HARMONY AND THE MUSIC OF THE SPHERES

MITTELLATEINISCHE STUDIEN UND TEXTE

HERAUSGEGEBEN

VON

PAUL GERHARD SCHMIDT

BAND XXX

HARMONY AND THE MUSIC OF THE SPHERES

The Ars Musica *in Ninth-Century Commentaries on Martianus Capella*

BY

MARIKEN TEEUWEN

BRILL
LEIDEN · BOSTON · KÖLN
2002

Illustration on the cover: picture of *Harmonia* in an eleventh-century manuscript from Milan, *Florence, Biblioteca Medicea Laurenziana, San Marco 190*, f. 108v.

This book is printed on acid-free paper.

Die Deutsche Bibliothek – CIP-Einheitsaufnahme

Harmony and the music of the spheres : The ars musica in ninth-century commentaries on Martianus Capella / by Mariken Teeuwen. – Leiden ; Boston ; Köln : Brill, 2002
(Mittellateinische Studien und Texte ; Bd. 30)
ISBN 90-04-12525-6

Library of Congress Cataloging-in-Publication Data

Library of Congress Cataloging-in-Publication Data is also available

ISSN 0076-9754
ISBN 90 04 12525 6

Cover design: Cédilles / Studio Cursief, Amsterdam

PRINTED IN THE NETHERLANDS

CONTENTS

ACKNOWLEDGEMENTS

This book is a slightly revised version of a Ph.D. thesis submitted to the Faculty of Arts of the University of Utrecht in December 2000. The research for this book has been sponsored for four years by the Foundation for Literature, Music and Drama Research (Stichting LMT) of the Netherlands Organization for Scientific Research (NWO), and partly also by the Royal Netherlands Academy of Arts and Sciences (KNAW). I thank both for their trust placed in me. I am also very grateful to the editor for accepting the result in this prestigious series.

The writing of a dissertation is a lonely affair, that paradoxically can not be completed without the help of many others. During the six years that I worked on this book, I have received support from many institutions and many individuals, who helped me conquer the inevitable obstacles. The completion of this book has depended to a large degree on their enthusiasm and support.

The research was carried out at the Faculty of Arts of the University of Utrecht, where I was so fortunate as to fall under the wings of Grada de Vries, secretary of the Department of Medieval Studies. Her practical support has been more important than she herself realises, and she created just the right working atmosphere.

All members of staff of manuscript departments in the libraries I visited were very helpful: not only did they trust me to study their valuable Carolingian manuscripts, they also helped me find my way into their systematic catalogues. I am very grateful, therefore, to the librarians of the reading rooms of the University Library in Leiden, the Bibliothèque Municipale in Besançon, the Bibliothèque Nationale de France in Paris, the Parker Library of the Corpus Christi College in Cambridge, the Bodleian Library in Oxford, and the Biblioteca Apostolica Vaticana in Vatican City.

In October 1998, this study travelled with me to a new job and new surroundings, and I had to fit it in next to a new research project I took upon me at the Constantijn Huygens Institute in The Hague, subsidiary of the Royal Netherlands Academy of Arts and Sciences. It would have taken me much longer to finish if my colleagues, especially Marijke Gumbert-Hepp and Olga Weijers, had not given me the elbow room I needed. I want to thank them for allowing me as much time as seemed necessary.

Furthermore, this book could not have been written without the generous help and interest I have experienced from many individuals. First and foremost I must mention my supervisors, who encouraged me in my work from a very early stage onward. I studied with both professor Arpad Orbán and professor Kees Vellekoop from the undergraduate level onwards, and have always benefited greatly from their expertise, encouragement and assistance.

As a research assistant, I was lucky to meet professor Calvin Bower (University of Notre Dame), who, despite the distance of an entire ocean between us, never ceased to encourage me and who showed me a warm hospitality. Professor Peter Gumbert (Leiden University) read previous versions of the second chapter of this book with his usual scrutiny and kept me from many mistakes and inconsistencies. For those that remain I am, of course, solely responsible.

Two colleagues have read the entire manuscript: Maria Sherwood-Smith, who painstakingly corrected my English and removed many Dutchisms, and Lia Couwenberg, with whom I shared not only an office for four years but also all my daily problems, and who has become a very dear friend.

Finally, I would like to thank my parents: Nel en Wil, who may never have guessed that they would raise two *doctores* from the Faculty of Arts. Perhaps all those years of taking me and my brother to museums and interesting places all over Europe did bear fruit after all. My two children, Sara and Jappe, did not contribute to this study in a particularly efficient way, but they were invaluable for putting my own work into perspective and conquering setbacks, just by their cheerful existence. My husband, Ab de Jong, never ceased to stimulate me to persevere. He read many previous versions of the chapters in this book and improved upon them greatly with his insightful comments. He managed to keep up an interest in a subject as drairy as Carolingian music theory, and I am profoundly grateful for that.

LIST OF PLATES

Cover:

Firenze, Bibl.Med.Laur., San Marco 190, fol. 108v

Chapter 2:

Chapter 3:

INTRODUCTION

In the ninth century, Martianus Capella's *De nuptiis Philologiae et Mercurii* was one of the late-antique texts read with scrupulous vigour by the intellectual elite. There remain today approximately 50 manuscripts that bear witness to this interest; most of them are conscientiously annotated, accumulating in their margins information from all kinds of sources in tiny script. These annotations form a rich source for the actual interests of the learned elite, and for the reception of classical learning in the so-called Carolingian renaissance. They provide insight into the precise aspects which formed the attraction of the work, which, as has often been stated, seems — to our modern taste — to have been immensely overrated. And perhaps the aspects that made it so attractive to its ninth-century readers are precisely those that made it so unattractive in the eyes of modern Classicists.

In *De nuptiis,* Martianus attempted to mix work with pleasure, to make us gain knowledge, while entertaining us with mythology and satire. It is, in other words, a marriage between story and learning, which is, in a way, what the title of the work already hints at. The Latin used by Martianus is a text-book example of mannerism: it is full of *hapax legomena*, employs many intricate grammatical constructions and rarely used poetical metres, and it is full of Greek vocabulary. In between the lines of its poems and satirical passages, it refers to a whole world of Greek myths. Apart from this, it treats the seven liberal arts, Grammar, Dialectic, Rhetoric, Geometry, Arithmetic, Astronomy and Music, in a nutshell. They appear as beautiful maidens before Mercury and Philology, and offer them their knowledge as an elaborate wedding gift. The result is — in the eyes of many modern readers — a rather monstrous match, but in the eyes of the Carolingian scholars, it was a *Fundgrube* of ancient knowledge.

Music, the *ars* treated in the ninth and last book of *De nuptiis*, occupies a special place within the order of the *artes liberales*. She is abundantly present not only in the ninth book, which is entirely devoted to her, but also in the first two mythological books, which set out the story which frames the work. For music is the quadrivial *ars* which holds the key to the universe. It is not only present on earth, but also in the divine order of creation itself; it encompasses the knowledge of the basic ratios according to which the cosmos is created. Music is the earthly reflection of these ratios. The same ratios that rule the cosmos are present in the proportions of the consonances of octave, fifth and fourth.

This approach to music as a quadrivial *ars* was not new, of course, and the learning on the subject gathered by Martianus in book IX, is little more than a rather sketchy summary of ancient learning. But it was new in the eyes of the Carolingian scholars who read and commented upon the text. They had another, much more powerful and much more complete, summary of ancient knowledge on the subject in *De institutione musica* of Boethius, and they had other late antique texts with more or less the same material at their disposal, but *De nuptiis* was certainly one of the texts that shaped their own thoughts on the *ars musica*.

In the present study, this early encounter with music as an art ruled by number will be opened up and analysed. It is divided into two basic parts: a study, and an edition of one particular gloss tradition on Martianus' musical material.

In the large number of glosses, several traditions have been recognized. Two corpora of glosses are ascribed with certainty to Carolingian scholars: one to the famous Irish scholar John the Scot Eriugene, who makes his appearance in Carolingian history as a scholar at the court of Charles the Bald, and the other to a scholar of a later generation, Remigius of Auxerre, who was trained in Auxerre, but, according to some sources, taught on Martianus Capella in Paris at the very end of the ninth century. Both these corpora of glosses have been edited by Cora Lutz, who made a great contribution to our present state of knowledge on the subject. There is, however, a third corpus of glosses, for which I have chosen the name the Anonymous commentary. It is even earlier than that of John the Scot, and even larger possibly than the commentary of Remigius, which occupies two sizeable volumes in its modern edition. It is the most enigmatic of the three, for its author has not been identified, in spite of the considerable amount of scholarship (carried out by great scholars at that, such as Claudio

Leonardi, Jean Préaux and Edouard Jeauneau) which has been devoted to the subject. A small part of it has been edited by Cora Lutz, and has been ascribed at first to an Irish bishop and scholar called Dunchad, and later to the more famous schoolmaster Martin of Laon, but both identifications have been shown to be doubtful.

Being so early, viz. from the second third of the ninth century, and so complete, it is also a text that can provide us with insight on many aspects of the intellectual life of the period. The annotations reveal a great deal about particular interests of the time, the books that were actually consulted in understanding and explaining such a difficult text, and the working method in getting access to Martianus' rich text.

The physical aspects of the annotated manuscripts, i.e., the hands that worked in them, the way text and glosses were laid out, the way layers of glosses were copied in the margins, etc., are also an important source of information. They may tell us more about the nature of glosses as texts and about their possible function. The codicological and paleographical details, moreover, may enlighten us about the geography of the reception of *De nuptiis*, and about its dissemination.

It is this Anonymous commentary, then, which stands at the center of the present study. It contains, first of all, a partial edition of the commentary: the glosses on book IX and on the musical material in books I and II. Furthermore the contents of the commentary on the subject of music will be analyzed and put into context, and relationships with contemporary texts on music theory will be explored.

The present study is set out as follows. The first chapter provides an introduction to the work and its context. First, the state of research on Martianus Capella himself and the work itself is briefly described. The question of its date, especially, has never been solved satisfactorily. Secondly, more complete overviews are given of the present state of research on the pre-Carolingian and Carolingian reception of the work. From these overviews it will appear that there are many gaps to be filled in our present knowledge. We know virtually nothing, for example, of the transmission of the text in the sixth to eighth centuries. We are uncertain of the actual form of the commentary written by John the Scot. Present scholarship remained, so far, unable to provide satisfactory answers to the problems of the identification of the author of the Anonymous commentary, the exact date of its origin or the specific location in a certain cultural center or intellectual circle. Furthermore, there are many manuscripts that contain glosses that have never been studied at all, and that have yet to be classified in one of the traditions we know, or in possible further traditions.

The second chapter is dedicated to the manuscripts themselves. First, since there remain today almost fifty ninth-century manuscripts, the need to make a strict selection from them was felt. The main criteria of this selection process are described in the first section of the second chapter. Eight manuscripts were finally chosen for the edition of the relevant gloss material: four manuscripts were considered of primary importance, and four were used to check readings, and solve problems caused by damages, corruptions or errors in the other four. Codicological and paleographical descriptions were made of these manuscripts. At the end of the second chapter, the conclusions of the manuscript evidence are drawn. The relationship between the manuscripts is sketched, and a tentative attempt is made to bring questions of authorship and intellectual context somewhat closer to a solution.

In the third chapter, which concludes the first part of the study, the focus shifts to the real subject of the study: the musical material in the earliest commentaries on *De nuptiis*. In this chapter the antique and late-antique treatises on music which were available to ninth-century scholars are introduced. Secondly, a close comparison is made between the Anonymous corpus of glosses on Martianus Capella's musical material, and the so-called *Glossa maior in Institutionem musicam Boethii*, a body of Carolingian and later glosses that date back to the same period as the Anonymous commentary on *De nuptiis*. With these two corpora of glosses we have the very earliest sources for the history of the development of musical thinking in the Carolingian period. These two sources are here studied in relation to each other for the very first time, for the *Glossa maior* has only recently been edited by Michael Bernhard and Calvin M. Bower, and the relevant parts of the Anonymous commentary on *De nuptiis* have only been edited in this study.

A second major theme in the body of thought on music in the Anonymous commentary on *De nuptiis* is the theme referred to in the title of the present study: the harmony of the heavenly spheres. In other words, the idea that the movements of the planets make the universe sound harmonious. This *topos* was prominent in both ancient and medieval speculations on the order of the cosmos, and the third section of chapter three develops this theme as it occurs in the Anonymous commentary in more detail. The sheer amount of gloss material on this theme bears witness to its importance in the eyes of the Carolingian commentators. Particularly long glosses are devoted to the musical ratios that rule the cosmos, and the sounds that might result. First, the context of Martianus Capella's view on the music of the cosmos is sketched. Then, the material on this theme in the Anonymous commentary is collected and analyzed, and it is held against the light of another thinker on the same subject, namely John the Scot Eriugene. For in one particularly long gloss on *De nuptiis*, this Irish scholar expresses a unique idea of the music of the cosmos and the physical laws according to which it sounds harmonious.

The sections that follow are essentially inventories of gloss material pertaining to the more technical aspects of the discipline of music. In accordance with Martianus' classification of subjects within the *ars musica*, five collections of gloss material have been made on, respectively, notes, intervals, tetrachords, *genera* (enharmonic, chromatic, diatonic), and modes (Dorian, Lydian, etc.). These collections of material have been translated and provided with comments, and, where possible, compared with contemporary or slightly later material, which may have been influenced by Martianus Capella's work or by the work of the glossator.

The last section of the third chapter, finally, treats those aspects of the oldest commentaries on *De nuptiis* that touch upon musical practice. There are, for example, several manuscripts in which notational signs are added to some of Martianus' poems. Secondly, there is a reference to the nature of the sequence being a song without words, which deserves closer attention. And finally, there is a reference to polyphonic singing in one of the glosses of John the Scot Eriugene which is analyzed in more detail.

The second part of this study contains the edition of relevant material from the Anonymous commentary, with the editorial principles and list of manuscripts. From book I and II in *De nuptiis* several sections were chosen that relate to music. In sections 11-12, for example, Martianus speaks of the secret and holy grove of Apollo, which resounds in perfect harmony. The vibrating trunks and branches of its trees sing together in consonance, and form an allegorical image of the music of the planetary spheres. From these sections *all* annotations have been edited, not only those that pertain specifically to music. This was felt necessary to have a clear picture of the relations between the manuscripts and their gloss collections.

Furthermore, the *accessus ad auctorem*, which certainly was part of the Anonymous commentary, even though it has not been copied in all the manuscripts containing the Anonymous tradition, was edited. This text was thought to be an especially helpful instrument in identifying the Anonymous commentary in manuscripts that were not studied here.

From book IX the whole of the corpus of glosses has been edited.

PART ONE

CHAPTER ONE

MARTIANUS CAPELLA AND THE SEVEN LIBERAL ARTS

Introduction: *Martianus Capella's* De nuptiis Philologiae et Mercurii

The focus of the present book is not on Martianus Capella himself and his encyclopedic work on the seven liberal arts, which is in itself a field of research in which many debates are yet to be decided. What is at the centre of attention here is the ninth-century reception of his work, or — more precisely — the early reception of materials on music in *De nuptiis*. The present chapter, therefore, is intended as a short introduction to author and text. Since a discussion of these subjects in their full complexity is not directly relevant to the present study, I shall give only brief outlines of the problems.[1]

1. *The author*

All that is known of Martianus Capella is found in *De nuptiis* itself.[2] In the manuscripts his full name is attested as Martianus Min(n)e(i)us Felix Capella. In book VI, the personification of the

---

[1] These are based mainly on the comprehensive study of W.H. STAHL & R. JOHNSON, *Martianus Capella and the seven liberal arts*, vol. I, which, despite its shortcomings, is still the best overview to date. (See the quite harsh criticism by D. SHANZER, in her review of the work in *Beiträge zur Geschichte der deutschen Sprache und Literatur* 104 (1982), pp. 110-117: (p. 117) "It is unfortunate that *scatet vitiis* must be the verdict on this book, especially since the translation is the first complete one ever.") For a more recent introduction to the work see also B. PABST, *Prosimetrum*, pp. 105-133, 211-228, 234-249. For a full analysis of its content and structure see S. GREBE, *Martianus Capella.*

[2] On the biography of Martianus Capella see, among others, M. CAPPUYNS, *Capella (Martianus)*, cols. 836-838; W.H. STAHL, *Martianus Capella and the Seven Liberal Arts* I, pp. 9-20; D. SHANZER, *A Philosophical and Literary Commentary*, pp. 1-3; S. GREBE, *Martianus Capella*, pp. 11-15.

literary genre herself, Satura, addresses him as *Felix meus*,[3] in book VIII as *Felix, vel Capella*[4] and in the final poem as *Felix Capella*.[5] Two of the earliest witnesses, Fulgentius and Cassiodorus, also mention him by this name. It is only in Gregory of Tours' *Historia Francorum* (10.31) that his first name, Martianus, is used for the first time, and the ninth-century commentators follow suit.

The author had a son, also called Martianus, whom he addresses in the second paragraph of the encyclopedia and to whom he dedicates the work in the final poem. Martianus describes himself as an old man,[6] with a "head, sprinkled with whitening hairs."[7] He lived in Carthage; in the manuscript tradition the words "Afer Carthaginiensis" are added to his name, and he refers to himself as an "*alumnus* of the beautiful city of Elissa" (Carthage, city of Dido).[8]

A wide range of professions have been suggested for the author: farmer, lawyer, rhetor, teacher of rhetoric, and even proconsul of Carthage. The two extremes, farmer and proconsul, have been rejected in recent scholarship,[9] and it is now considered likely that

---

[3] VI.576. The references are to the latest edition of the text by J. WILLIS. Translations are generally taken from W.H. STAHL, *Martianus Capella and the Seven Liberal Arts* II.

[4] VIII.806: *Felix, vel Capella vel quisquis es, non minus sensus quam nominis pecudalis* (Felix, or Capella, or whoever you are, with a sense to match the beast's whose name you bear).

[5] IX.999.

[6] The final dedicatory poem to his son opens with the line: *Habes anilem, Martiane, fabulam,* (Here you have the tale, Martianus, of an old man,); IX.997.

[7] *respersum capillis albicantibus verticem*, I.2.

[8] IX.999. It has been suggested that he was born in Madaura, to the southwest of Carthage, but there is no corroborative evidence for such an assumption. See for a full treatment M. CAPPUYNS, *Capella*, col. 837 and W.H. STAHL, *Martianus Capella and the Seven Liberal Arts* I, p. 12.

[9] On "farmer", see H. PARKER, *The seven liberal arts*, where he suggested that Martianus Capella was "a poor literary farmer" (p. 443). See Shanzer's refutation in *A Philosophical and Literary Commentary*, p. 2 n. 2: she places the references to rustic surroundings that lack culture in the context of literary commonplaces, used as a *captatio benevolentiae*. She mentions several parallels, among others in Gellius and Favonius Eulogius. On "proconsul", see W.H. STAHL, *Martianus Capella and the Seven Liberal Arts* I, pp. 16-17. The assumption was based on one corrupt line in the final poem of *De nuptiis*: *proconsulari verba dantem culmini* (IX.999, p. 385.14). (The manuscripts give many variant readings for *verba dantem*, such as:

he was a lawyer, rhetor or teacher. Stahl drew attention to the technical legal vocabulary Martianus uses in *De nuptiis* and suggested that he was a lawyer, but the evidence to this effect is not very convincing.[10] That he was a rhetor in a more general sense has been argued several times; the fact that the techniques of arguing cases get the majority of the attention in the book on rhetoric appears to point in this direction. The very undertaking of writing a handbook on the seven liberal arts, finally, may suggest that he was a teacher of rhetoric.[11]

The extravagancy of his style was seen, especially in earlier scholarship, as a sure sign of Martianus' lack of education, or as a crudeness originating in aspirations too high for his actual educational level. In recent scholarship, however, it has been shown that his style is deliberately chosen, and that it could only have been chosen by a highly educated person.[12]

2. *The story*

The encyclopedia consists of two parts: an introductory story in two books and seven books on the liberal arts.[13] The framing story, the marriage of Mercury and Philology, is told in the first and second books and referred to mainly at the beginning and end of each of the books on the individual arts. In the context of the story, the arts are personified as *dotales virgines* (VIII.810), seven beautiful and learned maidens constituting the dowry of the bridal couple. They enter the heavenly *convivium* and one by one they share their knowledge with Mercury, Philology and the assembly of the gods: Grammar in book III, Dialectic in book IV, Rhetoric

---

bonbinatore(m), bombinat ore, bobinatore.) Stahl stresses that it remains unsupported by any other evidence.

[10] W.H. STAHL, *Martianus Capella and the Seven Liberal Arts* I, pp. 19-20.

[11] *Ibid.*, pp. 17-18. For exact references see notes 40-44.

[12] See, for example, D. SHANZER, *A Philosophical and Literary Commentary*, pp. 3-4 (sections "Style" and "Knowledge of Greek"). She concludes: "Martianus is clearly a well-read author who writes deliberately convoluted and "artistic" Latin. This, coupled with the Greek and the interest in the *disciplinae*, suggests a highly educated and possibly academic background." See also F. LEMOINE, *Martianus Capella: a Literary Reevaluation*.

[13] See also S. GREBE, *Martianus Capella*, pp. 22-37.

in book V, Geometry in book VI, Arithmetic in book VII, Astronomy in book VIII and, finally, Harmony in book IX. Martianus thus establishes a set number of seven liberal arts, which differs from Varro's canon of the nine arts. He does mention Varro's two other arts who are waiting in the line of virgins, Medicine and Architecture, but these are sent away: there is not enough time to introduce nine arts at the wedding and the ones that concern themselves most with earthly things are dismissed (IX.891).

The story opens with Mercury's search for a suitable wife. Several options (Sophia, Mantice, Psyche) are considered and rejected when Virtus gives him the advice to consult Apollo. Apollo proposes Philology, a most learned, but mortal woman. Mercury and his companions set out to seek permission from Jupiter, and on their journey they are attended by the Muses and accompanied by the music of the heavenly spheres. Pallas urges Jupiter to convene a large group of married gods and goddesses, in order to decide upon the matter in a democratic way. A colourful company gathers and approves the marriage on condition that Philology will be accepted into the circle of the immortals.

In the second book, Philology is the main character. She rejoices at her fate, but is also worried about whether she will make a good wife for such a powerful god as Mercury. Jiggling with numbers, she finds out that the numerical values of their names (3 for him and 4 for her) make the perfect number 7, and rests reassured. On the basis of this numerical symbolism, the marriage between Philology and Mercury has, both in early and in later times, been taken to symbolize the union of eloquence and learning, the arts of the *trivium* and *quadrivium*, but also of secular and divine knowledge.[14] Her mother, Phronesis, dresses her for the wedding,

---

[14] G. NUCHELMANS, *Philologie et son marriage avec Mercure*. In book II, Martianus explains the perfection of the marriage with numerical symbolism: the value of the letters of Mercury's secret name (θωύθ) amounts to four (the quadrivial arts), that of Philology to three (the trivial arts). Together they make the number seven: the seven liberal arts and absolute knowledge. (II.101-109). See also W.H. STAHL, *Martianus Capella and the Seven Liberal Arts* II, p. 35, n. 9; G. MATHON, *Les formes et la signification de la pédagogie des arts libéraux*; B. PABST, *Prosimetrum*, pp. 108-109; S. GREBE, *Martianus Capella*, pp. 841-846; and the study by H.J. WESTRA, *The Allegorical Interpretation of Myth*, which focuses on the medieval reception of antique myths.

the Muses sing her praises, the Virtues bring her greetings and the three Graces kiss her. Athanasia, the daughter of Apotheosis, comes down to guide her through the heavenly spheres. Philology, though willing to leave the earth and ascend to heaven, finds herself to be too heavy because of the weight of her worldly learning. She regurgitates a multitude of books and loose leaves, written in all kinds of languages and alphabets, which are gathered eagerly by the Arts, Disciplines and Muses. Apotheosis offers her a potion of immortality from an egg-shaped cup; only then is she ready to ascend. She is carried in a palanquin by Labour, Love, Attention and Wakefulness, while Juno guides them on their way through the heavens and planetary spheres. Above the heavens, in the palace of Jupiter, the entire company of gods, goddesses, demigods and -goddesses, immortals and heroes awaits her. There Mercury and Philology are offered their wedding gift: the seven liberal arts.

The seven books that follow are each dedicated to one liberal art; first the arts of the *trivium*, then those of the *quadrivium*. They each explain their own discipline and are put to the test by the gods. Each has all-encompassing knowledge of its own art and the most learned mortal scholars are their students. Architectura and Medicina, the most earthly arts, are excluded, as are the seven magical arts, though — to the horror of Venus, who is already irritated by the tediousness of so much learning — these latter are later promoted by Apollo and Pallas and enter the heavenly *convivium* in book IX (891-897). They will be allowed to offer their knowledge to the bridal couple on another occasion.

Grammar explains her discipline in book III.[15] She defines her subject, discusses the letters of the alphabet exhaustively, covering their possible positions in words and their pronunciation. Further, she treats syllables and, in an introductory manner, the declension of nouns and the conjugation of verbs. Dialectic's treatise (book IV) reads as a mixture of introductory material on logic and on rhetoric; much of it is directly based on Aristotle's philosophical

[15] For a more elaborate treatment of this book, its contents, sources and influence, see W.H. STAHL, *Martianus Capella and the Seven Liberal Arts*, I, pp. 99-104.

works.[16] In book V, Rhetoric presents her own *ars*; Martianus uses Cicero's *De inventione* as a prime source for this treatise.[17] Book VI differs somewhat in character from the previous three: where the treatises of the trivial arts were mainly devoted to definitions, axioms and theoretical subdivisions, Geometry provides us in her book with a "conspectus of the known world", based mainly on Pliny and Solinus.[18] In Book VII, Arithmetic speaks on the subject of number. For this book, Martianus ultimately draws on two sources, Nicomachus and Euclid, be it through a number of intermediate sources.[19] Book VIII, on Astronomy, is not only the shortest book of the nine, it is also incomplete; the closing scene, in which Martianus usually returns to his framing story, is missing. It is, however, also generally considered as the "best" book in terms of content; as Stahl puts it, it is "the best of Martianus' quadrivium books and, considering its small compass, the best-proportioned and most satisfactory treatise on astronomy in the extant Latin literature before the Greco-Arabic revival."[20] Book nine is again a standard treatise on music, with definitions, classifications and theoretical explanations; it is a compilation of an unknown source on antique music theory (Varro?) and Aristides Quintilianus' *On music*.[21] Martianus gives ample space to the introduction of the maiden Harmonia, who is described as follows:[22]

---

[16] See *ibid.*, pp. 104-115, a section of Stahl's book which is written by E.L. BURGE.

[17] *Ibid.*, pp. 115-121.

[18] *Ibid.*, p. 127, see also pp. 126-148.

[19] See *ibid.*, pp. 149-170.

[20] *Ibid.*, p. 174; on Harmony, see pp. 171-201. See also the most recent analysis of this book and its reception in the early Middle Ages by B.S. EASTWOOD, *Astronomical Images and Planetary Theory*.

[21] For a full treatment of the position of this music treatise in the context of antique and late-antique learning on the subject see L. CRISTANTE, *Martiani Capellae Liber IX*. This study is laid out as an edition with translation, and a running commentary on the text. In the elaborate and especially helpful introduction section 2.4 (pp. 67-72) Martianus' sources are discussed.

[22] Martianus seems to have been the first to have personified the *artes*, and, in some of the manuscripts of his text, pictures are drawn of each of these maidens. A well-known example is *Firenze, BML, San Marco 190*. See also, for further references, K.-A. WIRTH, *Eine illustrierte Martianus-Capella-Handschrift*; and F. MÜTHERICH, *'De Rhetorica'*.

> [909] Harmony walked along between Phoebus and Pallas, a lofty figure, whose melodious head was adorned with ornaments of glittering gold. Her garment was stiff with incised and laminated gold and it tinkled softly and soothingly with every measured step and movement of her body. Her radiant mother the Paphian — who followed her closely — though she too moved with graceful measure and balanced steps, could scarcely match the gait of her daughter. In her right hand Harmony bore what appeared to be a shield, circular over-all, with many inner circles, the whole interwoven with remarkable configurations. The encompassing circles of this shield were attuned to each other, and from the circular chords there poured forth a concord of all the modes. From her left hand the maiden held, suspended at equal length, several small models of theatrical instruments, wrought of gold.[23]

Harmonia is the last of the arts to speak, and when she is finished the story is quickly brought to an end. She leads Mercury and Philology to the bridal chamber. The work is concluded by a final autobiographical poem, in which the author contemplates his own work. Not entirely to his own satisfaction, it has become a curious mix of serious scholarship, poetic fiction and trivialities. He humbly begs his reader's indulgence for his musings.

3. *Date of the text*

The date of the text is still a matter of much debate. Stahl and Shanzer, who both treat the discussion on dating *De nuptiis* in detail, provide an overview of the state of research on the question — Stahl up to the seventies and Shanzer up to and including much of the eighties; their summaries form the basis for the present review.[24]

The earliest students of the text — Shanzer mentions Vossius, Fabricius, Saxius and Funccius — dated the work ca. 470 A.D.: they

---

[23] Transl. W.H. STAHL, *Martianus Capella and the Seven Liberal Arts* II, pp. 352-353.

[24] W.H. STAHL, *Martianus Capella and the Seven Liberal Arts* I, pp. 12-15; D. SHANZER, *A Philosophical and Literary Commentary*, pp. 5-17. See also the most recent overview of S. GREBE, pp. 16-21.

linked the decadent literary style of the work to this period.[25] The first scholar to make a critical edition of the text, Eyssenhardt, found several passages in the text that could be interpreted as leads for assigning a date to *De nuptiis*.[26] First of all, in the final poem of the work he found a reference to the office of proconsul — an office which supposedly disappeared after 439 A.D., when the Vandal king Gaiseric conquered Carthage. Moreover, he interpreted a passage in book VI, *Carthago, inclita pridem armis, nunc felicitate reverenda*[27] (Carthage, once famous for her arms, now to be respected in her *felicitas*), as an indication that the text was written before the Vandal occupation of 439. These two passages thus gave him arguments for a *terminus ante quem* of 439 A.D. For the determination of a *terminus post quem* Eyssenhardt pointed to another passage in book VI.637, where he interpreted the words *ipsa caput gentium Roma* [...] *quamdiu viguit* ... (this city Rome, the capital of a people [...] as long as she thrived ...) as an indication that — in the eyes of the author — Rome was no longer thriving, and thus as evidence of its decline. In addition, Cappuyns suggested that it should be interpreted as an implicit lament for Alaric's sack of Rome in 410.[28] The argument received additional support from the fact that the words *quamdiu viguit* appeared to be a deliberate alteration of Pliny, *Natural History* 3.38, where Rome is simply mentioned as *terrarum caput*. Shanzer, however, rightly pointed out that the passage does not necessarily refer to this historical event,

---

[25] The one exception appears to have been Caspar Barthius, who claimed to possess a *Vita Martiani* which placed the author in the middle of the third century. As Shanzer has shown, however, this *Vita*, for the existence of which we have no other evidence than Barthius' claim, should at best be considered a playful falsification: D. SHANZER, *A Philosophical and Literary Commentary*, p. 5. See also J. PRÉAUX, *Securus Melior Felix*, p. 120 n. 2.

[26] In his dissertation (1861) Eyssenhardt argued for a different, even earlier dating: between 284 and 330 A.D. For this dating he based himself on the usage of the name of Byzantium (VI.657) for the city that was proclaimed capital and renamed as Constantinople by Emperor Constantine in 330. This passage, however, is cited from Solinus, *Collectanea rerum memorabilium*, 10.17, and thus loses evidential value. Nevertheless, in the edition of 1925 Dick adopted the early dating and used an aspect of Martianus' style as an additional argument: his use of so many different metres.

[27] VI.669.

[28] M. CAPPUYNS, *Capella (Martianus)*, col. 843.

and could in fact refer to many events or situations: Alaric's sack of Rome in 410, but equally, for instance, Gaiseric's sack in 455, or religious decline in general in fourth-century Rome.[29]

Thus Eyssenhardt and Cappuyns together set the borders of the period in which the text was written at 410-439 A.D. This dating proved to be highly influential and was accepted by many scholars, among them Willis and Préaux.[30] Stahl also accepted the dating, although he emphasized several times that the proposed *termini* are based on circumstantial evidence only.[31] The general acceptance of the dating was not altered by the fact that both Willis and Préaux in their later articles voiced doubts about the certainty with which the *termini* were settled upon. In 1973, Willis dated the text ca. 450,[32] Préaux dated it *post* 455 in 1975, and in 1978, he declared himself even less certain.[33] Neither Willis nor Préaux gave valid arguments for their shifts to later dates. Presumably for want of a good alternative, most scholars continued to use Cappuyns' dating between 410 and 439. Shanzer was the first to reconsider the entire dating question in her study of book I.

In her deconstruction of Eyssenhardt's and Cappuyns' arguments, she points to the fact that there is evidence for the continuation of the office of *proconsul* during the Vandal reign. In her view, the word *felicitas*, used in VI.669 to describe the circumstances in which Carthage found itself, should be read to mean prosperity in general, 'happiness' with respect to temporal goods, and not necessarily — as Cappuyns and Stahl suggested — specific, historically identifiable 'happy conditions.' The passage then also accords with a much later period in the fifth century, i.e. between 470 and 490. Shanzer

[29] D. SHANZER, *A Philosophical and Literary Commentary*, p. 5 n. 17.

[30] J. WILLIS in his dissertation, *Martianus Capella and his Early Commentators*; J. PRÉAUX, *Un nouveau texte sur la Vénus androgyne*, p. 484 n. 1.

[31] W.H. STAHL, *Martianus Capella and the Seven Liberal Arts* I, pp. 13-15. See, e.g., on the phrase *quamdiu viguit*: "On the other hand, placing all the weight of a *terminus post quem* on this expression of a florid writer is admittedly not secure, since the phrase does not necessarily imply that Rome was vigorous right up to the fall." On the *terminus ante quem* based on the office of *proconsul*: "This assumption is altogether likely, though not absolutely certain."

[32] J. WILLIS, *Martianus und die mittelalterliche Schulbildung*, p. 165.

[33] J. PRÉAUX, *Jean Scot et Martin de Laon en face du De Nuptiis*, pp. 162-163; and *id.* in *Les manuscrits principaux*, p. 76.

reassesses the *termini ante* and *post quem* and concludes that the *terminus ante quem* cannot be pushed back any further than the late fifth century.[34] She also argues that there may be evidence to push the *terminus post quem* forward, to a period later than or during the lifetime of Augustine (354-430). In his *Retractationes*, Augustine refers to a Varronian or some other, looser canon of nine liberal arts, while Martianus, who was acquainted with the canon of nine but excluded the arts Medicine and Architecture, seems deliberately to have peparted from the well-known canon to establish a new one of seven arts in his *De nuptiis*. The new format proved to be very influential; Boethius, Isidore and Cassiodorus, for example, used it in their works. The fact that Augustine does not use it could consequently serve as evidence for a chronology in which Augustine precedes Martianus, or in which Martianus and Augustine were contemporaries. This latter option is in fact preferred by Shanzer. On the one hand she uses the argument of the canon of the seven liberal arts as "a good reason for dating Martianus's work later than the lifetime of Augustine",[35] on the other hand she finds evidence which suggests that *De nuptiis* could have served as a source for Augustine's *De civitate dei*. Furthermore, there is the problem of the absence of any reference to Christianity in *De nuptiis*. Cappuyns noticed it, but did not conclude that Martianus was not Christian, since the entire field of secular science is a matter in which Christianity does not play a role in fifth-century Africa: it stands entirely in the Neo-Platonic tradition.[36] Shanzer, on the contrary, presents Martianus as belonging to late antique paganism *and* suggests that he not only knew Christian doctrines, but wrote his work as a polemic against them.[37] She uses this suggestion as

---

[34] This dating is based on Fulgentius's acquaintance with Martianus, as shown in the *Sermones Antiqui*. She accepts Helm's dating of Fulgentius as a contemporary of Dracontius in the reign of Gunthamund (484-496). *A Philosophical and Literary Commentary*, pp. 8-13.

[35] *Ibid.*, p. 16.

[36] M. CAPPUYNS, *Capella (Martianus)*, col. 838; W.H. STAHL, *Martianus Capella and the Seven Liberal Arts* I, p. 6.

[37] She defends this argument not only in *A Philosophical and Literary Commentary*, but also in her review of W.H. STAHL, pp. 110-111.

additional evidence for a dating at the end of the fifth century.[38]

There is, however, no positive evidence for Shanzer's reconstruction. Augustine is not mentioned in Martianus Capella's encyclopedia, nor can he be identified as a source. To what extent the dating of *De nuptiis* is a matter of interpretation is apparent from the fact that in the same year in which Shanzer published her reconstruction, two other students of the text, Cameron and Schievenin, published theirs. Cameron suggested two new leads for the dating of the text. First, he analysed the style of an editorial subscription, added to the text in the sixth century by a corrector, Securus Melior Felix.[39] Secondly, he proposed that the name of this corrector, *Felix*, might suggest a blood relation between the two, viz. of grandfather and grandson. In that case, he concludes, "the *De nuptiis* makes most sense in the second or third decade of the fifth century, the age of Rutilius Namatianus, Servius, and Macrobius."[40] On the basis of a re-interpretation of the autobiographical elements in the final poem, Schievenien argued that the office of proconsul provides a *terminus ante quem* in 429. For, although the title of proconsul did not disappear under the reign of the Vandals, its meaning shifted in such a way that it cannot be reconciled with the particular context.[41] On the other

---

[38] D. SHANZER, *A Philosophical and Literary Commentary*, pp. 16-17. S. GREBE (*Martianus Capella*, pp. 20-21) accepts the later dating. It is confusing, however, that Shanzer mentions the canon of the seven liberal arts of Martianus as "a good reason for dating Martianus's work later than the lifetime of Augustine" on page 16, but on the same page calls *De nuptiis* "the perfect source for Augustine's polemic in the earlier books of the *De Civitate Dei*" — thus leaving the reader in doubt whether she dates Martianus before or after Augustine's lifetime. On p. 17 she suggests that they were contemporaries and, since they both lived in Carthage, she considers it even "virtually impossible that they would not have known each other."

[39] On this subscription see the next section of this chapter.

[10] A. CAMERON, *Martianus and his first editor*.

[41] R. SCHIEVENIN, *Marziano Capella e il 'proconsulare culmen'*. In this article he sketches the image of Martianus as an old man, whose ambitions are frustrated, who is impoverished and who has become an outcast from the official circle of intellectuals because of his outdated ideals of a revival of the ancient pagan culture. To my mind, however, there is no solid base for this reading of his words. That this interpretation can be read into the text does not have evidential value; other meanings can also easily be read into the text.

hand, the supposed *terminus post quem*, 410 (Alaric's sack of Rome), was, according to him, not substantiated by conclusive evidence. In conclusion, he finds it wise to stick to the date derived from Martianus' use of material from Iamblichus, i.e. 330 A.D. Most recently, Pabst returned to the first *termini* proposed by Eyssenhardt and Cappuyns, viz. 410-439, and rejected Shanzer's conclusions.[42] His point of view can be taken as representative for recent views on the dating problem: the old *termini* of 410-439 A.D. are still generally used for the dating of *De nuptiis*.

De nuptiis *in the pre-Carolingian period*

From the sixth to the eighth centuries, the history of *De nuptiis* is shrouded in darkness. There are hardly any direct traces of the text, and there is very little indirect evidence of knowledge of the text. In modern sholarship the earliest history of Martianus' encyclopedia is heavily discussed and speculated upon, but there is only a small basis of evidence which can serve as the basis for such speculation.

The earliest evidence for dissemination of the work can be found in Fulgentius the Mythographer's book *Expositio sermonum antiquorum*, where Martianus is mentioned as a source:

> Celibatum dici voluerunt virginitatis abstinentiam, unde et *Felix Capella in libro de Nuptiis Mercuri et Philologiae* ait: "Placuitque Minervae pellere celibatum."[43]

Although the dating of Fulgentius himself is also under discussion, there seems to be some general agreement upon a dating in the

---

[42] B. PABST, *Prosimetrum*, p. 105 and n. 38.

[43] FULGENTIUS, *Expositio sermonum antiquorum* 45, ed. U. PIZZANI, p. 40 (see also his translation and comment on pp. 41, 173-174). The citation is located in *De nuptiis* I.5, but is very corrupt. Nevertheless, the name of the author and the title are solid evidence for dissemination of the work, though perhaps in some other form. Willis gives a list of parallels between *De nuptiis* and passages from Fulgentius' works in his unpublished dissertation, *Martianus Capella and his early commentators*, "II. Influence up to the time of Johannes Scotus."

second half of the fifth century.[44]

Whether Boethius (ca. 480-524/25) had any knowledge of *De nuptiis* is, again, a matter of disagreement. Although the two writers of *prosimetrum*, Martianus and Boethius, are often asociated with each other, there seems to be no positive evidence which suggests that Boethius was familiar with the work. Willis even pointed to one piece of negative evidence in *De institutione musica* I.25.26: when Boethius enumerates the Greek tone names with their Latin translations, he mentions Albinus as his source, and not Martianus Capella, who also gives a Latin translation for the same set of note names in IX.931.[45] The Albinus translation, however, differs slightly from the Latin names used by Martianus, who translates the Greek word *diezeugmenon* with the standard *divisarum* rather than *disiunctarum*. The translator of Boethius' text, Calvin M. Bower, argues that this is precisely the reason Boethius mentions Albinus: by citing this author, he can introduce a deviating set of Latin note names. Boethius himself, moreover, uses the standard set of names (which concurs with Martianus' set) as well: in his fourth book.[46] In addition, Shanzer pointed to the fact that Boethius might have been influenced by Martianus in his later work, *De Consolatione Philosophiae*. Due to the superficial likeness of these two works (the outline of Menippean satire, the use of *prosimetrum*) they are often

---

[44] Shanzer mentions a dating "surprisingly enough as early as the middle of the fourth century" (G. PENNISI, *Fulgenzio e la "Expositio Sermonum Antiquorum"*, Firenze 1963, pp. 19-27), but is convinced by Helm's argument, who argues that Fulgentius and Dracontius (484-496) were contemporaries. She refers to R. HELM, *Der Bischof Fulgentius und der Mythograph*, in: *Rheinisches Museum* 54 (1899), p. 120.

[45] J. WILLIS, *Martianus Capella and his Early Commentators*, p. 20. The passage in question from *De institutione musica* I.26 runs as follows: *Albinus autem earum nomina Latina oratione ita interpretatus est, ut hypatas principales vocaret, mesas medias, synemmenas coniunctas, diezeugmenas disiunctas, hyperboleas excellentes.* ("Albinus translated the names of these strings into the Latin language; thus he called the notes of the hypaton tetrachord *principales* ["principal"], those of the meson tetrachord *mediae* ["middle"], those of the synemmenon tetrachord *coniuncti* ["conjunct"], those of the diezeugmenon tetrachord *disjuncti* ["disjunct"], and those of the hyperboleon tetrachord *excellentes* ["high"].") Transl. C.M. BOWER, *Fundamentals of Music*, p. 46.

[46] C.M. BOWER, *Fundamentals of Music*, p. 46 n. 125. See also *De institutione musica* IV.3, where the naming of musical notes in Greek and Latin is recapitulated, and the Greek notational symbols are introduced.

associated with each other, but convincing evidence to prove concrete reliance of one upon the other has not (yet) been found.[47] Parker rejected the possibility that Martianus influenced Boethius' choice of style.[48] In his dissertation Willis kept the possibility of Martianian influence upon Boethius open, but in personal correspondence with Shanzer he expressed a negative view.[49]

The first datable piece of evidence that points to an active reception of *De nuptiis* is found in the manuscripts themselves. In more than 20 manuscripts[50] a subscription survives of Securus Melior Felix, a Roman rhetorician, who claims to have corrected the text from very corrupted exemplars:

> Securus Melior Felix, vir spectabilis, comes consistorianus, rhetor Urbis Romae, ex mendosissimis exemplaribus emendabam contra legente Deuterio, scolastico, discipulo meo, Romae, ad portam Capenam, consulatu Paulini, viri clarissimi, sub V nonarum Martiarum, Christo adiuvante.[51]
> (I, Securus Melior Felix, man of eminence, Aid in Council, rhetorician of the City of Rome, corrected <this text> from very faulty exemplars in consultation with Deuterius the scholar, a student of mine, in Rome, near the gate of Capena, under the consulship of the

---

[47] L. BIELER, one of the most recent editors, lists references to Martianus in his *fontes et similia*, but according to Shanzer (*A Philosophical and Literary Commentary*, p. 11 and n. 53), "in many of the cases there is a third possible source which Bieler has not mentioned, or the idea is a commonplace. They may be drawn from Martianus, but they are not the best evidence for Boethius's acquaintance with the work." (Reference to L. BIELER (ed.), *Anicii Manlii Severini Boethii Philosophiae consolatio*, (CCSL 94), Turnhout 1957, p. 116.)

[48] H. PARKER, *The seven liberal arts*, p. 453. His denial of any traces of Martianian influence in the works of Boethius is, however, based on a reasoning which cannot be supported in the eyes of modern scholarship: a deep loathing for Martianus' literary style and a high esteem for that of Boethius.

[49] W.H. STAHL, *Martianus Capella and the Seven Liberal Arts*, I, p. 57; D. SHANZER, *A Philosophical and Literary Commentary*, p. 11 n. 52.

[50] There seems to be some confusion as to the exact number of manuscripts containing the subscription. D. SHANZER speaks of 25 (*A Philosophical and Literary Commentary*, p. 8); C. LEONARDI counts 23 (*Nota introduttiva per un'indagine*, p. 279 n. 6, and *I codici di Marziano Capella* I, p. 446 n. 8); J. PRÉAUX 26 (he adds 2 to Leonardi's list of 23 in *Securus Melior Felix*, p. 103, and mentions another newly discovered one in *Les manuscrits principaux*, pp. 77-78).

[51] Cited after J. PRÉAUX, *Les manuscrits principaux*, p. 77.

most honorable Paulinus, on the third of March, with the help of Christ.)

The name of Paulinus in this subscription refers either to Flavius Decius Paulinus Junior, whose period of consulship included the third of March 534, or to Flavius Paulinus Senior, which would point to the date 3 March 498. It is generally agreed upon that the subscription should be dated in 534.[52]

In an article on the most important manuscripts of *De nuptiis*, Préaux convincingly showed that it is the revision of Securus Melior Felix that lies at the basis of all other texts. He classified the earliest manuscripts in three categories, reflecting different stages in the copying activity. The first category consists of the six most important manuscripts that reflect the oldest state of the text, which can be detected by errors originating from the *scriptio continua* of the sixth-century exemplar. In these manuscripts words were separated in the wrong place — which was often corrected by a later scribe — or they were misread and miscopied. Before the middle of the ninth century, there must have been exemplars with corrected word-division, which in their turn were copied, no longer showing any traces of the *scriptio continua* in the earliest exemplar.

Préaux reconstructed the earliest history of the text as follows.[53] Shortly before 534, *De nuptiis* was copied in Carthage and transported to Rome. In 533 the political situation in Carthage changed radically, due to the victory of Belisarius (general under Justinian) over the Vandals. Now that Carthage was set free from the Vandal occupation, new life was breathed into its contacts with Rome, including the cultural ones. In this same period the works of, for example, Valerius Maximus and Pomponius Mela were revised by Flavius Rusticus Helpidius Domnulus in Ravenna. This is the

---

[52] D. SHANZER, *A Philosophical and Literary Commentary*, pp. 8-13; A. CAMERON, *Martianus and his first editor*. I would like to stress, however, that the subscription is generally copied at the end of book I, in some cases at the end of book II. (LEONARDI, *I codici* I, p. 447 n. 15.) This might suggest that, although the seven books on the arts did exist in the sixth century, as is shown by the completeness of the earliest manuscripts, active reception of these same books remains unproven and should possibly be dated later, i.e. in the eighth or even ninth century.

[53] *Les manuscrits principaux*, esp. pp. 80-82.

context in which Securus Melior Felix's subscription should be situated.

Cassiodorus (ca. 485 - ca. 580) wrote in his *Institutiones* that he had heard of the existence of Martianus' encyclopedia of the seven liberal arts, but had been unable to find a copy:

> Audivimus etiam Felicem Capellam aliqua de disciplinis scripsisse deflorata, ne talibus litteris fratrum simplicitas linqueretur ignara; quae tamen ad manus nostras adhuc minime pervenire potuerunt.[54]
> (We heard that Felix Capella wrote an anthology on the disciplines, so that the ignorance of our brothers would not remain unacquainted with this kind of subject; we have, however, so far been unable to lay our hands on it.)

It should not surprise us, then, that in his treatment of the arts, Cassiodorus shows no knowledge of *De nuptiis* and depends largely on works of Augustine and Boethius.

Gregory of Tours (538/9 - 594) mentions Martianus and his work on the *septem disciplinae* in the epilogue to his *Historia Francorum*:

> Quod si te, sacerdos Dei, quicumque es, Martianus noster septem disciplinis erudiit, id est, si te in grammaticis docuit legere, in dialecticis altercationum propositiones advertere, in rhetoricis genera metrorum agnoscere, in geometricis terrarum linearumque mensuras colligere, in astrologiis cursus siderum contemplare, in arithmeticis numerorum partes colligere, in armoniis sonorum modulationes suavium accentuum carminibus concrepare: si in his omnibus ita fueris exercitatus, ut tibi stylus noster sit rusticus, nec sic quoque, deprecor, ut avellas quae scripsi.[55]
> (For, if our Martianus educated you, priest of God, whoever you are, in the seven arts — that is to say: if he taught you to read in his Grammatica, to apply arguments in discussions in his Dialectica, to recognize species of metre in his Rhetorica, to calculate the measures of countries and lines in his Geometrica, to contemplate on the course of the stars in his Astrologia, to add up parts of numbers in his Arithmetica, to sing the modulations of tones in songs of sweet intonations in his Harmonia — if you are so practised in all these

---

[54] *Institutiones* 2.2.17, ed. R.A.B. MYNORS, Oxford 1963, p. 130.

[55] *Historia Francorum* 10.31, ed. W. ARNDT & B. KRUSCH, MGH Scriptores Rerum Merovingicarum I, p. 449.

things, that our pen seems crude to you, even then, I pray you, do not tear up what I have written.)

But, although Gregory suggests familiarity with the work with the adjective "noster" and correctly associates Martianus' style with everything but simplicity, there is little evidence that he in fact used *De nuptiis* as a source of knowledge. Gregory mentions the arts in a different order, Martianus does not teach reading in *De grammatica* and the subject of metrics is treated only briefly in *De rhetorica* (III.519-522), whereas in *De harmonia* it gets a more ample treatment (IX.966-995).[56]

Both Willis and Manitius[57] pointed to parallels between *De nuptiis* and the anonymous sixth-century work of Welsh or Irish origin *Hisperica Famina*, which shows a "Martianus-like" love for strange Latinity. But one of the most recent students of that text, Michael Herren, does not consider this eccentric language evidence for Martianian influence. Discussing the problem of the genre and function of the *Hisperica Famina*, he says that the idea that it "constitutes some kind of encyclopedia, modelled after the *De Nuptiis* of Martianus Capella, can scarcely be maintained, if only because there is no conclusive evidence that the faminators knew Martianus".[58]

There is no consensus on whether Isidore of Seville had read and used *De nuptiis* for his encyclopedic work *Etymologiae*. In his dissertation, Willis published a list of parallels, and Fontaine finds evidence for the thesis that Isidore used *De nuptiis* as a source in a way which he describes as "discret mais suivi".[59] Leonardi, however, denies Isidore's usage of *De nuptiis*, except for book VII of the work, *De arithmetica*, which he had at hand when he wrote

---

[56] See also E. AUERBACH on this passage: *Literatursprache und Publikum*, p. 79-80.

[57] J. WILLIS, *Martianus Capella and his Early Commentators*, p. 24; M. MANITIUS, *Geschichte der lateinischen Literatur* I, pp. 156-158.

[58] M.W. HERREN, *The Hisperica Famina*, vol. I, p. 13.

[59] J. FONTAINE, *Isidore de Séville et la culture classique dans l'Espagne wisigothique*, vol. 2, p. 858. In the *Index locorum*, pp. 969-970, more evidence is summed up. Manitius treats the relation between Isidore and Martianus in *Geschichte der lateinischen Literatur* I, p. 57.

his *Liber de numeris*.[60]

Manitius and Willis also argued in favour of the possibility of Martianian influence in Bede's works,[61] but Stahl pointed out that the instances they cite can also be found in Pliny's *Natural History*.[62] Thus there is no indisputable proof that Martianus was read and studied in the insular world in the pre-Carolingian period.

In conclusion it can be said that the fate of *De nuptiis* during the sixth to eighth centuries remains unknown. The earliest manuscripts contain only small parts of the text, and are dated in the eighth century or the first decade of the ninth century. They are described by Leonardi in his census of manuscripts. The oldest one, *Karlsruhe, BL, Reich. Fragm. 136*, is dated in the late eighth century, probably written in a continental scriptorium but in Anglosaxon minuscule. It contains part of book III (*Grammatica*), sections 313-319, and could have been part of a grammatical miscellany. The fragment survived in the binding of *Reich. 199*, a ninth- or tenth-century manuscript containing Bede and Hraban Maur.[63] The second one, a manuscript Bischoff dated in the late eighth or early ninth century, also contains material from the book on Grammar: *Würzburg, UB, M.p.misc.f. 5a*.[64] It is located in a West or Southwest German scriptorium and written in a minuscule with insular features. It contains the second book of Cassiodorus' *Institutiones*, works on the computation of the date of Easter by Julius Severianus and excerpts of Martianus Capella's third book, sections 300-309 and 321-324.

Apart from these two fragments, however, there is no evidence of knowledge of *De nuptiis* at all in the late eighth century and first decades of the ninth century. There is one poem, attributed to

---

[60] C. LEONARDI, *I codici* I, p. 461 n. 94; and *id.*, *Intorno al Liber de numeris di Isidoro de Siviglia*.

[61] J. WILLIS, *Martianus Capella and his Early Commentators*, p. 25; M. MANITIUS, *Geschichte der lateinischen Literatur* I, p. 77.

[62] W.H. STAHL, *Martianus Capella and the Seven Liberal Arts* I, p. 61.

[63] *I codici* II, No. 75 (p. 57). See E.A. LOWE, *Codices Latinorum Antiqui* VIII, No. 1129. In his bibliography Leonardi also mentions A. HOLDER, *Die Reichenauer Handschriften* II, Leipzig, Berlin 1941, pp. 583-586.

[64] *I codici* II, No. 240 (pp. 495-496). See E.A. LOWE, *Codices Latinorum Antiqui* IX, Nos. 1401-1402.

Walahfrid Strabo, which could serve as evidence for his familiarity with the text: its language breathes Martianus Capella, so to speak. This would bring us to the time of Louis the Pious.[65] A letter of Prudentius, bishop of Troyes, to John the Scot, dated in 851, is the first certain evidence to break the silence. But these pieces of evidence take us well into the Carolingian period.

De nuptiis *in the Carolingian period*

1. *Carolingian manuscripts: Leonardi's census*

In 1959/1960, Claudio Leonardi published a large article in three parts on the manuscripts of Martianus Capella's *De nuptiis*, parts of the text, abbreviated versions of or commentaries on the encyclopedia.[66] The first part of the article is presented as an introduction to remaining parts. It provides an extensive background on questions of the date of *De nuptiis*, the earliest history of the text, Carolingian evidence for reception and dissemination, new waves of interest in the encyclopedia in the 12th and 14th century and the changed character of this interest, and the function of the text in the different periods. The second and third

---

[65] It is a poem in which a student praises his master, full of more or less explicit references to *De nuptiis*: (ed. E. DÜMMLER, MGH Poetae I, Carm. 5,50)

Laudent concordes sueto te more sorores,
Est quorum numerus bis quater atque semel,
Quae, testante mytho, volitantes alite multo
Auricomo comites intonuere polis.
Usibus una meis tantum restabat in arvis,
Quam vexit cygnus tardus ad istud opus.
(May the Sisters in concord praise you in the usual way,
Whose number is twice four plus one,
Who, according to the myth, fluttering with great speed,
Thunder in the heavens together with the Golden Haired.
Only one remained in the ploughed fields I used,
And a slow swan goaded her to this work.)

B. PABST, however, doubts the attribution, and argues that the poem seems to fit more in the context of the 940's, the time in which the reception of Martianus Capella was at its zenith. (*Prosimetrum*, pp. 215-217.)

[66] *I codici* I&II.

part are devoted to the real issue; they form a list of manuscripts from the 8th to the 16th century, provided with a short description of date and place (where possible), content and bibliography. As the details of the various commentaries and individual manuscripts will be treated elsewhere in this study, this section will focus on the place of Leonardi's census in modern scholarship on Martianus Capella, his methodological remarks and his general conclusions on the function and dissemination of the text.

Collecting and studying manuscripts of the encyclopedia throughout the Middle Ages, Leonardi was soon led to the conclusion that the work reached a peak of popularity in the ninth century. There remain today 49 ninth-century manuscripts, most of which contain a complete encyclopedia with its nine books and are provided with glosses and alternative readings.[67] The large-scale activity of copying, correcting and glossing the text has obscured the history of the text; commentaries, corpora of glosses and versions of the text itself have become inextricably entangled. The composition of a clear stemma of either text or glosses proved to be impossible.

Instead of trying to unravel the knot of corrupted and intermingled texts, as previous scholars had done, Leonardi wanted to throw light on the history of *De nuptiis* and its reception by means of studying the external features of the manuscripts themselves. Not in vain, as he concludes,[68] because as it turned out he was able to establish the place of origin of many manuscripts. Two methodological remarks precede the presentation of general conclusions, drawn from the census of manuscripts. First, Leonardi states that he payed special attention to manuscripts that come from a non-monastic background, such as manuscripts produced for, or in, cathedral centres or chapters. Since these sources were preserved in more quiet surroundings and were less exposed to study, correction and travelling, Leonardi suggested that they were less contaminated and delivered a purer textual tradition. Secondly, he put the manuscripts that can be related to certain individuals at the

---

[67] To contextualize this number, one should know that there are only two fragments of the text from the eighth century, 22 tenth-century, and 26 eleventh-century manuscripts.

[68] *I codici* I, p. 451.

centre of his investigation, and hoped thus to be able to study how the emendations and solutions for textual problems of Carolingian masters were disseminated to different parts of the Frankish empire.[69]

In his reconstruction of the transmission of *De nuptiis* Leonardi argues that in the course of the ninth century the encyclopedia gradually acquired the status of a school text. At first this is apparent in the production of compilations and the gathering of different readings in one codex, then it is shown by the activities of the ninth-century commentators, i.e. John the Scot Eriugene, Martin of Laon and Remigius of Auxerre. The point in time where Martianus' handbook of the arts seems to have become a standard schoolbook for basic knowledge in secular science is dated by Leonardi before the middle of the century, ca. 830-850.[70] The peak of glossing activity is dated in the reign of Charles the Bald (840-877). According to Leonardi, the first and second generation after Alcuin were the ones that revived the interest in Martianus Capella, seeing his work as the ideal textbook for Alcuin's educational programme: an introduction on a somewhat higher level to the seven liberal arts, surrounded by ancient mythology and full of cosmological remarks. Nevertheless it was the third generation, of Martin of Laon and John the Scot, which composed unified sets of glosses.

The first two books of the encyclopedia, the frame-story, usually received the highest amount of glosses. The books on the trivial and quadrivial arts sometimes found their place in topical collections.[71]

---

[69] I found both preset focusses of little value to the present research, since none of the manuscripts at the centre of this investigation can either be tied to a cathedral centre or chapter, or to certain individuals.

[70] I do not agree with this conclusion. See on this the conclusion of Chapter 2 below, and see also my forthcoming article *Ninth-century glosses*. I would date the entry of *De nuptiis* into the school curriculum at the end of the ninth or beginning of the tenth century, when Remigius of Auxerre composed his commentary.

[71] Leonardi mentions, for example, eleven collections containing Book IV (*Dialectica*) and texts on dialectic from Porphyry, Boethius, Ps.-Apuleius, Ps.-Augustin; two combinations of Book VII (*Arithmetica*) with Boethius' *De institutione arithmetica*; five miscellanies of book VIII (*Astronomia*), Hyginus and Macrobius' commentary on Scipio's dream; the gathering in one codex of texts on music by Augustine, Boethius and Martianus (Book IX, *Harmonia*) together with

In the ninth century, however, there is evidence to suggest an aim for complete versions of the work, supplied with alternative readings and glosses from several corpora.[72] The manuscript *Paris, BnF, lat. 12960* is unique in that it strived to unite three ninth-century commentaries.[73]

As for the dissemination of *De nuptiis*, Leonardi proposed the following reconstruction. Basing himself on the insular characteristics of what he thought to be two of the oldest manuscripts, *Leiden, UB, BPL 88* and *Cambridge, Corpus Christi College Library, Ms. 153*, and on the Irish origin of two of the most prominent commentators, i.e. John the Scot Eriugene and Martin of Laon, he suggested that the text was preserved in the insular world during the sixth to eighth centuries and was brought to the continent in the eighth century. In view of the problem of the apparent absence of direct traces of the text in the insular world,[74] and detailed studies on the two manuscripts mentioned,[75] this reconstruction can no longer be upheld. Nevertheless, it has often been argued that Irish scholars on the continent were among the first to draw attention to the work.[76] Leonardi locates the earliest

---

the ninth-century anonymous treatise *Musica Enchiriadis* as in the manuscripts *Bamberg, SB, Class. 9; Ivrea, BC, Ms. 84; Paris, BnF, lat. 7200. I codici* I, pp. 467-468, nn. 126-131.

[72] Throughout the Middle Ages Leonardi counts few more than 50 manuscripts containing the complete work with its nine books, of which 23 are to be dated in the ninth and tenth centuries. *I codici* I, pp. 462-463, nn. 103-105.

[73] *I codici* I, p. 463.

[74] Cf. section 2 (*De nuptiis* in the pre-Carolingian period) above.

[75] Bishop concluded that ms. *CCC 153* was modelled after a continental exemplum —though with Irish features. He noticed that there is only one glossing hand with a definite Irish origin to detect in ms. *BPL 88*. (T.A.M. BISHOP, *The Corpus Martianus Capella*, cf. the detailed description in Chapter 2 below). Leonardi argued the same in *Glosse eriugeniane a Marziano Capella*. Préaux does not mention either of these as a parent of a string of other manuscripts (cf. *Les manuscrits principaux*, pp. 78-80).

[76] See, e.g., C. LUTZ, *Martianus Capella*, in: *CTC* II, p. 368; É. JEAUNEAU, *Les écoles de Laon et d'Auxerre*, pp. 503-506; J. PRÉAUX, *Les manuscrits principaux*, pp. 87-88 and 92-93; J. CONTRENI, *The Irish "colony" at Laon*. Contreni stresses here, however, the dynamic character of the concentration of Irish scholars in the orbit with Laon at its centre. Although he is "impressed, first of all, by the number of Irishmen who can be traced to this one cultural centre," he finds it "difficult to isolate a specifically Irish element or quality in the cultural life of Laon." Therefore

(ninth- and tenth-century) manuscripts in Corbie,[77] Lorsch,[78] Auxerre,[79] Fleury,[80] Saint-Oyan,[81] Reims,[82] Soissons,[83] Paris,[84] Sankt Gallen,[85] Reichenau,[86] Trier,[87] Corvey,[88] Freising,[89] Montecassino,[90] Cologne,[91] Chartres[92] and Ivrea.[93] The centre of production seems to be Northeastern Francia, the area in between Corbie, Laon, Reims and Auxerre, possibly centred around Charles the Bald himself. From there, the text and glosses spread throughout the Frankish Empire. A separate centre of study seems to have been located farther East, in Sankt Gallen and Reichenau.

---

he speaks rather of a full assimilation and integration of Irish scholarship into a "new European culture." (p. 67)

[77] *St. Petersburg, Class.lat. F.V.10; Paris, BnF, lat. 8670, lat. 13955* and *lat. 12960.*

[78] *Vaticano, BAV, Pal.lat. 1577*; probably *Wien, NB, Ms. 266* and *Leiden, UB, BPL 36.*

[79] *Leiden, UB, Voss.lat.F. 48; Paris, BnF, Nouv.Acq. lat. 1615; Vaticano, BAV, Reg.lat. 1535.* To this list, Guy LOBRICHON has added several manuscripts which were produced or kept in Auxerre in the ninth century: *Bamberg, SB, Class. 39; Bern, BB, Ms. 56B*; and *Oxford, BL, Laud.lat. 118.*

[80] *Orléans, BM, Mss. 261* and *263;* the manuscripts *Paris, BnF, Nouv.Acq. lat. 1615; Vaticano, BAV, Reg.lat. 598* and *Orléans, BM, Mss. 191* and *309* became later part of the library of Fleury.

[81] *Besançon, BM, Ms. 594.*

[82] A large part of *Leiden, UB, BPL 88; Vaticano, BAV, Vat.Lat. 645*; tenth-century glosses in *Bamberg, SB, Class. 39.*

[83] *Paris, BnF, lat. 8669* and *lat. 7200.*

[84] *Paris, BnF, lat. 13026.*

[85] *Sankt Gallen, SB, Ms. 381* and *872*; *Napoli, BN, IV.G.68* and *Bruxelles, BR, Ms. 9565-9566.*

[86] *Karlsruhe, BL, Reich. 73* and probably *Reich. 137.* Préaux, however, expresses doubt on the localisation of *Reich. 73* in Reichenau. This manuscript, which was made during the reign of Louis the Pious and is one of the oldest sources, has been ascribed to the region of Fleury and South-Germany. *Les manuscrits principaux*, p. 81 and n. 18.

[87] *Trier, BP, Ms. 100.*

[88] *Wolfenbüttel, HAB, 132 Gud.lat.2°.*

[89] *Kremsmünster, SB, Fragm. I.20* and *München, BS, lat. 13084.*

[90] *Montecassino, BA, Ms. 332*; *Napoli, NB, IV.D.22*^bis^ is supposed to be a younger copy of a manuscript from Montecassino.

[91] *Köln, DB, Ms. 193*; and probably *London, BL, Harl. 2685.*

[92] *Chartres, BM, Ms. 103* and *105*, the first of which was destroyed in World War II.

[93] *Ivrea, BC, Ms. 84.*

Although Préaux, who studied the history of the text in great detail,[94] also dated the earliest stages of reception in the reign of Louis the Pious, with a peak during the reign of Charles the Bald, he disagreed with Leonardi on several other points. Préaux, first of all, did not believe that the insular world was the starting point of the dissemination of the manuscripts. In his opinion, the 'Scotti', who had settled in the valley of the Rhine (Mainz) and in the Frankish empire (Laon, Auxerre, Ferrières, Soissons, Fleury, Corbie and Reims), were driving forces behind the popularity of *De nuptiis*, but the oldest exempla from which the texts were copied go back to a sixth-century, Roman manuscript in *scriptio continua*, written in rustic capitals. Of the two manuscripts Leonardi mentioned as "belonging to the group of oldest ones" — *Cambridge, CCC, Ms. 153* and *Leiden, UB, BPL 88* —, the first one is not mentioned at all in Préaux' paper on the most important manuscripts of Martianus Capella, while the other is classified as a manuscript of the fourth or fifth generation.[95] There are six manuscripts in which the *scriptio continua* of the oldest exemplum has left its mark: *Karlsruhe, BL, Reichenau 73; Vaticano, BAV, Reg.lat. 1987; London, BM, Harl. 2685; Paris, BnF, lat. 8670; Bamberg, SB, Class. 39* and *Vaticano, BAV, Reg.lat. 1535*. Opinions differ on the place of origin of the first three manuscripts. *Reich. 73* has been located alternatively in South Germany and in the area around Fleury.[96] The Vatican manuscript *Reg.lat. 1987* has been attributed to Northeastern France by Bischoff, but has been placed in the Loire valley or in Laon by others.[97] Leonardi locates the Paris manuscript *lat. 8670*, in Corbie,[98] and the one that is now in Bamberg (*Class. 39*) in the Loire valley (Fleury, Ferrières, Auxerre). The origin of the second Vatican manuscript, *Reg.lat. 1535*, probably also lies in Auxerre.[99]

According to Préaux's reconstruction of the textual history of *De nuptiis*, the oldest exemplum in *scriptio continua* was first

---

[94] J. PRÉAUX, *Les manuscrits principaux.*

[95] J. PRÉAUX, *Les manuscrits principaux*, pp. 78-80.

[96] *Ibid.*, p. 81 n. 18.

[97] See the detailed description in Chapter 2.

[98] *Les manuscrits principaux*, p. 78. Bischoff located this manuscript in Corbie.

[99] *Ibid.*, p. 78. He does not give a possible localization for the London manuscript, *Harl. 2685*.

rediscovered and copied in the region Fleury - Ferrières - Auxerre. These copies were transmitted to Corbie and other monastic centres in the vicinity to be studied and copied there. Préaux freely speculates on all kinds of possibilities: the famous Lupus of Ferrières could have been involved in the first stages of the study of *De nuptiis*, or Hadoard of Corbie; there could have been a collaboration between John the Scot and abbot Wenebertus of Schüttern; a role could have been played by Martin of Laon, Fergus and Probus — two *Scotti* from Laon —, Muridac, Heiric and Haimo of Auxerre. The uncertainty ends, finally, with the teachings of Remigius of Auxerre, who composed a substantial commentary that was so widespread that his version of the text and his readings were clearly taken as a standard in the course of the tenth century.

Comparing the models of textual history as presented by Leonardi and Préaux, we can conclude that both assign an active role to the Irish or Scottish scholars who settled in the Frankish kingdom during the eighth and ninth century. Both view the period of the reigns of Louis the Pious and Charles the Bald as the peak of copying and glossing activity. Both, furthermore, acknowledge the region Corbie - Laon - Auxerre as a central region from which the text was distributed to the far corners of the kingdom. Leonardi, however, suggested that the earliest phases of the history of *De nuptiis*, are to be sought in the insular world, but based himself on shaky grounds. Préaux, on the other hand, divided the history of the text into several stages, which can be likened to concentric circles moving farther and farther away from a single late antique or Merovingian exemplum in *scriptio continua*. The exemplum is lost to us, but six manuscripts of its closest circle survived. The earliest reception can thus be located in the Loire valley, and disseminated from there to Corbie. Corbie, in turn, came to play an active role in the further dissemination of *De nuptiis*.

2. *The ninth-century gloss collections: The Anonymous corpus of glosses*

2.1 *The manuscripts*

The author of the ninth-century commentary generally considered as the oldest is still unknown. In 1944, Lutz edited the glosses under the name of Dunchad, but she based her edition on a single, incomplete source: *Paris, BnF, lat. 12960*, a late ninth-century miscellaneous manuscript written in the monastery St. Pierre in Corbie.[100] In this manuscript, the glosses are copied in a running form, and they concern only the last part of book II, the whole of book IV (*Dialectica*) and the first third of book V (*Rhetorica*). She was under the impression that *Paris, lat. 12960* was the only remaining witness of the commentary. In 1953,[101] however, Jean Préaux added several manuscripts to her one-item list and he continued to add more until his death in 1978. Furthermore, he re-opened the debate on authorship: he rejected Dunchad, and proposed the ninth-century Irish scholar Martin of Laon as author.[102]

In 1953, he presented a list of five ninth-century manuscripts which contained glosses from the Martin of Laon commentary in an interlinear and marginal form, that is, with glosses written in the margins and interlinear spaces of the text of *De nuptiis*: *Leiden, UB, BPL 87, BPL 88* and *Voss.lat.F. 48*; *Oxford, BL, Laud.lat. 118*; and *Paris, BnF, lat. 8670*. He signalled the manuscripts *Leiden, BPL 88* and *Voss.lat.F. 48* as the best witnesses of the commentary. In 1956, he added the manuscript *Besançon, BM, Ms. 594* to this list,[103] and in 1959, *Vaticano, BAV, Reg.lat. 1987*; he considered this last manuscript an even better witness of the commentary than the two Leiden manuscripts.[104] He also noticed that the glosses on book IX in the Leiden manuscript *BPL 88* do not belong to the Martin of Laon commentary and should be attributed to John the Scot.

First in an article,[105] and then in his census of Martianus Capella manuscripts, Leonardi briefly described the manuscripts mentioned above[106] and added some of his own: *Firenze, BML, San Marco 190* and *St. Petersburg, NL, Class.lat. F.V.10*; *Napoli, BN,*

---

[100] C.E. LUTZ (ed.), *Dunchad Glossae in Martianum.*

[101] J. PRÉAUX, *Le commentaire de Martin de Laon.*

[102] See on this the next section: *Debate on authorship.*

[103] J. PRÉAUX, *Un nouveau manuscrit de Saint-Gall*, p. 225 n. 11.

[104] J. PRÉAUX, *Deux manuscrits gantois.*

[105] C. LEONARDI, *Raterio e Marziano Capella*, p. 84 n. 3.

[106] *I codici di Marziano Capella* II, Nos. 83, 84, 89, 144, 161, 20 and 210 (following the above used order).

IV.G.68,[107] *Vaticano, Reg.lat. 1535* and *Vat.lat. 8501* and possibly *Orléans, BM, Ms. 191*.[108] The manuscript *Trier, Bibl. des Priesterseminars, Ms. 100* was also added to this list, but only after some confusion: Leonardi identified the glosses as belonging to the commentary of John the Scot,[109] Lutz disagreed and listed the manuscript among manuscripts with the Martin of Laon commentary.[110] In the same article, however, she wrote that the glosses "come from an earlier commentary which has not yet been identified."[111] In 1974, McDonough published his analysis of the manuscript and concluded that it was a witness of the commentary attributed to Martin of Laon.[112]

Of this list of fifteen manuscripts, twelve are dated around the middle or in the second half of the ninth century. The three exceptions are *St. Petersburg, NL, Class.lat. F.V.10* (a tenth-century manuscript from Corbie), *Firenze, BML, San Marco 190* (an eleventh-century manuscript from Milan), and *Vaticano, BAV, Vat. 8501* (a fifteenth-century miscellany containing glosses from the Martin of Laon commentary on book I and II). In only two manuscripts — *Paris, BnF, lat. 12960* and *Orléans, BM, Ms. 191* — the commentary was copied in a running form, that is independent of Martianus' text. In the thirteen others, the glosses are presented in a marginal and interlinear form, next to the text of *De nuptiis*. Most of them, again, contain the complete encyclopedia with its nine books, with the exception of the two copies of the commentary in running form (*Paris, BnF, lat. 12960* and *Orléans, BM, Ms. 191*[113]), the fifteenth-century manuscript *Vaticano, BAV, Vat. 8501* (Book I-II), *Napoli,*

---

[107] A manuscript that was also mentioned by Lutz in her edition of the commentary: *Dunchad Glossae in Martianum*, 'Appendix 2: Note on a Manuscript of Martianus Capella,' pp. 60-62.

[108] C. LEONARDI, *I codici* I, p. 455 n. 59 and p. 463 n. 109; and II, Nos. 60, 92, 131, 208, 220 and 134 (again following the order used above).

[109] *Ibid.*, No. 189.

[110] *Martianus Capella*, in: *CTC* II, p. 370.

[111] *Ibid.*, p. 371.

[112] C. MCDONOUGH, *Trier, Bibliothek des Priesterseminars Ms. 100*, p. 57 n. 8. In this article, however, McDonough compares the glosses of the Trier manuscript with the two manuscripts kept in Cambridge, containing an unidentified commentary.

[113] Cf. J. PRÉAUX, *Le manuscrit d'Avranches 240*, pp. 137-138.

BN, IV.G.68 (a miscellany from St. Gallen, containing only parts of book IV) and *Trier, Bibl. des Priesterseminars, Ms. 100* (Book I-III with glosses).

As far as it has been possible to pronounce upon the localisation of the manuscripts, the region of Northeastern France (Corbie-Soissons-Reims-Auxerre-Fleury) seems to have been the centre of the production of manuscripts with this commentary. The manuscripts *Paris, BnF, lat. 12960, lat. 8670* and *St. Petersburg, NL, Class.lat. F.V.10* have been attributed to Corbie with some certainty, and Préaux suggested Corbie as the place of origin for *Leiden, UB, BPL 87* also.[114] Historical evidence tied both *Vaticano, BAV, Reg.lat. 1987* and *Leiden, UB, BPL 88* to Corbie or Reims.[115] Préaux located the manuscripts *Oxford, BL, Laud.lat. 118* and *Besançon, BM, Ms. 594* in the region Auxerre-Soissons.[116] Two manuscripts, *Leiden, UB, Voss.lat.F. 48* and *Vaticano, BAV, Reg.lat. 1535*, were associated with the monastery St. Germain in Auxerre.[117] For *Orléans, BM, Ms. 191* the monastery St. Benoit sur Loire in Fleury has been suggested as place of origin.[118] *Napoli, BN, IV.G.68*, finally, is linked to St. Gallen by a possession mark, and is thus the only mansucript of the commentary localised outside the French region.[119]

2.2 Debate on authorship

The corpus of glosses has been attributed to different authors. Lutz attributed it to the Irish bishop Dunchad,[120] of whom we know

---

[114] B. BISCHOFF, *Hadoardus*; J. PRÉAUX, *Les manuscrits principaux*, p. 79.

[115] J. PRÉAUX, *Deux manuscrits gantois*. See also their descriptions in Chapter 2.

[116] J. PRÉAUX, *Les manuscrits principaux*, pp. 78-79. See their detailed descriptions in Chapter 2.

[117] For *Leiden, UB, Voss.lat. F. 48*, see the description in Chapter 2. See also J. PRÉAUX, *Les manuscrits principaux*, p. 79.

[118] C. LEONARDI, *I codici* II, pp. 415-416.

[119] In *Les manuscrits principaux*, J. PRÉAUX also links the text version of *De nuptiis* in manuscript *Trier, Bibliothek des Priesterseminars, Ms. 100* to the region Lake of Constance-Sankt Gallen.

[120] Lutz pays almost no attention to problems of identification in her edition. She does, however, treat the debate on authorship in appendix 3, 'Dunchad,' of her edition of the Eriugenian corpus; *Iohannis Scotti Annotationes in Marcianum*, pp. 229-235.

only that he once taught at the monastery of Saint Remi in Reims. In this she followed similar suggestions by Traube and Manitius[121], who pointed to a note on an inserted folio in the manuscript *London, Brit.Mus., Reg. 15.A.33*, which reads:

> Commentum Duncaht Pontificis Hiberniensis quod contulit suis discipulis Monasterii Sancti Remigii docens super astrologia Capellae Varronis Martiani.
> (Commentary of Dunchad, bishop from Ireland, which he collected for his pupils at the monastery of Saint Remi, when he taught on the subject of the astronomy of Capella Varro Martianus.)

The passage is followed by notes on the computation of time. The annotation was written on a folio that was later inserted in the manuscript,[122] but Dunchad was nevertheless generally accepted as the author of the commentary. Nine years later, Préaux denied the attribution and proposed a new candidate: Martin of Laon.

In an article on the manuscripts which supposedly contained the commentary,[123] he observed that the notes on the computation of time, attributed to Dunchad in *London, Brit.Mus., Reg. 15.A.33*, did not occur in any of these manuscripts and were, moreover, written in a different style. Préaux concluded that Dunchad probably wrote a commentary on book VIII on astronomy, but that the commentary edited by Lutz under his name could not have been his work. In a preliminary exploration of the commentary, he made the following observations. First of all, the glosses of 'Dunchad' and those of John the Scot are significantly different. The commentary of Remigius of Auxerre, on the other hand, seems quite similar,

---

[121] L. TRAUBE, *Computus Helperici*, pp. 103-104; M. MANITIUS, *Zur karolingischen Literatur*, p. 57; *id.*, *Geschichte der lateinischen Literatur* I, pp. 255-256; *id.*, in: *Didaskaleion* I (1912), pp. 138-172.

[122] This fact was missed by M. ESPOSITO, *Note on a ninth-century commentary*. He was immediately corrected by R. FLOWER, who also added that the folium in question was furthermore written in a different hand: *Irish commentaries*, pp. 566-567.

[123] J. PRÉAUX, *Le commentaire de Martin de Laon*. The manuscripts in question are *Leiden, UB, BPL 87* and *BPL 88*; *Oxford, BL, Laud.lat. 118*; *Paris, BnF, lat. 8670.* To this list he added in a postscript *Leiden, UB, Voss.Lat.F. 48*—a manuscript which he had not seen when writing his article, but which he considered of great importance for the reconstruction and edition of the commentary.

although Remigius, who does mention the name of John the Scot as his source at several instances, never mentions the name of another ninth-century schoolmaster. The anonymous commentator and Remigius focus their attention on the same kind of information: etymologies, grammar, history, geography and mythology. Unlike John the Scot, they seem to have little interest in philosophical issues. When the anonymous commentary and that of John the Scot offer contradictory interpretations or explanations, Remigius tends to prefer the anonymous commentator. These observations led Préaux to the conclusion that the anonymous scholar had to be someone close to Remigius, such as his own teachers: Martin of Laon or Heiric of Auxerre. Préaux further argued that the inconsistency found in the spelling of the word "dia-/dyalectica" in the anonymous commentary could never have been made by Heiric; in his *Collectanea* Heiric makes a distinction in the meanings of "dia" (i.e., *ex*) and "dya" (i.e., *duo*), whereas the anonymous commentator uses both spellings indeterminately. Heiric is thus dismissed as a possible author. This argument, however, is invalid, since in the anonymous commentary itself a gloss occurs in which the same distinction is argued:

> [30.18-9] Dia si per *i* nostrum scibitur, nihil aliud nisi ex significat, sed per *y* grecum, duo ostendit.
> (When *dia* is written with our *i*, it means nothing but *ex*, but when it is written with a Greek *y*, it means two.)

Préaux noted a number of signs that point to Martin of Laon as the author. For example, in the commentary there is a special interest in Greek vocabulary, and we still have a Greek-Latin glossary in the autograph of Martin of Laon (*Laon, BM, Ms. 444*). At the heart of the argument, Préaux points to parallels between anonymously delivered notes on the names of the nine Muses in this Greek-Latin glossary (which are surrounded by material that can be attributed with certainty to Martin of Laon), and a long gloss on the same subject in the anonymous commentary to Martianus Capella. The notes go back to Fulgentius the Mythographer and contain references to Martianus Capella. These striking parallels give Préaux enough evidence to replace 'Dunchad' with 'Martin of Laon'.

Préaux's hypothesis was accepted by many,[124] but the scholar who studied Carolingian Laon, its manuscripts and its masters in the greatest detail, John J. Contreni, did not agree.[125] In 1976 Contreni published two papers in which he cast doubt on the authorship of Martin of Laon, the famous ninth-century schoolmaster of Laon.[126] In both he argued that Préaux made a good case for the *possibility* of Martin's authorship, but collected only weak, circumstantial *evidence*. The parallels that occur between the Laon manuscript *444* and the commentary on *De nuptiis* are indeed striking, but they could equally indicate that a certain treatment of the ultimate source, Fulgentius the Mythographer, was known to both writers; that the writer of the note on the name of the muses knew and cited the gloss on the same subject from the *De nuptiis* commentary; or that both writings ultimately go back to the same source. It remains uncertain whether Martin of Laon is the author of both. Furthermore, not one of the Martianus Capella manuscripts collected by Leonardi is located with certainty in Laon. It seems clear that there was glossing activity in the region of northeastern France (Auxerre-Soissons-Laon-Corbie), but there are many other possible authors within this regional restriction: Heiric of Auxerre, Lupus of Ferrières, Haimo of Auxerre, Muridac (an Irish grammarian of Auxerre), John the Scot's student Winibertus, or Dunchad again.[127]

Although the discussion still finds itself in this stalemate — the one camp unable to produce stronger arguments, the other unable to point to one probable author —, the name of the ninth-century schoolmaster seems to have stuck to the commentary. I have decided, for the sake of clarity, to refer to the commentary as

---

[124] See, *inter multos alios*, C. LEONARDI, *I codici* I, pp. 463; *id.*, *I commenti altomedievali ai classici pagani*; and W.H. STAHL, *Martianus Capella and the Seven Liberal Arts* I, pp. 63-64.

[125] See his comprehensive study: *The Cathedral School of Laon*.

[126] *Three Carolingian Texts*, pp. 808-813; and *A note on the attribution*, pp. 451-452. See also his article *The Irish in the Western Carolingian Empire*.

[127] This point of view was accepted by many. See, for example, D. GANZ, *Heiric d'Auxerre glossateur du Liber glossarum*, p. 301; B. PABST, *Prosimetrum*, p. 218.

"Anonymous" in this study.[128] My own position in the discussion will be stated and clarified in the conclusion of Chapter 2 on the manuscripts of the Anonymous commentary.

2.3 *Character of the commentary*

Even from the incomplete version edited by Lutz it is clear that the size of the commentary is impressive.[129] It contains many short glosses, such as identifications, definitions, synonyms and etymologies, as well as a considerable number of lengthy explanations, excursions and rephrasings. The commentator makes remarks on rhetorical figures, such as litotes, hypallage and synecdoche, on grammar and on syntax. Lutz was wrong when she wrote that "although the study of metrics was an important aspect of grammar, no mention is made of the metres of the various poetic interludes,"[130] for the margins of most manuscripts with this commentary contain metrical identifications, usually written in small capitals and sometimes highlighted with coloured ink (red, yellow or green). This phenomenon has led Shanzer to speculate on a link to Lupus of Ferrières, who specialised in metrical annotations; he wrote a treatise on the metres of Boethius' *De consolatione Philosophiae* and metres have been identified in his autograph in the margins of a Prudentius manuscript.[131] Furthermore, identifications of the speaking or acting characters in the story are sometimes added in (poetic) passages, and are also usually written in capitals and/or coloured ink. Another feature of the commentary is its emphasis on and careful explanation of Greek vocabulary. Besides these focuses on grammar, syntax, and vocabulary, the commentator seems particularly interested in mythology and allegory.

In her introduction to the edition, Lutz listed the sources used in the commentary on part of book II, book IV and part of book

---

[128] I am aware of the fact that, in a different context, this apparently neutral solution can also cause confusion. There are a yet unknown number of further anonymous commentaries on *De nuptiis* from the later Middle Ages. On the other hand, it seemed equally misleading to speak of the "pseudo-Dunchad" or "pseudo-Martin" commentary, since the possibility of authorship of both is still kept open.

[129] *Dunchad Glossae in Martianum*, p. xiii.

[130] *Ibid.*, pp. xiii-xiv.

[131] D. SHANZER, *Review article*, p. 72 n. 60.

V.[132] These were not easy to identify, especially in the incomplete form of manuscript *Paris, BnF, lat. 12960*. The commentator rarely refers explicitly to a source, and a large number of the glosses could also go back to Martianus Capella's text itself. Additional sources that are definitely used in the Anonymous commentary are Macrobius' *Commentarium in Somnium Scipionis*, Virgil's *Aeneis*, Solinus' *Collectanea*, and Aulus Gellius' *Noctes Atticae*. The sources for mythological material are Servius' commentary on Virgil, Fulgentius (to a larger extent even than Lutz suggested),[133] Augustine and the second Vatican Mythographer. The astronomical glosses seem to reach back to "Bede, Macrobius and indirectly Pliny,"[134] whereas Isidore and Cassiodorus may have been used for glosses on rhetoric. Lutz further notes the commentator's supposed familiarity with Porphyry's *Εἰσαγωγή*, and with Aristotle's *Categories* and *De interpretatione*, probably — as she notes — "through the medium of the Pseudo-Augustinian commentaries and Boethius' commentaries."[135]

Lutz describes the tone of the commentary as didactic, but I would rather characterize it as exploratory. A lemma is usually followed by *scilicet, vel, id est* or *quasi*, quite frequently by *subauditur*, which do perhaps lend a didactic tone to the work. Lutz furthermore pointed to a preference on the part of the commentator for clauses with a particular structure (ending with the subject), and clauses smoothly following a certain prose rhythm, which might serve a mnemotechnic purpose. On the other hand, the interpretation of the text at hand is not the only task the commentator set himself. Whenever possible, the commentator associatively gathers and compares material from different antique and late antique sources; he seems anxious to compose a file of knowledge which is as complete as possible.

---

[132] C.E. LUTZ (ed.), *Dunchad Glossae in Martianum*, pp. xxiii-xxvi.
[133] *Ibid.*, p. xxv.
[134] *Ibid.*, p. xxv.
[135] *Ibid.*, p. xxvi.

3. *The ninth-century gloss collections: The Annotationes of John the Scot Eriugene*

3.1 *The manuscripts*

The second ninth-century author to whom a corpus of glosses is attributed is John the Scot Eriugene, an Irish scholar who spent part of his life at the royal court of Charles the Bald. Cora Lutz was the first to edit the entire collection of glosses attributed to him in 1939, under the title *Annotationes in Marcianum*. She based her edition entirely upon the single manuscript known to her: *Paris, BnF, lat. 12960*, a late ninth-century miscellany of commentaries, among others on Martianus Capella, written in the monastery St. Pierre in Corbie.[136] A few years after Lutz's edition, Labowsky discovered a second manuscript of John's commentary in a late ninth- or early tenth-century manuscript, now *Oxford, Bodl.Lib., Auct.T.2.19*, probably written in the monastery St. Vincent in Metz.[137] She noted that, while from the second book of the encyclopedia onwards the Metz manuscript could be easily identified with the commentary edited by Lutz, book I differs considerably in the presentation and selection of the lemmata. Jeauneau made this substantially different version of John's teachings on book I available in an edition.[138]

To this list of two manuscripts with John's commentary in a running form Lutz later added a number of manuscripts containing interlinear and marginal glosses.[139] She mentions three manuscripts in which John's commentary is added to the last four books (the quadrivial arts) of *De nuptiis*, while the first five books (frame-story

---

[136] She based her edition of the "Dunchad commentary" on the same source. C.E. LUTZ (ed.), *Iohannis Scotti: Annotationes in Marcianum*. See for a short description of the manuscript pp. xxviii-xxx of the "Introduction". Folios 47r-115v contain the collected glosses of John the Scot. Lutz's edition was severely criticised by J. WILLIS in his unpublished dissertation *Martianus Capella and his ninth-century commentators* — in my opinion his criticisms are only partly justified, as Lutz is editing one corrupt and lacunous manuscript and did not have the opportunity to compare her readings with others. It is true, however, that the edition contains erroneous readings and should be used with caution.

[137] L. LABOWSKY, *A new version of Scotus Eriugena's Commentary on Martianus Capella*.

[138] É. JEAUNEAU, *Le commentaire érigénien*.

[139] C.E. LUTZ, *Martianus Capella*, in: *CTC* II, p. 371.

and trivial arts) are provided with the Remigius commentary, i.e. *Bern, Burgerbibl., 331; Paris, BnF, lat. 8675* and *Wien, NB, 3222*. In his census of manuscripts, Leonardi identified glosses from John in the manuscript *Trier, Bibl. des Priesterseminars 100,*[140] but Lutz and McDonough rejected this identification.[141] The contents of two other manuscripts, *Vaticano, BAV, Pal.lat. 1577* (fols. 1-83) and *Wien, NB, 266* (fols. 1-150), are disputed: Leonardi believed that they contained John's glosses, whereas Lutz attributed the glosses in these manuscripts to Remigius of Auxerre. Lastly, Leonardi identified glosses on book IX in the manuscript *Leiden, UB, BPL 88* (fols. 168-181)[142] as belonging to John's commentary.[143]

3.2 *Debate on authenticity*

The most important evidence for the existence of a commentary on Martianus Capella by John the Scot was found in the introduction written by Prudentius, bishop of Troyes, to his treatise *De praedestinatione contra Johannem Scotum* (851). In the process of phrasing his refutation of John the Scot's treatise on predestination, Prudentius states that John, seeking the guidance of the works of pagan rhetoricians like Martianus Capella instead of the Scriptures and church fathers, at last became desperately entangled in the labyrinth of their mendacious fiction.[144]

In 1973, Schrimpf published a paper on the authenticity of John the Scot's *Annotationes in Martianum* in the proceedings of the

---

[140] C. LEONARDI, *I codici* II, pp. 454-455, 498.

[141] Lutz believed them to come from an earlier, not yet identified commentary (in: *CTC* II, p. 371); McDonough identified them as similar to the Dunchad/Martin of Laon-commentary (C. MCDONOUGH, *Trier, Bibliothek des Priesterseminars Ms. 100*, pp. 56-58).

[142] Not fols. 3-181 as Lutz wrote in the *CTC*, for the glosses added to book I-VIII (fols. 3-167) are not from John's commentary.

[143] *Glosse eriugeniane a Marziano Capella*. In this paper he promised to publish the glosses, but he seems not to have completed this undertaking so far.

[144] PL CVX, c. 1294: Nam ille tuus Capella, exceptis aliis, vel maxime te in hunc labyrinthum induxisse creditur, cujus meditatione magis quam veritati evangelicae animum appulisti. Quin etiam cum legeres beati Augustini libros, quos *De Civitate Dei* adversus paganorum fallacissimas falsissimasque opiniones mirabili affluentia digessit, invenisti eum posuisse ac destruxisse quaedam ex libris Varronis, quibus, quoniam Capellae tuo consona videbantur, potius assentiri quam veridici Augustini allegationibus fidem adhibere delegisti.

colloquium *The mind of Eriugena*.[145] In this article he addresses two questions: 1) the extent to which the original glosses on Martianus Capella of John the Scot are transmitted to us, and 2) whether we should distinguish two versions of his glosses to book I, i.e. a first and a second revised version. For these two questions, which he describes as the focal points of the discussion on authenticity, he provides overviews of the research done and opinions expressed so far. I will discuss these overviews in brief.

Although there were earlier suggestions on the authorship of John for a corpus of glosses in manuscript *Paris, BnF, lat. 12960*, a ninth-century manuscript from Corbie, [146] it was Hauréau who first gathered evidence for this attribution in 1862.[147] He identified the anonymously transmitted glosses as the work of John the Scot, by pointing to several parallels between glosses of Remigius in which he explicitly cites or refers to glosses of John the Scot and glosses found in the Corbie manuscript. Hauréau edited the glosses he needed as evidence for his hypothesis. More than forty years later, Rand picked up the discussion. In an article published in 1906,[148] he confirmed the identification and gave more examples of parallels of this sort. In 1912-13, Manitius edited a selection of glosses to book I-III.[149] Twenty years later, however, he was the first to express doubts about the straightforwardness of the identification, arguing that the corpus found in the Corbie manuscript could in any case not be the complete corpus of John's annotations on Martianus, for Remigius cites several glosses of John that cannot be found in the Corbie manuscript.[150] Lutz does not express any doubts on the authenticity of the entire corpus of glosses in her edition, but later she was convinced by Rand into changing her mind. First in 1940[151] and then in 1941,[152] Rand

[145] G. SCHRIMPF, *Zur Frage der Authentizität unserer Texte von Johannes Scottus*.

[146] Lutz refers to H. Floss, editor of the works of John the Scot in 1853 (cf. PL 122, col. xviii-xvi), who pleaded even then for further research on the Paris manuscripts, since he suspected the author could be identified as John the Scot (*Annotationes*, "Introduction", p. x).

[147] B. HAURÉAU, *Commentaire de Jean Scot Érigène*.

[148] E.K. RAND, *Die Glossen des Johannes*, pp. 81-82.

[149] In: *Didaskaleion* 1 (1912), pp. 157-172; 2 (1913), pp. 43-61.

[150] In: *Neues Archiv* 49 (1930-1932), p. 173.

[151] *How much of the Annotationes in Marcianum is the work of John the Scot?*

took, on the basis of stylistic evidence, the confusing position that book I, IV (*Dialectica*) and VIII (*Astronomia*) could not be part of the "Originalfassung", and that only book I should be considered as a reflection of the authentic form of John's teaching — a position which contradicts itself with regard to book I. In her elaboration of this view,[153] Lutz states that neither the style, nor the underlying method of glossing, nor the content of the glosses provide reliable criteria to distinguish between the work of Remigius and John. Through a close comparison of both authors, she reaches the conclusion that in the Corbie manuscript the glosses added to the quadrivial books (VI-IX) do not hand down the teachings of John, but give a short version of Remigius' commentary, probably compiled by one of his students. She even doubts if John ever glossed these quadrivial books. The glosses to books IV and V are again not John's teachings, but a compilation of them, probably written by one of his students. In her introduction to the edition of Remigius' commentary to *De nuptiis*, she argues that John's corpus, as edited by her in 1939, is an authentic version of his glosses with regard to book I. The rest is either severely abbreviated — in some cases to the point that the sense is distorted or lost —, or intermingled with material from Remigius' commentary.

The history of research on the second point, that is, the possibility of the existence of a first and a second revised version of John's commentary, has its starting point in 1943, when Labowsky discovered a second manuscript with the glosses of John the Scot on *De nuptiis*: *Oxford, Bodleian Library, Auct.T.2.19*, a manuscript located in Metz and dated at the end of the ninth, or beginning of the tenth century.[154] From book II onwards, the two known manuscripts (Corbie and Metz) concur reasonably well, but the glosses to book I are very different. They are similar in the way they interpret the text and the manner of argumentation, but differ, for example, in choice of lemmata. Labowsky suggested that the Metz manuscript contains a first version, which John revised and on

---

[152] A review of Lutz's edition in: *Speculum* 16 (1941), pp. 129-131.

[153] C.E. LUTZ, *The commentary of Remigius of Auxerre on Martianus Capella*, p. 140.

[154] L. LABOWSKY, *A new version of Scotus Eriugena's Commentary on Martianus Capella*.

which he elaborated to turn it into a second version, i.e. the Corbie version. In his unpublished dissertation of 1952, Willis attacked this point of view,[155] arguing that the Metz version does not contain John's glosses, but a compilation, in which his glosses are intermingled with glosses from other commentaries. Liebeschütz pointed in a different direction to explain the differences in the glosses to book I.[156] To his mind the Metz version is the first version of John's teachings, written before 847, that is, written before the attack of Prudentius on his "Capella-mindedness" and use of pagan sources of knowledge. In the second version of his commentary, written between 855 and 860, John rephrases his glosses to book I, to meet Prudentius' objections. This hypothesis found the hesitant approval of, among others, Mathon and Préaux.[157]

Schrimpf, however, proposed a different solution for the differences in the two versions. Carefully reconsidering the text, he calls attention to the didactic, rather than literary, character of the work, to which therefore literary criteria of style and composition do not apply. John's commentary was not composed as a running commentary, a unified set of glosses, but as a collection of glosses on individual words, phrases and ideas. It is in this didactic character that Schrimpf seeks an explanation for the differences in content. His reconstruction of the history of the text, which he himself calls a working hypothesis, contains four steps. First, the writing down of John's teachings in the form of interlinear and marginal glosses; second, — a possible, but not necessary step — the making of a copy in a running form; third, the composition of several distinctive excerpts; and fourth, the copying of these excerpts. To his mind, both the Corbie and the Metz manuscripts are texts that originated in the fourth phase. They evidence the state of the text in a running, excerpted, mixed, and (through copying) corrupted form. This hypothesis found favour in the eyes of

[155] J.A. WILLIS, *Martianus Capella and his Early Commentators*.

[156] *Texterklärung und Weltdeutung bei Johannes Eriugena*; *Zur Geschichte der Erklärung des Martianus Capella bei Eriugena*; *John Eriugena and his cosmological interpretation of Martianus Capella*, pp. 576-578.

[157] G. MATHON, *Jean Scot Érigène et le problème de l'âme universelle*, p. 364; J. PRÉAUX, *Les formes et la signification de la pédagogie des arts libéraux*, pp. 52-53.

Leonardi, although he continued to emphasize the aspect of mixture of the three ninth-century corpora.[158]

This is where the debate on authenticity of the glosses attributed to John the Scot ends. When the Irishman's commentary is described in recent research, usually no doubts about the authenticity of the work are voiced. And in the present study there will also be references to John's commentary on the ninth book, without taking the doubts ventured by Rand and Lutz into consideration. For their solution, the division of the commentary into parts which are definitely, possibly or probably not written by John the Scot, seems to lead us nowhere. In John's commentary on book IX, there is no new evidence which suggests that he could not have been the author, just as there is no evidence to prove that he was. Nevertheless, the style of the commentary seems to show enough unity to assume that it was composed by one and the same author. It is important, furthermore, that in contemporary and later texts John the Scot was credited with the composition of a commentary, even though its transmission is evidently problematic.

3.3 *Character of the commentary*

Most authors who have carried out research into the Martianus glosses of John the Scot have stressed its unique character. Its focus is not only on those aspects typical for Carolingian commentaries in general: philology, textual criticism and alternative readings, nor does it abound in synonyms, explanations on grammar or etymologies. Apart from these subjects, it places special emphasis on the philosophical ideas in Martianus' text. Whenever the author has the chance, he speculates on the Platonic idea of the soul, the physics of the cosmos, the planetary movements and the central importance of the sun. As Liebeschütz puts it, he seems to be interested in the "cosmological theology" of *De nuptiis*;[159] the link, present in this late antique encyclopedia, between the divine characters in the story and their cosmological meaning. This speculative, philosophical accent was noticed by John's contemporaries but did not find followers until the twelfth century.

---

[158] *I commenti altomedievali ai classici pagani*, p. 485; *Glosse eriugeniane a Marziano Capella*, pp. 181-182.

[159] *Zur Geschichte der Erklärung des Martianus Capella bei Eriugena*, p. 134.

Prudentius, as Liebeschütz pointed out, clearly linked John's extraordinary cosmological ideas with his reading of Martianus.[160]

His isolation is also reflected in the sources he uses, of which Lutz drew up a list in her edition of the Corbie version of his commentary.[161] It seemed clear to her that he had access to an (unknown) earlier commentary or commentaries,[162] and she drew up a list of parallel passages that attest the close connection between the Anonymous commentary and John's glosses.[163] Furthermore, there is evidence that John frequently used the Neoplatonic metaphysics of Calcidius' *Commentary on the Timaeus* and Macrobius' *Commentary on Scipio's Dream* to clarify Martianus' cosmological system. Pliny seems to be his source in questions of the mechanics of the moving planets and celestial harmony. Lutz noted glosses referring to Priscian's *De sideribus*, Aratus, Bede's *De rerum natura*, Pseudo-Augustine's *Decem categoriae*, Boethius' commentaries on Porphyrius, on the *Decem categoriae* and *De interpretatione* of Aristotle, Isidore of Seville's *Etymologiae* and probably Cassiodorus' *Institutiones*. On rhetoric, John used Cicero's *De inventione* and possibly *De oratore* or *Brutus* of the same author. For the mythological material in *De nuptiis*, his list of sources is abundantly rich: "Fulgentius, Hyginus, Servius' commentaries on Virgil, Macrobius' *Saturnalia*, Mythographus Primus, St Augustine's *De civitate dei*, as well as Isidore of Seville's *Etymologiae*, are all used a number of times."[164] Etymologies and lexicographical glosses mostly come from Isidore, and the works of Solinus, Iuvencus, Virgil and Ovid are cited by Lutz as sources for isolated glosses. Remarkably, she is only able to point to two citations from the Bible and almost no mention of any Patristic literature — the only exception is Augustine's *De civitate dei*.

Style is another aspect of John's commentary Lutz treats briefly

---

[160] *The place of the Martianus glossae in the development of Eriugena's thought*, p. 52.

[161] *Annotationes*, pp. xx-xxv.

[162] *Ibid.*, pp. xx-xxi.

[163] *Ibid.*, Appendix III, "Dunchad", pp. 229-235; pp. 232-233. G. SCHRIMPF noted that the list would undoubtedly become much longer when the Anonymous commentary is properly edited in its entirety; *Zur Frage der Authentizität*, pp. 134-135.

[164] C. LUTZ, *Annotationes*, pp. xxiv-xxv.

in her introduction.[165] She notes a special interest in, and fair knowledge of, Greek vocabulary, a special appreciation of 'strange' words (*hapax legomena* and neologisms), a certain dexterity in the forming of new words, especially in the fairly straightforward translation of Greek words.

4. *The ninth-century gloss collections: The Cambridge glosses*

4.1 *The manuscripts*

In the *Catalogus Translationum et Commentatorium* Lutz acknowledges a third commentary to *De nuptiis* dated in the second or third quarter of the ninth century, which she refers to as the "Anonymus Cantabrigiensis."[166] There are two witnesses for this batch of glosses: *Cambridge, Corpus Christi College, Ms. 153* (a ninth-century manuscript, *CCC 153*) and *Cambridge, Corpus Christi College, Ms. 330* (which she dates in the eleventh century, *CCC 330*). The glosses cover all nine books of the encyclopedia and, Lutz concluded, have a probable insular origin: this is suggested by the type of script used and the presence of Old Welsh glosses in manuscript 153. She seems, however, to have overlooked an important and detailed study of these sources by Bishop, who reached different conclusions.[167]

Bishop studied the two manuscripts in an attempt to establish the sources drawn upon in England in the first half of the tenth century, a period of the revival of Anglo-Latin letters. He pointed to a number of Welsh manuscripts which must be included in the list of manuscripts linked to this revival, and among those are the two manuscripts with the "Anonymus Cantabrigiensis" commentary. In the late ninth- or early tenth-century Welsh manuscript *CCC 153* he discovers the hand of an English scribe of the mid-tenth century, writing holograph square minuscule, who filled in lacunae and appended an eclectic commentary in a running form. This same commentary survived in *CCC 330*, a manuscript which according to Bishop is continental and probably dates from the late ninth century, rather than the eleventh.

[165] *Ibid.*, pp. xxv-xxx.

[166] C.E. LUTZ, *CTC* II, pp. 371-372.

[167] T.A.M. BISHOP, *The Corpus Martianus Capella.*

Both manuscripts go back to the same exemplum, which was written in Carolingian minuscule. On the basis of a detailed analysis of the writing errors, Bishop concluded that the model manuscript contained a number of continental abbreviations *and* many abbreviations that could indicate Irish influence. These "Irish" characteristics, however, were already common on the continent in the second half of the ninth century, as he himself adds. In *CCC 153* en *CCC 330* it is possible, therefore, to observe the problems of transmission of a manuscript, written in a Carolingian, continental monastery with Irish elements, by a tenth-century English scribe. The result is a highly corrupted text containing many scribal errors, of which Bishop gives numerous examples.[168]

Apart from the body of glosses in a running form attached to the text by the English scribe, Bishop detects in *CCC 153* marginal and interlinear glosses which he attributes to one of the writing hands at work elsewhere in the manuscript, an insular hand of the second half of the ninth century. The glosses, however, which are eclectic in nature, will have to await discussion in a future study.

4.2 *Character of the commentary*

While Lutz made no remark on the content of the commentary, Bishop explained in a footnote that he restricted himself to a brief investigation of the sources, since Préaux intended to edit the batch of glosses added to Martianus Capella in *CCC 153*.[169] The plan seems never to have been realized. Nevertheless, despite its brevity Bishop's examination is accurate and revealing. His only goal was — as he explains in the same footnote — to "suggest an approximate date for the compilation." He observed that the glosses "include (i) some glosses taken *verbatim* from John the Scot, (ii) numerous glosses taken from 'Dunchad', (iii) many lexicographical glosses and a number of longer ones which I (i.e. Bishop) have not attempted to attribute, (iv) apparently nothing from Rémy of Auxerre, whose work circulated in his lifetime." Although he did not make it explicit, Bishop thus seems to suggest an eclectic commentary, drawing from the previous commentaries of John the Scot and the Anonymous commentary. Perhaps some kind of dictionary

[168] *Ibid.*, pp. 270-271.
[169] *Ibid.*, p. 257 n. 1

(*glossarium*), which should be dated prior to Remigius' commentary, was used as well. There has not been any further research into this compilation of older material.

5. *The commentary of Remigius of Auxerre*

In the Benedictine monastery St. Germain at Auxerre, Remigius, born after 841, studied with Heiric of Auxerre, a teacher who in his turn was educated by the great scholar Lupus of Ferrières who studied with John the Scot, and who was esteemed so highly that Charles the Bald sent his son Lothaire to be taught by him.[170] The dates of Remigius' life are not entirely clear. He became schoolmaster at Auxerre, probably upon his teacher's death, between 875-877, or in 883.[171] At some point between 882 and 900 (the period that Fulco was archbishop of Rheims), he and a fellow-scholar, Hucbald of Saint-Amand, were called to Rheims by Fulco to breathe new life into the cathedral school after the destruction by the Normans. Lutz dated this event in 883, but a later date (893) was argued by Mabillon and defended by Jacobsen.[172] After Fulco was murdered (900), Remigius seems to have settled in Paris, where he was one of the first to establish a school. Among his pupils, around 901-902, was Odo of Cluny, whose *vita* mentions Remigius' teaching in the liberal arts in general and specifically dialectic and music.[173] It has been assumed that it was during this late period of

---

[170] Most details of Remigius' life and work are taken from the thorough introduction of C.E. LUTZ in her edition of Remigius' commentary on Martianus Capella: *Remigii Autissiodorensis commentum in Martianum Capellam I*, pp. 1-62, but see also several contributions in D. IOGNA-PRAT, C. JEUDY, G. LOBRICHON (eds.), *L'école carolingienne d'Auxerre*, pp. 373-453.

[171] J.J. CONTRENI, *The Cathedral School of Laon*, p. 148. See pp. 459-460 in *L'École carolingienne d'Auxerre*, where the controversies concerning the dates of Remigius' life are discussed by C. JEUDY.

[172] C. Jeudy refers to P.C. JACOBSEN, *Flodoard von Reims. Sein Leben und seine Dichtung 'De triumphis Christi'*, (Mittellateinische Studien und Texte 10), Leiden, Köln 1978, pp. 6-8 (*non vidi*). C. JEUDY, in *L'École carolingienne d'Auxerre*, p. 459.

[173] *Vita sancti Odonis - scripta a Joanne monacho, ejus discipulo* (PL 133 c. 45): "Deinde [Odo] apud Parisium dialectica musicaque a Remigio doctissimo viro est instructus" (Subsequently [Odo] was taught on the subjects of dialectic and music in Paris by the most learned man Remigius). And, a little further (PL 133 c. 52): "His diebus abiit Parisius, ibique dialecticam sancti Augustini Deodato filio suo

Remigius' life that he taught Martianus Capella. In 908 or shortly after, Remigius died and his name is mentioned in the obituary of the cathedral chapel at Auxerre:

> VI. Non. Maii. Obiit Remigius monachus et egregius doctor.[174]
> (On the second of May Remigius, monk and eminent teacher, died.)

5.1 *The manuscripts*

There are 76 manuscripts containing Remigius' commentary, in part or entirely, in a running or in marginal and interlinear form, abridged, excerpted or complete. The manuscript tradition is, in other words, much too large to treat here in full detail. The data on Remigius manuscripts as gathered by C. LEONARDI (*I codici*) and C. LUTZ (*CTC* II, pp. 372-376), reproduced and completed by the research-group at Auxerre (*L'École carolingienne d'Auxerre*, pp. 489-494), are here represented in Figure 1.

Lutz considers Remigius' commentary "a good example of a "living" text";[175] it was not considered fixed and was subjected to wilful alterations, perhaps even by Remigius himself. When copying the commentary scribes added their own information and omitted information which they considered less interesting or superfluous. From the very beginning of the textual tradition, as is also clear from Fig. 1, even the form of the commentary was open to variation. Lutz distinguishes three distinct forms in which it occurred: "(1) on manuscripts of Martianus Capella, written on the top, bottom, and sides of the pages, (2) as a continuous treatise, though divided, with short sections of the text of Martianus Capella headed TEXTUS, each followed by the appropriate glosses of Remigius headed COMMENTUM, (3) as a running commentary with relatively short *lemmata* accompanied by glosses and explanatory material."[176] She considers this last form, which is by far the most

---

missam perlegit, et Martianum in liberalibus artibus frequenter lecticavit; praeceptorem quippe in omnibus habuit Remigium" (In these days he left for Paris, and there he read through the dialectic of Augustine, which he sent to his son Deodatus, and he often read Martianus on the liberal arts; as his teacher in all these things he had Remigius).

[174] Cited after C. JEUDY, *L'École carolingienne d'Auxerre*, p. 460.

[175] *Remigii Autissiodorensis* I, p. 50.

[176] *Ibid.*, p. 50.

	Commentary in running form	Commentary alternating with text	Text with glosses	Books I-II in abridged form	Excerpts	Sum:
c. 9			1		1	2
c. 9-10			2			2
c. 10	10	1				11
c. 10-11	3		2			5
c. 11	6		5		2	13
c. 11-12	1	1				2
c. 12	8	2	4	1	1	16
c. 12-13	1	1	4			6
c. 13		1	5		1	7
c. 14	1		1	1		3
c.15	2	2	2		1	7
c. 16	2					2
Sum:	34	8	26	2	6	76

Figure 1: Manuscripts of Remigius' commentary on Martianus Capella, classified per centruy, as summarized in *L'Ecole carolingienne d'Auxerre*, "Liste d'oeuvres III.17, *Commentarius in MARTIANUM CAPELLAM*," pp. 489-494.

common one, closest to the original format.

Although the commentary on the first five books of the encyclopedia (mythological introduction and *trivium*) is substantially the same in all manuscripts, there remain two different versions for the quadrivial books. Lutz divided the manuscripts into two families, A and B, and studied the relation between the two in detail.[177] Her conclusion was that Remigius was indeed the author of version A, but that B (a condensed version of A) must have been written by one of his pupils. She edited the A-text, using only manuscripts which preserve the entire text. Two manuscripts, *London, British Library, Royal Library 15 A 33* (*L*) and *Paris, Bibliothèque Nationale, lat. 8786* (*P*), formed the basis of her edition, to which she added the variants, additions and corrections of three more manuscripts in the apparatus: *Paris, Bibliothèque Nationale, lat. 14754* (II); *Paris, Bibliothèque Ste. Geneviève 1041-1042* (*G*) and *Cesena, Biblioteca Malatestiana, Plut. 16 cod. I* (*C*).

L was written in the Benedictine abbey of St. Remigius at Rheims in the beginning of the tenth century, *P* may have been copied at St. Germain d'Auxerre in the first half of the same century. Both manuscripts represent the commentary in a running form, marking the *lemmata* with capital letters and sometimes punctuation signs. II is a twelfth-century source, written in France. It contains an anathema which locates the book in the abbey of St. Victor at Paris. It also contains the glosses in a continuous form, but the distinction between *lemma* and gloss is marked differently: the *lemma* is usually written on a separate line. *G* and *C*, finally, are laid out differently: they contain the commentary alternating with the complete text of *De nuptiis*. *G* is a very large work in two volumes, dated ca 1200 and localized in France.[178] *C* is an early fifteenth-century manuscript, probably written at Cesena for the founder of the library, Malatesta Novello dei Malatesti.

The only ninth-century source containing the A-version of

---

[177] *The commentary of Remigius of Auxerre on Martianus Capella.*

[178] Lutz dates the work in the thirteenth century and locates it at the monastery Ste. Geneviève. The Auxerre research group, however, seems to be more certain about the date (ca. 1200), and less certain about the location. They say the manuscript is French and contains the library catalogue of a yet unidentified French library. *L'École carolingienne d'Auxerre*, p. 492.

Remigius' commentary is now *Bamberg, SB, Class. 39*, a manuscript from the second half of the ninth century with contemporary or slightly later glosses, among which Bischoff recognized ones written in an early tenth century hand.[179] It has been located in Fleury. Two early sources are classified in the B-family: *Bern, BB, Ms. 56B* (mentioned by Lutz as a manuscript from the ninth/tenth century)[180] and *Angers, BM, Ms. 277* (not mentioned by Lutz but noticed by the Auxerre research group and dated in the fourth quarter of the ninth century).[181]

5.2 *Character of the commentary*

The main trait of the commentary of Remigius is its completeness.[182] Remigius seems to have tried to merge all earlier comments and all relevant scholarship together, and gave the commentary a new air of running text by incorporating Martianus' text into his own paraphrases. While the commentary of John the Scot is cited seven times by Remigius, the Anonymous corpus of glosses, which is used to a much larger extent, is not mentioned at all. This can only be explained, to my mind, by the supposition that for Remigius, the Anonymous commentary was an integral part of the text tradition of *De nuptiis*. The commentary shows a strong interest in almost every aspect of *De nuptiis*: it includes many glosses on textual criticism, grammar, lexicography, vocabulary and etymologies, but also long treatises on the allegorical interpretation

---

179 This is reported by the Auxerre group, in: *L'École carolingienne d'Auxerre*, p. 492, and by D. SHANZER, *Review article*, p. 68.

180 Strangely, this manuscript is not entered in the list in *L'École carolingienne d'Auxerre*, although elsewhere it is mentioned as the best source for family B (p. 391). There, the source is dated in the tenth century and located in Auxerre or Fleury.

181 This manuscript contains abstracts from Remigius' commentary on book VI. It is located by B. Bischoff in the region of Rheims. Cf. *L'École carolingienne d'Auxerre*, p. 494.

182 See C.E. LUTZ, *Remigii Autissiodorensis* I, pp. 17-39; *id.*, *The Commentary of Remigius of Auxerre on Martianus Capella*. In the latter, she mainly goes into the relation between the two versions of Remigius' commentary and the *Annotationes* of John the Scot. This description, however, is seriously flawed by the absence of a proper edition of the Anonymous commentary. Her conclusions can no longer be upheld, and the whole matter should be re-examined on the basis of the material from the Anonymous commentary.

of the *dramatis personae* and of the story as a whole, on Greek myths and their "hidden" truths.

The list of sources consulted by Remigius or by the earlier commentators (who are, in turn, cited by Remigius), is impressively long.[183] Passages are quoted from the classical poets Simonides, Valerius Soranus, Terence, Horace, Virgil, Persius, Lucan, Juvenal, Statius and Prudentius, and from the prose of Varro, Cicero, Pliny, Priscian, Solinus, Macrobius, Servius, Fulgentius, and Chalcidius. References are made to Homer, Plato, Aristotle and Orosius. The Vulgate is cited, as are the Christian writers Claudianus Mamertus, St. Gregory and St. Augustine. Furthermore, Lutz identified unacknowledged borrowings from Pliny, St. Augustine, Servius, Festus, Isidore, Macrobius, Fulgentius, Hyginus, Solinus, Mythographus I and II. Many of these references, however, must have come from intermediary sources, of which, as a complete edition would undoubtedly show, the Anonymous corpus of glosses is the most important one. Nevertheless, the commentary as a whole is an impressive accumulation of ancient knowledge.

It is generally assumed that Remigius composed his commentary for school purposes, and that he completed it for his teaching in Paris shortly after 900. The tone of the commentary is pedagogical, and Lutz characterized its effect as "that of the oral presentation of a professor's class notes on a text which presents difficulties of content and interpretation."[184] From the manuscript tradition it is also clear that by this time, *De nuptiis* had come to be included in the school curriculum. The commentary of Remigius, in its turn, became the basis for many 10th-century and later commentaries in Latin and also in the vernacular.[185]

---

[183] *Ibid.*, pp. 22-23.

[184] *Remigii Autissiodorensis* I, p. 24.

[185] Cf. C.E. LUTZ, *Remigii Autissiodorensis* I, pp. 40-49; *id.*, *CTC* II, pp. 376-380, B. PABST, *Prosimetrum*, pp. 244-245.

De nuptiis *in later centuries*[186]

The popularity of *De nuptiis* did not die after the Carolingian period, but it did change in character. In the ninth century both the introductory books with the story of Philology and Mercury's marriage and the seven books on the arts received intensive study. In the later centuries, the manuscript evidence and the surviving commentaries show a concentration on the allegorical story of the encyclopedia, which is related in the first two books, and a dwindling of interest in the more technical treatises of the arts in books III-IX. More often than not, the manuscripts contain nothing but the allegory, or separate books (mostly book eight, on Astronomy)[187] are incorporated into a thematical collection.

From the tenth century there remains a commentary and partial translation from Notker of St. Gall. His commentary contains the first two books of *De nuptiis*, and is an almost verbatim translation of Remigius' commentary into Old High German. Furthermore, it has been shown that he used the Anonymous commentary and that of John the Scot.[188]

Much work remains to be done on the later commentaries from the 12th, 13th and 14th centuries. Only a few of them are properly researched and edited.[189] Lutz compiled a tentative list of commentaries in her article for the *Catalogus Translationum et Commentatorium*, but stressed the fact that the listing was indeed tentative. A substantial number of the commentaries have been

---

[186] Since this aspect of research falls outside the scope of the present study, I will only touch briefly on the subject. For a complete overview see C. LUTZ, *CTC* II, pp. 367-370 and 376-381; B. PABST, *Prosimetrum*, pp. 244-249. Furthermore, see W.H. STAHL, *To a better understanding of Martianus Capella*; and the introductions and bibliographies of H.J. WESTRA's editions: *The commentary attributed to Bernardus Silvestris*; *The Berlin Commentary*. Westra refers to the studies of, among others, E. JEAUNEAU (esp. *Note sur l'École de Chartres*), and P. DRONKE (esp. *Fabula*).

[187] See on this several studies of B.S. EASTWOOD, of which the most recent is *Astronomical Images and Planetary Theory*.

[188] A.K. DOLCH, *Notker Studien*, pp. 297-386. The text has been edited by J.C. KING, *Notker der Deutsche. Martianus Capella, De nuptiis Philologiae et Mercurii*, (Die Werke Notkers des Deutschen, Neue Ausgabe 4), Tübingen 1979.

[189] Haijo Westra edited the commentary attributed to Bernardus Silvestris, and the so-called Anonymus Berolinensis. See references above.

attributed to scholars from the "School of Chartres", although this very concept has been under fire for some years now. William of Conches, Thierry of Chartres and John of Salisbury all show concern for the philosophical implications of the marriage between Mercury and Philology,[190] and commentaries have also been attributed to Bernardus Silvestris and Alexander Neckam. Many poems and prosimetrical works, furthermore, reveal that their authors were inspired by the strange language and colourful content of *De nuptiis*: the *Anticlaudianus* of Alanus of Lille, for example, or the *Bataille des sept arts* of Henri d'Andeli.[191]

Lutz lists the following commentaries. First of all, she introduces the commentary on *De nuptiis* attributed to Bernardus Silvestris. Since her article, the attribution has been doubted, but the commentary is undoubtedly strongly connected with the intellectual life of the milieu of the School of Chartres in the 12th century.[192] The text survives in only one unfinished copy from the early 13th century, and breaks off in section 37 of book one. Originally it must have covered books I and II.[193] Two more commentaries from the same intellectual circle have been attributed to Alexander Neckam and Thierry of Chartres.[194] The former covers book I and II, the latter book VIII (on Astronomy). There is one further commentary on just book VIII: the so-called *Anonymus Londinensis*, whose text survives in a 14th-century manuscript.[195] Further, Lutz lists four *Anonymi* (*Anonymus Vaticanus, Zwettlensis, Berolinensis* and *Florentinus*) who commented upon the first two books of the encyclopedia, and who all made intensive use of the commentary of

---

[190] Cf. G. NUCHELMANS, *Philologie et son marriage avec Mercure.*

[191] See for many more examples the thorough and exhaustive study of B. PABST, *Prosimetrum*, esp. Part V, "Die prosimetrische Literatur des lateinischen Mittelalters." (pp. 309-1048)

[192] Because of its many parallels with a commentary on Virgil's *Eneid*, which was formerly attributed to Bernardus Silvestris, the commentary on *De nuptiis* was also ascribed to him. The most recent editors of the *Eneid*-commentary, however, doubt the attribution, which makes the attribution of the commentary on *De nuptiis* to Bernardus Silvestris equally uncertain. H.J. WESTRA, *The commentary attributed to Bernardus Silvestris*, pp. 7-17.

[193] *Ibid.*, pp. 11.

[194] C. LUTZ, *CTC* II, No. 8 and 9 respectively.

[195] *Ibid.*, No. 11.

Remigius of Auxerre.[196] Two of these, the *Anonymi Zwettlensis* and *Berolinensis*, have subsequently been studied by Haijo Westra, who sketched the relation between the two commentaries as follows. The *Zwettlensis* commentary actually consists of three different parts, one of which is an incomplete version of the *Berolinensis* commentary. The other two are abridged versions of Remigius' commentary, with significant additions from Augustine, Fulgentius and Eusebius. Furthermore, the *Berolinensis* commentary shows relations in content and method to the commentary attributed to Bernardus Silvestris, although a close inspection also shows incompatible elements.[197]

Lastly, Lutz mentions a 16th-century commentary on the first two books of *De nuptiis* by Johannes Dubravius, bishop of Olomouc from 1546 to 1553.[198] She concludes her article with a long list of manuscripts with unidentified commentaries. In *CTC III* she added one more 16th-century commentary to her list: a commentary on book III (on Grammar) by Joannes Rhagius Aesticampianus, published in 1507.[199]

---

[196] *Ibid.*, No. 10, 12, 13 and 14.
[197] Cf. H.J. WESTRA, *The Berlin Commentary*, pp. xxxv-xxxix.
[198] C. LUTZ, *CTC* II, No. 15.
[199] *CTC* III, pp. 449-451.

CHAPTER TWO

THE MANUSCRIPT TRADITION OF THE ANONYMOUS COMMENTARY

Selection

The many problems that remain unsolved concerning the authorship of the corpora of glosses and the assessment of their size leave us with no other option than to carefully consider each manuscript individually.[1] This, however, seems useful only for those manuscripts which are likely to contain relevant material for the present research, that is, those manuscripts that contain ninth-century reflections on Martianus Capella's body of thought on music. Music is, of course, one of the seven arts Martianus presents in his encyclopedia, and the ninth book in which *Harmony* treats her own art is treated in full in the present study. But in addition she is abundantly present in the two mythological books, narrating the story of Mercury's and Philology's marriage. There are, for example, several passages which refer to the theme of the harmony of the heavenly spheres, the idea that the planets, orbiting the earth, produce tones because of their movements and that they sound together in perfect harmony. Music, furthermore, is presented as an art of numbers, revealing the heavenly order to humankind in a set of numerical ratios which lie at the basis of the perfect harmonies of octave, fifth and fourth. These themes prompted the writing of glosses in the ninth century, and it is on the ninth-century reception of these themes that the present book also focuses.

The number of manuscripts containing (parts of) *De nuptiis* is considerable. Even to deal with just the ninth-century material in full would be a research-project of many years, for there remain today 49 ninth-century manuscripts. In order to keep the mass of

---

[1] C. LEONARDI stressed this same point on several occasions: *I codici* I, p. 450; *I commenti altomedievali ai classici pagani*, pp. 459-462; *Martianus Capella et Jean Scot*, pp. 198-200.

textual evidence manageable, a selection according to strict textual criteria was therefore felt to be unavoidable. The starting point for this selection is provided by Leonardi's census of manuscripts.[2] For the present research, a preliminary selection from his list was carried out on the basis of three criteria:

a. dating of the manuscript in the Carolingian period (9th-11th century);
b. presence of the three books that contain material on music, i.e. books I, II and IX;[3]
c. presence of glosses in (at least) these three books.

The selection based on these criteria left us with a list of 21 manuscripts that qualified for closer inspection; these are listed in the figure below (Figure 1), together with Leonardi's numbers and a summary of the relevant information in his short descriptions.[4]

---

[2] C. LEONARDI, *I codici di Marziano Capella* I-II.

[3] One could argue that book III, on Grammar, should also have been included in this study, since, especially in the early Middle Ages, reflections on music often followed the same principles as reflections on grammar. See on this, among others, M. BIELITZ, *Musik und Grammatik*, pp. 20-36, 47-67; C.M. BOWER, *The Role of Boethius*, p. 164; *id.*, *The grammatical model*; M.-E. DUCHEZ, *Description grammaticale et description arithmétique*; *id.*, *Le savoir théorico-musicale carolingiens*, pp. 555-558. I am sure that a systematic study of the glosses on book III would indeed have yielded even further interesting material, but this subject will have to await discussion in a future study.

[4] Leonardi's data, especially where dating and location are concerned, are partly outdated and have in some cases been corrected in recent scholarship. For example, brief descriptions of the first four manuscripts are given in B. Bischoff's recently published catalogue of ninth-century manuscripts (*Katalog der festländischen Handschriften*). He locates *Bamberg Class. 39* in Fleury or its surroundings (with a question mark), and dates it in the third quarter of the ninth century (211, pp. 47-48). He dates *Bern 56B* in the first half of the tenth century, and claims its origin to be French (498, pp. 105-106). He dates *Besançon 594* in the third quarter (629, p. 134), and *Bruxelles 9565-9566* in the fourth quarter of the ninth century (731, pp. 157-158). Since Leonardi's census was my starting point, however, I chose to present his information in this table. The manuscripts that are found, after a narrower selection, to be relevant for the present study, will be treated amply further on in this chapter. In this detailed treatment, Leonardi's data will be completed and corrected by reviewing more recent scholarship on the subject.

L	signature	origin, dating	glosses
8	Bamberg, Staatl.Bibl., Class. 39	French, c. 9	contemporary and 10th century glosses, Reims, unidentified
15	Bern, Sadtbibl., 56 B	?, c. 9-10	Remigian corpus
20	Besançon, BM, 594 (*Be*)	French (St.-Oyan?), c. 9	many contemporary glosses, Anonymous corpus
26	Bruxelles, Bibl.Royale, 9565-9566	region Bodensee, c. 9	many glosses, slightly later, unidentified
28	Cambridge, Parker Library (Corpus Christi Coll.), 153	Welsh/English, c. 9	unidentified contemporary and later glosses
37	Chartres, BM, 103	?, c. 9-10	unidentified glosses
60	Firenze, BML, San Marco 190	Milan, c. 11	glosses from the Anonymous corpus
80	Köln, Dombibl., 193	?, c. 9-11	contemporary and later glosses, unidentified
82	Leiden, UB, BPL 36 (*Leb*)	Lorsch?, c. 9	contemporary glosses, unidentified
83	Leiden, UB, BPL 87 (*Lea*)	?, c. 9	glosses from the Anonymous corpus
84	Leiden, UB, BPL 88 (*Lb*)	Reims?, c. 9	glosses from the Anonymous corpus and (book IX) John the Scot's commentary
89	Leiden, UB, Voss.Lat.F. 48 (*La*)	Auxerre?, c. 9	many glosses from the Anonymous corpus
125	München, Bayer.Staatsbibl., lat. 14729	St. Gall?, c. 9-10	unidentified glosses in contemporary and later hands
144	Oxford, BL, Laud.lat. 118 (*Oxf*)	French, c. 9	contemporary and later glosses, Anonymous corpus

159	Paris, BnF, lat. 7900A	French, c. 10	glosses from the Remigian corpus
160	Paris, BnF, lat. 8669	French (Soissons?), c. 9	unidentified contemporary and later glosses
161	Paris, BnF, lat. 8670	Corbie, c. 9	glosses from the Anonymous corpus
162	Paris, BnF, lat. 8671 (*Par*)	?, c. 9	unidentified contemporary glosses
92	St. Petersburg, Nat.Libr., Class.Lat.F.V.10	Corbie, c. 10	glosses from the Anonymous corpus
208	Vaticano, BAV, Reg.lat. 1535	Auxerre, c. 9	glosses from the Anonymous corpus
210	Vaticano, BAV, Reg.lat. 1987 (*Va*)	Reims?, c. 9	many glosses from the Anonymous corpus

Figure 1

These 21 manuscripts were studied on microfilm, with the following questions in mind:

a. What is the quantity of glosses on music in books I, II and IX?
b. To what extent does the manuscript contain glosses that are still unedited, or — in other words — how much new material is present in each manuscript?[5]
c. Is it possible to recognize groups of manuscripts, of which the individual members are clearly related to each other?

On the basis of this study the following observations were made.

First, a focus on the oldest commentaries, that of John the Scot

[5] On this question one must add that each manuscript has its own, unique collection of glosses. One can, of course, detect likenesses between glossed manuscripts to differing extents, but there are no two manuscripts with the exact same collection of glosses. This implies much for the status, or authoritative quality, of glosses as text — see on this A.C. DIONISOTTI, *On the Nature and Transmission of Latin Glossaries*.

and the enigmatic Anonymous corpus attributed first to Dunchad and then to Martin of Laon, appeared to offer the best possibilities for research. Three factors determined this choice: a) the fact that the focus of the present book is on music and it will therefore address parts of books I and II and book IX in its entirety; b) a strict focus on the earliest period of active reception, which lies in the ninth century; c) the presence of new and unedited material in the manuscripts.

For the corpus of Remigius of Auxerre a close analysis of its earliest sources seemed to be a matter of less importance. For although Remigius' commentary differs per manuscript, as is the case in the other gloss-collections, the variations here are in general minor and the outline is clearer. The two manuscripts from the preliminary selection which contain glosses from the commentary of Remigius, i.e. *Paris lat. 7900 A*, *Bern 56B*, are therefore considered to be of secondary importance for the present research. They both represent the so-called B-version of Remigius' commentary — that is, the version Lutz chose not to edit —, and the merit of their 'new' material should be analyzed, but falls outside the scope of the present study. A third manuscript, *München Clm. 14729*, contains what seems to be an abstract of Remigian glosses and is therefore also not discussed in this study.

The ninth-century corpus generally attributed to Martin of Laon is only partly edited under the name 'Dunchad.' It was copied on a significantly larger scale than the commentary of John the Scot. It remains unclear how large this corpus actually is, but there is enough evidence to assume a comprehensive commentary of all books. A large amount of yet unedited and unseen material is hidden away in the margins of its manuscripts. The commentary of John the Scot has been made available in an edition by Lutz, but the present research revealed two sources which add new material to that edition, precisely for the ninth book on music.

Secondly, a considerable number of the Carolingian manuscripts contain a mixed collection of glosses. In four manuscripts (*Paris lat. 8669; Paris lat. 8670; Vaticano Reg.lat. 1535; Köln 193*) we can detect glosses from the Anonymous corpus in a first, older layer of glosses, contemporary with the manuscript, and additions from the Remigian corpus in a hand or hands that worked approximately one generation later. In two of these, *Paris lat. 8669* and *Vaticano Reg.lat. 1535*, glosses which can be identified neither with Remigius' glosses, nor with the Anonymous ones, but which sometimes resemble John's glosses, are interspersed in the two layers.

Mixtures of the Anonymous commentary and John's commentary are found in three manuscripts: *Paris lat. 8671; Leiden BPL 88; Vaticano Reg.lat. 1987.* The latter two are closely related and contain a unique kind of mixture; they give a version of the Anonymous commentary in books I-VIII of *De nuptiis*, but in book IX a version of John's glosses that differs considerably from Lutz's edition. *Paris lat. 8671* is more difficult to categorize, since it contains only a few glosses. It appears to consist of a selection from or different version of the Anonymous commentary with some additions from John's commentary. Three manuscripts (*Leiden BPL 87; Leiden Voss.Lat.F. 48; Besançon 594*) contain what is here cautiously considered to be a "purer" version of the Anonymous commentary.

More complex mixtures of glosses from different collections — not only the three already mentioned, but also unidentifiable ones — are found in the manuscripts *Leiden BPL 36*, *Cambridge 153* and *Oxford Laud.lat. 118.*

The relations between the manuscripts are, unfortunately, more difficult to analyze than a simple description of their contents might suggest. Any two manuscripts, after all, may contain glosses from the same corpus and at the same time be wholly different. An important factor in assessing the relationship between the individual sources is the *size* of the gloss collection — a factor which also played a decisive role in the selection of the most valuable manuscripts for this research. For if, because of damaged or lacunal transmission, a particular manuscript does not contain the relevant passages in the books I, II or IX, it is evidently useless for our purposes. After consultation of the microfilms and comparison of their glosses with the available editions eleven manuscripts were excluded from the selection for reasons of scarcity of glosses or defective transmission.

First, *Chartres 103* was not just damaged in World War II, as Leonardi mentioned in his census, but virtually destroyed. There remain today nothing but a few "fragments incomplets" and the manuscript had not been reproduced photographically before the war.[6]

Secondly, there are five manuscripts which have either been transmitted defectively or do not contain glosses for some of the relevant books. In *St. Petersburg Class.Lat.F. V.10* the first few leaves are missing (including relevant passages on music in book I), only two leaves of book II have survived, and book IX does not contain glosses. *Bruxelles 9565-9566* contains only glosses in book I, not in book II or IX. Moreover the text is incomplete: book IX breaks of in section IX.973. There are no glosses on book IX in *Firenze San Marco 190*, *München Clm. 14729* and *Paris lat. 7900A* (which is also lacunal in its transmission of book IX).

Thirdly, there is a group of five manuscripts which are excluded from the narrow selection for a more subjective reason: they were felt to contain less *relevant* gloss material. These are *Bamberg Class. 39* (the first two books contain only a small number of glosses, book IX is only glossed in the introductory narrative), *Köln 193* (which is also sparsely glossed in book I and II and contains even less glosses in book IX), *Leiden BPL 87* (which forms such a close parallel to *Leiden BPL 88* for books I-II that it can be disregarded for that reason, and has almost no glosses to book IX), *Paris lat. 8669* (the glosses on book IX cover less than one page in transcription), and *Vaticano Reg.lat. 1535* (the glosses on book IX are mainly corrections to the text or variant readings).

If we combine all these observations — concentrating on the oldest commentaries and evaluating the amount of new and relevant material in the gloss-collections per manuscript —, nine manuscripts remain from the initial list of 21. These can be divided into four

[6] Letter from the head of the manuscript-department, Ms. Chr. Pollin, 18 February 1994. J. Smits van Waesberghe, however, took photographs of book IX before World War II for his private collection. These are now kept on microfilm in his archive in the Department of Music of the University of Amsterdam. (I thank Eddie Vetter for lending the film to me.) The film reveals that the manuscript indeed contained glosses from the Anonymous corpus, and that it forms a fairly close parallel to *Be* and *La*. Unfortunately, the glosses only cover the introduction of *Harmonia*, and not her treatment of the technical aspects of the *ars musica*. They stop at section 920. Furthermore, the text itself is incomplete: it breaks off in section 930.

groups, which are introduced briefly below.

1. *The Anonymous corpus*

There are three manuscripts, *Leiden Voss.Lat.F 48* (*La*), *Besançon 594* (*Be*) and *Oxford Laud.lat. 118* (*Oxf*), that seem to contain a 'pure' version of the Anonymous corpus of glosses attributed to Dunchad or Martin of Laon. These sources are located in Auxerre (?), St.-Oyan (?) and the region Auxerre-Soissons (?) respectively.[7] Of these three *La* appears to be the oldest: ca 850. *Be* is dated shortly after 850 and *Oxf* in the second half of the ninth century. The first two manuscripts of this group, *La* and *Be*, are closely related and seem to be the best sources, at least quantitatively. *Oxf* has the same amount of material for book I-II, but only about one sixth of the material for book IX. It was therefore decided to concentrate on the material found in *La* and *Be*, and use *Oxf* as a background source, turning to it for checking, correcting and contrasting the glosses from the two main manuscripts.

2. *The Anonymous corpus* and *John the Scot*

Two manuscripts, *Leiden BPL 88* (*Lb*) and *Vaticano Reg.lat. 1987* (*Va*), contain a unique and idiosyncratic mixture of commentaries: they have a version of the Anonymous corpus for the first eight books of *De nuptiis*, but for book IX, *On harmony*, they give John the Scot's material — in a different version than the one edited by Lutz.[8] The two manuscripts are located in the same region: *Lb* in Reims (?) and *Va* in Northeastern France (Reims?). They are dated

[7] The dating and localization of *La* and *Be* will be treated in full in the second part of this chapter, in which these two manuscripts are described in detail. For the dating and localization of *Oxf*, cf. among others J. PRÉAUX, *Les manuscrits principaux*, pp. 78-79, and its description in the third part of this chapter.

[8] As mentioned before, the manuscript *Leiden BPL 87* is a close pendant of *Leiden BPL 88*, but has hardly any glosses to book IX. Literal parallels and similarities in details of the lay-out of these two manuscripts make clear, however, that *Leiden BPL 87* should be considered as a member of the same group.

in the second half of the ninth century,[9] and are both rich in material. Préaux argued that *Va* contains a less contaminated version of text and commentary.[10]

3. *An eclectic (insular?) corpus*

The manuscript *Cambridge, Corpus Christi College 153* contains a highly eclectic commentary. The manuscript originated in Wales, following a continental model.[11] In the tenth century it was kept in England, where more glosses were added. The Welsh part of the manuscript is dated at the end of the 9th century, beginning of the 10th; the English part in the middle of the 10th century. The last leaves of the manuscript contain a batch of *collectae glossae*, which are also found (without the text of Martianus Capella) in *Cambridge, Corpus Christi College 330* — a manuscript that was not included in the selection and has not been studied in detail.

This 'group of one' will not be at the centre of interest in the present study. It serves as a background source, to contrast the ninth-century continental commentaries to this earliest survivor of insular reception. Since it is its sole survivor, is seems to offer few possibilities for general conclusions.

4. *Mixed corpora*

Most ninth-century manuscripts carry glosses brought together from several sources. Three of these, *Leiden BPL 36* (*Leb*), *Paris lat. 8670* and *Paris lat. 8671* (*Par*), provide us with the most interesting material. As a group, however, this category of manuscripts is not as coherent as the other three, and this is reflected also in the differences in localization. *Leiden BPL 36* has a mixture of glosses from the Anonymous commentary and some unique, still unidentified glosses, *Paris lat. 8670* a mixture of Anonymous and

---

[9] *Leiden BPL 87* is situated in the same region: Eastern France (?), and dated in the same period; cf. among others J. PRÉAUX, *Les manuscrits principaux*, p. 79; C. LEONARDI, *I codici* II, p. 62. The dating and localization of *Leiden BPL 88* and *Vaticano Reg.lat. 1987* will be treated in detail in the second part of this chapter.

[10] J. PRÉAUX, *Deux manuscrits gantois*, p. 17.

[11] Bishop studied this manuscript in great detail: T.A.M. BISHOP, *The Corpus Martianus Capella*.

Remigian glosses, and *Paris lat. 8671* a mixture of Anonymous glosses and material from the commentary of John the Scot. *Paris lat. 8670* is the oldest of the three manuscripts: it is dated ca 850. *Paris lat. 8671* and *Leiden BPL 36* are both dated in the second half of the 9th century. The origin of *Paris lat. 8670* is posited in Corbie[12] — seen as a center of Martianus-study and dissemination —, *Leiden BPL 36* in Lorsch (?).[13] On the origin of *Paris lat. 8671* opinions differ considerably: Auxerre (?) and Bodensee (?) are proposed by Préaux, and Lorsch can also be added to the list of possible places of origin.[14]

Quantitatively spoken these sources are not as rich in material as those of groups 1 and 2. Poorest is *Paris lat. 8670*, which was therefore put on a sidetrack in the present study.

Since the glosses added to the books I-II in the manuscripts of this group largely overlap with material found in the other groups, they are considered to be of less relevance. Only the glosses to book IX from this group will therefore be used, as a background source for the material from groups 1 and 2.

At the end of this selection process there thus remained four manuscripts of great importance: *La*, *Be*, *Va* and *Lb*. It was decided to edit the glosses from these sources in the present book. Four other manuscripts, *Oxf*, *Lea*, *Leb* and *Par*, were found to be of next-to-primary importance and were used as background sources, to check the readings in the four most important manuscripts and to contextualize the glosses. The outcome of the selection is represented in Figure 2 below.

---

12 Cf. among others C. LEONARDI, *I codici* II, p. 437; B. BISCHOFF, *Hadoardus*, p. 53; J. PRÉAUX, *Les manuscrits principaux*, pp. 78, 97, 121; D. GANZ, *Corbie in the Carolingian Renaissance*, p. 152.

13 Cf., among others, C. LEONARDI, *I codici* I, p. 452; B. BISCHOFF, *Lorsch im Spiegel seiner Handschriften*, pp. 45, 96-97.

14 Préaux suggested both Auxerre (?) and Bodensee region (?), in *Les manuscrits principaux*, pp. 79 and 125 respectively. There are, however, several characteristics in the script which Bischoff mentioned in his description of the style of Lorsch. Moreover, these pieces of paleographical evidence are confirmed by textual evidence, for the corpus of glosses found in *Paris lat. 8671* appears to be related to the one found in *Leiden BPL 36*, i.e. the Leiden manuscript that contains hands that are localized in Lorsch by Bischoff.

For the study of the content of this material, it was checked against, supplemented by and contrasted with the glosses edited by C.E. Lutz (*Dunchad*, *Johannes Scotus*, *Remigius of Auxerre*) and E. Jeauneau (*Johannes Scotus*, B-version, book I). Furthermore it was analyzed in the context of content-related contemporary material, such as the ninth-century glosses to Boethius' *De institutione musica*, available in the editions of M. Bernhard and C.M. Bower.[15]

One of the many things that remain unexplored in this study is the comparison with and analysis of the material in relation to that found in the eclectic insular commentary in the Cambridge manuscripts (group 3), and in the unedited B-version of Remigius' commentary, as found in, for example, *Paris lat. 7900A* and *Bern 56B*.

[15] Cf. C.E. LUTZ (ed.), *Dunchad glossae in Martianum*; *id.*, *Iohannis Scotti Annontationes in Marcianum*; *id.*, *Remigii Autissiodorensis Commentum in Martianum Capellam, Liber I-II, Liber III-IX*; E. JEAUNEAU, *Quatre thèmes érigéniens*; M. BERNHARD & C.M. BOWER (ed.), *Glossa maior in institutionem musicam Boethii.*

	Glosses on musical passages in books I and II, Anonymous corpus		
group 1	*La*	*Be*	(*Oxf*)
group 2	*Va*	*Lb*	(*Lea*)

	Glosses in book IX, Anonymous corpus	
group 1	*La*	*Be*
group 4	(*Leb*)	(*Par*)

	Glosses in book IX, B-version of Corpus John the Scot	
group 2	*Va*	*Lb*

Be = Besançon, Bibliothèque Municipale, Ms. 594
La = Leiden, Universiteitsbibliotheek, Vossianus Latinus Folio 48
Lb = Leiden, Universiteitsbibliotheek, BPL 88
Lea = Leiden, Universiteitsbibliotheek, BPL 87
Leb = Leiden, Universiteitsbibliotheek, BPL 36
Oxf = Oxford, Bodleian Library, Laud.lat. 118
Par = Paris, Bibliothèque Nationale de France, lat. 8671
Va = Città del Vaticano, Biblioteca Apostolica Vaticana, Reg.lat. 1987

Figure 2

Plate 1: *Leiden, UB, Voss.Lat.F. 48* (La), fol. 2r. (Cf. pp. 88-98) Opening page of *De nuptiis*, Book I.

The beginning of Martianus Capella's *De nuptiis* can be seen on this page, preceded by a header in uncial letters. The main text is written in small Carolingian minusucle, leaving ample space between the writing lines for glosses. Narrow columns for longer glosses surround the text and the outer margins are used for even more glosses.

Three different hands can be recognized in the glosses. The scribe who originally glossed the text used the gloss columns or the margins for this purpose. His glosses are usually neatly tied to their lemmata with reference symbols. A second, also contemporary scribe wrote his glosses with a thinner pen and darker ink, mainly between the lines, but also in the margins.

Thirdly, roughly between fifty and a hundred years later, Ratherius of Verona (880-974) highlighted unusual words from the text by repeating them in majuscules in the margin: EGERIMION, for example, in the lower margin.

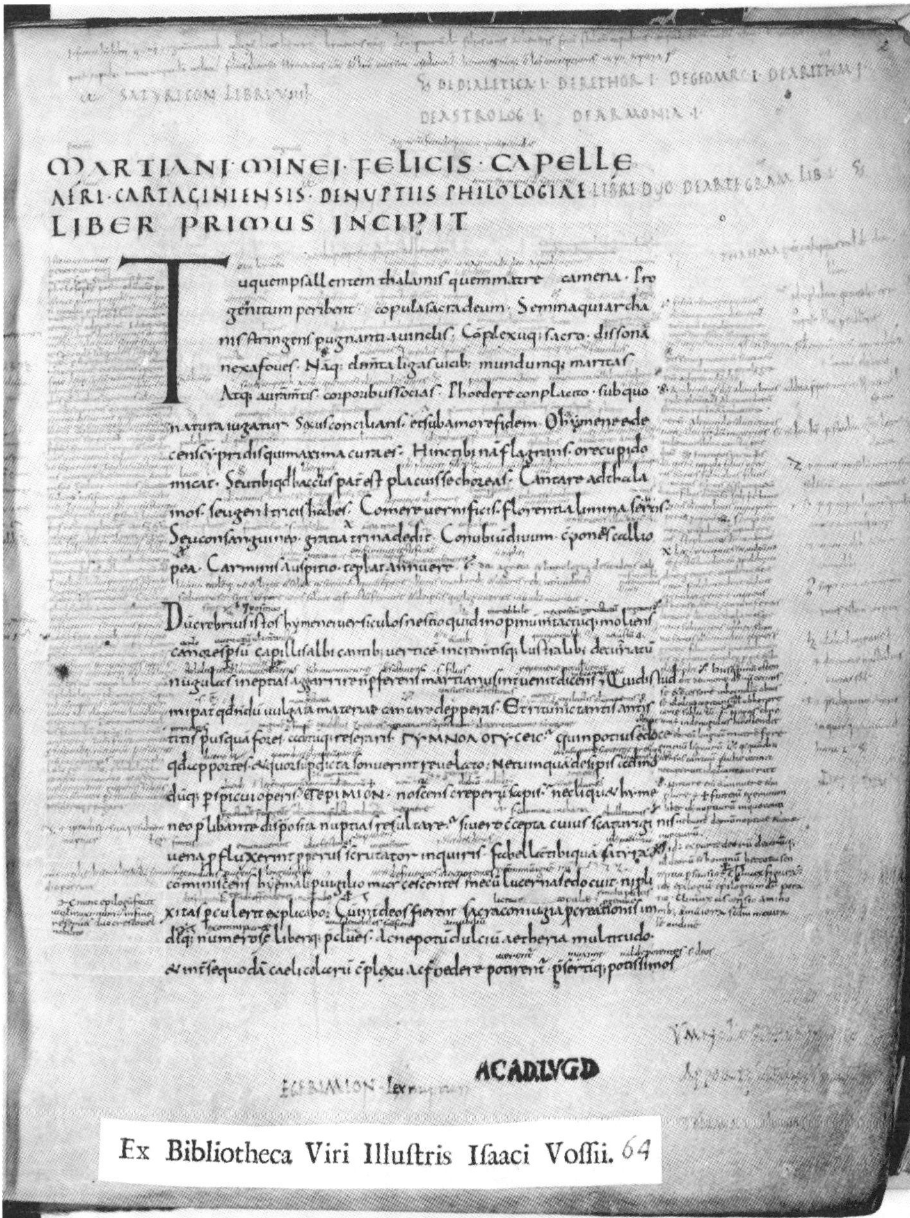
MARTIANI MINEI FELICIS CAPELLE
AFRI CARTAGINIENSIS DE NUPTIIS PHILOLOGIAE
LIBER PRIMUS INCIPIT
ACADLVGD
Ex Bibliotheca Viri Illustris Isaaci Vossii. 64

Plate 1

Plate 2: *Besançon, BM, Ms. 594* (Be), fol. 16r. (Cf. pp. 98-103)
Last page of Book II, beginning of Book III.

Since the parchment of the opening pages of *Besançon 594* has darkened so much and the ink has faded, the opening page of the manuscript could not be reproduced. Nevertheless, it will be clear that *Leiden, Voss.Lat.F. 48* and *Besançon 594* resemble each other closely in script and lay-out.

The page shown has the same widely spaced lines, with relatively small Carolingian minuscules on the writing lines, leaving ample space for the glossator to write comments above the words of the text. A narrow column for glosses was marked off, but was never actually used for this purpose. It was incorporated in the writing space of the main text, and the glosses are restricted to the outer margins and the interlinear spaces.

The glosses seem to have been written by one hand. The glossator used a set of letters very similar to that of the text.

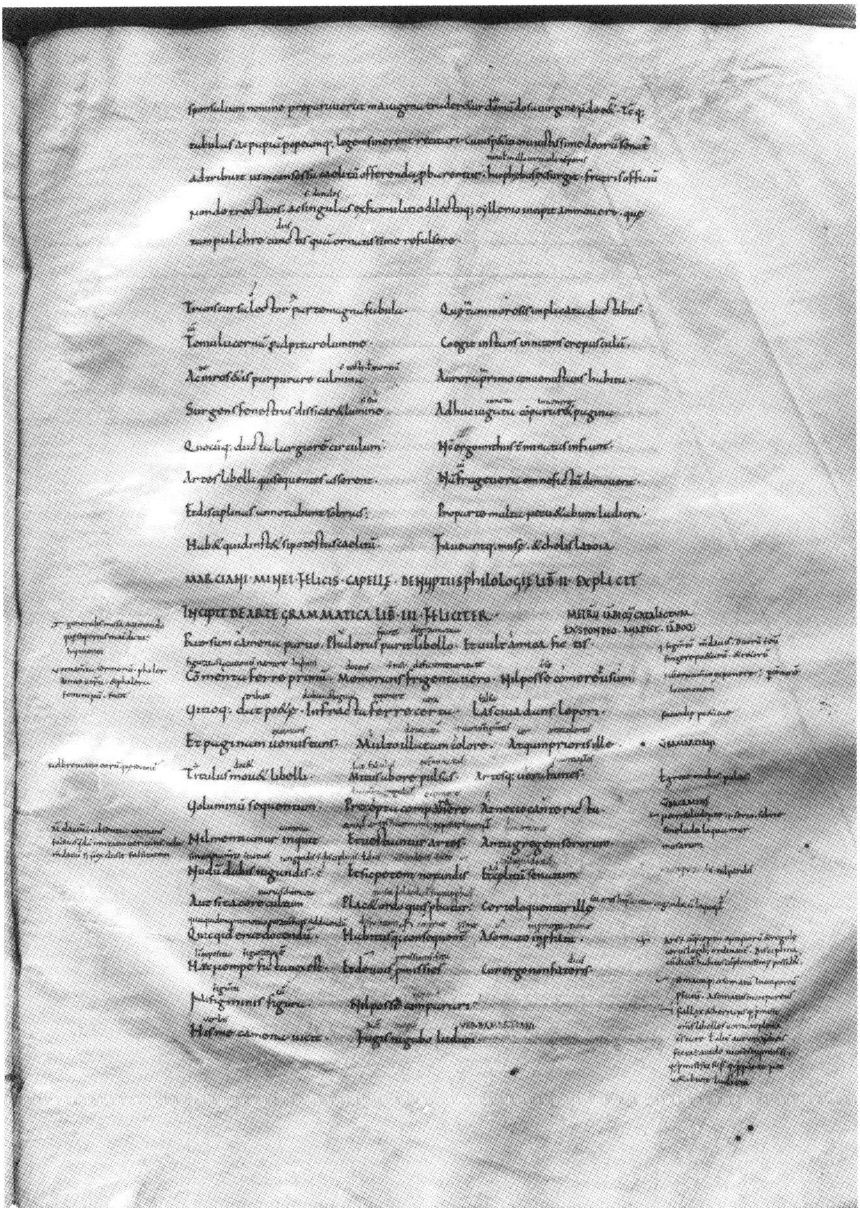

Plate 2

Plate 3: *Vaticano, BAV, Reg.lat. 1987* (Va), fol. 2r. (Cf. pp. 103-117) Opening page of *De nuptiis*, Book I.

The opening page of *Vaticano, Reg.lat. 1987* shows a decapitated text: the first five lines of the opening poem appear on a single leaf, added to the manuscript at a later stage.

Many different hands worked in this manuscript, even though its appearance is relatively uniform. Striking are the clubbed ascenders, x's with long descenders pointing to the left, and an oddly shaped rt-ligature (e.g. *sertis* in line 8, or *apportes* in line 17).

The lay-out is straightforward in one column, and allows little room for glosses. Nevertheless, glosses were very carefully written in the margins and between the lines, in a tiny, very regular Carolingian minuscule. The glossator is very precise in tying lemma and gloss together with a reference sign. Often, the reference signs used are similar to those used in *Leiden, BPL 88* (*Lb*, Plate 4). The ink he used is light brown and has occasionally almost faded.

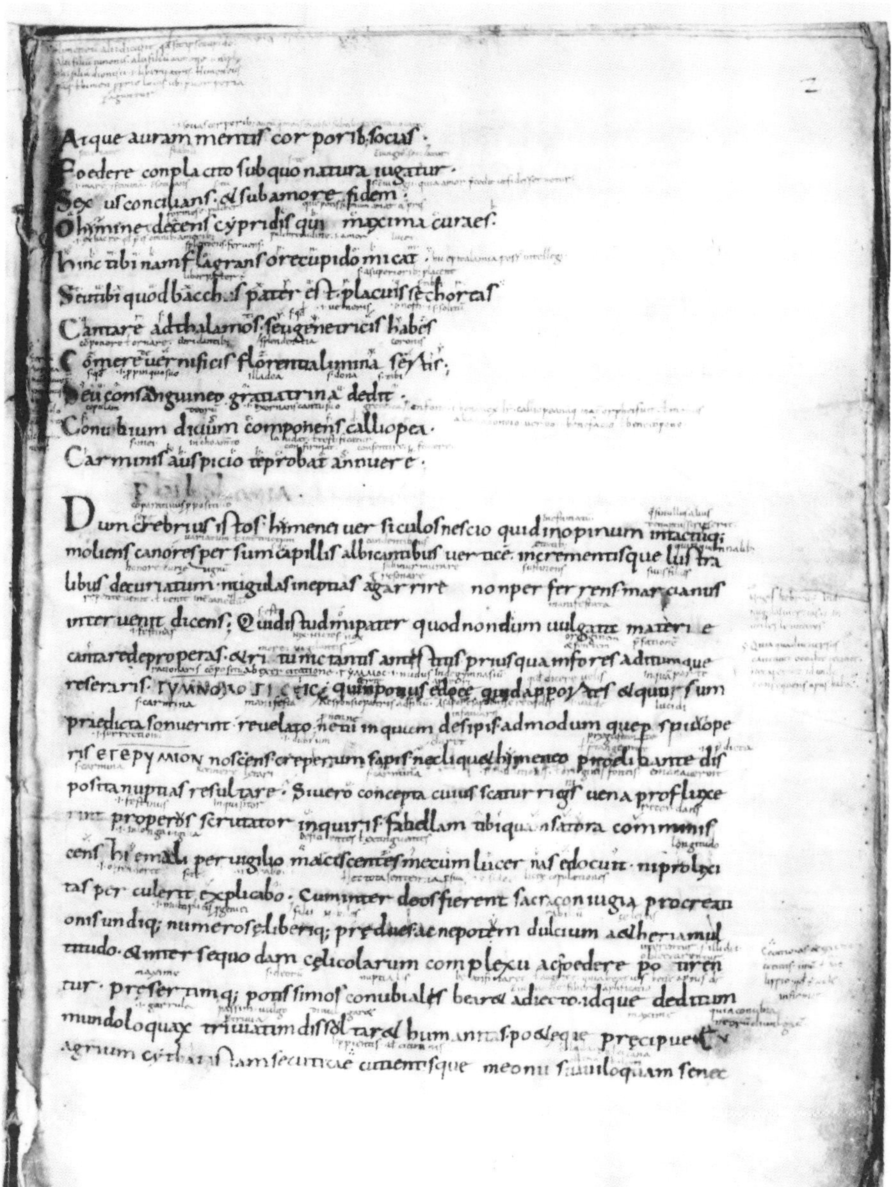
Atque auram mentis corporibus socias.
Foedere conplacito sub quo natura iugatur.
Sexus concilians et sub amore fidem.
O hymine decens cypridis qui maxima cura es.
hinc tibi nam flagrans ore cupido micat.
Seu tibi quod bacchus pater est placuisse choreas
Cantare ad thalamos seu genetricis habes
Comere vernificis florentia limina sertis.
Seu consanguineo gratia trina dedit.
Conubium divum componens calliopea.
Carminis auspicio te probat annuere.
Dum crebrius istos himenei versiculos nescio quid inopinum intactumque
moliens canores per sum capillis albicantibus verticem incrementisque lustra
libus decuriatum nugulas ineptas aggarrire non perferens marcianus
interuenit dicens. Quid istud mi pater quod nondum vulgatae materiae
cantare deproperas et ritu nictantis antistitis priusquam fores aditumque
reseraris ΥΜΝΟΝ ΑΕΙΔΕΙΣ quin potius edoce quid apportes et quae sunt

Plate 3

Plate 4: *Leiden, UB, BPL 88* (Lb), fol. 3r. (Cf. pp. 117-126)
Opening page of *De nuptiis*, Book I.

This plate shows folio 3 recto, originally probably the first folium of the manuscript, and the opening page of Martianus Capella's *De nuptiis*. Even though no separate gloss columns are provided in this codex, its lay-out is careful and spacious, with ample room for glosses.

The text is written in several hands, but has a relatively uniform appearance throughout the manuscript. The r's are tilted to the right and sometimes descend below the line, ascenders are lightly clubbed, and x's have long descenders to the left. The letters are somewhat suspended above the lines. The colour of the ink varies between an almost faded light brown and nearly black.

Several layers of glosses can be detected, but two scribes contemporary with the manuscript are responsible for the main glossing activity. The first layer of glosses is written in a tiny, very regular Carolingian minuscule, both in the margins and in the interlinear spaces. The scribe is precise in tying lemma and gloss together with a reference sign. Often, the reference signs used are similar to those used in *Vaticano, Reg.Lat. 1987* (*Va*, Plate 3). A second scribe executed his glosses in darker ink, and seemingly with less care. This hand can also be found emending and filling in lacunae in the text, and he may also have been responsible for the headers that identify the number of the book and its contents, which are added throughout the codex.

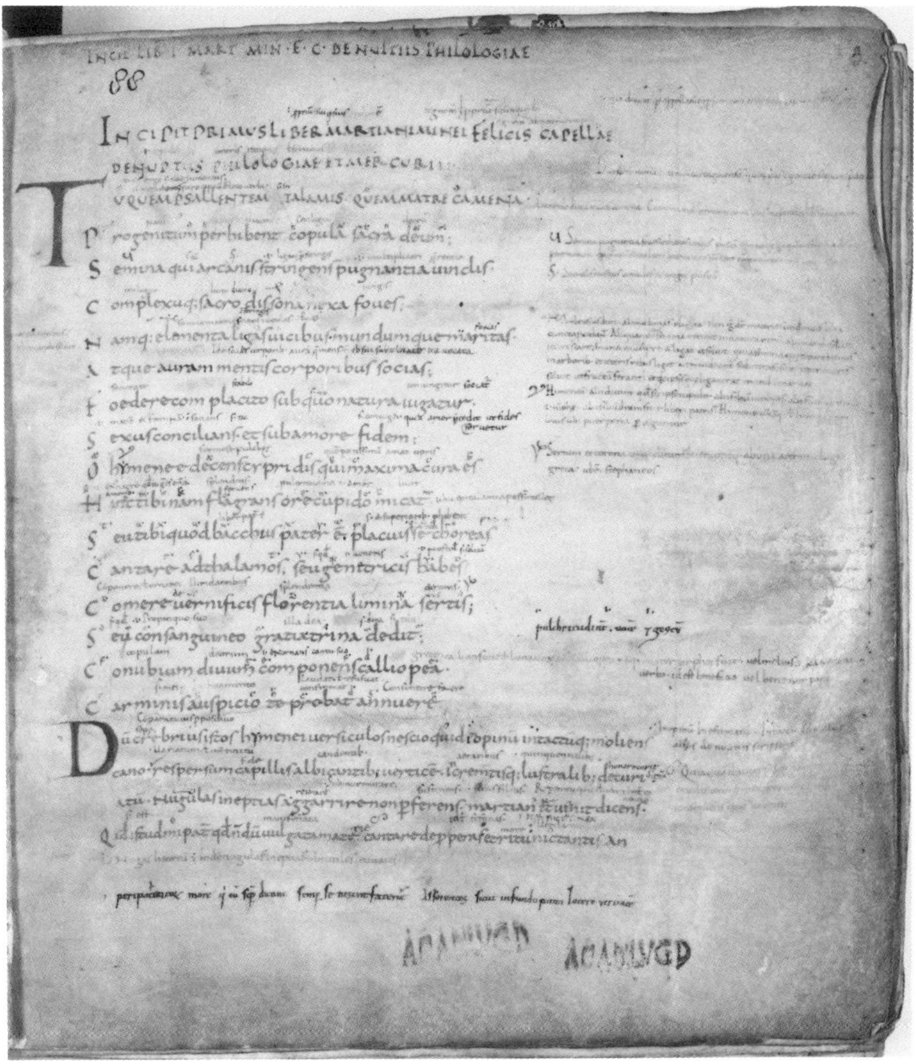

Plate 4

Plate 5: *Oxford, Bodl.Lib., Laud.lat. 118* (Oxf), fol. 1r. (Cf. pp. 127-131)
Opening page of *De nuptiis*, Book I.

The appearance of this manuscript resembles that of *Leiden, Voss.Lat.Fol. 48* (*La*, Plate 1). The lay-out is even more spacious, leaving ample room for glosses. Headers with uncial and rustic capitals, often used in deliberate alternation, nicely structure the text. The characteristics of the script that suggest an early dating of *La* (odd ligatures and "old" letter forms), however, are absent from this manuscript. It has been suggested that this manuscript originates from the same scriptorium (viz. St.-Germain of Auxerre), but is of a later date.

In the glosses, again, mainly two hands are at work, who are distinguishable primarily by the differences in the colour ink they use. They are contemporary with the main text. The highest amount of glosses is found in Books I and II, and after these two their number quickly decreases. The content of the glosses, furthermore, shows little resemblance to *La*: it is an idiosyncratic mixture of glosses from the Anonymous corpus and John the Scot with unique ones.

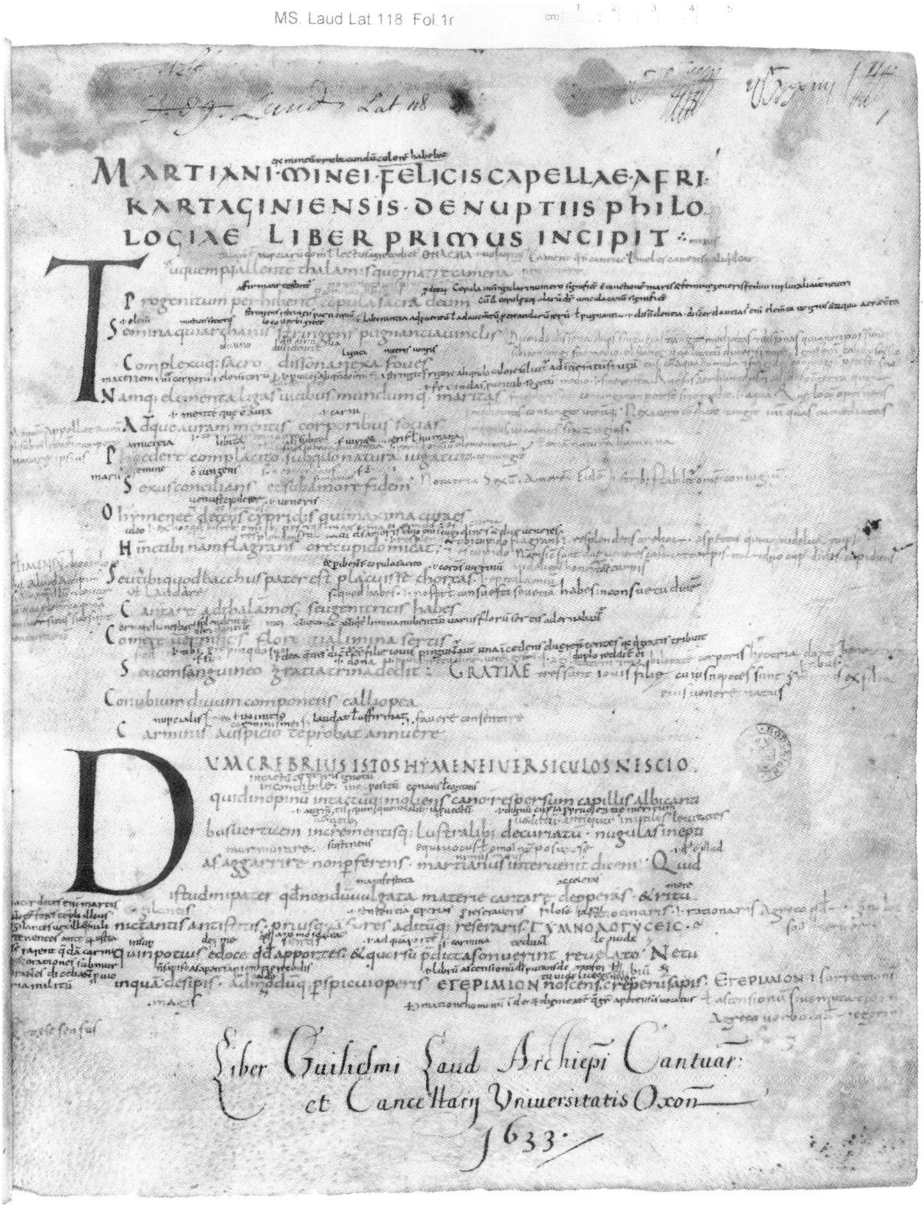
MS. Laud Lat 118 Fol 1r
MARTIANI MINEI FELICIS CAPELLAE AFRI KARTAGINIENSIS DE NUPTIIS PHILOLOGIAE LIBER PRIMUS INCIPIT
Liber Guilielmi Laud Archiepi Cantuar: et Cancellarij Universitatis Oxon 1633

Plate 5

Plate 6: *Leiden, UB, BPL 87* (Lea), fol. 2v. (Cf. pp. 131-135)
Opening page of *De nuptiis*, Book I.

The opening page of *De nuptiis* is found on folio 2 verso, originally the verso side of the first folium of the manuscript. On folio 2 recto, the *accessus ad auctorem* of the Anonymous commentary is found.

The codex starts out as a careful copy of *De nuptiis* with the Anonymous corpus of glosses, written in elegant Carolingian minuscule and with carefully executed headers and subtitles in red rustic capital. Further on in the codex, however, the scribes become sloppier and careless, not only with their script but also with the text itself, which shows more and more omissions. The glossing activity quickly decreases after the second book. The last book is incomplete.

The glosses added to the first two books are closely parallel to those found in *Vaticano, Reg.lat. 1987* (Plate 3) and *Leiden, BPL 88* (Plate 4). A comparison between Plate 3 and this plate shows that even the reference signs used to tie lemma and gloss together are copied.

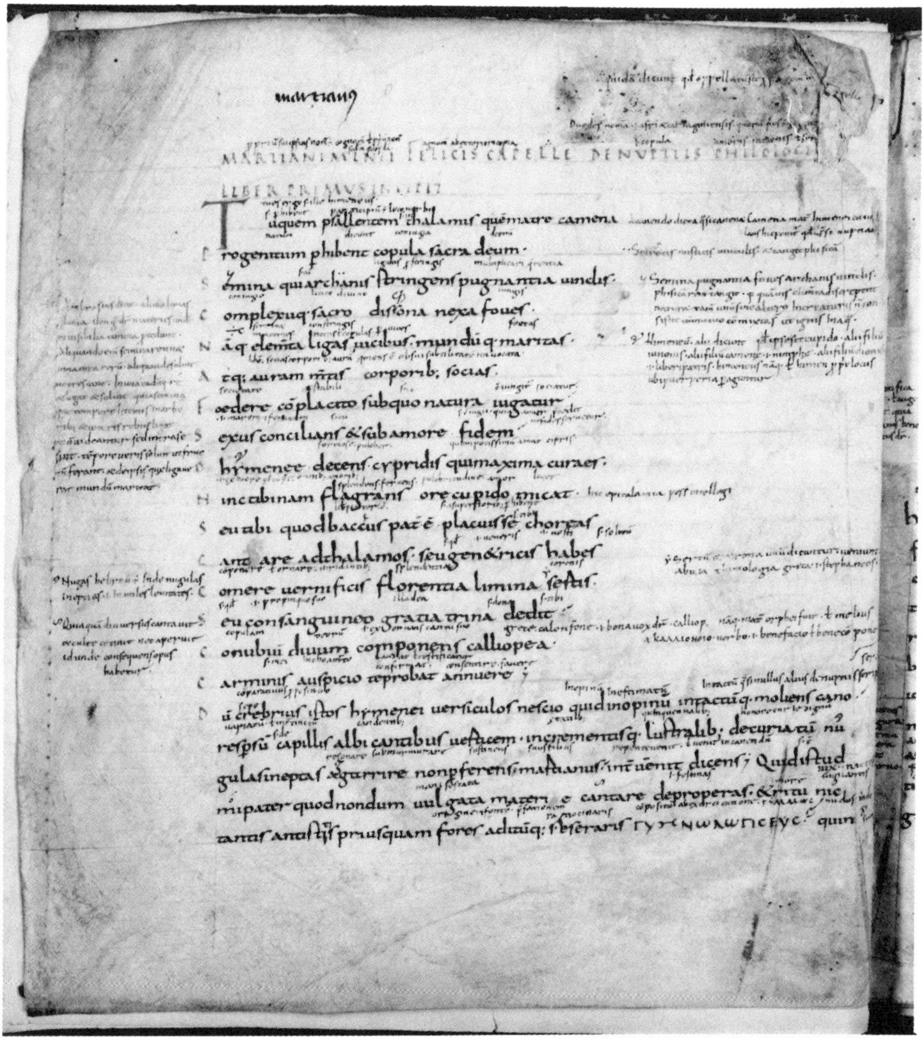

Plate 6

Plate 7: *Leiden UB, BPL 36* (Leb), fol. 115r. (Cf. pp. 135-140)
Opening page of Book IX, *De musica.*

A range of hands can be distinguished in this codex, varying in date from the third to the fourth quarter of the ninth century. This particular plate shows a script to be dated in the fourth quarter.

The glosses are executed mainly by two hands, distinguishable by the colour ink they used. Both hands are active especially in the two mythological books (I-II) and in the quadrivial books (VI-IX). Bischoff pointed to several characteristics which point to Lorsch, but none of these appear on this particular page, which shows the end of book VIII (Astronomy) and the beginning of book IX (Music). Red ink is used for the *explicit/incipit* and for initial letters. The lay-out of the book is straightforward in one column.

The glosses can be identified as a selective and deviating version of the Anonymous corpus of glosses, and is relatively closely parallel to *Paris, lat. 8671* (Plate 8).

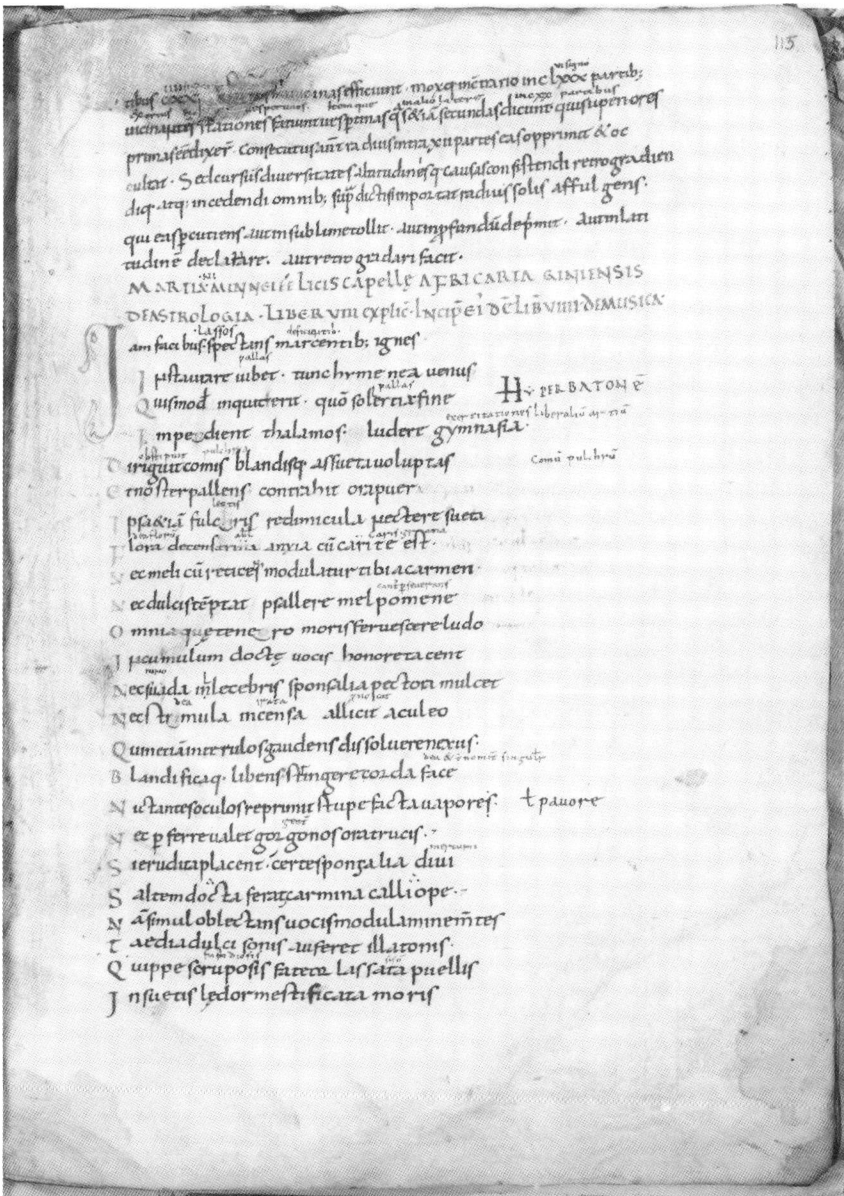

Plate 7

Plate 8: *Paris, BnF, lat. 8671* (Par), fol. 73r. (Cf. pp. 140-145)
Opening page of Book IX, *De musica*.

Several hands were at work in this late ninth-century manuscript, some using abbreviations characteristic for the writing school of Lorsch. (These, however, do not appear on this particular page.) The glosses are also written by several different hands, ranging in time from the late ninth to the early tenth century.

The text is not very carefully copied. The lay-out is plain, the script of both text and glosses is not very polished. There are many lacunae in the text.

This plate shows the end of book VIII (Astronomy) and the opening of book IX (Music). The script is late ninth- or perhaps tenth-century. A contemporary layer of glosses can be identified as a highly selective and deviating version of the Anonymous commentary, which is relatively closely parallel to that found in *Leiden, BPL 36* (Plate 7). There are, however, also many differences in detail.

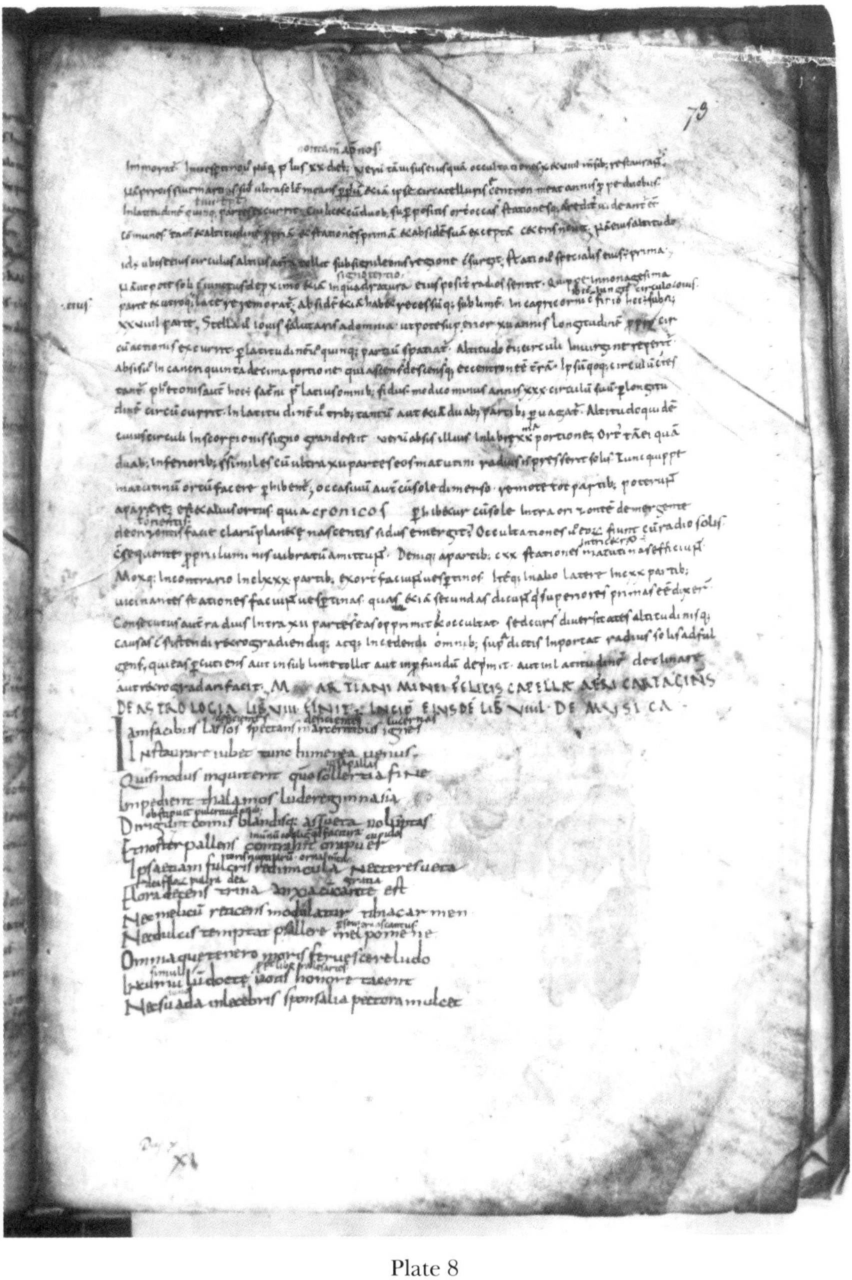

Plate 8

The four most important manuscripts

The second part of this chapter consists of an introduction in greater detail to the four most relevant manuscripts. The descriptions presented in this chapter are perhaps more detailed than is strictly necessary for the central problem of the present research: the reception of Martianus' account of classical musical theory in the Carolingian period. Nevertheless, I remain convinced that a detailed study of the individual sources is the best approach to the study of Martianus reception. Observations which appear irrelevant for our purposes, may very well be the key to answers to other questions.

The manuscript descriptions are based on (earlier descriptions in) secondary literature and on my own observations.

1. *Leiden, Universiteitsbibliotheek, Voss.Lat.F. 48* (La) (Plate 1)

This parchment manuscript was written ca 850, probably in Auxerre. It contains the entire encyclopedia *De nuptiis Philologiae et Mercurii* of Martianus Capella, heavily glossed in contemporary and later hands, and some miscellaneous additions.

Bibliography:

C. LEONARDI, *Raterio e Marziano Capella*

The author draws attention to the autograph of Ratherius of Verona on the fly-leaves and in the margins of the manuscript. He studies the dedication formula of archdeacon Heliseus of Auxerre (f. 50v) and the chronology on f. 95r.

C. LEONARDI, *I codici* I, pp. 452, 462-463; II, pp. 67-68

Leonardi discusses the dedication formula of Heliseus of Auxerre and briefly describes how the manuscript went through the hands of Ratherius of Verona, Petau, Christina of Sweden and Vossius

G. GLAUCHE, *Schullektüre im Mittelalter*, p. 45

Glauche adds that this manuscript, together with 11 others, witnesses the special interest of the "Irish colony" in *De nuptiis*. He suggests a concentration of manuscripts and manuscript production with the commentary of the learned Irishman Martin of Laon in Northeastern France or the surroundings of Laon.

K.A. DE MEYIER, *Voss.Lat.F. 48*

An excellent short description of the manuscript, on which I relied especially in the description of its contents.

J. PRÉAUX, *Les manuscrits principaux*, pp. 79, 101, 123

Préaux dates the manuscript in 847, but it is not clear why. He mentions the fact that the manuscript was given to Saint-Germain of Auxerre shortly after its origin. He suggests that this manuscript is close to the sixth-century archetype, because it carries the correction note of *Securus Melior Felix*. He ranges it, however, under the category of 'second generation' Martianus manuscripts; the many mistakes due to the *scriptio continua* of the archetype are corrected in *La*.

D. SHANZER, *Review article*

Shanzer repeats the conclusions of the above-mentioned authors and adds that in the stemma of manuscripts, this manuscript is related to *Be*, *Paris, BnF lat. 8669*, and *Oxf*

In other papers, e.g. L. HOLTZ, *L'école d'Auxerre*, in: *L'école carolingienne d'Auxerre*, pp. 131-146, esp. 138; and the unedited list of Carolingian manuscripts in Auxerre of G. LOBRICHON, the same information is repeated.

The manuscript consists of 12 quires of eight leaves (4 *bifolia*). The first quire lacks one leaf (which was probably originally blank[16]) the rest is perfectly regular: $\underline{1}^8$(-2) [1-7], $\underline{2}^8$-$\underline{12}^8$ [8-95].[17] All quires except the twelfth carry an original quire signature: *Q i* - *Q xi*. These are written in the middle of the lower margin of the last verso of each quire, with the exception of the signature of the first quire which is written on the first (originally blank) recto-side of the quire (f 1r - i; f 15v - ii; f 23v - iii; f 31v - iiii; f 39v - v; f 47v - vi; f 55v - vii; f 63v - viii; f 71v - viiii; f 79v - x; f 87v - xi).

The foliation is modern, written in the upper right of each recto-side in pencil.

The leaves measure ca 310 x 250 mm. That this closely resembles the original size of the manuscript can be concluded from the fact

---

[16] The text of *De nuptiis* only starts on f 2r, and the first folium contains later annotations (a sermon by Ratherius of Verona, and a 15th-century index of the work on *rasura*).

[17] This formula for the description of collation follows the model of J.P. Gumbert. The quires are numbered and underlined; the number of leaves per quire as it was originally planned is represented in superscript. For example: $\underline{1}^8$ means that quire number 1 was conceived as a quire of 8 leafs. The irregularities in the quire are mentioned between round brackets. For example: (-2) means that the second leaf is missing from the quire, $(+2^*)$ that a leaf is added after the second leaf of the quire. The (modern) foliation is given in square brackets.

that the complete text, including the glosses in the margins, has been preserved and was nowhere damaged by cutting. On f 80r in the lower margin there are drawings of two signs of the zodiac, *Sagittarius* and *Leo*, of which the body of the archer and the legs of the lion fall off the page. This, however, seems to have been the case in the original lay-out of the drawings; it is not the result of cutting.

The parchment is of medium quality. The thickness of the leaves shows some variation, but this is not extensive. There are two obvious exceptions to the general thickness of the leaves: the third quire is made of thicker and stiffer parchment, and the middle bifolium of the fifth quire is thinner. The colour of the parchment is reasonably constant, although hair- and flesh-sides are easily recognizable.[18] All quires are arranged according to the same pattern: HFHF.[19] The last quire is dampstained and shows most signs of wear.

The leaves contain 28 or 29 lines (10 lines measure ca 63 mm), ruled with a hard point, mostly above topline.[20] The lay-out of the page differs per group of quires. The ruling for 1-6 can be described by the formula[21] 11-1:J/0/1-1:C/F, 22<32><128>68 x 37<210>63 mm — to put it in words: a narrow column for glosses (32 mm), a broader one for the text of *De nuptiis* (128 mm). On the

---

[18] Markings of the skin are still visible on ff. 56r, 58r, 59v-60r, 63v, 75v-76r, 78r, 93v-94r.

[19] In this formula the nature of the subsequent *recto*-sides is shown; thus, in *La*, the recto-sides are subsequently hair- and flesh-sides of a hide. There is an exception, however in the third quire: f 25r is (by error) not a flesh-side, but a hair-side.

[20] Exceptions are folios 8r-9v, 15r-v, 16r-21v, 22v-23v, where there are 28 lines, below topline.

[21] The formulas are modeled on formulas of D. Muzerelle (and J.P. Gumbert), described in J.M.M. HERMANS & G.C. HUISMAN, *De descriptione codicum*, pp. 34bis-34sexies. A moderately altered version was recently published: D. MUZERELLE, *Une méthode de notation symbolique*. The formula consists of four zones, separated from each other by a slash (/), of letters (which correspond to a certain length of a rule), and of numbers (which correspond to the number of lines within a certain zone of the formula). The first zone describes vertical lines; the second, horizontal lines in the margin (that is: not the lines that are used for text); the third, the lines that project into the margins, "through lines"; and the fourth, writing lines.

first nineteen folia, the gloss column is already incidentally used for text, and from f 20 onwards the column is still present in the ruling pattern, but not used as such. The glosses are written in the inner and outer margins. In quire 7 the narrow column has disappeared and the outer margin is marked off with a double line (1-2:J/0/1-1:C/J, 22<153>5>70 x 39<209>60 mm). In quires 8-12, a single instead of a double line was used to mark off the writing space: 1-1:J/0/1-1:C/J, 25<158>68 x 45<204>60 mm. Prickings for the vertical lines are found next to the first and last text-line, for the horizontal lines, on the outer vertical line. From the fact that the prickings for the lines are not found on the edges, and that the horizontal and vertical projecting lines do not reach the edges of the leaves, one can conclude that pricking was done by use of a template. The ruling was exercised on the hair-sides and per two bifolia, folded open — as one can conclude from the irregularities and mistakes, e.g. in the last quire.[22] Thus, the general pattern of furrows and ridges is:[23] **>** > **>** > ¦ < **<** < **<**.[24] In quires 2 and 9, however, deviations from this general pattern can be noticed: in 2 the two inner bifolia, and in 9 the two outer bifolia were (accidentally) folded the other way around: 2 **>** > < **<** ¦ **>** > < **<**; 9 < **<** **>** > ¦ < **<** **>** >.

The script is Carolingian minuscule with cc-a's (alongside closed a's), clubbed ascenders, ligatures of -nt, -rt, (-st and -ct), et-ligature

---

[22] This is consistent with the method named "Old Style" by E.K. RAND in *How many leaves at a time?*, cf. p. 77: "The method that I have termed Old Style consists in the ruling of two or four leaves at a time on a hair side, after they have been arranged with hair-sides confronting hair-sides and flesh-sides confronting flesh-sides (Rule I). The Old Style is regularly employed from the end of the eighth century down to about 820 or somewhat later, when the New Style came in." In the New Style, which was practised from the middle of the century onwards, the ruling is equally carried out on two leaves at a time on hair side, but only after they have been arranged with the flesh-side of the upper leaf on the hair-side of the lower leaf. Subsequently the leaves were re-assembled in accordance with "Rule I". This yields a different pattern of furrows and ridges: the Old Style pattern is **>** > **>** >; the New Style pattern is **>** < **>** <.

[23] I.e. each opening of the codex shows either furrows facing furrows, or ridges facing ridges.

[24] Bold angular brackets are used for the leaf that is ruled, normal angular brackets for the leaf in which this ruling shows through.

within a word or even in between two words (e.g. *sensificar&onis*, f 84r l.26) and relatively wide line spacing. The scribe uses an open g, the insular abbreviation for *est*, round and straight d's. For majuscules both uncial letters and *capitalis rustica* are used (often in a deliberate alternation). The transmission of Greek words, written in capitals, is quite corrupt; the scribe intermingles Greek and Latin letters.[25] A clear change of hands is not recognizable, but the script slowly grows smaller (relatively wider spaced), more pointed and less round. When one leafs through the manuscript, the pages look less and less full towards the end, which is not only due to a diminishing number of glosses. The letters are largest on f 3 — measuring slightly more than 3 mm —, but from f 13 onwards their corpora are never higher than 2 mm.

In the glosses, however, three different scribes can be recognized. First, the scribe who originally glossed the text (probably simultaneously with the text of *De nuptiis*) and who used for this purpose the narrow column for glosses or the margins. The scribe uses a round, relatively broad letter in a slightly faded brown ink, and a quite broad pen. The glosses are usually neatly tied to the gloss with a reference symbol. Secondly, a contemporary or slightly later layer of glosses is added to the text with a thinner pen, a smaller sized Carolingian minuscule and darker ink. This scribe scribbled his glosses mostly between the lines, but also in the margins. He is less careful in the placement of his glosses and in using reference signs for tying gloss to lemma. He uses more round d's and more *e-caudatae* (instead of *ae*) than the first hand. The glosses from these two hands can be identified as belonging to the Anonymous corpus.[26] One of these two (probably the first) writes short passages in Tironian shorthand (e.g. f 2r, inner margin; f 4r, between the lines above line 21 and 23). Thirdly, the work of a later scribe can be observed, who did not actually comment upon the text, but highlighted words by repeating them in majuscules in the margin. These words are usually not keywords in the text — as one would expect —, which could serve as a kind of index, but can be characterized as 'strange' or unusual words, which the student of

[25] E.g. f.86r l.22: ΠΑΡΙΠΑΤΗΥΡΑΤΟΝ (*παρυπάτη ὑπάτων*); f.87v l.10 OKYΠYCNOY (*ὀξύπυκνοι*).

[26] J. PRÉAUX, *Le commentaire de Martin de Laon*.

the text would perhaps have wanted to add to his vocabulary. Only an occasional longer gloss of this hand appears in the margin (e.g. f 9r). This third hand has been identified by Leonardi as the autograph of Ratherius of Verona (ca 880-974), who also wrote a sermon on f 1r and a short letter in the lower margin of f 95v.[27]

The punctuation is original, sober and sparing. Majuscules are used for *explicits/incipits*, and (at random) at the beginning of verses, passages and 'sentences.' Prose and verse are generally distinguished by writing the verses in two columns (for which separate verticals are sometimes drawn) and by beginning each verse in principle with a majuscule. The only rubrication in the entire manuscript is found in the *explicit/incipit* of book II/III — the originally red ink now rather black or silver-coloured. There are a few initials that stand out because of their tallness: the initial letter of book I is five lines tall, of book II three lines. The other books have initial letters of one or two lines high, as do some prose or verse passages.

The manuscript is not decorated. There are two drawings on f 80r of two signs of the zodiac, *Sagittarius* and *Leo*, but these function, to my mind, more as commentary to book VIII (*De astronomia*) than as decoration.

The history of the manuscript is full of mysteries. Although there seem to be many clues for origin and date, they are in fact no more than riddles, as, for example, in the case of the chronology found on f 95r:

> Ab origine mundi usque ad nativitatem noe sunt anni mille LVI. Inde usque ad nativitatem abraham fiunt anni DCCCXCII. Ab origine mundi usque ad abraham colliguntur anni mille DCCCXLVIII. Inde usque ad exitum filiorum israel de terra aegypti anni II milia CCCCXLVIII. Inde usque ad edificationem domus a salomone fiunt anni CCCCLXXX et stetit illa domus annos CCCCX. Inde usque ad edificationem templi a zorobabel et zra (*sic, lege Ezra*) anni LXX et stetit illa CCCCXXX annos. Ab eversione vero templi quæ facti (*sic*)

[27] C. LEONARDI, *Raterio e Marziano Capella*. B. Bischoff confirmed the attribution in a letter to F. WEIGLE, who edited Ratherius' letter in *Die Briefe des Bischofs Rather von Verona*, MGH, *Briefe der deutschen Kaiserzeit* I, pp. 48-49 (N° 9a).

est a tito usque in presentem annum qui est annus incarnationis domini DCCCXV,[28] fiunt anni DCCLXX et[29] VI. Ab origine mundi usque in presentem annum IIII milia DLXXV.
(From the origin of the world to the birth of Noah there are 1056 years. From there to the birth of Abraham 892 years. From the origin of the world to Abraham 1848 years are counted. From there to the exodus of the sons of Israel out of the land of Egypt 2448 years. From there to the construction of the temple by Solomon there are 480 years, and this temple lasted 410 years. From there to the construction of the temple by Zorobabel and Ezra (there are) 70 years, and this temple lasted 430 years. But from the destruction of the temple, which was caused by Titus, until the present year, which is the year of our Lord 815, there are 776 years. From the origin of the world until the present year (there are) 4575 years.)

In this enigmatic chronology the numbers do not add up: 892 plus 1056 is not 1848, but 1948; the destruction of the temple by Titus in the year 70 A.D. is neither 776 years away from the year 815 (corrected version), nor from 915 (original version). In short: it actually gives us no hold for dating the manuscript. It is highly unlikely that the manuscript was written in 815 — all paleographical evidence points to a later date. Possibly the chronology was copied out of an exemplar from the year 815, but that does not prove anything for the text of *De nuptiis*, let alone for the glosses added to it.

There is, however, another clue for the dating of this manuscript, hidden in the margin of f 50v. Along the edge of this folio the text *HUNC LIBRUM DEDIT HELISEUS ARCHIDIACONUS SANCTO GERMANO PRO VITA ETERNA* (archdeacon Heliseus gave this book to St. Germain for eternal life) is written, from bottom to top and left to right, in rustic capitals. Heliseus can be traced back to the late 10th-, early 11th-century obituary list of the St. Stephanus cathedral of Auxerre. He died on 20 December,[30] but the list

---

[28] Actually, there are four C's, but the fourth one is deleted by expunction.

[29] et *superscripta*.

[30] "Obiit Eliseus archidiaconus, qui dedit fratribus res proprietatis suae in villa quae dicitur Gratiacus, arpennos vinearum novem, propter remedium peccatorum suorum suique memoriam recolendam." (Archdeacon Heliseus died, who gave the brothers his possessions in the village called Gratiacus, nine arpents of vineyard (1 arpent equals 120 x 120 Roman feet), for the healing of his sins and to revive his

unfortunately does not mention of which year. He is also found, however, as one of the signatories to a decree at the council of Douzy in 871.[31] Shortly before or after then, he must have given this book to the abbey of Saint-Germain in Auxerre. It remains unclear where the manuscript was actually written, but shortly after its origin (to conclude from paleographical evidence) it was in the possession of the monks of Saint-Germain of Auxerre — possibly it was also written in their scriptorium.

In the tenth century, the manuscript came into the hands of Ratherius of Verona (890-974), whose autograph is found in the manuscript. In his childhood Ratherius was oblated to the monastery of Lobbes, which — at that moment — was closely tied to the famous school of Liege,[32] possessed a rich collection of books and was therefore able to educate its pupils well. Ratherius, however, led a turbulent life, in which he undertook many travels. Three times he became bishop of Verona, but every time he was sabotaged, expelled, imprisoned or exiled by the local aristocracy. From 953 to 955 he held the episcopal see of Liege and, hence, was abbot of Lobbes, supported by Bruno, the brother of Otto I. In 955 he became a victim of a conspiracy of Balderic, bishop of Utrecht, and Rotbert, the archbishop of Trier. The nephew of the bishop of Utrecht, also called Balderic, became the new bishop of Liege and Ratherius was again exiled. With the support of Bruno he became abbot of a little monastery in Aulne, in the vicinity of Lobbes, from 955 to 960. He became bishop of Verona for the third time in 961, only to be expelled again after 6 years; with his sharp tongue and pen he made too many enemies and was forced to leave Verona in 967. For some time he drifted from monastery to monastery and he died in Aulne in 974.

---

remembrance.) Cited after C. LEONARDI, *Raterio e Marziano Capella*, p. 80.

[31] C.J. HEFELE & H. LECLERQ, *Histoire des conciles* IV, Paris 1911, pp. 619-635. "Heliseus archidiaconus, ad vicem mei Christiani praesulis ecclesiae S. Stephani Antisiodorensis, subscripsi" (I, archdeacon Heliseus, in the city of Christianus, head of the church of St. Stephanus of Auxerre, signed.) Christianus was bishop of Auxerre from 860 until his death in 869 — cf. GAMS, *Series Episcoporum Ecclesiae Catholicae*, Regensburg 1873, p. 502.

[32] Although the monastery actually fell within the diocese of Cambrai, from 889 its abbots where also bishop of Liege. This situation would last until 990, when Heriger became abbot and Notker held the episcopal see of Liege.

It will be clear that Ratherius' notes in the margins of *La* hold no unequivocal clues to the whereabouts of the manuscript in his time. It is unlikely that the manuscript was present in the library of Lobbes, for *De nuptiis* is mentioned in neither of the two catalogues of the library that have survived.[33]

From the tenth century onwards the history of the manuscript eludes us completely. On folio 20v a note in a thirteenth-century script says *Guido de Lidols* and *Petrus*. On folio 53r again *Petrus et qui con illo* — perhaps in the same hand. The meaning of these notes, however, is unclear. Only at the end of the sixteenth century the codex reappears as part of the collection of Paul Petau (1558-1614): his signature *D.32* is written on one of the paper fly-leaves. At his death the collection came into the possession of his son, Alexander Petau, who sold it to the Swedish queen Christina in 1650. The manuscript then came into the hands of Isaac Vossius and via Gerardus Vossius it became part of the collection of the University Library of Leiden, where it was given its present signature: *Vossianus Latinus Folio 48*. A *schidula* with his name and a stamp of the University Library can be found in the lower margin of f 2r.

La once had a parchment cover with a French charter on the inside. This cover is now separated from the manuscript and preserved in a box with signature *BPL 2513*. Its present binding is a 19th-century parchment back with cardboard boards. In the gutters of the quires six sewing stations can be seen, spaced (measured down from the head of the spine) 21-40-66-66-64-33 mm.[34] There are six paper fly-leaves at the beginning and three at the end of the book.

The manuscript contains:

f 1r Originally blank, in the tenth century filled in with a sermon of Ratherius of Verona (autograph): *CUM*

[33] By order of Folcuin, abbot from 965-990, a list was made of the books the monastery possessed, edited by J. WARICHEZ in *L'Abbaye de Lobbes*, pp. 254-256. In 1049 abbot Hugo had a more extensive inventory made, ed. F. DOLBEAU, *Un nouveau catalogue de Lobbes*, I: pp. 3-36, II: pp. 191-248.

[34] Older sewing stations can be seen in some of the gutters, but I remained unable to reconstruct the original binding.

POTENTUM ALIQUIS AD PENITENTIAM ACCEDIT DE QUO IGNORATUR UTRUM CORDE AN PRETEXTU TANTUM UT ASSOLET HOC AGAT HUMILITATIS. HEC ORATIO SUPER EUM POST PSALMOS ET LETANIAM FUNDATUR.

f 1v On *rasura* an index in 15th-century French cursive script:

liber satiricorum libri viii
item de dyaletica liber i
item de rethorica liber i
item de geometria liber i
item de arismetica liber i
item de astrologia liber i
item de armonia liber i[35]

f 2r -91v Martianus Capella, *De nuptiis Philologiae et Mercurii*, books I-IX
I 2r-11r; II 11r-19r; III 19r-30v; IV 30v-40r; V 40r-52v; VI 52v-66r; VII 66r-74v; VIII 74v-82r; IX 82r-91v.

f 92r Figures to book VI (*On geometry*) and VIII (*On astronomy*)

-92v Pseudo-Cyprianus, *De xii abusivis saeculi*,[36] prologue: *Duodecim abusiva sunt seculi. hoc est sapiens sine operibus ... dei iudicium rotatur.*

f 93r -94v Pseudo-Hieronymus, *Commentarius in canticum Deborae*:[37] *INCIPIT CANTICUM DEBORAE. Barach vir Deborae prophetidis fuisse traditur ... Et si queritur cur vinum non dederit. respondendum est quia domus recabh vinum non bibebant. sicut habes in iheremia propheta.*

f 94v -95r (l.28) Anonymous commentary on days of fasting: *haec dicit dominus exercituum. Ieiunium quarti. Ieiunium quinti. Ieiunium septimi. Ieiunium decimi erit domui iuda gaudium et letitia ... et vasa domus domini in babilone translata fuerunt.*

(f 95r, l.22) List of the names of the months in Hebrew: *Nomina mensuum hebreorum. Primus mensis martius aebraicae nisan vocatur ... Ianuarius sebbath. febroarius*

[35] The 15th-century reader apparently did not notice book III, *On grammar*.

[36] De Meyier refers to the edition of S. HELLMANN, *Texte und Untersuchungen* .. 34. 1, 1909, 32.1-9.

[37] PL 23, col. 1383C-1390C.

adar.

(l. 25) Chronology:

Ab origine mundi usque ad nativitatem noe ... Ab origine mundi usque in presentem annum III milia CLXXV.

f 95v Figures to book IX (*On harmony*), enumeration of note names; note to book 4 (*On dialectic*): *predicativus. UTRUM UTILIS SIT DIALECTICA. Omnis disciplina utilis est ... MIXTUS. UTRUM UTILIS EST DISPUTANDI SCIENTIA. Si bene disputare utile est ... omne iustum honestum ...*

Lower margin: letter of Ratherius of Verona (autograph):[38]

Obsecro karissimam dominationis vestrae paternitatem ... et ego propterea X missas vobis faciam cantare.

2. *Besançon, Bibliothèque Municipale, Ms. 594* (Be) (Plate 2)

A parchment manuscript dated in the third quarter of the ninth century, located in Saint-Oyan, abbey Saint-Claude (?). It contains the entire encyclopedia *De nuptiis Philologiae et Mercurii* of Martianus Capella and is for the greater part heavily glossed, apparently in only one contemporary hand. The glosses form a close parallel to those found in *La*.

Bibliography:

A. CASTAN, *La bibliothèque de l'abbaye de St.-Claude du Jura*, p. 325 n. 2; *id.*, *Catalogue général*, pp. 352-353

Castan dates the manuscript in the middle of the ninth century and notes that the script of this manuscript closely resembles those manuscripts that were produced by order of Manno, who taught at Laon and later became provost of Saint-Claude in the Jura. He observes that many of the marginal glosses seem to be written in the same hand as the text.

S. TAFEL, *The Lyons Scriptorium* II, pp. 49-50

Tafel also mentions Manno: a ct-ligature characteristic for his hand is also recognizable in *Be*. A Martianus Capella manuscript is mentioned in the 11th-century catalogue of St.-Oyan. It is, however, described as a "codex continens Martiani libros novem, habens in fine Porfiri librum ad Constantinum imperatorem". But nowadays the codex, which still

---

[38] Ed. F. WEIGLE, *Die Briefe des Bischofs Rather von Verona*, pp. 48-49, N° 9a.

has its original ninth-century sewing (!), only contains Martianus Capella.

J. PRÉAUX, *Un nouveau manuscrit de St. Gall*, p. 225 n. 11
Préaux is the first to identify the corpus of glosses in *Be* as being a representative of the Anonymous corpus, which he attributes to Martin of Laon.

C. LEONARDI, *I codici* I, pp. 452, 462-463; II, pp. 13-14
Concerning the place of origin Leonardi mentions St.-Oyan — the manuscript was there, at least, shortly after its origin. He also recognizes the Anonymous corpus in the glosses, and observes that it was written by two different hands.

G. GLAUCHE, *Schullektüre im Mittelalter*, p. 45.
Glauche adds that this manuscript, together with 11 others, bears witness to the special interest of the "Irish colony" in *De nuptiis*. He suggests a concentration of manuscripts and manuscript production with the commentary of the learned Irishman Martin of Laon in Northeastern France or the surroundings of Laon.

J. PRÉAUX, *Les manuscrits principaux*, pp. 79, 123
Préaux dates the manuscript in the second or third quarter of the 9th century and mentions that the manuscript was present in St.-Oyan shortly after its origin. *Be* is not as close to the 6th-century archetype as *La*; it is removed one step further.

D. SHANZER, *Review article*
She repeats the conclusions of the above mentioned authors and adds that in the stemma of manuscripts, this manuscript is related to *Paris, BnF, lat. 8669*, *Oxf* and *La*.

The same information is repeated in B. BISCHOFF, *Katalog der festländischen Handschriften*, 629, pp. 157-158.

The manuscript consists of 11 quires: 1-3^{8} [1-24], 4^{10} [25-34], 5^{6} (-1, -4) [35-38], 6-10^{8} [39-78], 11^{6} (-6) [79-83]. It is not clear why quires 4 and 5 deviate from the pattern of regular quaternios. For the last quire (11) a smaller gathering was obviously used because not many leaves would be needed for the remainder of the text. All quires except the first and eleventh carry an original quire signature in the lower margin of the last verso of each quire: f 16v - ii; f 24v - iii; f 34v - iiii; f 38v - v; f 46v - vi; f 54v - vii; f 62v - viii; f 70v - viiii; f 78v - x.

The foliation is modern, written in the upper right of each recto in ink. On f 2r, 37r and 53r there are second folionumbers in pencil. In the first quire the bifolia, which are now loose, have been re-arranged and numbered in the wrong order. Nowadays the second bifolium is placed in the middle of the gathering, so that the leaves are now in the order 1342¦7568. To read the leaves in the correct order, one should read (following the modern pagination) f 1r-v, then 4r-v, 2r-v, 3r-v, 6r-v, 7r-v, 5r-v and finally 8r-v. Originally the leaves were bound together correctly, as can be concluded from the HF-arrangement and marks left by binding thread in the gutter of bifolium 3^6 (originally 4^5).

The leaves measure ca 335 x 255 mm. This must be close to the original size of the manuscript, for the margins are quite large, there are no signs of cutting and in quire 4 (the *quinio*) the edges of the actual skin are visible (bifolia 26^33 and 28^31).

The parchment is of medium quality. Hair- and flesh-sides are in general easily recognizable.[39] The thickness of the leaves is fairly constant. The quires of 8 leaves are arranged according to the pattern HFHF. In the first quire, the present pattern is HHFF, but restoring the original order, the pattern is HFHF. In the two *terniones* and the *quinio* the arrangement is the same, taking the irregularities of the gatherings into account. The manuscript does not have a cover, which caused darkening of the outer leaves and fading of the ink. Both from the front and from the back many wormholes penetrate into the manuscript, as far as f 37 and f 58 respectively. Moisture damaged the manuscript, especially the last quire. On several leaves in the margins white spots can be seen, perhaps some kind of fungus or the remains of inserted damp and sticky paper leaves. Folium 36 (half of a bifolium, fastened only by a stub) is torn and almost loose.

The leaves contain 29 lines, with exception of the first bifolium of the first quire (f 1^8), where 30 lines are ruled, but the 30th line is not used. Ten lines measure ca 90-94 mm. Two ruling types are used in this manuscript: the quires 1, 2, 4, 5, 6 and 11 have one column, no horizontal projecting through lines and only two vertical projecting through lines: 1-1/0/0/J, ca. 60<134>62 x 22<264>52 mm (quires 1 and 2) or ca. 22<264>52 x 36<158>62 mm (quires 4-6 and 11). The quires 3, 7, 8, 9 and 10

---

[39] Markings of the skin are clearly visible on 12v-13r, 20v-21r, 29r, 37r, 40v-41r, 42v-43r, 45r, 46v-47r, 52v-53r, 54v-55r, 65r, 67r, 73r, 76v, 79r, 80v-81r, 82v-83r.

have in lay-out a narrow column (probably originally meant for glosses) next to a broad column, marked off by an extra vertical through line, but this narrow column has never actually been used; the writing lines stretch from the first to the third vertical through line. This ruling-type can be described by the formula 11-1/0/0/J, ca. 35 < 17-136 > 65 x 22 < 264 > 52 mm. In both types the prickings for the vertical lines are high in the upper, and low in the lower margin. Pricking for the writing lines are along the edges of the leaves. Ruling is executed by a hard point, generally on the hair-side of each bifolium, laid open. There are two different ruling-patterns for the *quaterniones*: > < < > ¦ < > > < in quire 1 (having restored the original order of the bifolia);[40] > < > < ¦ > < > < in the quires 2, 3, 6-10. The two *terniones* (5 and 11) follow the same pattern: > < > ¦ < > <. The *quinio* (4) is again ruled according to the same pattern as the *terniones*, but a *binio*, ruled per two open bifolia, is added to it: > > > < > ¦ < > < < <.

The script is Carolingian minuscule, with cc-a's alongside two-storeyed a's and lightly clubbed, almost vertical ascenders. U's are round or sometimes v-shaped, g's have large tails which are nearly closed and reach backwards. Capital N's and U's sometimes have descenders. The line-spacing is very wide; letters of 1-1.5 mm (with long ascenders) in lines of 9-9.5 mm. The copyist must have reckoned with interlineair glosses. The script is regular, but there are several transitions where the letter seems to grow smaller, more packed together and written with a thinner pen.[41] The scribe uses a fine *capitalis rustica* (with capital and uncial (round) D) for (a.o.) *explicits/incipits* and is able to write nice, regular Greek letters.

The glosses are written in a contemporary hand (or perhaps hands) very similar to the text-hand. In a regular and small script (letters of ca 1 mm) many glosses are written in the books I, II, IV and VII, slightly fewer in the other books.[42] There are no clear

---

[40] Here the ruling of the inner pair of leaves has been executed on the flesh-side.

[41] Ff 16v-17r, 25r-25v, 34v-35r, 59r-59v. There may be a change of hands on f 48v.

[42] In general the first and last couple of pages of each book are heavily glossed, but in between the density of glosses decreases.

transitions of one scribe to another, nor distinguishable layers of glosses. The commentator sometimes does, and sometimes does not use symbols to tie gloss and lemma together.

The original punctuation is sober and scarce and uses only two symbols: an elevated dot (pause) and an elevated dot with an ascending stroke (question mark). Further on in the manuscript (e.g. f 25v) *puncti elevati* (semicolons) and *puncti versus* ("inverted semicolons") are used as well, which may have been added by a slightly later scribe. Majuscules are used for *explicits/incipits*,[43] words in Greek, subtitles, refrains of poems (e.g. f 10v), definitions of metres, at the beginning of verse lines or sentences. The majuscules used at the beginning of books, or of passages of poetry or prose are in general somewhat larger: 1.5 or 2 lines high. There are no rubrications or decorations.

We have very little information on the history of *Be*. There are no traces of usage other than the ninth-century ones, nor are there any notes or stamps which reveal former possessors — the only stamp is the one of the Bibliothèque Municipale of Besançon on f 1r and 83v. Castan and Tafel[44] both linked the manuscript to Manno, teacher at Laon under Charles the Bald and then provost of Saint-Claude in St.-Oyan. His hand can be recognized by a characteristic ct-ligature in his *ex voto*,[45] and closely resembles the hand of the scribe of *Be*. Martianus Capella's encyclopedia is mentioned in an 11th-century library catalogue of St.-Oyan and this title can perhaps be identified as the manuscript now kept in Besançon. In the catalogue, however, the codex is described as a codex containing Martianus Capella *and* Porphyry, whereas *Be* only contains Martianus Capella. Moreover, since the original sewing has been preserved, it is unlikely that the manuscript has ever contained

---

[43] For the explicit of book I/incipit of book II three lines were left open, but it was never actually written (f 8v).

[44] A. CASTAN, *La bibliothèque de l'abbaye de St.-Claude du Jura*, p. 325 n.2, p. 355; *id.*, *Catalogue général des manuscrits des bibliothèques publiques de France*, pp. 352-353; S. TAFEL, *The Lyons Scriptorium* II, pp. 49-50.

[45] "Voto bonae memoriae Mannonis liber ad sepulchrum sancti Augendi oblatus"; cited after S. TAFEL, *The Lyons Scriptorium*.

more texts than this one.

The manuscript does not have a cover. At present it is kept in a modern box made to fit it. The first quire is loose, but its sewing stations concur with the other quires. The other ten quires are sewn together and still have their original herringbone sewing (!). In the gutter of the quires five stations can be seen, corresponding to three thongs and headbands, spaced (measured down from the rim) 25-39-96-100-33 mm. The headbands are fastened on a piece of leather, which is first folded back and then fastened to the quires on the edges of the spine (see my sketch in Figure 3 below).

The manuscript contains Martianus Capella, *De nuptiis Philologiae et Mercurii*, books I-IX: I 1r-8v; II 8v-16r; III 16r-27r; IV 27r-35r; V 35r-43v; VI 43v-57r; VII 57r-66r; VIII 66r-73v; IX 74r-83v.
On f 83v: figure to book VIII (*De astronomia*), made by use of a pair of compasses.

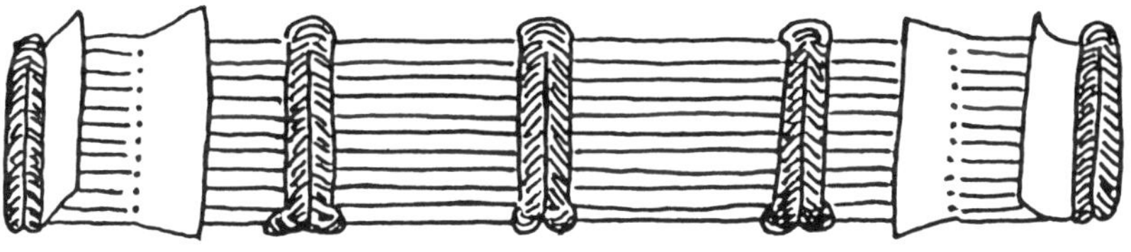

Figure 3

3. *Vaticano, Biblioteca Apostolica Vaticana, Reg.lat. 1987* (Va) (Pl. 3)

A ninth-century parchment manuscript containing the entire encyclopedia *De nuptiis Philologiae et Mercurii*, with many contemporary and a few later glosses. It is in fact composed of two parts (book I-II and book III-IX) that originally did not belong together. The joining of these two parts, however, must have taken place at a very early stage, probably already in the ninth century. Both parts are located in Northern or Northeastern France, the first part may have been written in Reims.

Bibliography:

E.M. BANNISTER, *Monumenti Vaticani*, p. 101 col. 2

The author draws attention to the two songs on f 1r-v, which were written in the 11th century and to which musical notation was added.

J. PRÉAUX, *Deux manuscrits gantois*

Préaux observes the 11th-century additions to this and a Leiden manuscript (*Lb*), which reveal that the manuscript at that time was kept at St. Peter's abbey in Ghent. They are written in a hand that can be identified as belonging to the school of abbot Wichard († 1058). He suggests that the monastery asked its sister-abbeys, i.e. Corbie or Reims, for books to enhance the educational and cultural level. One of these two, probably Reims, could then have been the place where the manuscript was written. He also notes that the manuscript is in fact a compilation: books I-II are of a different origin than books III-VIII. The script of books I-II could indeed be located in Reims, while that of the second part may resemble that of Tours, according to Bernhard Bischoff.

C. LEONARDI, *I codici* I, 446, 462-464; II, pp. 470-472

Leonardi notes the 'Wichardian' additions and the fact that the manuscript seems to have been compiled from two different manuscripts. The joining of these parts, however, must have taken place in an early stage. For the glosses (Anonymous corpus) were written throughout the encyclopedia in only one hand (or possibly two very similar ones) in the ninth century.

G. GLAUCHE, *Schullektüre im Mittelalter*, p. 45

Because of the presence of 'Martinus of Laon's' corpus of glosses, this manuscript is also included in Glauche's list of sources which point to a special interest of the "Irish colony" in *De nuptiis*.

S. CORBIN, *The neumes of the Martianus Capella manuscripts*

Corbin studied the 11th-century neumes on ff 1r-v and identified them as Lorrainian notation.

J. PRÉAUX, *Les manuscrits principaux*, pp. 78, 101-102, 122-123

Since *Va* contains the correction note of *Securus Melior Felix* and reflects — through many visibly corrected word divisions — the *scriptio continua* of its sixth-century archetype, Préaux lists it as one of the three most important manuscripts. He dates the codex in the second quarter of the ninth century, and locates it in North(eastern) France.

D. SHANZER, *Review article*

She repeats the conclusions of the above-mentioned authors and adds that in the stemma of manuscripts, this manuscript is related to *Vaticano Reg.lat. 1535*, *London Harl. 2685*, *Bamberg 39*, *Paris lat. 8670* and *Karlsruhe Reich. 73*.

J.C. FRAKES, *Remigius of Auxerre, Eriugena, and the Greco-Latin circumstantiae-formula of accessus ad auctores*, p. 241 n. 46

Frakes cites part of the *accessus ad auctorem* found in both *Va* and *Lb*, to illustrate an "Irish predilection for the accessus".

The results of these studies are repeated in, among others, C.E. LUTZ, *CTC* II, p. 370; L.D. REYNOLDS, *Texts and Transmission*, p. 246; B.S. EASTWOOD, *The chaster path of Venus*, pp. 147, 150, 152, 154; and M.-E. DUCHEZ, *Jean Scot Érigène premier lecteur*, p. 175. In *RISM, Theory of Music* the manuscript is wrongly dated to the 12th century (!).

The manuscript consists of 20 quires, 17 of 8 leaves, 2 of 3 and 1 of 10. The first two quires are in fact *bifolia*, to which single leaves are added: $\underline{1}^{2}$ (+0*) [1-3] and $\underline{2}$(+0*) [4-6] (Figure 4). The actual text of Martianus Capella's *De nuptiis* begins on f 1v, which contains five lines of the opening poem; the poem is continued on bifolium 2^3. Quires $\underline{3}$ and $\underline{4}$ are perfectly regular: $\underline{3}$-$\underline{4}^{8}$ [7-22]. With these four quires, books I and II are complete.

Quire $\underline{5}$ lacks the first leaf: $\underline{5}^{8}$ (-1) [23-29]. This quire opens with book III, On Grammar, and is written in a different script. The first leaf was perhaps cut off in order to make the joining of the two parts from different origins less conspicuous. Quires $\underline{6}$, $\underline{7}$ and $\underline{8}$ are again regular: $\underline{6}$-$\underline{8}^{8}$ [30-53]. Quire $\underline{9}$ is a *quaternio* with two singletons: $\underline{9}^{8}$ (3|6) [54-61]. Quire $\underline{10}$ is, again, a regular *quaternio*: $\underline{10}^{8}$ [62-69]. A smaller leaf, on which longer glosses were written, was added after the sixth leaf. The quires $\underline{11}$-$\underline{19}$ are also regular: $\underline{11}$-$\underline{19}^{8}$ [70-141]. Quire $\underline{20}$ has five leaves: $\underline{20}^{4}$ (+4*) [142-146] (Figure 5). The text of the ninth and last book of *De nuptiis* actually ends at the bottom of f 145v, and 146 contains annotations of a later date. Perhaps, then, the quire was planned as a binio, in which 142^145 were singletons; f 146 is probably a later addition.

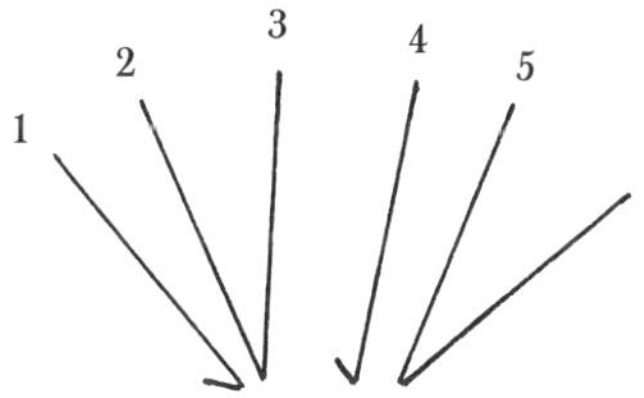

Figure 4 — Quire $\underline{1}$ and $\underline{2}$

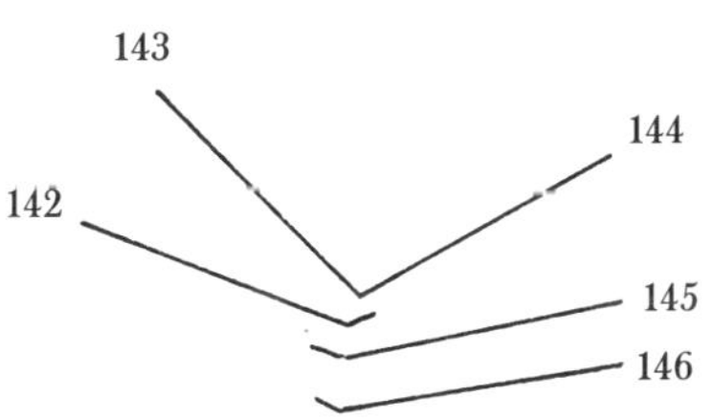

Figure 5 — Quire $\underline{20}$

Formula: 1^{2} (+0^{*}) [1-3]
2^{2} (+0^{*}) [4-6]
3-4^{8} [7-22]

5^{8} (-1) [23-29]
6-8^{8} [30-53]
9^{8} (3¦6) [54-61]
10^{8} [62-69]
11-19^{8} [70-141]
20^{4} (+4^{*}) [142-146]

Quires 3 and 4 both carry an original quire-signature in the lower margin of their last versos: *F* (14v) and *G* (22v). The quires 9-19 carry the signatures *E-P*, also in the lower margin of their last versos: *E* (61v), *F* (69v), *G* (77v), *H* (85v), *I* (93v), *K* (101v), *L* (109v), *M* (117v), *N* (125v) *O* (133v) and *P* (141v). The implications of these quire-signatures will be treated below.

The leaves measure ca 295 x 225 mm. They are numbered in ink by a modern hand, in the upper right of each recto. There are no traces of an original foliation. Although sometimes the edges of the actual skin and most prickings alongside the edges of the leaves are visible, there is evidence that at some point small strips were cut from the upper and lower margins, for some of the marginal glosses are no longer complete (e.g. f 44v-45r, f 135, f 145).

The parchment is of medium quality, the thickness not very constant and the hair- and flesh-sides are sometimes clearly (e.g. f 16v-17r), sometimes hardly (e.g. quires 7 and 17) recognizable. The quires of 8 leaves are all arranged according to the pattern HFHF (3-20). The first two quires have the patterns /HF¦H (1) and /FH¦F (2). Ruling is always executed on the hair-side. There are many edges, tears and holes in the leaves of this manuscript, some restored with new parchment, some not. Edges are seen, for example, at ff 16, 49 (repaired), 59, 76; holes in ff 17, 57, 60, 61, 63, 67, 70, 87, 104, 114, 116, 131, 134, 141, 142 and 143. Pieces have been cut off the margins of ff 83, 117, 135 and 145; ff 17 and 119 have been torn. From the back wormholes penetrate into the manuscript as far as f 107. The last quire is more worn than the others: dark stains obscure the text.

The lay-out of the pages is always in one column, but otherwise there is little regularity. The number of lines per page varies from 23 to 31 and six different types of ruling with a hard point are used.

The first folium (quire 1, f 1) has 25 lines. Its lay-out does not provide an inner margin: the lines stretch from the gutter of the bifolium to the outer margin, marked off with a double vertical through line. The upper- and lower margins are marked off with double horizontal through lines: 0-2/0/2-2/C, measurements ca 170>43 x 28<202>64 mm. Ten lines measure ca 85 mm, the text is written above topline. Prickings for these lines must have been cut off. The next five leaves (quire 1 and 2, ff 2-6) have 29 lines, above topline (ca 87 mm per 10), a single vertical through line marking off the inner margin and a double one marking off the outer margin. The first three and last two writing-lines are projecting into the margins: 1-2/0/3-2C/J, ca 7<165>8>32 x 28<230>35 mm. Prickings for the lines are found along the outer edge of the leaf, and those for the vertical through lines on the first horizontal. Ruling is executed on the hair-side, according to the pattern < < (ff 2-3) < < | < (ff 4-6).

Quire 3 has 4 vertical through lines (pricked on the first and last writing line) and two horizontal ones, both at the top of the leaf: 2-2/0/2-0/J (ca 11<7<162>7>25 x 24<245>26 mm). Its 30 text-lines (above topline, 85 mm per 10) are pricked alongside the outer edge of the leaf and ruled on the hair-sides per two bifolia, in what has been called by E.K. Rand "Old Style": > > > > | < < < <.

The 4th quire — that is, the first quire of the part which supposedly has a different origin — again has 4 vertical through lines but no horizontal ones: 2-2:E/0/0/J (ca 7<6<171>7>34 x 25<233>36 mm). It has 29 lines (above topline, 84 mm per 10) and is pricked and ruled according to the same pattern.

Quire 5 is different again: it has the same number of vertical and horizontal through lines as quire 3 (2-2/0/1-0/J, ca 12<8<164>10>34 x 20<241>32 mm), but 29 lines (above topline, 86 mm per ten) instead of 30, and these are pricked along the outer vertical through line, instead of along the edge of the leaf. The last two leaves (ff 28-29) have 28 lines instead of 29. They were ruled per two folia, and not per two bifolia, as the pattern of ridges and furrows reveals: / > > > | > > > >.

Quires 6, 7 and 8 have the same lay-out as quire 5, but they are ruled per four bifolia: > > > > ¦ < < < <. Quire 6 has 29 lines, above topline, 7 and 8 28. These last two quires also have different measurements: 9<7<166>10>32 x 17<231>47 mm.

For quire 9, again a different scheme was used: double vertical through lines and single horizontal ones above and below the text-lines, 28 lines (above topline). The formulae for this quire are 2-2/0/1-1/J. The two singletons in this quire do not form a proper pair in the pattern of furrows and ridges, for this yields the irregular pattern: > > < < ¦ > < < <.

Quire 10 is laid out as 5-8, has 28 lines and the same method of ruling per four bifolia as 6-8: > > > > ¦ < < < <. Not only the ruling was done per four bifolia, but also the pricking, as one can conclude from a slanting line of holes for the rules in all the leaves of quire 10. After f 67 (the 6th leaf of the quire), a smaller leaf is inserted for a long gloss with its own lay-out: no margins are created and it has only writing lines, 48 in number, above topline, 46 mm per 10.

Quire 11 is laid out with two double vertical projecting lines and one horizontal one at the top (as quire 5-8 and 10): 2-2/0/1-0/J. It has 28 lines, which are ruled per two bifolia: > > > > ¦ < < < <.

In quire 12 the one horizontal through line disappears: 2-2/0/0-0/J, it has 28 lines and is ruled per four bifolia: > > > > ¦ < < < <.

The quires 13 to 19 have this same lay-out with two double vertical through lines and no horizontal ones: 2-2/0/0-0/J. All these quires with the exception of quire 19 have 28 lines, above topline; quire 19 has 29 lines. In the quires 13 to 17 the rules are not, however, pricked along the fourth vertical through line (as in the quires 6-12 and 18), but the third. In quire 19 the prickings must have been placed along the edge of the leaf and cut off, for they no longer remain. Quire 13 was ruled per two folia (the bifolia being already folded): > > > > ¦ > > > >, just as quire 5. The quires 16 and 17 are ruled per two bifolia (as 3-5 and 11): > > > > ¦ < < < <; 14, 15, 18 and 19 per four bifolia (as 6-8, 10 and 12): > > > > ¦ < < < <.

In quire 20, finally, two different ruling-patterns are used: ff 142-144 have 31 lines (height per 10: 78 mm), two double vertical through lines and no horizontal ones: 2-2/0/0-0/J. The measurements, however, are different: 10<6<179>8>13 x 30<224>41 mm.

quire	lay-out	number of lines	ruling per 2 or 4 bif.	pricking hor.	pricking vert.
1^{2}	1-2/3-2	29		[edge]	1st line
2^{2}	1-2/3-2	29		[edge]	1st line
3^{8}	2-2/2-0	30	2	edge	1st/last l.
4^{8}	2-2/0-0	29	2	edge	1st/last l.
5^{8}	2-2/1-0	29	2	edge	1st/last l.
6^{8}	2-2/1-0	29	4	4th vert.	1st/last l.
7^{8}	2-2/1-0	28	4	4th vert.	1st/last l.
8^{8}	2-2/1-0	28	4	4th vert.	1st/last l.
9^{8}	2-2/1-1	28	2	4th vert.	1st/last l.
10^{8}	2-2/1-0	28	4	4th vert.	1st/last l.
11^{8}	2-2/1-0	28	2	4th vert.	1st/last l.
12^{8}	2-2/0-0	28	4	4th vert.	1st/last l.
13^{8}	2-2/0-0	28	2	3rd vert.	1st/last l.
14^{8}	2-2/0-0	28	4	3rd vert.	1st/last l.
15^{8}	2-2/0-0	28	4	3rd vert.	1st/last l.
16^{8}	2-2/0-0	28	2	3rd vert.	1st/last l.
17^{8}	2-2/0-0	28	2	3rd vert.	1st/last l.
18^{8}	2-2/0-0	28	4	4th vert.	1st/last l.
19^{8}	2-2/0-0	29	4	[edge]	1st/last l.
20^{4}	2-2/0-0	31		[edge]	1st/last l.

Figure 6

Ff 145-146 seem to have been cut to size for this manuscript, but originally laid out for a larger codex: on f 145 29 lines are used for text, although more lines can be seen in the lower margin. The text on f 145 (the whole leaf is used for the last section of *De nuptiis*: the *explicit* stands at the very bottom of f 145v) does not fit the ruling-pattern; f 146 shows that there are in fact prickings for 23 lines (height per 10: 95 mm). This same leaf reveals two double vertical (pricked in the upper and lower margins) and two double horizontal through lines.

From this long description of different types of lay-out used in the Vatican manuscript *Reg.lat. 1987* (which I have tried to capture more systematically in Figure 6 above) the following conclusions can be drawn. First of all, it is clear that the general hair-flesh arrangement is HFHF (Rand's Rule I), and that ruling is always executed on the hair-side. Where the prickings for these ruling-patterns are not found along the edges of the leaves, a template must have been used.

Secondly, two main types of ruling are used: per two and per four bifolia. In the quires 3, 4, 11, 16, 17 and 20 pairs of bifolia are ruled,[46] in 6-8, 10, 14, 15, 18 and 19 whole quires.

Thirdly, there are two main types of lay-out: with double vertical through lines on both sides of the text and either with or without one horizontal through line at the top of the leaf. The quires 4 and 12-20 correspond with each other in lay-out: 2-2/0-0. The quires 5-8, 10 and 11 follow the other pattern: 2-2/1-0.

Fourthly, the main numbers of lines are either 28 (7-18) or 29 (1, 2, 4-6, 19).

These observations lead to the conclusion that there is no clear system behind all these variations. The variations in the four aspects mentioned do not occur in coherence with each other, but at random. This and the fact that there are two different series of quire-signatures suggests that the codex in its present form is a composition, as was noted already by Préaux and Leonardi.[47]

[46] In the quires 5 and 13 per two folia.

[47] J. Préaux, *Deux manuscrits gantois*, pp. 19-20; C. Leonardi, *I codici* II, pp. 470-471. Their descriptions of the collation of the manuscript, however, do not correspond with each other, nor do they correspond with my own observations.

The actual break occurs between quires $\underline{1}$-$\underline{4}$ on the one hand (containing the text of book I-II), and quires $\underline{5}$-$\underline{20}$ on the other (book III-IX). The quire-signatures of $\underline{3}$ and $\underline{4}$, *F* (14v) and *G* (22v) respectively, make it clear that these were originally the 6th and 7th quire in a no longer existing codex. The first two quires, thus, are a ninth-century addition (ff 2-3) and an eleventh-century addition (f 1). Préaux suggested that they consist of pieces of what originally must have been quire *E*, but since ff 5^6 form a bifolium, this is not possible. Préaux furthermore suggested that quires $\underline{1}$ to $\underline{4}$ were originally ff 33-40 (*E*), ff 41-48 (*F*) and ff 49-56 (*G*), and that the text of *De nuptiis* originally opened on f 35v. With this, however, he implies that in this original manuscript a text covering the leaves of four quires (*A-D*) must have preceded *De nuptiis*.

Quires $\underline{5}$-$\underline{8}$ are the quires *A-D* of a second manuscript, but these quires are not numbered. The quires $\underline{9}$-$\underline{19}$, however, carry the signatures *E-P*. These then must have constituted a manuscript which contained only the books on the arts (III-IX). Although it cannot be proved by means of quire-signatures, it seems to me that quire $\underline{20}$ (a binio with the last sections of book IX) might originally have been part of the same source as the quires $\underline{1}$-$\underline{4}$; this, at least, is suggested by its script, which appears to be slightly older and resembles the first four quires more closely.[48]

To conclude: the first four quires (and perhaps the last quire) of this manuscript seem to stem from an older manuscript than the quires containing the books on the artes. If this manuscript is indeed a joining of two already existing, but incomplete or otherwise

---

Leonardi counts only 19 quires, overlooking the fact that the first six pages actually consist of two quires. He describes these folia as a *quaternio*, from which the first two pages are missing ($\underline{1}^{8}$ (-1,-2)). Furthermore, he does not notice the irregularities in quire $\underline{9}$. Préaux, on the other hand, does describe the first two quires correctly ($\underline{1}^{2}$ (+0^{1}) [1-3] and $\underline{2}^{2}$ (+0^{1}) [4-6]), but makes things more complicated than they actually are for the quires $\underline{5}$-$\underline{7}$. In his opinion, quire $\underline{5}$ is a *ternio* ($\underline{5}^{6}$ [23-28]), in mine (and Leonardi's) a *quaternio* ($\underline{5}^{8}$ (-1) [23-29]). Furthermore, he describes quire $\underline{6}$ as a *quinio* ($\underline{6}^{10}$ [29-38]), $\underline{7}$ as a *ternio* ($\underline{7}^{6}$ [39-44]), while according to my (and Leonardi's) observations these quires are regular *quaterniones* ($\underline{6}^{8}$-$\underline{7}^{8}$ [30-45]).

[48] On the other hand, there is no break or overlap in the text in the joining of quires $\underline{19}$ and $\underline{20}$.

unsuitable manuscripts, it must have been a joining of 1) an older manuscript containing at least the first two books of the encyclopedia, preceded by another text that filled four quires, and 2) a newer one, containing only the books on the seven liberal arts from *De nuptiis*. Perhaps an old, severely damaged manuscript (quires 1-5 and perhaps 20?) were taken as the starting point, and quires 5-20 (or 19) were newly created to complete it. We can be certain, in any case, that the joining of these two parts took place in the ninth century, for the glossing of the whole was done in one ninth-century hand.

The text is written in Carolingian minuscule. As was already suggested in the evaluation of the different ruling-patterns, many different hands worked on this codex, which is moreover composed of parts from different manuscripts. There are many certain and even more possible changes of scribes, writing letters which vary in height between 1.5 and 3 mm. A few of the possible breaks are listed here:

f 1r-v	A later addition to this manuscript, written in two eleventh-century hands, which can be identified as belonging to the school of abbot Wichard of Ghent.
f 4r; l.5	From f 4r (quire 2) onwards the letter seems to be written in a more vertical direction; the scribe uses a different kind of capital A.
f 22v-23r	This scribe uses a more open g, cc-a next to two-storeyed a. The cauda of the e-caudata is almost closed.
f 41v;l.23-f 42r;l.10	A thinner pen is used, which is possibly the one of the glossator.
f 43r	The letters become sloppier, lopsided and irregular.
f 50v	The scribe uses tidier, more regular script.
f 102v; l.2 and 21	The lines 3-20 are perhaps written on *rasura*; the letters grow larger and sloppier at first, then smaller and more regular.
f 120r; l.7	At the beginning of book eight a small, rather flat script takes over, which uses ligatures that

	could perhaps be dated slightly later, but which uses cc-a's as well.
f 132v	The script is written with a thinner pen, on slightly undulating lines.
f 138v	The script becomes increasingly irregular and sloppy.
f 142r	From here (quire 20) a small, tidy and regular letter is written, almost vertical and with slightly clubbed ascenders.

Some of these hands use ligatures that could point to a dating in the earlier 9th century, such as an nt-ligature (e.g. f 70v; l.12), ligatures with an r that stretches beneath the ruling: rt (e.g. f 96v; l.21) and cre (e.g. f 4r; l.9). The number of corrected word divisions is striking; they are mainly executed by the hand that wrote the text. This could be a second characteristic which points in the direction of an early dating of at least part of the manuscript, for it suggests a close relation with the (sixth-century) original in *scriptio continua*. Another notable feature occurs in the Greek words: these are most often marked with a line over the Greek capital letters (e.g. f 2r; l.17, 19).[49]

The glosses are almost entirely written in one and the same hand,[50] or in two hands that resemble each other closely. In the glosses one can observe variations in ink-colour, from brown to black, but it is not possible to point to clear changes of hands. The glossator sometimes shapes his glosses in the form of a V. There are a few additions from later (11th- and 12th-century) hands, e.g. on f 6v, f 28r-v, ff 71v-72r, ff 86v-90v. The letters of the glossator(s) measure only 1 mm, his g's are 3-shaped with open tails, his r protrudes beneath the (imaginary) line and he has a characteristic preference for giving serifs to letters such as i, n, m. The books I, II, VIII and IX are most densely glossed. Sometimes a hand that closely resembles the one that wrote the text bridges small lacunae in the margins. F 1r-v, finally, is written in two eleventh-century

[49] This characteristic, however, occurs only in the first book.

[50] This was the main reason for Préaux to consider this source superior to the Leiden manuscript *BPL 88*: *Deux manuscrits gantois*, p. 17.

hands, which belong to the school of abbot Wichard of Ghent. These same hands can be recognized on the last folia.

On the first folium of the manuscript the poem *Versus de astris et signis caeli* and the refrain of the poem *Scande caeli templa* are written, in an 11th-century, 'Wichardian' hand. To these songs musical notation is added, which was first studied by Bannister in his *Monumenti Vaticani de Paleografia Musicale Latina*,[51] and more recently by Corbin.[52] The neumes are in Lorrainian notation and since the text of the song was obviously written in Ghent, the musical notation may very well share this place of origin. There are no specific characteristics that suggest otherwise.

Both original punctuation and punctuation added by later scribes are found in the manuscript. Apart from the usual medial and low points (with or without commas, generally added by the glossator), the scribes use question marks. An addition that is probably later is the small upside down triangle consisting of three points or two points and a comma, used for longer rests (e.g. at the end of a book) or in between verse-lines (e.g. f 120v). The scribes write majuscules for subtitles and *explicit/incipits*, mainly using rustic capitals and occasionally uncial E, M, D, and F (protruding beneath the line).[53] Coloured majuscules are used for the first *explicit/incipit*, sometimes for the initial letters of verses or prose-passages (e.g. f 2r, 6r). The glossator writes key-words and subtitles in rustic capital.

Some of the majuscules, as already mentioned, are decorated with colour. The majuscules on ff 2r and 7r are written in red ink; the bows of some majuscules are coloured with red (ff 115r-v, 120v, 124v, 125v) or green (f 6r) ink. On f 1r a figure of the planets is drawn in red and green ink, and in the title of the poem below, these same colours are also used.

---

[51] Leipzig 1913, p. 101. He locates the notation in Northeastern France, but observes characteristics that to his mind belong to the notational tradition of Laon or St. Gallen. Corbin, however, does not share this view.

[52] *The neumes of the Martianus Capella manuscripts*, esp. p. 6.

[53] The *incipit* for book II is written in uncial script, in red ink (f 12v). The other incipits are in black ink, in rather clumsy uncial script (ff 22v, 79r, 102r, 120r), in rustic capitals (f 41v) or in a mixture of uncial and rustic capitals (ff 23r, 57v-58r, 132v). Only the last *explicit* (f 144v) is written in the first beautiful and regular rustic capital with uncial M.

Unfortunately, there are few sources which reveal the history of *Va*. As was argued before, it consists of two parts (books I-II & III-VIII), but these were joined already in the ninth century — before the main layer of glosses was written. In the eleventh century the manuscript was present in the library of the monastery of St. Peter in Ghent, as is clear from the presence of an 11th-century hand from the writing school of abbot Wichard (988/994-1058). On f 146v one can decipher a later confirmation of ownership:[54]

> Liber Sancti Petri gandensis ecclesie. Servanti benedictio tollenti maledictio. Qui folium ex eo tulerit vel curtaverit anathema sit
> (Book of the church of Saint Peter of Ghent. A blessing to the keeper, a curse to the thief. May he who took or cut a folium from it be damned.)

Préaux speculatively suggested that the book was given to the monastery of St. Peter by one of its sister-abbeys, Corbie or Reims.[55] In the sixteenth century the codex was obtained by the humanist scholar Franciscus Nansius (1525-1595), who left his mark of ownership on f 1r.[56] Later, the manuscript became part of the collection of the Swedish queen Christina, who — as is well known — donated a large part of her book-collection to the Vatican. Thus it entered the Vatican collection in 1690, receiving its present

[54] Partly erased, mentioned both by C. LEONARDI (*I codici* II, p. 471) and J. PRÉAUX (*Deux manuscrits gantois*, p. 15), whom I cite here.

[55] *Deux manuscrits gantois*, p. 20. In *Les manuscrits principaux*, however, he mentions a possible relationship with Reichenau, for which he unfortunately does not give any evidence. He tentatively suggests that both *Va* and a manuscript that in text-tradition is closely related, *Karlsruhe Reich. 73*, are copies of one and the same model, which could have remained in Reichenau in the first half of the ninth century. This model could have been studied and copied by Lupus of Ferrières, his fellow-students, and 'Irish' scholars from Mainz. The Vatican manuscript would then become the 'French' model for a new generation of manuscripts, the one now remaining in Karlsruhe, the 'German' model. This reconstruction would shed light on the fact that the text tradition of these two manuscripts is relatively similar, but the corrections and additions are different. Préaux notes, furthermore, that the corrections of the 'French' Carolingian philologist have been more widely disseminated.

[56] This possession-mark is partly erased as well and has become undecipherable to me, but Préaux reads *ἐκ τῶν τοῦ μεγάλου Francisci Nansii* (out of the <books> of the great Franciscus Nansius). *Deux manuscrits gantois*, p. 17.

signature *Reginense Latinus 1987.* A stamp on f 146v confirms this.

Apart from the ninth-century traces of usage, there are some 11th- or early 12th-century hands recognizable in the manuscript, in which occasional additions or marginal remarks were written.

The ninth-century manuscript today has modern boards, covered with reddish brown leather with a golden ornamental edge. In their present form the boards measure 306 x 224 mm and the spine shows five raised bands (spaced 50-50-45-47-48-66 mm).[57] On the spine, the signature is printed in gold letters: *Reg 1987.* The codex has red-and-yellow silk sewn headbands. The spine is — especially at the head — quite gravely affected by worm-holes.[58] It has two paper fly-leafs, both empty.

The manuscript contains:

f 1r-v Additions in an eleventh-century hand: a figure of the planetary orbits, made by use of a pair of compasses; the song *Versus de astris et signis caeli* (in several manuscripts attributed to 'Priscianus Grammaticus')[59] and the refrain *Scande caeli templa* (Martianus Capella, book II.117 ff.), both with musical notation.

f 1v The first 5 verses of *De nuptiis*, book I in the same eleventh-century hand.

f 2r-145v Martianus Capella, *De nuptiis Philologiae et Mercurii*, in a ninth-century hand:
books I.6-II: I 1v-12v; II 12v-22v;

---

[57] I was unable to establish whether the sewing was still original or not. It seems unlikely that the sewing is original, since the codex consists of two originally different parts, but I did not see any clear traces of old, no longer used sewing stations in the gutters of the quires.

[58] When I visited the Vatican Library (May 1996) the manuscript had already been transferred to the restoration-department, waiting to be repaired. I would like to express my gratitude to the staff of the manuscript department for making it possible for me to study the manuscript in spite of its being on the waiting list for restoration.

[59] A. RIESE & F. BUECHELER, *Anthologia latina sive poesis latinae supplementum* I, Carmina in codicibus scripta II, Leipzig 1906, N° 679, pp. 154-155. Identified by C. LEONARDI in *I codici*, where he refers to H. WALTHER, *Initia carminum ac versuum medii aevi posterioris latinorum*, Göttingen 1959-1969, N° 310.

books III-IX: III 23r-41v; IV 41v-57v; V 58r-79r; VI 79r-102r; VII 102r-120r; VIII 120r-132v; IX 132v-145v.

f 146r-v Additions in the same, eleventh-century hand: on the recto a Greek alphabet with numerical values, figure of the proportions of the intervals *bisdiapason, diapason, diapente, diatessaron* and *tonus*; on verso a partly erased possession-mark of St. Peter in Ghent and a stamp of the *Biblioteca Apostolica Vaticana.*

4. *Leiden, Universiteitsbibliotheek, BPL 88* (Lb) (Plate 4)

A ninth-century parchment manuscript of *De nuptiis Philologiae et Mercurii* of Martianus Capella, heavily glossed especially in the books I, II and IX. The last two quires, containing the text of book IX, were originally part of another manuscript or were especially made to complete an already existing, incomplete one.[60] They contain a different corpus of glosses: in the books I-VIII a version is found of the Anonymous corpus, in book IX of John the Scot's commentary. The joining together of these parts must have taken place already in the ninth century. The last two quires may have been written in Reims.

Bibliography:

J. PRÉAUX, *Le commentaire de Martin de Laon*

Préaux dates the manuscript in the second half of the 9th century and identifies its commentary as a version of the Anonymous corpus, which he attributes to Martin of Laon. Speculating on the presence of 11th-century annotations of ownership of the abbey of St. Peter in Ghent, he suggests that the manuscript was originally written in St.-Omer or Laon, i.e. the two places to which the monks of St. Peter fled during hostile invasions in 880.

A. VERHULST, *L'activité et la calligraphie du 'scriptorium' de l'abbaye Saint-Pierre-au-Mont-Blandin de Gand*

Verhulst mentions the *ex libris* and *anathema*-formula of St. Peter's abbey in Ghent and describes with great precision the Wichardian hand writing on the added bifolium 1^2 and the last leaf of the manuscript.

[60] In the last two quires a chaotic mixture of lay-outs is found, which may suggest that these were made from left-overs of already pricked and ruled, but otherwise unused bifolia.

J. PRÉAUX, *Deux manuscrits gantois*
Préaux again studies the 11th-century additions in 'Wichardian' hand, which are found in this and the above described Vatican manuscript (*Va*). He now suggests a different place of origin. Speculating that the monastery asked her sister abbeys, i.e. Corbie or Reims, for books to enhance the educational and cultural level, he identifies one of these two as the probable place of origin. On the basis of a paleographical study Bischoff suggested a likeness to the script of Tours in the beginning of this manuscript, but from book V onwards a localization in Reims. Préaux argues for a break later in the manuscript using codicological as well as internal evidence; he considers book I-VIII as one, book IX as a second part.

C. LEONARDI, *I codici* I, pp. 451-453, 457, 462-464; II,pp. 62-63
Leonardi establishes the same compilation of two parts: book I-VIII and book IX. The joining together of these two parts must have taken place in the ninth century, for a small, Carolingian hand writes glosses throughout the entire encyclopedia. Referring to Bischoff, he locates the second part of the manuscript in Reims.

G. GLAUCHE, *Schullektüre im Mittelalter*, p. 45
Because of the presence of 'Martinus of Laon's' corpus of glosses, this manuscript is also in Glauche's list of sources which point to a special interest of the "Irish colony" in *De nuptiis*.

T.A.M. BISHOP, *Autographa of John the Scot*
Although Bishop never mentions the signature of *Lb*, he is in fact writing about it in this paper, in which he is concerned with the problem of two hands, I^1 and I^2, both identified as the autograph of John the Scot.[61] I^2 appears as one of the glossators in the ninth book of *Lb*. His conclusion is that I^2 is not the hand of John the Scot, since this hand shows "nothing peculiarly scholarly; nothing at all to show a moment-to-moment contact between the acts of thinking and writing" (p. 93).)

C. LEONARDI, *Glosse eriugeniane a Marziano Capella*
Leonardi adds to his earlier observations that there are three layers of glosses in the ninth book of *Lb*:[62] E — a hand of Irish origin, which

[61] L. TRAUBE, preface to E.K. RAND, *Iohannes Scottus*, in: *Quellen und Untersuchungen* I (1906), vii-ix; *id.*, *Autographa des Iohannes Scottus*, in: *Paläographische Forschungen* V (1912), (*non vidi*). Rand denied the identification in the article *The supposed autographa of John the Scot*.

[62] In the short summary that precedes his paper Leonardi states that he envisages it as a preliminary study for a new, critical edition of the B-version of John's commentary to book IX. This project, however, was never realized.

writes glosses to the narrative part and the last poem; A — a contemporary continental hand, writing glosses to the music-theoretical part of book IX; and R — a slightly later (but still ninth-century) hand which can also be recognized in the other part of this manuscript (books I-VIII). A and E write a version of John the Scot's commentary. He remains unable to identify the glosses of R.

J. PRÉAUX, *Les manuscrits principaux*, pp. 79, 101, 123

Préaux dates *Lb* in the third of fourth quarter of the 9th century, locates it in Auxerre (?), without further references (p. 79). Further on he mentions either the region Auxerre-Corbie, or Reims as places of origin (p. 101). *Lb* is a source for the Anonymous commentary, here attributed to Martin of Laon, of a second generation, following earlier sources such as *La* and *Be*.

J.C. FRAKES, *Remigius of Auxerre, Eriugena, and the Greco-Latin circumstantiae-formula of accessus ad auctores*, p. 241 n. 46

Frakes cites part of the *accessus ad auctorem* found in both *Lb* and *Va*, to illustrate an "Irish predilection for the accessus".

The results of these studies are repeated in, among others, C.E. LUTZ, *CTC* II, p. 370; J.D. REYNOLDS, *Texts and Transmission*, p. 246; B.S. EASTWOOD, *The chaster path of Venus*, pp. 147, 150, 152, 154; and M.-E. DUCHEZ, *Jean Scot Érigène premier lecteur*, p. 175.

The manuscript consists of 23 quires; 22 of eight leaves, 1 of four (quire 21) and 1 of six (quire 23). In quire 1 a bifolium, which was added to the originally eight leaves, was numbered in the modern foliation: 1^{8} [1-10]. To quire 13 a single, smaller leaf is added after the sixth leaf in the quire, which was also numbered in the modern foliation (13^{8} [99-107]). For the rest the collation of this manuscript is regular.

Formula: 1^{8} [1-10]
$2\text{-}12^{8}$ [11-98]
13^{8} [99-107]
$14\text{-}20^{8}$ [108-155]
21^{4} [164-167]

22^{8} [168-175]
23^{6} [176-181]

All quires but the last carry an original quire-signature, written in the lower margin of their last versos: *I* (10v), *II* (18v), *III* (26v), *IIII*

(34v), *V* (42v), *VI* (50v), *VII* (58v), *VIII* (66v), *VIIII* (74v), *X* (82v), *XI* (90v), *XII* (98v), *XIII* (107v); *XIIII* (115v) *XV* (123v), *XVI* (131v), *XVII* (139v), *XVIII* (147v), *XVIIII* (155v), *XX* (163v), *XXI* (167v). Quire 22 carries (by error?)[63] the number *XXI* (175v).

The foliation is modern, written in the upper right corner of each recto in pencil.

The leaves measure ca 253 x 235 mm. This must closely resemble the original size of at least the first part of the manuscript (books I-VIII), for the marginal glosses are all complete. Only on two folia in quire 23 are glosses in the lower margin damaged by cutting (ff 179r, 180v), which suggests that this part of the codex came originally from a slightly larger one.

The parchment is in general of fine quality. The leaves are relatively thick and stiff, some of the flesh-sides being perfectly white, some of the hair-sides perfectly smooth (e.g. quires 5 and 6).[64] The added bifolium (ff 1^2) and the last two quires are clearly of a lesser quality, thinner and more worn. The parchment has very few holes[65] and the margins of only a few leaves show the edge of the skin.[66] All quires of 8 leaves are arranged according to the pattern HFHF (2-12, 14-20, 22), and the quires of lesser leaves are arranged in the same manner, taking the number of their leaves into account: quire 21 has HF|HF, and quire 23 HFH|FHF.

[63] Leonardi sees in this irregularity in the quire-numbering another argument for his thesis that the last two quires of BPL 88 originally belonged to a different manuscript. Although I do agree with his conclusion, I doubt whether the irregular quire-numbering found in 22 can serve as evidence, for the number has exactly the same lay-out as the other ones. This makes me suspect that the quire-numbering was added after the joining together of the two parts, just like two of the layers of glosses and the headers, uniformly added throughout the manuscript, and that the irregularity is simply an error.

[64] There are, however, also folia that show the markings of the skin, e.g., ff 52v, 54v, 83r, 88v, 90v, all hair-sides of quire 12.

[65] Only two in the main part of the manuscript (book I-VIII): ff 119, 127; five in its additions (bifolium 1-2 and book IX): ff 2, 169, 171, 176.

[66] A bad spot in f 89, and possibly the missing edges of ff 158, 159, 160, and 161.

Ruling is executed with a hard point per folium on the hair-side. The dominant pattern of furrows and ridges is > < > < ¦ > < > < (1-20, 22).[67] The *binio* and *ternio* show the same pattern in fewer leaves: > < ¦ > < (21), > < > ¦ < > < (23). The pattern for the lay-out of the page is uniform from quire 1 to quire 21: double vertical through lines on either side of the text, pricked in the upper and lower margins; single horizontal through lines as the first and last textline, pricked in the outer margin; 21 rules, above topline, pricked along the edge of the leaves: 2-2/0/1-1/J, 26<9<135>9>55 x 32<170>50 mm; 21 lines, above topline, height per ten ca 83 mm.

There are only two exceptions: on f 55 the lower horizontal through line stops at the third vertical one (2-2/0/1A-1C/J) and on the bifolium 75^82 both horizontal through lines stop at the fourth vertical one (2-2/0/1-1:C/J).

In the last two quires five different lay-outs are found. In 22 the first and fourth bifolium (i.e. 168^175 and 171^172) have two vertical through lines on either side of the text, one horizontal through line at the head of the text, none below; 2-2/0/1-0/J. A double row of prickings can be seen along the edge of the first bifolium, a single row on the fourth one. The measurements are ca 25<8<142>7>53 x 27<196>29 mm, ten lines are ca 68 mm high. The second and third bifolium, however, have 2 horizontal through lines above, one below the text; 2-2/0/2-1/J, 22<7<142>7>54 x 28<195>28 mm. Again a double row of prickings is found on the one (169^174), a single row on the other bifolium (170^173).

In 23 three different types of lay-out occur in three bifolia:

* ff 176^181: 2-1/0/2-1/J, 19-21A,[68] a single row of prickings along the edge of the leaf, 30 lines, atl (ca 69 mm per ten), 23<7<145>52 x 25<198>22 mm.
* ff 177^180: 2(+2)-2(+2)/0/5-1/J,[69] a double row of prickings

[67] Both the added bifolium in quire 1 and the added folium in quire 13 remained unruled: 00> < > < ¦ > < > < (1), > < > < ¦ > <0> < (13).

[68] The ruling of lines 19-21 reaches the edge of the page.

[69] I counted four vertical projecting lines on either side of the text, but for only two of these, prickings were made in the upper and lower margins.

along the edge of the leaf, 31 lines, atl (ca 68 mm per ten), 25<8<140>7>52 x 21<205>28 mm.

* ff 178^179: 2-2/0/3-2/J, 19-21A, a single row of prickings along the edge of the leaf, 30 lines, atl (ca 69 mm per ten), 20<7<143>8>52 x 25<200>28 mm.

The many different kinds of lay-out in these two quires may suggest that these were newly composed in order to complete an incomplete or damaged manuscript. These bifolia were, perhaps, prepared for other manuscripts, but not yet used.

The text is written by several scribes in Carolingian minuscule.[70] The general appearance, however, is relatively uniform, with a small, upright letter (ca 2 mm high), 3-shaped g's, x's with long descenders to the left, r's which are tilted to the right and sometimes descend below the line, uncial capital M's and D's (besides rustic capital ones), and lightly clubbed ascenders. The letters are somewhat suspended above the lines. Some of the scribes carefully executed high-quality, very regular letters (e.g. ff 3-19r), others more of a scribbling, wavering script (e.g. ff 46v-48r). Striking are the differences in ink-colour, which varies between an almost faded light brown and nearly black.

In the glosses, several hands can be detected. The main glossing-activity of the first part of the manuscript took place contemporarily with or shortly after its making, by two contemporary glossators. The one who wrote the first layer of glosses wrote his additions in a tiny, very regular Carolingian minuscule of only 1 mm high. He used both the margins and the space between the lines and is precise in tying lemma and gloss together with a sign. The ink he used has a light-brown colour and has occasionally almost faded. The second layer of glosses must have been added only a few years later, for the script of this second

[70] The script changes slowly, and it is difficult to point to the exact places where a change of hands occurs, but I suspect breaks in ff 19r-19v, 32r-32v, 46v, 48r, 54r, 59r, 61v, 71r, 71v, 87v, 167r-168v. The last two quires (ff 168r-181v) are clearly written in a different hand, using long ascenders and descenders (without clubbing), closed g's, one-storeyed a's, st-ligatures. The scribe spaces his letters less widely, which, combined with the larger number of writing lines, makes the page look very full.

glossator closely resembles that of the first one. His glosses are executed in darker ink, seemingly with less care, for the script does not look as regular as that of the first layer. This hand can also be found emending and filling in lacunae in the text. Throughout the codex headers are added which identify the number of the book (verso) and its content (recto). In the margins one sometimes finds subtitles written in rustic capitals, which were probably written by the second glossator. The headers, which look very similar, may also have been written by him.

These two glossators seem to have worked their way through the manuscript, including the additional last two quires containing book IX. They are especially active in books I, II, IV, and at the beginning of VIII. In books I, II and VIII the two hands are approximately equally active, whereas in III, V, VI and VII hand 2 is by far the most active.

In book IX, finally, the first hand disappears after the first page and the second hand appears to have been working together with a third one. This third hand shows many Irish characteristics, such as a high, open e, a low and open a, an r with a relatively long descender and a straight T-shaped t. His letters are low, broad, widely spaced and look as if written in haste. The script was once identified as the autograph of John the Scot Eriugene, then as I^2, the hand of a student of John.[71]

In addition to the three hands, contemporary with the manuscript, several notes of later hands can be detected. First of all the Wichardian hands can be observed, who copied large schemes on the topic of Dialectica on ff 1v and 2r, a figure representing the planets and the zodiac on f 2r, an *accessus ad auctorem*, the refrain *Scande celi templa* with musical notation[72] and a Greek alphabet with numerical values on f 2v. They have also been active in some of the books of the encyclopedia, as can be seen e.g. on ff 63r and

[71] T.A.M. BISHOP, *Autographa of John the Scot*, pp. 90-92. For a description of this insular hand, cf. C. LEONARDI, *Glosse eriugeniane a Marziano Capella*, pp. 175-178.

[72] In her study of musical notation in Martianus Capella manuscripts, Corbin did not look at *Lb*. She did, however, look at the 11th-century neumes on f 1r-v in the Vatican manuscript *Va*, which are an exact parallel to those in *Lb*. The neumes can thus be identified as of Lorrainian origin. S. CORBIN, *The neumes of the Martianus Capella manuscripts*.

78r. Several later readers of this manuscript wrote their occasional notes in the margins of *Lb*, e.g. on ff 18v, 22r-v, 33v, 43r, 84r, 115r.

The punctuation is partly original, partly added by the (second?) glossator. Majuscules are used for *explicits/incipits*, at the beginning of verses, passages and "sentences", for titles and subtitles, headers, refrains, etc. They are written in rustic capital, with occasional uncial M's and D's. Initial letters of 1.5-3 lines high are found at the head of books or prose-passages after verses. In book IX, the initial letters of the verses stand out because of their placement between the double vertical through lines. Colours are used to mark beginnings of books, poems and prose-passages. There are several titles and initials in red ink,[73] and passages with capital letters in brown ink, coloured with ochre.[74] Sometimes the two colours are found together, as for example on ff 96r-96v, where red capital letters are marked with ochre.

Again there are few sources which reveal the history of this manuscript. As was noted above, the codex contains parts from two different manuscripts, bound together in the ninth century. Bischoff noted that the script of book I-IV resembled that of Tours, and suggested that the second part could have been written in Reims.[75] Although Préaux argued for a break later in the manuscript — i.e. from book IX onwards —, he accepted the localization in Reims. Like *Va*, *Lb* was part of the library of St. Peter's abbey in Ghent in the eleventh century, as can be concluded from the presence of 'Wichardian' script in the added bifolium and the *anathema* on f 2v (the same as that found in *Va* on f 146v):

> Liber s. petri gandensis ecclesiae. Servanti benedictio. Tollenti maledictio. Qui folium ex eo tulerit. vel curtaverit. anathema sit.
> (Book of the church of Saint Peter of Ghent. A blessing to the keeper, a curse to the thief. May he who took or cut a folium from it be damned.)

---

[73] Ff 3r, 22v-24r, 24v-26v, 27r-31v, 40r-v, 79v-80r, 96r-v. Some of these red majuscules are written in red ink, others are redrawn with red ink (ff 24v-26v, 27r-31v, 40r-v), perhaps when the second or third layer of glosses was added to the manuscript.

[74] Ff 3r, 8r-v, 9r, 10v, 11r, 13r, 17v, 18r, 58v, 59r, 60v, 61r, 96r-v.

[75] He suggested this in a personal letter (29 December 1957) to J. PRÉAUX, who repeated his view in *Deux manuscrits gantois*, pp. 18-19.

In an early article Préaux voiced the hypothesis that the manuscript could have been written in St.-Omer or Laon, two places to which the monks of St. Peter's fled for hostile Norman invasions in the 880's.[76] In later articles, however, he opted for a second hypothesis, which dates the arrival of the manuscript in St. Peter's abbey in Ghent in the period when Wichard was abbot, i.e. 1034/35-1058, and seeks the place of origin among its sister abbeys, Corbie or Reims. Because the two ways of establishing the origin of the manuscript, i.e. Bischoff's hypothesis based on paleographical evidence and Préaux' hypothesis based on historical evidence, partly concur, Reims is now generally accepted as the place of origin of the second part, that is, the last two quires, of the manuscript.[77]

Again like *Va*, the codex was in the sixteenth century obtained by the humanist scholar Franciscus Nansius, from whose heirs the university library of Leiden bought the manuscript ca. 1600. There it received the signature *Biblioteca Publica Latina 88*. A *schidula* with this signature is attached to the spine of the codex and a stamp of the library can be seen in the lower margin of f 3r.

Today, the codex has a parchment spine and cardboard covers, covered with marbled paper. On the inside, paper leaves are glued to the covers. Two paper fly-leaves open the book, the second one watermarked with a bunch of grapes. The codex has five double cords (spaced 55-42-44-48-42-36 mm). The sewing is not original, for several now unused sewing stations can be seen in the gutters of the quires.[78] In the back of the codex, no paper fly-leaves are present,

---

[76] *Le commentaire de Martin de Laon*, p. 457: "Or je crois qu'il est permis de supposer avec assez de vraisemblance que cet important codex de 182 folios [...] ne provient sûrement pas de Saint-Pierre de Gand, dont l'histoire intellectuelle n'est guère brillante. D'autre part il n'est pas sans intérêt de noter ici que lors des invasions normandes de 880, Gand servit de quartiers d'hiver aux pirates et que dès 846 les moines de Saint-Bavon et de Saint-Pierre se réfugièrent à Saint-Omer, puis à Laon [...]. On peut raisonnablement penser que le *Leidensis 88* arriva à Gand lors du retour des moines après les invasions [...]."

[77] Each of the two localizations is, however, a hypothesis; neither is in itself based on conclusive evidence. The localization should therefore be taken for what it is: a hypothesis.

[78] I have been unable to reconstruct the sewing-method in its original form.

but there is a parchment folium which carries the signature *BPL 88* and the number 182 (i.e. the 182nd folium of the codex). On it is a Latin text in a fourteenth-century cursiva.[79] Its measurements do not concur with the manuscript: it is part of a bifolium, from which the first leaf is cut to a strip of only 35 mm wide, the second measures 200 (width) x 240 (height) mm. Its prickings are enigmatic: they run along the edge of the strip and form a rectangle on the edge of the preserved folium. Perhaps another piece of parchment was sewn onto it to make it large enough to serve as a cover. The leaf is now fixed to the sewing with one piece of string.

The manuscript contains:

	two paper fly-leaves;
1r	empty, offset of a strip of a missal, used to strengthen the binding;[80]
1v-2r	ninth-century annotations: schemes highlighting the *ars dialectica*, a figure of the planets and the zodiac, made by use of a pair of compasses;
2v	eleventh-century annotations in 'Wichardian' hands: an *accessus ad auctorem* from the Anonymous corpus, the refrain *Scande celi* with neumes, a Greek alphabet with numerical values, the *anathema* of St. Peter's abbey, Ghent;
3r-181v	Martianus Capella, *De nuptiis Philologiae et Mercurii*, with many glosses taken from the Anonymous corpus (books I-VIII) and John the Scot's commentary (book IX). I 3r-18v; II 18v-34r; III 34r-58v; IV 59r-80r; V 80r-106r; VI 106v-133r; VII 133v-152v; VIII 152v-167v; (in a different script and with different glosses) IX 168r-181v;
182r-v	bifolium cut to size with a Latin text in fourteenth-century cursiva, perhaps once used as a cover.

[79] The text has been identified by one of the librarians as "b. Bernardi, Orationem ad cor b. Virginis". The librarian refers to BPL 90 and BPL 94, in which leaves of the same manuscript can be found.

[80] One of the annotations on the fly-leaves identifies fragments from the same missal in BPL 48 and BPL 120, both of which also belonged to Franciscus Nansius. The fragment itself is now filed away in box BPL 2514 (B:78).

Four manuscripts of secondary relevance

In this third section, four manuscripts are described which are of secondary importance. They are considered reliable witnesses of the Anonymous commentary, but only for part of the whole corpus. In *Oxford, Laud.lat. 118* and *Leiden, BPL 87*, for example, the glosses in book I and II are closely related to the Anonymous commentary as found in the four manuscripts described above, while in *Leiden, BPL 36* and in *Paris, lat. 8671* the closest parallels are found for the Anonymous commentary on book IX, which is otherwise only found in *Besançon, 594* (*Be*) and *Leiden, VLF 48* (*La*) Furthermore, the testimony of the four manuscripts described below at times proved invaluable for solving problems caused by damage, corruptions or errors in the transmission of the glosses.

1. *Oxford, Bodleian Library, Laud.lat. 118* (Oxf) (Plate 5)

This parchment manuscript was written in the third quarter of the ninth century, possibly in Auxerre (?). It contains the entire encyclopedia of Martianus Capella, with many contemporary glosses, especially for books I and II, and some miscellaneous additions.

Bibliography:

R. KLIBANSKY, *The Rock of Parmenides*, p. 181
Klibansky remarks that *Oxf* contains a commentary which is closely related to the commentary attributed to Dunchad. He dates the manuscript in the ninth century, and argues that "It is obvious that in our case the gloss preserved in Ms. Laud.lat. 118 represents a stratum of material older than Scotus' commentary".

J. PRÉAUX, *Le commentaire de Martin de Laon*
The glosses added to this manuscript belong to the commentary attributed to Dunchad/Martin of Laon. The manuscript can be dated in the ninth century.

C. LEONARDI, *I codici* I, pp. 456, 462-463; II, p. 423
The manuscript seems to be written in one hand. The glosses are written either by the scribe of the text himself or by a contemporary scribe. They are added to books I-IV and VIII-IX and are numerous in books I-II. They belong to the Anonymous commentary. Occasionally later glosses are added.

S. CORBIN, *The Neumes of the Martianus Capella Manuscripts*

In *Oxf* neumes are added to several poetic passages. Corbin identifies the notation as Aquitanian and locates it in the area west of the Massif-Central (Limoges-Angoulême). She compares different versions of the melody added to the refrain "Scande caeli templa" found in four manuscripts, and comes to the conclusion that they are entirely independent. The only resemblance is the recitative character of the melodies in both *Oxf* and *Paris, lat. 8670.*

J. PRÉAUX, *Les manuscrits principaux*, pp. 79, 123

Préaux dates the manuscript in the second or third quarter of the ninth century. He locates *Oxf* in the region Auxerre-Soissons, together with three more manuscripts: *La*, *Be* and *Paris, lat. 8669.* He suggests that the manuscript might be related to the group of scholars surrounding Lupus of Ferrières, but does not give evidence for this. He categorizes *Oxf* in group C, that is just below the six manuscripts of major importance for the edition of *De nuptiis.*

D. SHANZER, *Review article*

Shanzer copies Préaux's dating and the grouping of the manuscripts *Oxf*, *La*, *Be* and *Paris, lat. 8669.* She remarks on pp. 72-73, n. 60: "T.A.M. Bishop kindly informs me that he has found one of the hands in F (=*Oxf*) in manuscripts connected with Lupus of Ferrières. It would be intriguing to know whether this manuscript, and perhaps the ancestor of the whole group, could be associated with Lupus. I note that it is primarily MSS of this group that carry metrical annotations [...], a speciality of Lupus, who wrote on the meters of Boethius *De consolatione philosophiae* and identified the meters in the Wolfenbüttel Prudentius (56, 18 Aug. 4° [...]), probably using the *Centimetrum* of Servius (also to be found in Laud.lat. 118 ff 90^{v}-93^{r})."

D. GANZ, *Heiric d'Auxerre glossateur du Liber glossarum*, in: *L'École carolingienne d'Auxerre*, pp. 297-312

Ganz notes that Heiric, the third master of the school of Auxerre, often quotes Martianus Capella in his *Liber glossarum*. He listed e.g. the word *crotula*, which is the corrupt version of *crotalorum tinnitu* (II.133). This corruption is known to be present in only four Martianus-Capella manuscripts: *Oxf*, *La*, *Be* and *Paris, lat. 8669.* He suggests Heiric as a candidate for the authorship of the Anonymous commentary.

The manuscript consists of twelve quires of 8 folia: $\underline{1}$-$\underline{12}^{8}$ [1-95]. The quires are regular, but in the modern foliation, one leaf (of which the outer margin was cut) was not numbered. Between ff 35 and 36, a leaf that should have carried the number 36 is present. On the last verso of all quires except the last a quire signature is executed: $\underline{1}$ 8v:

i, 2 16v: ii, 3 24v: iii, 4 32v: iiii, 5 39v: v, 6 47v: vi, 7 55v: vii, 8 63v: viii, 9 71v: viiii, 10 79v: x, 11 87v: xi.

The leaves measure ca 288 x 237 mm.

The arrangement of the folia in all quires is HFHF|HFHF.

Prickings usually must have been executed along the edges of the leaves, for they no longer remain in the larger part of the quires. The first leaf of quire 1, however, is an exception: here prickings for the writing lines are visible along the third vertical through line. In the quires 2-12 almost all prickings, which were executed along the edges of the leaves, were cut off.[81]

Ruling was done with a hard point, according to the following patterns:

1-5: 11-1B1A/1-1:H/1H-0/J,
measurements ca 10<22<208>31>21 x 6<24<149>32>24 mm. 27 (1-4) or 28 (5) lines atl, ca 80 mm high per 10.

6-12: 11-11/0/1-1/J, measurements ca 10<25<162>32>3 x 16<252>24 mm. 31-34 lines[82] atl, ca 76-68 mm high per ten.

Ruling was always executed on the hair-sides of the folia, and carried out per two bifolia. There general patterns of ridges and furrows is < < < < | > > > > (1-6, 8-10, 12), but in 7 and 11 the arrangement is different:
< < > > | < < > > (7) and > > > > | < < < < (11).

The script is Carolingian minuscule and resembles the script found in *La*, a manuscript dated ca 850 and located with some probability in Auxerre. The text is written by several scribes, who use a fairly similar set of letters, which makes it difficult to establish the exact places of changes of hands. The ascenders are lightly clubbed, the scribes use many ligatures, closed and open g's, round and upright d's, closed and open a's. However, some of the characteristics of the script in *La* (unusual nt- and at-ligatures, the use of the &-ligature in between words) that specifically point to an early dating, are not

---

81 Noteworthy is that in quire 10 prickings can be seen along the edges which do not correspond to the actually executed ruling-pattern. This quire was perhaps prepared for a manuscript of a different size, but in that case its destination was changed before it was actually ruled.

82 31 lines: 11, bifolia 80^87, 81^86; 12. 32 lines: 11, biff 82^85, 83^84. 33 lines: 6, biff 40^47, 41^46; 7; 8; 9; 10. 34 lines: 6, biff 42^45, 43^44.

found in *Oxf*. The manuscript is therefore supposed to be of a somewhat later date. If the Leiden manuscript was indeed written in the monastery St.-Germain of Auxerre, *Oxf* could very well originate from the same scriptorium. In the Bodleian Library an unpublished list is kept, compiled by Guy Lobrichon, of manuscripts which were written or kept in St.-Germain of Auxerre in the ninth century. Three manuscripts which are part of my original, broad selection are listed: *La*, *Vaticano, Reg.lat. 1535* and *Oxf*.

In the glosses mainly two scribes are at work, contemporary with the scribes of the text. They use different ink-colours but are otherwise hard to distinguish, for their script is very similar. Comparing the content of the glosses, the parallel with *La* — the manuscript which resembles *Oxf* closely in palaeographical respect — is, surprisingly, not very close. There are parallels with the Anonymous corpus of glosses, but there are also parallels with the commentary of John the Scot. Furthermore, a fairly large number of the glosses seem to be unique. To books I and II, many glosses are added, but then the glossing activity decreases. In book IX only a few glosses are added. Among these, however, are some remarkable parallels with *La* and *Be*, which are not found in any of the other manuscripts.[83]

A remarkable aspect of this manuscript is the fact that one of Martianus' poems was provided with musical notation. Not only the refrain "Scande caeli templa" (II.117 ff) has neumes, but also the verse "Tandem laboris fructus" (II.120). The neumes are written in exactly the same ink as was used by one of the two contemporary glossators and are therefore supposed to be contemporary with the layers of glosses. The neumes are identified as Aquitanian point neumes — a type of musical notation that was common in Southern

[83] E.g. 357.4-5 Hanc quartam divisionem Boethius non ponit *BeLa*, *Oxf* f 84: hanc quartam divisionem boetius non ponit; 358.19-20 Non negat Boetius sonum dividi in duas partes, sed in aequas partes dividi non posse dicit *BeLa*, *Oxf* f 84r: <non negat> boethius sonum di<vidi> in duas partes, sed in ae<quas> partes dividi non poss<e dicit>; and 358.21-359.1 Diapente nihil est aliud nisi tonus additus ad diatessaron *BeLa*, *Oxf* f 84r: Diapente nihil est aliud <nisi> tonus additus ad diates<saro>n.

France and Northern Spain.[84] Provided that neumes and glosses are by one and the same hand, this suggests a location further south for the glossing of the manuscript, or at least for one of the scribes working on the manuscript.

In the manuscript red ink is sometimes used for *explicits/incipits*, for initials and initial verse letters. For capital letters both uncial and rustic capital letterforms are used, sometimes in deliberate alternation. Apart from the use of red ink, the manuscript is wholly undecorated.

Today the manuscript is leather-bound, the wooden covers covered with dark brown leather, five raised bands on the spine.

2. *Leiden, Universiteitsbibliotheek, BPL 87* (Lea) (Plate 6)

This ninth-century parchment manuscript contains the encyclopedia of Martianus Capella, with contemporary glosses. The ninth book of *De nuptiis* is incomplete; it ends at section IX.920 on f 134r.

Bibliography:

J. PRÉAUX, *Le commentaire de Martin de Laon*

The glosses added to this manuscript belong to the Anonymous commentary, attributed to Dunchad/Martin of Laon. The manuscript can be dated in the second half of the ninth century. In the tenth century, the codex was given to the Abbey of St. Adelbert in Egmond by Egbert, archbishop of Trier (977-993). The glosses are slightly younger than the text — late 9th-, or early 10th-century.

J. PRÉAUX, *Deux manuscrits gantois*, p. 17.

Préaux notes that the annotation of Securus Melior Felix was copied in *Lea*, at the end of book I.

C. LEONARDI, *I codici* I, pp. 446, 448, 450, 457, 462-463, and II, p. 62

The manuscript contains glosses from the Anonymous commentary. Leonardi dates it in the ninth century and remarks that the glosses were probably written by the same scribe that wrote the text. He notes some passages in Tironian shorthand in the margins. At the end of the tenth century the manuscript was given to the abbey of St. Adelbert in Egmond, where the manuscript was given the siglum P.IIII.10. It was

[84] For a more ample treatment of this aspect, see Chapter 3, section *Theory and practice*, 1.

donated to Leiden University Library by Hendrik Bredius.

G. GLAUCHE, *Schullektüre im Mittelalter*, p. 45
Together with ten more manuscripts which contain the *accessus* of the Anonymous commentary, *Lea* is attributed to Martin of Laon, a witness of a special interest in *De nuptiis* among the Irish scholars in Laon and the Loire-region (Northeastern France). Glauche further remarks that the manuscript was given to Egmond in the tenth century by archbishop Egbert of Trier, and that the codex is mentioned in the catalogue of the library of St. Adelbert in Egmond under number 17: *Item Martianus.*[85]

J. PRÉAUX, *Les manuscrits principaux*, pp. 79, 101, 123
Préaux dates the manuscript in the third or fourth quarter of the ninth century. He locates it in Corbie (?), but does not give any evidence to support this location. In a footnote (11bis) he mentions that Bischoff located the manuscript in Eastern France. To his mind, the manuscript is of secondary importance for an edition of *De nuptiis*, for it represents the text in a later stage. The glosses belong to the Anonymous commentary, which he attributes to Martin of Laon, and also reflect a second stage of the textual history. They are an (indirect) copy of the commentary as found in manuscripts like *La*, *Be* and *Va*.

B.S. EASTWOOD, *Plato and circumsolar planetary motion*, p. 11 and n. 14
There are six ninth-century manuscripts in which the orbits of Mercury and Venus are depicted as intersecting, epicyclic circles, with the sun located in their common middle region: *Lea*, *Lb*, *Leb*, *Paris lat. 8669*, *Par*, and *Va*.

Some of the above mentioned remarks are repeated in C.E. LUTZ, *CTC* II, p. 370; J.C. FRAKES, *Remigius of Auxerre, Eriugena and the Greco-Latin circumstantiae-formula of accessus ad auctores*, pp. 241, 253 n. 46.

The manuscript consists of 17 quires: $\underline{1}^{8}$ [1-9] — one leaf was later added before the first quire, and it has been numbered in the modern foliation —, $\underline{2}$-$\underline{16}^{8}$ [10-129], $\underline{17}^{4}$ (+4*) [130-134]. The first folio, which contains a possession note linking the manuscript to the monastery of Egmond, is a fifteenth-century addition. The last quire ($\underline{17}$) is written in another, tenth-century hand and seems to be a later addition as well. The manuscript is incomplete and the text ends in section IX.920.

Quires $\underline{3}$-$\underline{5}$ carry quire signatures: capital letters with Roman

[85] Edited with a commentary by J.P. GUMBERT, *De Egmondse boekenlijst*, pp. 151-157, esp. 154-155.

ciphers written underneath. These signatures are found in the lower margins of the last verso of each quire: 25v, *C iii*; 33v, *D iiii*; 41v, *E v*. There are traces of quire numbering according to a different system for the quires 6-13, although there are no signs which point to a break in the codex in any other respect. The same hands continue to work in the manuscript, there are no added leaves, no gaps or overlappings in content. In quire 6 the Roman cipher *i* is written at the bottom of the lower margin of its last verso (49v), in quire 10 a *v* (81v), in 12 a *vii* (97v), and in 13 a trace of what might very well be an *viii* (105v). The place of these signatures — at the bottom of the leaves — suggest that the signatures of the quires 7-9 and 11, and possibly also of 14-16, have been trimmed.

All *quaterniones* have been arranged according to the pattern HFHF¦HFHF, or in a similar fashion taking the number of leaves into account.

Ruling is executed in drypoint on the hairside of each folium, according to the pattern > < > < ¦ > < > <.

The lay-out of the quires 1-4 and 6-11 can be described in the formula 1-11/0/1-1/J-J'. The text lines are ca 81 mm high per 10, the gloss lines ca 40 mm per 10. The approximate sizes are 35<180>45 x 15<170>50>10 mm. It is in two columns, one for the text of Martianus Capella and an adjacent one for glosses. The writing spaces of both text and glosses are ruled, with 23 lines[86] for the text and twice as many for the glosses.[87] Extra lines for glosses are drawn in the upper (ca 5) and lower (8 or 9) margins. The three vertical through lines are pricked in the upper and lower margin, the two horizontal ones and the textlines on the third vertical through line. The prickings for gloss lines are made in the upper and lower margins and in between the prickings for the text lines, with a smaller sharp point than that used for the text lines.

In quire 5 pricking and ruling for the gloss lines are missing, and the lay-out is described in the formula 1-11/0/1-1/J.

In quire 12 the gloss lines are pricked, but not carried out. Quire 13 and part of 14 (ff 107, 111, 112) are again different: they have

---

[86] The exception is quire 14, which has 23 lines on the first, 24 on the other seven folia.

[87] The bifolium 29^30 in quire 4 is an exception: unlike the other leaves, it has no separate ruling or prickings for the gloss lines.

two vertical through lines, no second column, and no writing lines for glosses, but they do have additional prickings for gloss lines: 1-1/0/1-1:C/J. In 15 the same pattern is laid out, but gloss lines are only pricked in ff 114, 120, 121 and the lower margin of 115, drawn in the lower margins of 114 and 121. In the rest of 14 (ff 106, 108-110, 113), 16 and 17 (as far as one can establish in this worn quire) there are no writing lines for glosses[88] and three vertical through lines and two horizontal ones, in 14 with length B, in 16 and 17 with length A: 1-11/0/1A or 1B-1A or 1B/J.

The text is written by several scribes in Carolingian minuscule. *Capitalis rustica* is used for most of the subtitles. At the beginning of the codex these scribes wrote in an elegant script and took the effort of filling in *explicits/incipits* in red ink. In the course of the book, however, the scribes become sloppier and careless, not only with their script, but also with the text itself, which shows more and more omissions. In the last few quires subtitles and *explicits/incipits* are often not executed and the last book is incomplete.

This same tendency can be observed for the glossing activity. The first books have many glosses, executed neatly in mostly one and the same hand, but after book IV hardly any glosses are added. On f 2r, a contemporary scribe wrote the *accessus ad auctorem*. Remarkable are some oddly shaped *NOTA*'s in the margins and Tironian notes.[89] Furthermore it should be noted that the glosses added to the first books of *Lea* are a close parallel of those added to *Lb* and *Va* – not only in content, but even in lay-out. For with a frequency that excludes coincidence, the glossator of *Lea* used exactly the same signs to tie gloss to lemma.

Red ink is used quite often in the beginning of this manuscript, for subtitles, *explicits/incipits*, and for capital letters at the beginning of books, sections, verses or for refrains. The initial letter of book IV was coloured with yellow ink, but this is the only case in which coloured ink, other than red, was used.

Today the manuscript is leather-bound, with thick cardboard

---

[88] Prickings for these, however, are visible in 14.

[89] E.g. on ff 33r, 41v, 42r, 44v, 45r, 46v, 48v, 49r, 52v, 58v, 60r, 118r-v, 119r-v, sqq.

covers covered with red-brown leather and 5 raised bands on the spine. The original sewing was not preserved. On the inside of the covers marbled paper fly-leaves (black and bluish green) are stuck. Three more paper fly-leaves with modern annotations and clippings from catalogues are found at the beginning of the codex, two at the end.

Of the history of the manuscript little is known. According to a note of ownership on f 1v, the manuscript was given to Egmond by archbishop Egbert of Trier in the late tenth century; the codex is mentioned in the catalogue of the library of St. Adelbert in Egmond, where it was rebound by order of abbot Gerardus of Poelgeest in 1465. Via Hendrik Bredius, it became part of the University Library of Leiden.
The contents of the manuscript are:

f 1r-v Signature of unknown origin (*P.iiii.IO*) and note of ownership of Egmond:
Dominus egbertus archiepiscopus treverensis obtulit m[onasterio egmondensi] hunc librum quem venerabilis domnus gherardus de poelgheest abbas eiusdem loci religari fecit anno domini m°.cccc°.lxv°
(The lord Egbert, archbishop of Trier, offered this book to the monastery of Egmond, (this book) which the honourable dom Gherardus of Poelgheest, abbot of this same place, had rebound in the year 1465.)

f 2r *accessus* from the Anonymous commentary

ff 2v-134r Martianus Capella, *De nuptiis Philologiae et Mercurii*, books I-IX.920 (I 2v-15v, II 15v-27r, III 27r-45r, IV 45r-60v, V 60v-81v, VI 81v-104r, VII 104r-117v, VIII 117v-128v, IX (up to § 920) 128v-134r.)

3. *Leiden, Universiteitsbibliotheek, BPL 36* (Leb) (Plate 7)

Some of the hands in this ninth-century parchment manuscript are identified by Bischoff as representatives of the school of Lorsch. It contains the entire encyclopedia of Martianus Capella, with a corpus of glosses that remains unidentified.

Bibliography:

J. PRÉAUX, *Deux manuscrits gantois de Martianus Capella*, p. 17
Préaux lists *Leb* among manuscripts which have the correction-note of Securus Melior Felix at the end of book I.

C. LEONARDI, *I codici* I, pp. 446, 448, 450, 452, 457, 462, and II, pp. 61-62
Leonardi dates the manuscript in the ninth century and proposes a possible location in Lorsch (I, p. 452; II, p. 61). He remarks that the manuscript contains glosses in a contemporary hand and corrections and *variae lectiones* in a slightly later hand.

C.E. LUTZ, *CTC* II, p. 381
Lutz listed the ninth-century manuscript *Leb* among the manuscripts with "unidentified glosses".

B. BISCHOFF, *Lorsch im Spiegel seiner Handschriften*, pp. 45, 96-97[90]
Bischoff recognized in *Leb* a number of characteristics which can be related to the scriptorium of Lorsch. He briefly describes the manuscript in the section "Das Skriptorium seit etwa 860." He remarks that the history of the origin of the manuscript must have been an intricate one, for among the scribes are some hands which show the characteristics of Lorsch, and some which belong to other, unidentified schools. He argues that the sections ff 74-81 and 106r-127v are 10th-century additions, and that the glosses and corrections are written in several hands from the 9th and 10th century. Characteristic for Lorsch are, for example, the abbreviation hł (*hic* or *hoc lege*, e.g. f 21r) and a Tironian *hic* (e.g. f 20r).

J. PRÉAUX, *Les manuscrits principaux*, pp. 79, 122-123
Préaux dates the manuscript in the third or fourth quarter of the ninth century. He locates it in Auxerre (?), but does not give any evidence to support this location. The manuscript is of secondary importance for an edition of *De nuptiis*, for it represents the text in a later stage. Both *Lb* and *Leb* are, to his mind, good witnesses of the text as corrected and explained by philologists associated with the court of Charles the Bald.

B.S. EASTWOOD, *Plato and circumsolar planetary motion*, p. 11 and n. 14
There are six ninth-century manuscripts in which the orbits of Mercury and Venus are depicted as intersecting, epicyclic circles, with the sun located in their common middle region: *Leb*, *Lea*, *Lb*, *Paris lat. 8669*, *Par*, and *Va*.

---

[90] See also the second, revised edition: *Die Abtei Lorsch im Spiegel ihrer Handschriften*, pp. 54, 106-107.

The manuscript consists of 17 quires: 1^8 (+7*)[91] [1-9], 2-15^8 [10-121], 16^6 [122-127]. Quire 17 consists of three leaves, a single leaf and a bifolium. With exception of the last, all carry quire signatures in their lower margins. Some of the signatures are placed on the last verso of the quire (1-3), others on the first recto (9-16), and in some quires both the first and the last leaf were numbered (4-8; see figure 7 below).

The HF-arrangement of all quires is HFHF. Ruling is executed with a hard point on the hair side of the folia, usually according to the pattern > < > < ¦ > < > < (all quires except 3, 5 and 8; see fig. 7). In quire 3, 5 and 8, however, ruling was sometimes executed per pair of bifolia: in 3 and 5 the inner pair was ruled from the outside (> < > > ¦ < < > <), and in 8 two pairs were ruled from the outside (> > > > ¦ < < < <; see fig. 7).

The lay-out of the pages is in one column, with double vertical through lines on both sides of the writing space. The number and length of the horizontal through lines, however, differ. In general (all quires except 10-14), one through line is used at the top, and two at the bottom of the text: 2-2/0/1-2/J, with 32 writing lines,[92] height per 10 ca 76 mm. Measurements: ca 30<227-6> 47 x 20<6<145>7>47 mm.[93] Quires 10-12 and 14 have no horizontal through lines: 2-2/0/0-0/J, 32 lines, and approximately the same writing space. Quire 13, finally, has two horizontal through lines on either side of the text: 2-2/0/2-2/J.[94] The irregularities are captured in figure 7 below.

In quire 17, on the three folia with figures for the books *Dialectica*, *Arithmetica*, *Astronomia* and *Harmonia*, ruling can be seen, but a pattern is hard to detect.

---

[91] I remain unable to find an obvious explanation for the addition of this leaf. It has several peculiarities: the scribe uses letters which first grow and than shrink again to their normal size, but they seem to be written by the same hand. Perhaps the leaf was added to replace a damaged one, but this earlier leaf must already have been an addition to the original quire; the text does not overlap.

[92] Quire 16, however, has a deviating number of writing lines: 33 in stead of 32.

[93] Within these quires, however, there are some variations: 1 (e.g. ff 5, 6, 8): 2-2/0/1-1/J, and 9 (ff 69^70) 2-2/0/1A-1C/J.

[94] Ff 100^103 of quire 13 are different: 2-2/0/0-2/J.

Prickings for the vertical through lines can be seen on the first line, or last line but two, for the horizontal lines along the edges of the leaves.

The text is written by several scribes, in Carolingian minuscule. Bischoff recognized in some of them characteristics which can be related to the writing school of Lorsch, but does not locate the entire manuscript there. Furthermore, he believed that the sections ff 74-81 and 106-127 (that is, quires 10 and 14-16) were 10th-century additions.

q.	signatures		hor. through lines (≠ 1-2)	writing lines (≠32)	ruling pattern (≠ >< ><)
1		9v			
2		17v			
3		25v			> < > >
4	26r	33v			
5	34r	41v			> < > >
6	42r	49v			
7	50r	57v			
8	58r	65v			> > > >
9	66r				
10	74r		0-0		
11	82r		0-0		
12	90r		0-0		
13	98r		2-2		
14	106r		0-0		
15	114r				
16	122r			33	

Figure 7

This view, however, can no longer be upheld, for on these pages glosses are added which seem to have been written by the same ninth-century hands that appear throughout the rest of the manuscript.[95] The larger part of the glossing is carried out by two hands, contemporary with the manuscript. They can be distinguished only by the colour of the ink they used: one wrote in light brown ink, the other in darker ink. They use a very similar set of letters: a very small Carolingian minuscule. Both hands are active in all books, but especially in I, II, VI, VII, VIII and IX, that is in the first two mythological books, and in the books on the quadrivial arts. A third hand, which uses light brown ink and a larger letter, seems to add glosses casually where there are few. His lines of writing often slow a marked slant. He is especially active in book VII, on *Arithmetica*. Characteristics in the script of the glossators that point to Lorsch are, among other things, the abbreviations *hd* and *hł* (*hic/hoc deest* and *hic/hoc lege*) and some Tironian signs.

Comparing the glosses from this manuscript with the three main corpora, we can conclude that it contains glosses from the Anonymous corpus together with glosses which have not been copied elsewhere and which remain unidentified. The manuscript is most closely related to *Par*, a manuscript which also contains an idiosyncratic mixture of glosses from different corpora.

Red ink is used for *explicits/incipits*, for capital letters at the beginning of books, sections, verses or even sentences, the first lines of strophes or refrains. More and more red ink is used towards the end of the codex. The initial letter of book IX is executed in red ink and is even slightly decorated with pear-shaped drops hanging from the serif of the *J*(am).

The history of the manuscript is unclear. Notes on f 1r show that it was bought in Paris in February 1576, and given to the Library of Leiden in 1606 by Peter Schrijver (Petrus Scriverius): *Petrus Scriverius L.M.D. Bibliothecae Lugdunensi MDCVI ipsis eidibus*

[95] The additions may be of a later date than the rest of the codex: the script shows characteristics that may point to a slightly later date. But to date them in the tenth century is perhaps too bold.

octobris.

Today the manuscript has thick cardboard covers, covered with marbled paper. The parchment spine shows five bands, but these are not from the original binding; several older sewing stations can be seen in the gutters of the quires.

The contents of the manuscript are:

ff 1r-127v:	Martianus Capella, *De nuptiis Philologiae et Mercurii*, book I-IX (I, 1r-11v; II, 11v-22v; III, 22v-41r; IV, 41r-56r; V, 56v-74v; VI, 75r-90v; VII, 90v-104v; VIII, 104v-115r; IX, 115r-127v).
ff 128r-129v:	Figures to the books on Dialectic (128r), Arithmetic (128v), Astronomy (129r) and Music (129v).
f 130r:	Anonymous letter *Quid sit ceroma*, written in a tiny Carolingian minuscule (9th/10th century?) (ed. PL 96, cc. 1385-1388; MGH Epp. 6, p. 114)
f 130v:	empty
f 131r-v:	Old fly-leaf, written in 12th-century Gothic textualis and glued to the back cover: Gregory, *Moralia* 2.34/55-38/62.

4. *Paris, Bibliothèque Nationale de France, lat. 8671* (Par) (Plate 8)

A ninth-century manuscript with Martianus' encyclopedia *De nuptiis Philologiae et Mercurii*, with some lacunae. The glosses it contains are written in contemporary and later hands, and mainly added to the first few folia. Book IX, however, also contains a reasonable number of glosses.

Bibliography:

C. LEONARDI, *I codici* I, pp. 448, 462; and II, pp. 437-438

Leonardi dates the manuscript in the ninth century. It contains Martianus Capella's encyclopedia with several lacunae and some contemporary glosses and a few *variae lectiones.*

C.E. LUTZ, *CTC* II, p. 381

Lutz listed the ninth-century manuscript *Leb* among the manuscripts with "unidentified glosses".

J. PRÉAUX, *Les manuscrits principaux*, pp. 79, 122-123

Préaux dates the manuscript in the second or third quarter of the ninth century and locates it in Auxerre (?), without, however, giving evidence for this localization. He argues that this manuscript and the manuscripts *Bruxelles, 9565-9566* and *München, Clm. 14729*, show the state of the text in the scholarly milieu of the late ninth or early tenth century, in the Bodensee region.

B.S. EASTWOOD, *Plato and circumsolar planetary motion in the Middle Ages*, p. 11 and n. 14

There are six ninth-century manuscripts in which the orbits of Mercury and Venus are depicted as intersecting, epicyclic circles, with the sun located in their common middle region: *Par*, *Leb*, *Lea*, *Lb*, *Paris lat. 8669* and *Va*.

The collation of the manuscript is not very regular:[96]

$\underline{1}$-$\underline{2}^{8}$ [1-16]
$\underline{3}^{8}$ (4¦5) [17-24]
$\underline{4}^{6}$ [25-30]

$\underline{5}^{8}$ (2¦7, 3¦6) [31-38] (Figure 8)
$\underline{6}^{8}$ (-3-6) [39-42]
$\underline{7}^{8}$ (+$6^{8\,(2|7)}$) [43-58] (Figure 9)
$\underline{8}^{8}$ (-5-6) [59-64]
$\underline{9}^{8}$ (4¦5) [65-72]

$\underline{10}^{8}$ [73-80]
$\underline{11}^{4}$ (+4^{*}) [81-85]

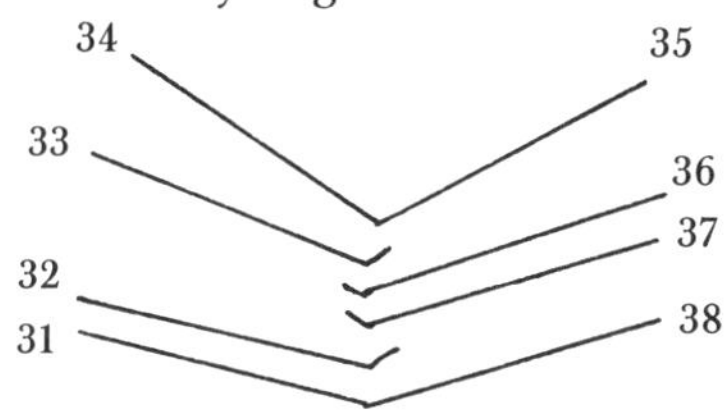

Figure 8

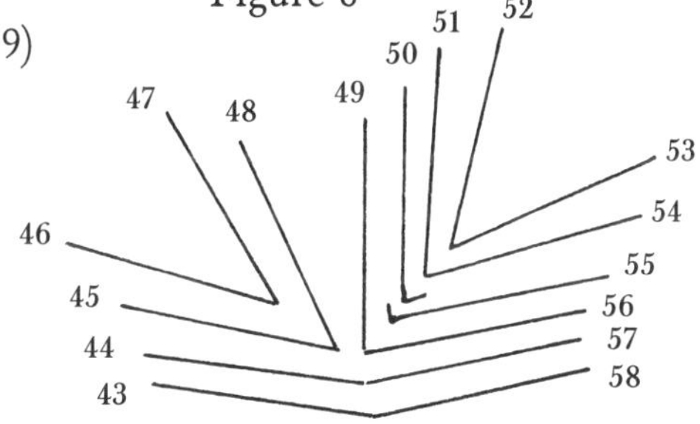

Figure 9

All quires except $\underline{1}$, $\underline{9}$ and $\underline{11}$ carry quire signatures, executed in the left corner of the lower margin of the first recto of each quire. The last quire is numbered on the last leaf. The tenth quire, which contained almost the entire eighth book, is lacking.[97]

The leaves are numbered in ink in a modern hand, in the upper

[96] It is even more chaotic than Leonardi noted in his short description (*I codici* II, 438). He took the third, fifth and ninth quires, and the eight folia added after the 6th page of quire $\underline{7}$, for regular *quaterniones*.

[97] A seventeenth-century hand wrote in the lower margin of 73r, i.e. on the first leaf of quire $\underline{10}$, "Deest x."

right corner of each recto. The leaves measure ca 340 x 225 mm, but some of the added leaves are smaller. The parchment is of mediocre quality, with many holes and discolourings. Especially quire 3 was severely damaged by damp.

The HF-arrangement and the patterns of furrows and ridges show many irregularities, which are shown in the table below. The lay-out varies per quire and sometimes even per folium. Ruling was carried out with a hard point, according to the following patterns (cf. fig. 10):

1-3: 1-1/0/1-1/J, measurements ca 37<255>48 x 13<165>47 mm. 37 (1-2) or 38 (3) lines, atl, 68-71 mm per 10.

4: 2E-1E-2E/0/1-1-1:B/H, measurements ca 34<238>28>46 x 16<7<125>7>23>50 mm. 40 (f 25) or 39 (ff 26-30) lines, atl, ca 69 mm per 10. In the original lay-out rules are provided for 36 lines, then an extra through line was added in the lower margin to make room for an extra set of three or four lines. The extra column (perhaps for glosses) provided in the outer margin was never actually used; its space was added to the writing space.

5: 1-1/0/1-1/J, measurements ca 28<277>38 x 22<175>38 mm. 39 lines, atl, ca 72 mm per 10.

6, 8: 1E-2F-1E/0/1E-1F-1E, measurements ca 37<230>21 x 22<128>7>26>45 mm. 39 lines, atl. ca 68 mm per 10. As in quire 4, the original lay-out of a page, providing for 36 lines and an extra column in the outer margin, has been adapted for this manuscript. An extra through line marks the space for three more lines, and the writing space stretches from the first to the fourth vertical through line.

7: The folia which form a regular *quaternario* (ff 43-48, 57-58) have the same lay-out as quire 6. The eight folia added after f 48 have other patterns: a) ff 49^56: 1-1/0/1-1:J/J, measurements ca 13<270>45 x 23<180>33 mm. 39 lines, atl, 68 mm per 10; b) ff 50 and 51^54: 1A-1B-11E/0/1-1:B/J, measurements ca 26<244>14 x 11<90>15<70>27 mm. There are two columns laid out, but they are not used. The lines run from the inner to the outer margins. F 51 has the same lay-out, but is inserted upside down; c) ff 52^53 and 55: 1-1/0/1-2/J, measurements ca 20<269<8>39 x 7<177>45 mm. 40 lines, atl, 73 mm per 10.

9: 1-1:E/0/1-1:E/J, measurements ca 30<271>45 x 6<186>35 mm. 38 lines, atl, ca 76 mm per 10.

10: In the lay-out of 10 again an extra column is marked off in the outer margin, this time by a double through line: 2-2-2/0/1-1/J. The scheme is never actually used: the lines run from the first to the last vertical through line. There are 38 lines, atl, ca 71 mm per 10.

11: 1-1:E/0/1-1:E/J, measurements ca 32<242>60 x 37<143>55 mm. 38 lines, atl, ca 64 mm per 10. On f 83 extra lines are drawn for figures.

q.	HF-arrangement (≠ HFHF)	lay-out (≠ 1-1/0/1-1)	nr. of lines	ruling pattern (≠ > < > <)
1	(?)		37	< < > <
2			37	< > > <
3	HHFF¦HHFF		38	> > > <
4		2-1-2/0/1-1-1	39-40	
5			39	
6		1-2-1/0/1-1-1	39	
7	¦HF$^{?HFH¦FH??}$HF	as in 6 ^{irregular}	39-40	¦ > <$^{> < > > ¦ < < < <}$ > <
8		as in 6	39	
9	HFHH¦FFHF		38	> < > >
10		2-2-2/0/1-1	38	
11			38	

Figure 10

The script is Carolingian minuscule. Several hands are at work in the manuscript, some using clubbed ascenders and characteristic abbreviations. Several hands, contemporary and later (10th-century), can be recognised in the glosses as well. One of those — who added glosses on f 1r — writes them in the shape of a triangle or rhomb, another uses Tironian shorthand (f 50r, column b, l 25), others again use the abbreviations *hd* and *hl* (*hic/hoc deest* and *hic/hoc lege*), which Bischoff thought characteristic for the writing school of Lorsch.

The glosses remain unidentified, which is largely due to the scarcity of glosses in the first two books. In book IX, however, a fairly large collection of glosses is added to the text. Comparison of these with the other glosses examined shows that *Par* contains a selective and deviating version of the Anonymous commentary. There are also some parallels with the commentary of John the Scot. A more important observation is that *Leb* and *Par* are relatively close parallels of each other, and since *Leb* is located in Lorsch, this constitutes extra evidence for a location in Lorsch of *Par*. In broad outline the two manuscripts concur with each other, but there are also many differences in detail.

In the manuscript majuscules are used for the first letters of verses, passages or sentences, and for the initial letters of books, refrains and *explicits/incipits*. The largest initial letters are up to 7 lines high. Red ink is used for some of the *explicits/incipits*,[98] initial letters,[99] refrains,[100] and figures.[101] Two of the initials are decorated with simple plaiting and animal heads in black ink.[102] On f 23r a sketch can be seen in light ink of a bearded man, barefooted, wearing a turban and with his hands put together as if contemplating something — a picture of an 'oriental' Martianus Capella as a philosopher?

The manuscript contains *De nuptiis Philologiae et Mercurii* of Martianus Capella, with some lacunas and some miscellanous additions: Book I, 1r-9r; II, 9r-17v; III, 17v-29r; a section from Bede's *De natura rerum*, 29v-30v; IV (up to 422.3), 31r-40v; V (from 474.21),[103] 41r-49v; VI, 49v-62v; VII, 62v-72r; VIII (sections 810-883 are missing, that is most of book VIII), 72r-73r; IX (sections 954-977 are missing), 73r-83r. On ff 83v-84r figures are drawn in

[98] Books I (f 1r), II (f 9r), III (f 17v), IV (f 31r) and VIII (f 72r).
[99] ff 5r, 31r
[100] ff 10v-12r
[101] f 83v
[102] f 9v, 6 lines high, and f 12v, 7 lines high.
[103] The end of book IV and the beginning of book V are missing.

addition to the books on *Geometrica* and *Astronomia*, on 84v the names of the tones and *modi* are listed. Folium 85r-v seems to have been part of a different Martianus Capella manuscript. It contains the sections IX.954-966, which are missing in the ninth book on ff 73r-83r, but it does not fill in the entire lacuna.

Today the manuscript is covered with parchment, and has four raised bands on the spine.

Conclusions and outlook

The stemma of Martianus Capella manuscripts is complex.[104] The text must have survived during the sixth, seventh and eighth centuries in a single manuscript, and is likely to have been copied in the late eighth century in pre-Carolingian *scriptio continua*.[105] When *De nuptiis* gained popularity in the first half of the ninth century, it was copied in many major and minor centres of scholarship of the Carolingian empire; today there are almost fifty ninth-century manuscripts of the text or parts of it extant. In the later ninth century and in the tenth, the tradition soon becomes "hopelessly contaminated",[106] but in the first half of the ninth century, the text is preserved in a series of manuscripts which are fairly close to the pre-Carolingian archetype.[107] It is remarkable that among these manuscripts there is abundant evidence of a kind of continuous contact among them, shown, for example, by the additions, corrections and alterations copied in these texts by second hands, but also by the corpora of glosses added to them. This is

---

[104] See on this J. PRÉAUX, *Les manuscrits principaux*; and esp. D. SHANZER, *Review article*, pp. 62-81 — in this article, she criticizes Willis' edition of the text, and especially his choice of manuscripts and description of their relations.

[105] On the basis of the nature of the errors, Shanzer argues that this late eighth-century manuscript was copied from a late antique manuscript "with a two-script format — some sort of minuscule for the prose and rustic capital for the verse." Furthermore, she adds: "Two major lacunae and many universal errors make it clear that all extant texts stem from this archetype." SHANZER, *Review article*, p. 63.

[106] These are Shanzer's words, SHANZER, *Review article*, p. 63.

[107] Préaux's "manuscrits principaux"; J. PRÉAUX, *Les manuscrits principaux* (pp. 78).

perhaps reflected by the tendency in modern scholarship to look for a single centre from which the text was disseminated; Préaux suggested Corbie.[108]

In general, one can conclude that the relations between the corpora of glosses are roughly in concurrence with the stemma of the actual text of Martianus Capella. Of the four most important manuscripts for the Anonymous corpus, Shanzer mentions three in her stemma (*Leiden, Voss.Lat.F. 48* = V; *Besançon, 594* = E; *Vaticano, Reg.lat. 1987* = X), and of the four manuscripts of secondary importance, two (*Oxford, Laud.lat. 118* = F; *Paris, lat. 8671* = M). In her stemma, these manuscripts are related in a way which is reflected by the relationships between the gloss-corpora. As far as this study could determine,[109] relations of these latter texts can be sketched as follows: a common archetype or perhaps archetypes[110] was the basic source for the four most important manuscripts. These, however, clearly divide into two pairs. *Leiden, Voss.Lat.F. 48* (*La*) and *Besançon, 594* (*Be*) are so closely related that I remain unable to determine which of the two could have been the original and which the copy.[111] The other pair consists of the Ghent-manuscripts *Leiden, BPL 88* (*Lb*) and *Vaticano, Reg.lat. 1987* (*Va*) — both of which contain the Anonymous commentary for books I-VIII, and a version of John the Scot's commentary for book IX. Errors in the lay-out of the glosses suggest that *Va* was copied from *Lb*, in which the hand I^2, a scholar from the circle of John the Scot, appears.[112] The four manuscripts of secondary importance

---

[108] *Les manuscrits principaux*, pp. 96-97.

[109] The conclusions are tentative: I did not study the whole of books I and II, and did not study books III-VIII at all. Moreover, all manuscripts which did not have any glosses for book IX were excluded from my selection. For a thorough analysis of the relationships between the different sources of the Anonymous commentary, one should start with a much more ample selection than my narrow one.

[110] Writing-errors and errors in lay-out prove that none of the manuscripts I studied contains an original text.

[111] It must be stressed, however, that the determination of the exact relationships was not the main purpose of this study. The focus of this book is on the reception of antique music theory in the Carolingian era.

[112] See on this my forthcoming article *Ninth-century glosses*; in this paper it is shown that errors in lay-out can be explained by positing the Leiden manuscript as the model, the Vatican manuscript as the copy.

are related to these sources in varying degrees of clarity. *Oxford, Laud.lat. 118* (*Oxf*), for example, is related to *La* and *Be*, but is idiosyncratic in many respects. *Leiden, BPL 87* (*Lea*) is closest to *Lb*, but is much less complete. The manuscripts *Paris, lat. 8671* and *Leiden, BPL 36* (*Leb*) are related to each other and to the common archetype, but it is unclear through how many intermediary sources. The following picture thus arises:

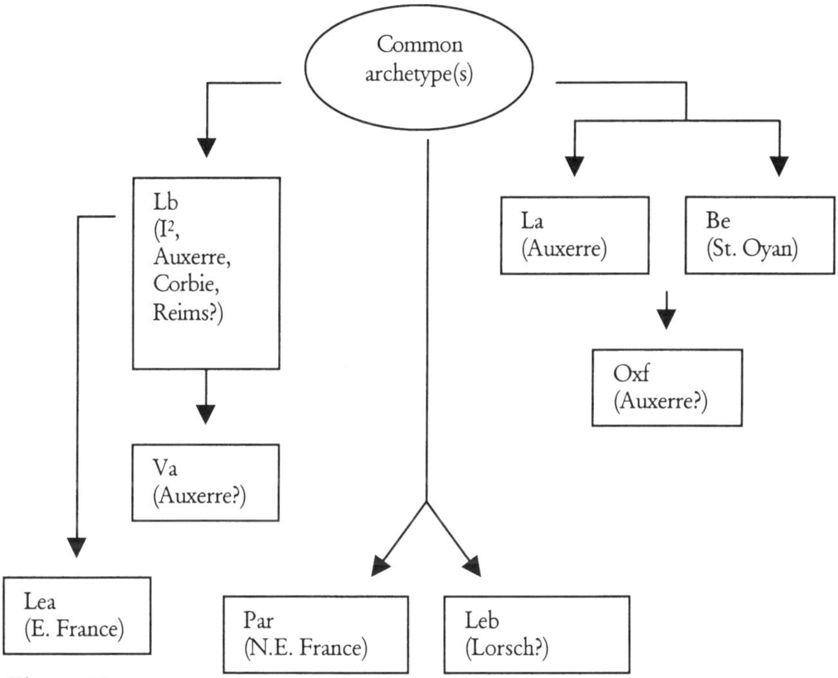

Figure 11

Most of the manuscripts stem from the central part of the Carolingian empire: Auxerre, Corbie, Reims, North, Eastern or North Eastern France, and, further to the East, Lorsch, have all been suggested on the basis of their paleaographical characteristics. Both the place of origin and the author of the Anonymous commentary remain enigmatic, despite the respectable amount of scholarship produced on the subject.

Speculatively — and I cannot stress this word enough —, I would like to suggest a possible solution to the problems. First of all, it must be noted that the copies we have today of the Anonymous commentary are all copies of text and commentary together. Often the manuscripts have been laid out from their very beginning to contain both text and glosses; gloss columns are pricked and ruled in their quires. This strongly suggests that there was a more or less authoritative body of glosses, with a more or less fixed form.[113] The glosses, in other words, are no random annotations of individual scholars, but copies of an established text. It also suggests that text and commentary were, from the beginning of Martianus Capella study to the period of the dissemination of Remigius' commentary at least, seen as a unity. Secondly, at the same time, there are no two manuscripts which contain the same corpus of glosses. Even the manuscripts which are closest in relation, are different in many aspects: lemmata which are annotated in one manuscript are without glosses in the next, and the individual glosses are transmitted with a fair amount of variants. The fluidity of the transmission of the body of glosses under scrutiny in this study seems to be an essential part of its literary status.

In recent scholarship, this Anonymous corpus of glosses has been attributed to almost every influential scholar from the reigns of Louis the Pious and Charles the Bald: to Dunchad, and Martin of Laon of course, but links with Heiric of Auxerre, Lupus of Ferrières, the 'Irish' intellectual communities of Laon and Auxerre have also been suggested. Copies of *De nuptiis* with commentary appear, from the 850's onwards, throughout the Carolingian empire. Would it not be possible, then, to suggest an origin among a group of scholars collected at the courts of Louis the Pious and Charles the Bald?[114] An attribution to a group of scholars rather than to an individual author would explain why the commentary is never cited by reference to its author. Remigius of Auxerre, who copied large parts of the Anonymous commentary into his own, never mentions the author of it, and mentions John the Scot, on the other

---

[113] This point is especially proved by the body of figures, tables and drawings (mainly to the books on arithmetic, astronomy and music) which seem, somehow, to be part of the corpus of glosses, and which is consistently transmitted by a number of manuscripts. See, for example, on astronomical figures, the articles of B.S. EASTWOOD, esp. *Plato and circumsolar planetary motion*.

[114] In fact, one of the manuscripts which reflect an early stage of the text is, in Shanzer's stemma, attributed to the Palace School of Louis the Pious — *Karlsruhe, Reich. Aug. 73*; D. SHANZER, *Review article*, p. 77.

hand, many times when referring to his *Annotationes*. If my suggestion that the Anonymous commentary was, for Remigius, part of the transmission of *De nuptiis* holds true, it would have been unnecessary and even illogical for him to cite its author. It would also explain why the Anonymous commentary is almost completely integrated (silently) into his own, while he makes explicit references to the *Annotationes* of John the Scot. Secondly, the attribution to a group of scholars related to the Carolingian courts would also explain its simultaneous dissemination to different parts of the centre of the empire. Martin of Laon, Heiric of Auxerre, Lupus of Ferrières and even John the Scot,[115] could all have been part of the circle of scholars who first studied and explained the text and composed a body of glosses to explain it.

Furthermore, the character of the glosses also concurs with the hypothesis. They seem not to be designed for the class-room, but rather to reflect a gathering of as much knowledge as possible, mainly from antique and late antique sources, to reach a full understanding of the text. The commentary provides basic explanations on Martianus' often intricate syntaxis or manieristic grammatical constructions, but they also lead the reader associatively to other antique sources on, for example, the harmony of the spheres or numerical speculations (Macrobius, Calcidius, Boethius), on ancient mythology (Fulgentius, Servius), Greek vocabulary or etymological explanations (Isidore).

The existence of such scholarly collaborative ventures instigated by the court has already been proved in the field of theological issues — glosses on the Bible, the debate on predestination could perhaps also be mentioned —, but so far the study of secular texts has not been analyzed from this angle. Yet it is a fact that Charlemagne's successors, especially Louis the Pious and Charles the Bald, attracted scholars to their courts whom they consulted on a wide range of topics.[116] These court 'schools', if they can be given

[115] In the third section of Chapter 3, on the theme of the harmony of the spheres in the Anonymous commentary and that of John the Scot, evidence is presented which hints at a strong relationship between the two.

[116] There is a large bibliography on this subject. To cite just a few of the more recent titles (in chronological order): J.J. CONTRENI, *The Cathedral School of Laon from 850 to 930*; J. MARENBON, *From the Circle of Alcuin to the School of Auxerre*;

that name, were loosely-organised institutions, their members were transient; they were often dispersed to monasteries or cathedrals after a stay of several years at the court. It is also known that the libraries of these kings contained many classical and late antique texts, which were explored with great enthusiasm. In the words of John Contreni, the Carolingian leaders instigated "a well-organised programme which, given the authority behind it, could not help but achieve its goals."[117] To my mind, there is enough evidence to assume that Martianus Capella's *De nuptiis* was one of the texts which, on initiative of the court, were copied in the large monastic scriptoria of Carolingian Europe. On the other hand, the fluidity of the Anonymous corpus of glosses aptly illustrates the "differences in resources, talents and interests in the cultural landscape", of which Contreni also speaks.[118] The goal of a unitary culture was, in many respects, illusory, and the plurality of the gloss traditions show the vitality of learning in the Carolingian era.[119]

---

articles in D.L. WAGNER (ed.), *The Seven Liberal Arts in the Middle Ages*; M.W. HERREN (ed.), *The study of Greek in the West in the Early Middle Ages*; articles in the proceedings of the conference *Giovanni Scoto nel suo tempo: l'organizzazione del sapere in età carolingia*; P. RICHÉ, *Ecoles et enseignement dans le haut moyen âge*; M.T. GIBSON & J.L. NELSON (eds.), *Charles the Bald: Court and Kingdom*; D. GANZ, *Corbie in the Carolingian Renaissance*; D. IOGNA-PRAT, C. JEUDY & G. LOBRICHON (eds.), *L'école carolingienne d'Auxerre*; articles by D. BULLOUGH, collected in *Carolingian Renewal: Sources and Heritage*; J. CONTRENI, *Carolingian Learning, Masters and Manuscripts*; various articles in P.L. BUTZER & D. LOHRMANN (eds.), *Science in Western and Eastern Civilizations in Carolingian Times*; R. MCKITTERICK, *Books, Scribes and Learning in the Frankish Kingdoms*; R. MCKITTERICK (ed.), *Carolingian Culture: Emulation and Innovation*; the articles by B. BISCHOFF, collected and translated by M.M. GORMAN in *Manuscripts and Libraries in the Age of Charlemagne*; the contributions of J.J. CONTRENI and D. GANZ in *The New Cambridge Medieval History*, Vol. II, *c. 700 - c. 900*.

[117] J. CONTRENI, *The Carolingian Renaissance: Education and Literary Culture*, p. 711.

[118] *ibid.*, p. 712.

[119] This is also shown by the numerous letters exchanged by Carolingian scholars. See, for example, their edition in the 5 volumes of MGH, *Epistolae Karolini Aevi*, and the edition of the letters of Lupus by L. LEVILLAIN, *Loup de Ferrières, Correspondance*, (Vols. 1-2), Paris 1927-1935. See on the intellectual network also the articles by B. BISCHOFF on the Carolingian court libraries; P. DEPREUX, *Büchersuche und Büchertausch*; or M. DE JONG, *From Scolastici to Scioli.*

CHAPTER THREE

THE *ARS MUSICA* IN THE EARLIEST COMMENTARIES ON *DE NUPTIIS*

Introduction

In the present chapter, the focus will shift to the musical material in the earliest commentaries on *De nuptiis*. In order to evaluate the results of the detailed analysis of the individual glosses of the Anonymous commentary, it will be necessary to introduce in broad brushstrokes the intellectual context in which they were conceived and functioned.[1] On the one hand, there is the legacy of Greek music theory, which was handed down to the Middle Ages through Latin translations, compilations and encyclopedias — a legacy which was essentially Pythagorean in nature, explaining music in terms of mathematical ratios. On the other hand, the last decennia of the ninth century and the beginning of the tenth witnessed the birth of medieval writing on music theory, with Anonymous treatises and those attributed to Aurelian of Réôme, Regino of Prüm and Hucbald of Saint-Amand, in which first attempts were made to reflect systematically on the repertoire of the liturgy. Everyone who is familiar with both categories of sources knows that, on the one hand, a close relationship exists between the two, and, on the other hand, they appear scarcely related. The sources of musical thinking of the Middle Ages often faithfully reproduce central pieces of classical theory. The myth of Pythagoras, for example, who discovered the basic ratios of musical consonances when

---

[1] On this see, among others, N. PHILLIPS, *Classical and Late Latin Sources*; R. CROCKER, *Frankish Music Theory*; M. BERNHARD, *Überlieferung und Fortleben*; D. HILEY, *Plainchant and Early Music Theory*, in: *Western Plainchant*, pp. 442-447; S. RANKIN, *Carolingian Music*; C.M. BOWER, *Transmission of Ancient Theory* forthcoming in: *The Cambridge History of Music Theory*. I thank Calvin Bower for showing me his chapter before publication.

passing a blacksmith's shop, figures in many medieval treatises on music, as do speculations on the music of the planetary spheres, the Boethian threefold classification of music — *musica mundana, musica humana* and that which rests in musical instruments —, and classical myths on the power of music (*e.g.* Orpheus and Eurydice, Amphion and Arion). At the same time, however, the organisation and standardisation of Gregorian chant, pursued from the mid eighth century onwards,[2] was realized by means of intrinsically medieval principles, such as the eight church modes, neumatic and alphabetic notations, and the solmization system. It is important, therefore, to unravel the relation between ancient sources and their medieval inheritors with great care, to establish shifts of meaning on a terminological level and assess the importance of the ancient sources in the development of a theoretical frame for liturgical chant. With the earliest commentaries on the musical contents of *De nuptiis* we possess an important witness for the earliest reception of ancient music theory. Together with the *glossa maior* tradition on Boethius' *De institutione musica* it forms the only source for the nature of this reception.

1. *The legacy from Antiquity*

A number of texts from Antiquity were handed down to the Middle Ages, and, as their manuscript histories show, were in fact present in the cultural centres of the Carolingian empire.[3] All of them were firmly rooted in the Platonic tradition, which is particularly obvious from the works of the 4th-century author Calcidius, who translated and commented upon Plato's *Timaeus*, and the 5th-century author Macrobius, who wrote a commentary on Cicero's *Somnium Scipionis*. The text of Calcidius was central in the dissemination of the idea that the cosmos was created according to mathematical principles: the numerical ratios which created the World Soul are reflected in musical consonances. Of Calcidius' text three ninth-century copies are extant, all copied in

---

[2] Cf. S. RANKIN, *Carolingian Music*, pp. 275-279.
[3] N. PHILLIPS, *Classical and Late Latin Sources*.

Northern France.[4] The text was used as a source of knowledge by John the Scot and Remigius of Auxerre in their commentaries on *De nuptiis*, and cited extensively in the Anonymous treatises *Musica* and *Scolica enchiriadis*. The commentary on the *Somnium Scipionis* of Macrobius was studied from the early ninth century onwards: its astronomical contents are cited or paraphrased in the Anonymous commentary on *De nuptiis* and in that of John the Scot. Regino of Prüm, furthermore, cites large portions of the work on intervals and the effects of music.[5]

Augustine of Hippo (354-430), whose writings strongly represent the Platonic tradition which shaped him, wrote a treatise on music in six books, *De musica*.[6] In his work, music is again an embodiment of Pythagorean proportions. However, these proportions are not related to principles of pitch or consonance, but to quantitative verse instead, to rhythm and meter. The first five books of his treatise are dedicated to this subject, while in the sixth book the same ratios lead the soul to understanding the abstract beauty of the perfectly harmonized creation, and through this to knowledge of God. While Augustine's teachings on the subject of rhythm and meter remained relatively unimportant in the Middle Ages,[7] his sixth book was influential. John the Scot, for example, cites it twice in his principal philosophical work *Periphyseon*.[8] Augustine's treatise is nevertheless important in the

---

[4] Cf. *ibid.*, pp. 126-131.

[5] M. BERNHARD, *Überlieferung und Fortleben*, p. 19; *id.*, *Studien zur Epistola de armonica institutione des Regino von Prüm*, pp. 34-35.

[6] Cf. the article on Augustinus by F. HENTSCHEL in the new *MGG* (Personenteil I), Kassel etc. 1999, cols. 1169-1175; N. PHILLIPS, *Classical and Late Latin Sources*, pp. 120-126.

[7] In the Middle Ages, the discipline of rhythm and meter was no longer part of the *ars musica*, but of the *ars grammatica* instead. Medieval authors used handbooks on grammar and rhetoric, such as those by Priscianus, Donatus, Quintilianus, and Fortunatianus, to gain familiarity with the subject, rather than Augustine's *De musica*. Only in *Scolica enchiriadis* Augustine's teachings are applied to ninth-century chant. Cf. N. PHILLIPS, *Classical and Late Latin Sources*, p. 124.

[8] Cf. *ibid.*, pp. 124-126. Phillips furthermore points to two indirect witnesses for knowledge of Augustine's *De musica* in the ninth century: a reference in a poem by Alcuin and another in one of Lupus of Ferrière's letters.

present context, because partly on his authority the secular disciplines, the liberal arts in general and music in particular, were established as an integral part of Christian education.[9]

Martianus Capella's work espoused the same Platonic ideal of knowledge of God through knowledge of secular learning. The number of liberal arts which was to become canonical was established in his encyclopedia: seven maidens, personifications of the arts of the trivium (*grammatica, dialectica* and *rhetorica*) and quadrivium (*geometrica, arithmetica, astronomia* and *harmonia*) are presented to the bridal couple, Mercury and Philology. The bride and groom themselves symbolize the union of the secular sciences (Philology) and eloquence (Mercury), or, in a different interpretation, the union of secular learning and divine knowledge. Mortal Philology has to undertake a journey through the circles of heaven to reach a divine status, which is comparable to Cicero's account of the ascent of the soul to knowledge of its own immortality in the *Somnium Scipionis*. Martianus, moreover, gives the discipline of music a unique place among the seven liberal arts: the technical content of the *ars musica* is treated in the ninth book of the encyclopedia, but it is also abundantly present in the first two narrative books, where the numerical ratios of creation resound in the music of the sacred Apollonian grove and of the heavenly spheres. Lady *Harmonia* is the last maiden to present her knowledge to the wedded couple, but not because she is the least important: on the contrary, because of her special relationship with divine creation, hers is the art cherished most by the gods.

The technical contents of the ninth book of *De nuptiis* are quite basic: hardly anything can be found which is absent from Boethius' much more elaborate study on the fundamentals of music. The concepts of the discipline of music are introduced, and seven subjects are treated separately: pitches (*soni*), intervals (*spatia*), systems (*systemata*), modes of tetrachord division (*genera*), modes (Dorian, Lydian, Phrygian, etc.; *modi*), modulation (*commutatio*), and melodic composition (*modulatio*). Furthermore

[9] Not only on his authority, though, for Cassiodorus, Isidore and, of course, Martianus Capella were far more influential on this point.

the subject of quantitative verse is treated. The introductory tone of the treatise, however, which made *De nuptiis* as a source of learning inferior to Boethius' *De institutione musica*, made it at the same time easier for the Carolingian students of both texts to come to terms with the basic concepts of the ancient *ars musica*.[10]

Boethius (†524) is without doubt the most influential scholar read in the early Middle Ages, certainly when it comes to the discipline of music.[11] He envisioned a series of works on the four disciplines of the quadrivium, and realized two of them: *De institutione arithmetica* and *De institutione musica*. Using Greek writers of the Platonic and Pythagorean traditions such as Nicomachus, Ptolemy, Euclid, Plato and Aristotle, he wrote two dense treatises on number: *De institutione arithmetica* focused on number itself, and *De institutione musica* on number in ratios and proportions. The basic philosophy underlying Boethius' work was, again, that knowledge of these secular arts could lead man, who is, without knowledge, essentially a slave of his own senses, to abstract truth. And again, music occupies a special place among the liberal arts, for as the most sensual one it is able to influence human behaviour and to shape a man's character.[12] Furthermore, he preaches that the rational principles which govern music are truths revealed about the order of the cosmos — sensual experience is relatively unimportant. The subject of music is numerical essence, musical intervals are scientifically measured ratios and consonance is mathematical harmony.

Boethius' work as a whole deeply influenced medieval musical thinking, but there are a few aspects which proved particularly

---

[10] Hucbald prefers Martianus even explicitly over Boethius on the interpretation of Greek tone names: (*Musica*, 117a): "Interpretatio autem horum nominum succincte quidem a Boetio attacta, planius autem Martianus exsequitur." Cited from Y. CHARTIER, *L'oeuvre musicale d'Hucbald de Saint-Amand*, p. 192.

[11] See C.M. BOWER, *The Role of Boethius*; A. WHITE, *Boethius in the Medieval Quadrivium*; N. PHILLIPS, *Classical and Late Latin Sources*, pp. 103-108; and M. BERNHARD, *Überlieferung und Fortleben*, pp. 24-31.

[12] See J. SMITS VAN WAESBERGHE, *Die besondere Stellung der* ars musica *im Zeitalter der Karolinger*.

important. One of those is Boethius' use of diagrams, for example to illustrate the functioning of the Greek Perfect System. In these diagrams each horizontal line represents a given string or pitch. Another influential aspect is his employment of various letters to designate certain points in the division of the monochord — thus taking a major step in the development of a system in which a certain letter represents a certain pitch. These aspects formed the building blocks for the development of a music theory for the contemporary chant practice at the end of the ninth century and the beginning of the tenth.

The last two authors who shaped music-theoretical thinking of the ninth and tenth centuries are the early medieval encyclopedists Cassiodorus († ca. 583) and Isidore of Seville (†636).[13] For his monastic community at Vivarium Cassiodorus wrote a handbook of divine and secular studies: *Institutiones divinarum et humanarum litterarum*, a work designed to provide a summary of learning for the brothers of his community. In Isidore of Seville's *Etymologiarum seu originum libri XX* the liberal arts are treated briefly in books 1-3. These authors, though concerned mainly with the preservation and transmission of classical knowledge, are the first to introduce Christian examples in their texts: they discuss the musical instruments present in the Bible, and stress the centrality of singing in Christian worship. They thus begin to break down the boundaries between the ancient discipline of music and the living practice of liturgical music.

2. *The ninth- and early tenth-century sources*

The earliest medieval sources of musical thinking are without doubt the oldest gloss traditions found in manuscripts of Martianus Capella's *De nuptiis* and Boethius' *De institutione musica*. These notes, which fill the margins and interlinear spaces of many ninth-century manuscripts, are the very first witnesses of the encounter of Carolingian scholars with ancient music theory.

---

[13] Cf. N. PHILLIPS, *Classical and Late Latin Sources*, pp. 108-118; M. BERNHARD, *Überlieferung und Fortleben*, pp. 31-35.

There is evidence for the study of *De nuptiis* and *De institutione musica* from the early ninth century onwards. Both the oldest layer of the corpus of glosses on *De institutione musica* and the Anonymous corpus on *De nuptiis* can be dated in the second third of the ninth century. It is striking that both commentaries remain anonymous, which could perhaps lead us to assume that the corpora of glosses were the collective work of a group of scholars rather than of one author. Perhaps it was a group of scholars at the court of Louis the Pious, or of his son Charles the Bald, who first explored these texts, and wrote their findings down in the margins of the manuscripts kept in the court library.[14] The earliest witnesses of these commentaries are dated around the middle of the ninth century, and there is evidence (writing errors, erroneous lay-out, lack of *signes de renvoi*) which marks them as copies. The originals of these commentaries must, in other words, date back to the first half of the century. There is also evidence to suggest a related reception of the two: material from Boethius is used in the commentaries on *De nuptiis*, and vice versa.[15]

With respect to the nature of these glosses, it should first be noted that the scholars who wrote the glosses were no musicians; their primary concerns lie with the vocabulary found in these late antique texts,[16] etymological explanations of proper names, philology, and the philosophical frame-work, not with the music-theoretical contents. The author (or authors) of the commentary on Boethius' treatise, moreover, seems to have been fascinated by mathematics — a trait which is, of course, for a large part dictated by *De institutione musica* itself.[17] There are numerous comments on the ratios underlying the semitone, apotome (2187:2048) and comma (524,288:531,441). Moreover, in the earliest reception of Boethius' *De institutione musica* any reference

---

[14] See the forthcoming Volume 4 of the edition of the *Glossa maior in institutionem musicam Boethii*, and C.M. Bower's forthcoming article for the *Cambridge History of Music Theory*; see also his earlier publications, esp. *Die Wechselwirkung*.

[15] See the second section of this chapter.

[16] Martianus Capella with his manieristic Latin, his *hapax legomena* and his many Greek terms proved to be a particularly rich source in this respect.

[17] See on this C.M. BOWER, *Die Wechselwirkung*, pp. 174-178.

to liturgical practice is conspicuously absent.[18] It is only in the layers of glosses dated in the late ninth and early tenth centuries that examples from liturgical chant are used for the illustration of musical intervals. Perhaps, however, this too is partly dictated by *De institutione musica* itself. After all, the content of Boethius' treatise is *number*, not sounding music. The abstract mathematical concepts that govern sounding music are seen as a way of grasping the divine order of creation, and the *ars musica* is more a discipline of reason than of the senses. In any case, the same cannot be said of the commentaries on *De nuptiis*: here contemporary liturgical practice is present even in some of the earliest glosses, be it only on a few occasions.[19]

A separate tradition is fromed by the first medieval treatises on music theory. Triggered by the problems which arose from the attempts to unify the liturgical repertoire for the entire Carolingian empire, treatises were written which focused on contemporary chant practice and struggled to describe its organizing principles.

Five treatises are dated in the last decades of the ninth century and the first decades of the tenth. These are, in uncertain chronological order:

- Aurelianus of Réôme, *Musica disciplina*;
- the earliest portions of the Anonymous *Alia musica*;
- the also Anonymous treatises *Musica* and *Scolica enchiriadis*;
- Hucbald of Saint-Amand, *Musica*;
- and Regino of Prüm, *Epistola de armonica institutione*.

The textual tradition of these treatises is characterized by great complexities: sections of them have been preserved independently, the commentaries on Boethius or Martianus Capella are cited in the treatises without acknowledgement of the sources, and excerpts from Cassiodorus and Isidore are incorporated in the same manner. Moreover, the status of these treatises was such that scholars felt free to revise them, excerpt them or elaborate upon them during the ninth, tenth and

[18] *Ibid.*, pp. 182-183.

[19] See the last section of this chapter, *Theory and Practice*.

eleventh centuries.[20]

Aurelianus of Réôme wrote his *Musica disciplina* around the middle of the ninth century.[21] His main concern was a kind of modal theory *avant la lettre* and classification of individual chants from liturgical practice. These same concerns are treated in the anonymous treatise *Alia musica*.[22]

Two further anonymous treatises are *Musica* and *Scolica enchiriadis*,[23] which have been dated to the second half of the ninth century.[24] Their complex textual histories have been explored by H. Schmid and N. Phillips, who, tellingly, arrived at significantly different conclusions concerning the relationships between the extant manuscripts. The *enchiriadis* treatises make use of several ancient treatises, among them Calcidius' translation and commentary of Plato's *Timaeus*, Augustine's *De musica*, Boethius' *De institutione arithmetica* and *De institutione musica*, and Cassiodorus' *Institutiones*. Conspicuously absent as a source of ancient music theory is Martianus Capella's *De nuptiis*. In the recent translation of the treatises by R. Erickson, the treatises are introduced and characterized as follows:

> "The treatises known commonly today as *Musica enchiriadis* and *Scolica enchiriadis* are products of ninth-century reverence for and interest in the received wisdom of antiquity. But even more they

---

[20] Cf. N. PHILLIPS, *Classical and Late Latin Sources*, p. 102, n. 7: "It should be noted that the text traditions of each of these Carolingian treatises suggests that the texts we have received were reworked in varying degrees during the ninth or tenth centuries." In the same note, on the Aurelian text tradition, she cites L. GUSHEE (ed.), *Aureliani Reomensis Musica disciplina*, pp. 12-14: "[Aurelian's] contribution ... wielding of scissors and paste-pot," "almost total compilation."

[21] Cf. L. GUSHEE, *Aureliani*, esp. "Introductione", pp. 11-49.

[22] Cf. J. CHAILLEY (ed., transl. and intr.), *Alia musica. Traité de musique du ixe siècle.*

[23] H. SCHMID (ed.), *Musica et scolica enchiriadis una cum aliquibus tractatulis adiunctis*; N. PHILLIPS, *"Musica" and "Scolica enchiriadis": The Literary, Theoretical, and Musical Sources*; R. ERICKSON, (transl. and intr.), *Musica Enchiriadis and Scolica Enchiriadis.*

[24] It has also been suggested, however, that they could have been written in the middle of the ninth century, or even earlier. There is no evidence for a certain dating — either before, or after 850. N. PHILLIPS, *"Musica" and "Scolica enchiriadis"*, pp. 511-516.

> reflect concern for the training and education of singers engaged in daily worship. Indeed, the pragmatic stance of these documents, evident in the inclusion of exercises and of music from the repertory — features totally absent from the ancient formal treatments of music — testifies to Carolingian practicality."[25]

Their contents include the first fixed pitch notation, the earliest description of the polyphonic practice called *organum*, and the first technical description of modal theory based on final and ambitus.[26]

In Hucbald of Saint-Amand's *Musica*, the distance from the ancient treatises that focussed on numerical ratios becomes even larger.[27] Its character is described by its latest student, Yves Chartier, as "un cours pratique d'initiation au chant liturgique destiné à des débutants."[28] Nevertheless, Hucbald refers to the ancient authorities such as Boethius, Martianus Capella and Calcidius in the opening portions of his treatise. The most important contributions of the treatise are its new application of the ancient theory of tetrachords, the building blocks of the Greek Perfect System, to the four principal church tones (*protus, deuterus, tritus* and *tetrardus*), and its new notational system, influenced by Boethius' notational signs (*daseian notation*).

Regino of Prüm, finally, wrote his musical treatise in the form of a letter dedicated to archbishop Radbod of Trier in the early years of the tenth century.[29] The main body of his letter forms a compilation of Boethius, Martianus Capella, Macrobius, Calcidius, Cassiodorus and Fulgentius; a compilation, in other words, of almost the entire late antique tradition of music theory. On the other hand, his principal intention is, according to his own claims, the correction of intonation formulas of the

---

[25] R. ERICKSON, *Musica Enchiriadis and Scolica Enchiriadis*, p. xix.

[26] *Ibid.*, pp. xx-xxi.

[27] Y. CHARTIER, *L'Oeuvre musical d'Hucbald de Saint-Amand*, pp. 43-81.

[28] *ibid.*, p. 46.

[29] Cf., among others, H. HÜSCHEN, *Regino of Prüm*; Y. CHARTIER, *L'Epistola de armonia institutione de Réginon de Prüm*, diss. University of Ottawa 1965 (*non vidi*); *id.*, article "Regino of Prüm" in the *New Grove*; C.M. BOWER, *Natural and Artificial Music*; M. BERNHARD, *Studien zur Epistola de armonica institutione des Regino von Prüm.*

antiphons and responsories of Mass and Office. From his own experience with the liturgy as performed in the abbey of St. Maximin at Trier, where he composed his letter, he argues that some of these intonations were "degenerate". For this purpose, the treatise was followed by a comprehensive tonary with the incipits of a great number of antiphons, responsories etc., classified according to the eight modes.

The corpora of glosses on Boethius and Martianus Capella and these earliest treatises then form the sources for the history of the development of musical thinking in the Middle Ages, in which the Pythagorean tradition was incorporated in the systematization of the Gregorian repertoire. The earliest witnesses of medieval scholarship on ancient theory are the earliest corpora of glosses on Boethius' *De institutione musica* and Martianus Capella's *De nuptiis* — witnesses which we are now able to study in relation to each other for the first time, because of C.M. Bower's and M. Bernhard's edition of the *Glossa maior* on Boethius, and the edition of the Anonymous commentary on Martianus Capella provided in the present work. In this chapter such a study is made, and I have chosen to treat the subject of music in the glosses on Martianus Capella's *De nuptiis* thematically. First of all, the relation between the corpora of glosses on *De institutione musica* and *De nuptiis* will be analyzed in some detail, for this relationship has important implications for the yet unsolved problems of dating and authorship of the Anonymous corpus. Secondly, the theme of the Harmony of the Spheres is treated, which is central in the first two books of Martianus' encyclopedia and received particularly lengthy glosses of a more philosophical and speculative character. From these glosses it becomes clear that the arts of philosophy, mathematics, astronomy and music were basically unseparated in medieval scholarship; each of these disciplines is based on the same ratios, which reflect the order of creation itself. The third aspect treated in this chapter is the technical content of Martianus Capella's ninth book, and of the glosses attached to it. Five sections are devoted to this aspect: tones, intervals, *genera* (enharmonic, chromatic, diatonic), tetrachords, and modes (Dorian, Phrygian, Lydian, etc.). Finally, the few glosses which relate to contemporary musical practice are treated in the last section.

Boethius' De institutione musica *and Martianus Capella's* De nuptiis: *related reception of the two main sources on antique music theory*

1. *Introduction*

The two main works which brought antique music theory to the Middle Ages are Boethius' *De institutione musica* and the ninth book of Martianus Capella's *De nuptiis Philologiae et Mercurii.* These two works differ greatly in almost every aspect — their style, scope, degree of detail, sources, language etc. could hardly differ more. Through these two works, however, Carolingian scholars inherited a consistent body of music theoretical knowledge from Antiquity. A thorough investigation into the relation between the earliest reception of both works is therefore much needed. Moreover, because of the recent edition of the *Glossa maior* by Michael Bernhard and Calvin M. Bower,[30] that is, the glosses on Boethius which form a more or less coherent tradition and which date back to at least the second third of the ninth century, and my own edition in this study, a comparison becomes possible for the first time. Let us first, however, consider the present state of research on the subject, which is mainly based on the findings of three authors: first of all, the two editors of the *Glossa maior*, Michael Bernhard and Calvin M. Bower,[31] and secondly, a series of articles from the hand of Marie-Elisabeth

---

[30] M. BERNHARD & C.M. BOWER, *Glossa maior in Institutionem musicam Boethii*, I-III, München 1993-1998. The fourth volume with commentary and conclusions is forthcoming.

[31] See, *e.g.*, M. BERNHARD, *Glossen zur Arithmetik des Boethius*; *id.*, *Überlieferung und Fortleben*; *id.*, *Glosses on Boethius*; *id.*, *Boethius im mittelalterlichen Schulunterricht*; C.M. BOWER, *The Role of Boethius' "De institutione musica" in the Speculative Tradition of Western Musical Thought*; *id.*, *Boethius' De institutione musica: A Handlist of Manuscripts*; *id.*, *The grammatical model of musical understanding in the Middle Ages*; *id.*, *Die Wechselwirkung.*

Duchez.[32]

In her articles Duchez reconstructs the reception of antique music theory, and draws the following conclusions. The reception of Martianus Capella's *De nuptiis* and Boethius' *De institutione musica* caused a basic change in musical thinking. Besides a language-oriented approach to music — present in, for example, the works of late-antique authors such as Augustine, Cassiodorus and Isidore — an approach based on mathematical concepts came to life because of the study of the works of Boethius and Martianus Capella. Musical thinking became saturated with the mathematical proportions which form the basis of creation in (Neo)Platonic works, such as Plato's *Timaeus* and Calcidius' translation of that text, Macrobius' commentary on Cicero's *Dream of Scipio*, and the first two narrative books of *De nuptiis*.

Duchez sees the development of this process of the "mathematicization" of music in three stages. First, through reading and studying *De nuptiis*, ninth-century scholars (from the time of Charles the Bald) were introduced to abstract concepts such as tone, pitch, interval, etc. — unconnected with the language-oriented concepts of melody, phrase, melodic formula. The first encounter of these scholars with these concepts is reflected in the Anonymous corpus of glosses on *De nuptiis*. Secondly, because of their acquaintance with these concepts, Carolingian scholars were able to familiarize themselves with *De*

---

[32] M.-E. DUCHEZ, *Jean Scot Erigène premier lecteur du "De institutione musica" de Boèce?*; *id.*, *Description grammaticale et description arithmétique des phénomènes musicaux*; *id.*, *Vocabulaire musical et conceptualisation musico-scientifique*; *id.*, *L'émergence acoustico-musicale du terme "sonus" dans les commentaires carolingiens de Martianus Capella*; *id.*, *Le savoir théorico-musical carolingien dans les commentaires de Martianus Capella*. In this last article, she anticipates on a work in progress on the reception of Martianus' music theoretical material in the Carolingian period: *Imago mundi. La naissance de la théorie musicale occidentale dans les commentaires carolingiens de Martianus Capella et sa structuration par le "De institutione musica" de Boèce* (*ibid.*, p. 165, n.2). In a letter (13 October 1992), however, she kindly informed me that the work was intended to be her doctoral thesis, but that things turned out otherwise. She stated that her conclusions on the Carolingian material are of the same nature as those presented in her earlier articles. So far, she has not completed *Imago mundi*.

institutione musica, which is much more detailed and complex than Martianus' introduction to the art of music. This second stage is reflected in copies of and glosses on *De institutione musica*, and in John the Scot's commentary on *De nuptiis*, where he gives evidence of a thorough knowledge of both works. Thirdly, the gloss tradition on *De nuptiis* was transformed into a more complete and more superficial corpus at the same time by the compilatory activity of Remigius of Auxerre; the text entered the school curriculum and lost its innovatory influence.

Bernhard and Bower, editors of the *Glossa maior* on Boethius' *De institutione musica*, reached quite different conclusions on the history of Boethius reception. First of all, they do not see John the Scot as a pioneer and lonely intellectual star in this history. On the basis of thorough manuscript research, they reached the conclusion that *De institutione musica* was copied on a somewhat larger scale in Northern France from ca. 825 onwards to 860 — roughly the reign of Louis the Pious. This corresponds strikingly closely with the conclusions on the production of glossed manuscripts of *De nuptiis*, which was at its zenith around the middle of the ninth century in the same region. Eleven manuscripts of *De institutione musica* date back to the last third of the ninth century, and in six of these 9th-century glosses are found. Bower characterizes the earliest glosses in terms which are again remarkably equally true for the Anonymous corpus of glosses on *De nuptiis*. The character of the glosses is not primarily pedagogical; it is a learned commentary on the text, and does not only strive for a better understanding of the text, but also aims to elaborate upon it by seeking parallels in other (late) antique texts. The commentators of both texts draw their knowledge from a common *Fundgrube* of antique material on mythology and etymology.

If we follow Bower's conclusions, it seems very likely to presuppose a related reception of the two texts on antique music theory, rather than the phased one for which Duchez argued. In this chapter a comparison between the *Glossa maior* and the Anonymous commentary on *De nuptiis* will be made, in an attempt to shed new light on this question. First, however, Duchez' arguments and conclusions must be discussed, and the role of John the Scot in the reception of Boethius must be looked into. Then, the commentaries will speak for themselves — citations, shared sources, and shared figures and tables will be presented.

2. *John the Scot's* Annotationes in Marcianum *and Boethius'* De institutione musica

Duchez published her contribution to the International Eriugena Colloquium, held in Freiburg im Breisgau in 1979, under the title "Jean Scot Érigène premier lecteur du *De institutione musica* de Boèce?".[33] She argued in this article that the first traces of examination and knowledge of Boethius' work on music can be found in John the Scot's commentary on *De nuptiis*. The traces she detects are (as she herself stresses) not direct: no quotations from Boethius and no mentions of his name. Boethius' influence on John emerges from short, textual reminiscences, the analogous use of vocabulary, the use of words in the meaning that is given to them by Boethius (and by Boethius alone). Her strongest example of such a textual reminiscence is John's interpretation of the word *paginula*; he uses this word to explain a passage in *De nuptiis* II.138. In this passage, Philology is willing to leave the earth and ascend to heaven, but she is too heavy, because of the weight of her worldly learning. She then throws up a multitude of books in all kinds of languages and scripts (II.138). The books and pages are eagerly gathered by the young women who personify the Arts and Disciplines:

> [138] in aliis [voluminibus] quippe distinctae ad tonum ac deductae *paginae*, in aliis circuli lineaeque hemisphaeriaque cum trigonis et quadratis multiangulaeque formae pro theorematum vel elementorum diversitate formatae; dehinc pictura animalium membra multigenum in unam speciem complicabat.
> (In some of these books the *pages* were marked with and subdivided for notational signs; in others there were circles and straight lines and hemispheres, together with triangles and squares, and polygonal shapes drawn to suit the different theorems and elements; then a representation joined together the limbs of various animals into

[33] In: *Eriugena. Studien zu seinen Quellen*, pp 165-187.

particular species.)[34]

John the Scot, however, had a different manuscript version at his disposal, where the word *paginae* was corrupted into *paginulae*, and commented on it in the following words:

> [60,8] PAGINULAE dicuntur in musica ubi tropi in altitudinem ascendunt pro differentiam (sic) tonorum
> (PAGINULAE are called in music the place (*i.e.* on the page) where the modes rise for the differentiation of the pitches)[35]

The meaning of this sentence can only be understood in the light of Boethius' explanation of modal theory in book IV, chapter 15-17 of *De institutione musica*. In this explanation, Boethius makes use of two diagrams, which were, as he himself states, handed down from antiquity.[36] The theory involved concerns the transposition of systems of (for example) an octave from the lowest tone of the Perfect System (the *proslambanomenos*) to higher tones, by means of which modal transposition is accomplished. When a scale of an octave, with its succession of tone steps of whole and half tones characteristic for the Hyperdorian mode, is thus transposed from the *proslambanomenos* to — *e.g.* — the *hypate hypaton*, the next tone in the Perfect System, it forms the Hypophrygian mode.[37] The principle is illustrated by means of a large diagram, in fact a matrix of eight columns and eighteen rows. The eight columns contain the tones of the eight modes, while the eighteen rows contain the names of the eighteen tone names of the diatonic system.[38] In each cell, a notational sign is presented which corresponds with the tone in

---

[34] Trans. cf. W.H. STAHL, *Martianus Capella*, II, p. 47.

[35] ed. C.E. LUTZ, *Iohannis Scotti Annotationes*, p. 65.

[36] BOETHIUS, *De institutione musica*, IV,15.

[37] The explanation of Boethius on the way in which such a transposition should be accomplished is sparse, and the accompanying diagram contains significant differences between signs used in this text and those found in other sources. The basic principle of transposition, however, seems clear. Cf. C.M. BOWER, *Fundamentals of Music*, pp. 153-154, nn. 84-85.

[38] Cf. Fig. D.20 in C.M. BOWER, *Fundamentals of music*, p. 155. For a plate of a corresponding diagram in a ninth-century manuscript (*Paris, BnF, lat. 13020*, fol. 51r) see M.-E. DUCHEZ, *Jean Scot Érigène premier lecteur*, Planche II.

the right mode and right function. In a second diagram, the modes are presented horizontally, and in the cells notational signs for each tone of the Perfect System are written — thus creating a diagram with eighteen columns (for the eighteen tones of the diatonic system) and eight rows (for the eight modes), which are arranged like steps (see Plate 1 in this section).[39] In this second diagram it is shown how tones with the same pitch have different functions in the Perfect System, and how the modes relate to each other: *Descriptio continens modorum ordinem ac differentias* — "a diagram containing the disposition of and differences between the modes".[40] The term *paginula* is used by Boethius to refer to the little boxes in these diagrams, which contain the notational signs for the tones in question. And it is this meaning — Duchez keenly observes[41] — which John the Scot had in the back of his mind when explaining the passage.[42]

In her article, Duchez not only argues that John the Scot knew *De institutione musica* and integrated the Platonic and epistomological concepts of the work into his own philosophical system, but also argues that his knowledge of Boethius' *De institutione musica* is unique and can not be encountered in the other commentaries:

---

[39] Fig. D.21 in *Fundamentals of Music*, p. 156. See also M.-E. DUCHEZ, *Jean Scot Érigène premier lecteur*, where Planches II and III show examples of the same diagrams in a ninth-century manuscript (Paris, Bibl.Nat., lat. 13020).

[40] BOETHIUS, *De institutione musica*, IV, 16 (343,7). *Differentias* and not *differentiam*, as Duchez erroneously cites on p. 177. Transl. C.M. BOWER, *Fundamentals of Music*, p. 154.

[41] M.-E. DUCHEZ, *Jean Scot Érigène premier lecteur*, pp. 176-178.

[42] Duchez also notes that a hand which appeared in one of the earliest manuscripts of *De institutione musica*, identified as the hand of I^2 — a hand which belonged to a scholar "from the circle of John the Scot," also appears in a Martianus Capella manuscript: it is one of the hands which wrote glosses in the margins of book IX of *De nuptiis* in Leiden, University Library, BPL 88. In this manuscript a somewhat different version of John the Scot's commentary is found for book IX, while the other books carry glosses from the Anonymous corpus. (See the manuscript description in the previous chapter.) Cf. M.-E. DUCHEZ, *Jean Scot Érigène premier lecteur*, pp. 182-183; C.M. BOWER, *Die Wechselwirkung*, p. 166.

> Notons que la comparaison des gloses érigéniennes avec d'autres gloses du même passage, issues, au cours du IXe siècle, du même milieu érudit de Laon, montrent les différentes attitudes devant le *De musica*, de ceux qui avaient pris le *De nuptiis* comme base d'enseignement: après le glossateur des manuscrits Leyde B.P.L. 88 et Vatican Regina lat. 1987 — Martin de Laon, selon Préaux — qui ignore le *De musica*, Jean Scot semble en avoir une vue d'ensemble juste, je vais essayer de le montrer; mais à fin du siècle, Remi d'Auxerre, bien que connaissant effectivement le texte de Boèce sans, peut-être, l'avoir bien compris, ne fait dans son *Commentarium in Martianum Capella* aucun rapprochement du passage glosé avec la théorie boécienne.[43]

John the Scot, she thus argues, is the only ninth-century scholar to give evidence of a deeper and fuller understanding of Boethius' work. There is no evidence of knowledge of *De institutione musica* in glosses from the Anonymous tradition, and those Boethian traces found in the commentary of Remigius of Auxerre show less understanding. Moreover, they are to be dated substantially later than the traces of Boethian thought in John's commentary.

It seems obvious that Duchez' conclusions deserve closer scrutiny now that a more complete edition of the Anonymous commentary is available. The passage in which Duchez saw such philosophically and epistemologically complete adoptions of Boethian material into John's glosses on *De nuptiis* also received comments in the Anonymous commentary. They are presented below, side by side:

Anonymous commentary	John the Scot, *Annotationes*

42.19-1 SED DUM TALIA VIRGO UNDANTER EVOMERET, PUELLAE QUAMPLURES, QUARUM ARTES ALIAE, ALTERAE DICTAE SUNT DISCIPLINAE, SUBINDE, QUAE EX ORE VIRGO EFFUDERAT, COLLIGEBANT

42.19-1	60,3
TALIA s. volumina *LaLbVaLeaOxf*	SUBINDE frequenter. inter

[43] *Ibid.*, p. 175.

VIRGO Philologia *LbVaLea*
UNDANTER copiose *BeLaLbVaOxf*
QUARUM puellarum *LbVaLea*
ARTES Pos<t>quam ad perfectionem venitur *La*
  Quia latior est [p]ars quam disciplina *LbVaLea*
ALTERAE puellae *LbVa*
  Quia disciplina continetur in artibus *LbVaLea*
DISCIPLINAE Inter artem et disciplinam hoc interest, quod disciplina dicitur dum docetur, ars vero cum ad perfectam animi intelligentiam perducitur *BeLaLbVaLea*
SUBINDE actutum statim *BeLa*
  statim saepe *LbVa*
  statim *LeaOxf*
ORE suo *LaLbVa*
EFFUDERAT emiserat *LbVa*

artem et disciplinam hoc interest, quod quando discitur, disciplina vocatur, quando perfecta in habitu mentis est, ars nuncupatur. Ars dicta est ἀπὸ τῆς ἀρετῆς, hoc est a virtute.

42.21-2 IN SUUM UNAQUAEQUE ILLARUM NECESSARIUM USUM FACULTATEMQUE CORRIPIENS

42.21-2
IN SUUM UNAQUAEQUE ILLARUM NECESSARIUM "Quia non omnia possumus omnes" *BeLa*
  vel artibus vel disciplinis *LbVaLea*
  Bene dicit in suum necessarium usum et facultatem, quia dum advivimus non omnia capere possumus *La*
FACULTATEMQUE possibilitatem *LaLbVaOxf*
CORRIPIENS accipiens *LaLbVa*
  violenter capiens *Oxf*

60,5
FACULTATEM instrumenta. Facultas artis est omne instrumentum quo discitur et docetur.

42.22-3 IPSAE ETIAM MUSAE, PRAESERTIM URANIE CALLIOPEQUE, INNUMERA GREMIO CONGESSERE VOLUMINA

42.22-3
PRESERTIM maxime *LbVa*
URANIE nominativus Grecus *LbVaLea*
CALLIOPEQUE Per Uraniam ratio caelestium elementorum, per Calliopen, quae bona sonoritas dicitur, musica totius naturae, et celestis et terrestris, significatur *BeLaLbVaLea*
INNUMERA copiosa in GREMIO in sinu suo *BeLaLbVa*
plura *Oxf*
†CONCESSERE collegere *BeLaOxf*
collegerunt *LbVa*

60,5
Ideo URANIA ET CALLIOPE collegerunt libros pre ceteris quia doctiores erant. Praesul est Urania caelestis musicae, Calliopea artificialis musice praesul.

42.23-1 IN ALIIS QUIPPE DISTINCTAE AD TONUM AC DEDUCTAE PAGINAE

42.23-1
IN ALIIS QUIPPE s. voluminibus *BeLaOxf*
DISTINCTAE s. erant *BeLaLbVaLeaOxf*
AD TONUM usque ad tropum *BeLaOxf*
i. ad proprietatem uniuscuiusque toni *BeLaLbVaLeaOxf*
DEDUCTAE extentae *BeLaLbVaOxf*
PAGINAE Musicam tangit *Be*

60,8
PAGINULAE dicuntur in musica ubi tropi in altitudinem ascendunt pro differentiam tonorum.

43.1-3 IN ALIIS CIRCULI LINEAEQUE HEMISPHAERIAQUE CUM TRIGONIS ET QUADRATIS MULTIANGULAEQUE FORMAE PRO THEOREMATUM VEL ELEMENTORUM DIVERSITATE FORMATAE

43.1-3
IN ALIIS s. voluminibus *LaLbVaLeaOxf*
Geometricam tangit *Be*
Nunc tangit geometricam et astrologiam *LaLbVaOxf*
CIRCULI s. planetarum *BeLaOxf*

60,8
EMISPERIA semicirculi

60,10
TEOREMATUM speculationum, teorema proprie dicitur in geometria

planetarum cursus *LbVaLea*
LINEAEQUE celestes semitae *LaOxf*
Lineae sunt in caelo clxxxiii: duo hemispheria, unum superius a signifero usque ad polum articum, alterum vero inferius sub signifero usque ad antarticum i. usque ad austrinum polum *BeLaLbVaOxf*
CUM TRIGONIS Trigonus quasi tergona i. genua flectens, quia gonum dicunt Greci genua *BeLaLbVaLeaOxf*
MULTIANGULAEQUE Multiangulae sunt omnes figurae, a quadrato usque ad infinitum, ut sunt pentagoni, exagoni, eptagoni ceterique numeri in plures angulos ducti *BeLaLbVaLeaOxf*
THEOREMATUM Theorema est contemplatio rationis, cum lineae lineis iunguntur, anguli lateribus, latera angulis et anguli angulis et latera lateribus etcetera *BeLaLbVaLeaOxf*
contemplationum *Oxf*
ELEMENTORUM Initiorum, per quae incipiunt ipsae disciplinae *LbVaLea*
Elementa vero geometriae sunt punctus, linea, angulus *BeLaLbVaOxf*

quod elementa in astrologia; quod enim geometria fingit in geometricis figuris, hoc describit in naturalibus corporibus astrologia.

43.3-4 DEHINC PICTURA ANIMALIUM MEMBRA MULTIGENUM IN UNAM SPECIEM COMPLICABAT

43.3-4
DEHINC postea *Oxf*
PICTURA ANIMALIUM cum. Tangit cosmographiam *Be*
In similitudinem animalium erant factae *BeLa*
Cosmographiam tangit hic id est

60,11
IN UNAM SPECIEM hoc est in unam mundi imaginem. Ut enim una imagine totus mundus cum regionibus et animalibus in una quadam specie compraehenditur

descriptionem mundi *LaLbVaOxf*
IN UNAM SPECIEM in unam paginam *BeLa*
i. paginam *LbVaLeaOxf*
COMPLICABAT depingebat *BeLaOxf*
conpingebat *LbVaLea*

aspicientiumque oculis manifestatur, sic tota musica in una imagine describitur per tonos et tropos caeterosque modos et horum singula super notatos numeros et litteras et notas.

43.4-6 ERANT ETIAM LIBRI, QUI SONORUM MELA SIGNAQUE NUMERORUM ET CANTANDI QUAEDAM OPERA PRAEFEREBANT

43.4-6
ERANT ETIAM LIBRI Musicam tangit *LaLbVaLeaOxf*
SONORUM MELA dulcia *Be*
dulcedinem soni *La*
i. dulcedines sonorum *Oxf*
dulcedinem *LbVaLea*
SIGNAQUE NUMERORUM Proprie numerus dicitur concordia humanae vocis *BeLaLbVaLeaOxf*
OPERA studia *LbVa*
PRAEFEREBANT demonstrabant *BeLaLbVaOxf*

43.6-7 POSTQUAM IGITUR ILLAM BIBLIOTHECALEM COPIAM NIXA IMITUS VIRGO DIFFUDIT

43.6-7
[1]Omnia quae vomuit Philologia, ad significationem mundanorum studiorum posita sunt. [2]Humana enim studia aliquando fallunt, quia in hac vita artium veritas apparere non potest. [3]Ideoque inflantur quasi quodam superbiae humore illi, qui illis utuntur. [4]Per illud vero sorbillandum i. poculum, quod Philologiae dederat, veritas artium, quae solis immortalibus et veritatem contemplantibus manifestabitur, apertissime significatur *BeLaLbVaLeaOxf*

In this comparison it becomes clear, first of all, that the Anonymous commentary discussed Martianus' text at much greater length, and, secondly, that the two commentaries share some of the material. John, for example, calls Urania the *praesul caelistis musicae*, the leader of celestial music, and Calliope the *praesul artificialis musicae*, the leader of artificial — that is earthly? — music. In the Anonymous commentary, Urania is the *ratio caelestium elementorum*, the proportion of the celestial elements, while Calliope (*bona sonoritas* or "harmonious harmony" — an etymology probably taken from Fulgentius, *Mythologiae*, I.15 (*Fabula de novem musis*)) — stands for the *musica totius naturae, et celestis et terrestris*, all music, celestial and earthly. Neither of the two are in terminological concurrence with Boethius' division of music into *musica mundana* (cosmic music), *musica humana* (human music) and the music which rests in musical instruments,[44] but both roughly equate Urania with Boethius' *musica mundana*, and Calliope with Boethius' *musica humana* and "*instrumentalis*". There is no real difference there in the adoption of Boethian material.

Theorema, which, according to Duchez, is paraphrased with *speculatio* in a typically Boethian manner, also receives a very similar explanation in the Anonymous commentary: *contemplatio rationis* — a "theoretical contemplation". The only place in which John clearly refers to Boethian material where the Anonymous commentary does not is the gloss added to *paginula* described above. But this was not caused by John's deeper understanding of Boethius, as Duchez would have it; it was simply caused by a deviating manuscript tradition, in which the word *pagina* was transformed into *paginula*, and which must have evoked the association with Boethius' diagrams of modal transposition.

Furthermore, a diagram resembling that of Boethius is copied in several Martianus Capella manuscripts, two of which carry glosses from the Anonymous commentary, viz. *Leiden, University*

[44] BOETHIUS, *De institutione musica*, I,2.

Library, BPL 36 and *Köln, Dombibliothek, 193*.[45] This figure (see Plate 2 in this section) bears the title *Hi sunt tropi xv hoc est formae carminum qui peraguntur in cithara viginti octo cordarum* — "These are the fifteen modes, that is the shapes of songs which are performed on the cithara of twenty-eight strings." An accompanying text furthermore explains:

> Cordae sunt quae vergunt a fronte paginae in eius imam partem. Lineae vero quae porriguntur a leva in dexteram termini sunt troporum.
> (The strings are the (lines) which run from the head of the page to its bottom. The lines, on the other hand, which stretch from the left to the right, determine the limits of the modes.)

And:

> Mese ypodorii est adquisita eolii. Mese ypofrigii, adquisita iastii. Mese ypoeolii, adquisita lidii. Mese ypoiastii, adquisita yperdorii, Mese ypolidii, adquisita yperfrigii. Mese dorii, adquisita ypereolii. Mese frigii, adquisita yperiastii. Mese eolii, adquisita yperlidii.
> (The *mese* of the hypodorian mode, is the *adquisitus* of the aeolic mode. The *mese* of the hypophrygian mode, is the *adquisitus* of the iastian mode, ... (etc.))

The strings of the cithara, in other words, are here presented in columns, while the fifteen modes are presented in rows. The diagram does not, however, present all twenty-eight strings of the Perfect System, but has only eight cells per mode — that is, the first octave, from *proslambanomenos* (or *adquisitus*) to the *mese* in the diatonic genus. The eighth cell in each row represents the *mese*. Thus it becomes visible how the *mese* of one mode is the *adquisitus* of another, by following the column of the *mese* in one mode as far down as it goes. The figure is simpler than Boethius' diagram of transposition, because it contains only eight steps per mode instead of eighteen, and because it does not contain Greek notational signs. It does, however, make use of the same system.

[45] The first manuscript contains a mixed corpus of Anonymous, Remigian and unidentified glosses, the latter a mixture of glosses from the Anonymous tradition and later glosses from the commentary of Remigius of Auxerre. The figures are found on fol. 129^{v} and 201^{r-v} respectively. See on this also the section on "*Modi* and *Tropi*" in this chapter.

(See Plate 2)

The fact that this Boethian figure was copied in two ninth-century manuscripts with Martianus Capella's *De nuptiis* further weakens Duchez' hypothesis that John the Scot was the only ninth-century scholar to have had knowledge of Boethius' modal theory. In fact, when looking at music theoretical material in the Anonymous commentary, one must conclude that material from *De institutione musica* was readily and frequently used to explain Martianus' sometimes obscure explanations of antique music theory, and to identify the strange (Greek) vocabulary used by him.

3. *Boethian material in the Anonymous commentary and vice versa*

First of all, Boethius is mentioned by name in several glosses, and in many more Boethian material clearly forms the context of the Anonymous glosses. Thus, for example, Boethius is mentioned in comments added to the term *proslambanomenos*, the first tone of the Greek Perfect System, which is also called *prosmelodos* in *De institutione musica* I.20:

> [357.12-3] Secundum Boetium prosmelodos dicitur proslambanomenos. *BeLa*, Boetium] boeticum *Be*
> (According to Boethius the *proslambanomenos* is called *prosmelodos.*)

When Martianus speaks on the subject of semitones, the commentator remarks that according to Boethius, a tone cannot be divided into two equal semitones, which is treated at several instances in *De institutione musica* (e.g. I.16, ll 17-19: *Rursus tonus in aequa dividi non potest, cur autem, posterius liquebit; nunc hoc tantum nosse sufficiat, quod nunquam tonus in gemina aequa dividitur.* — Moreover, a tone cannot be divided into equal parts; why, however, will be explained later. For now it is sufficient to know only that a tone is never divided into two equal parts.[46])

> [358.19-20] Non negat Boetius sonum dividi in duas partes, sed in aequas partes dividi non posse dicit. *BeLa*

[46] Transl. C.M. BOWER, *Fundamentals of music*, p. 26. Boethius looks ahead at book III, 1-2.

(Boethius does not deny that a tone can be divided in two parts, but he says that it cannot be divided in two equal parts.)

Furthermore, there is a gloss added to the contradictory statements in *De nuptiis* on the number of tones. In section 931, where Martianus enumerates the names of the tones in Greek and Latin, he mentions 18 tones, but in section 940, this time following a different authority, he speaks of 28 tones. The commentator explains this, by referring to Boethius' enumeration of the tones in I.22 or to his presentation of these tones in a diagram:[47]

[362.7-8] Supra xviii posuit, nunc xxviii. Si volueris diatonicum tantum commemorare, tunc x et viii habebis. Si autem cromaticum vel enarmonium intromiseris, x supra habebis et erunt xxviii. Sub uno tamen nomine v cromatici et v enarmonici adduntur i. pro una corda. Sed subtiliter et obscure iste dixit. Boetius vero enucleatius patefecit *BeLa*, et erunt xxviii. *om. La*
(Above he presented 18 (notes) now 28. If you would want to enumerate just the diatonic (genus), you would have 18 (notes). But if you would put in the chromatics and enharmonics, you would have 10 (notes) extra, and there will be 28. Yet under one name 5 chromatics and 5 enharmonics are added, that is for one string. But he said this in a subtle and obscure manner. Boethius however made this clear in a more precise manner.)

In Martianus' sections on the power of music, Boethius' name is not mentioned in the glosses, but his influence is nevertheless apparent. Where Martianus speaks of the "Pythagoreans" who "assuaged the ferocity of men's spirits with pipes and strings and taught that there is a firmly binding relationship between souls and bodies" (IX.923),[48] the Anonymous commentator makes the association with a story in Boethius' *De institutione musica* I,1, in

[47] Cf. G. FRIEDLEIN, pp. 215-216; C.M. BOWER, *Fundamentals of Music*, Fig. A.18 (p. 44). An enumeration of the tones, as a matter of fact, is also found in some ninth-century Martianus Capella manuscripts, often in combination with the diagram which illustrates the relations between the different modes. (Cf. the section on "*Modi* and *tropi*" in this chapter.)

[48] "Pythagorei etiam docuerunt, ferociam animi tibiis aut fidibus mollientes, cum corporibus adhaerere nexum foedus animarum"; transl. W.H. STAHL, *Martianus Capella*, II, p. 357.

which a drunk and raving young man is brought to reason by someone playing the right kind of music to him:[49]

> Vulgatum quippe est, quam saepe iracundias cantilena represserit, quam multa vel in corporum vel in animorum affectionibus miranda perfecerit. Cui enim est illud ignotum, quod Pythagoras ebrium adulescentem Tauromenitanum subphrygii modi sono incitatum spondeo succinente reddiderit mitiorem et sui compotem? Nam cum scortum in rivalis domo esset clausum atque ille furens domum vellet amburere, cumque Pythagoras stellarum cursus, ut ei mos, nocturnus inspiceret, ubi intellexit, sono phrygii modi incitatum multis amicorum monitionibus a facinore noluisse desistere, mutari modum praecepit atque ita furentis animum adulescentis ad statum mentis pacatissimae temperavit.
> (It is common knowledge that song has many times calmed rages, and that it has often worked great wonders on the affections of bodies or minds. Who does not know that Pythagoras, by performing a spondee, restored a drunk adolescent of Taormina incited by the sound of the Phrygian mode to a calmer and more composed state? One night, when a whore was closeted in the house of a rival, this frenzied youth wanted to set fire to the house. Pythagoras, being a night owl, was contemplating the courses of the heavens (as was his custom) when he learned that this youth, incited by the sound of the Phrygian mode, would not desist from his action in response to the many warnings of his friends; he ordered that the mode be changed, thereby tempering the disposition of the frenzied youth to a state of absolute calm.)

In the Anonymous commentary, the same story, although quite different in detail, is referred to in the following words:

> [354.7-8] Hic videtur tangere quod legitur, quod quidam iuvenis dives temulentus in urbe quadam scortum adiit. Cuius domum — cum intrare nequiret — domum illius post reveniente (*sic, lege* reveniens) cum urbe flammis cremare [videbatur] conabatur. Unde universi commoti sunt. Quod cum quidam Pytagorae nunciasset, qui tunc forte astris intendebat, ille tibicinam vocavit. Que cum docte tibicinaret, non tamen iuvenis a ferocia sua cessaret, ipsae (*sic, lege* ipsi) Pytagoras fistulam mutari precepit adque aliut modulamen

[49] G. FRIEDLEIN, pp. 184-185; transl. C.M. BOWER, *Fundamentals of music*, p. 5.

> concini iussit. Quod cum actum fuisset, iuvenis insanire cessavit. *La* (This seems to touch on a story: a rich, drunk young man went to see a prostitute in a certain city. When he was unable to enter her house, after returning he tried to set fire to her house and the city. By which everybody was disturbed. And when somebody told this to Pythagoras, who at that time happened to be studying the stars, he called a female flute-player. And when she played the flute skilfully, but the young man did not cease his frenzy, Pythagoras advised her to change flute and ordered another melody to be played. And when that was accomplished, the young man ceased his insanity.)

Other implicit examples of Boethian influence in the ninth-century glosses on Martianus Capella are found in figures and tables, which appear both in manuscripts of *De nuptiis* and of *De institutione musica*.[50] The diagrams, for example, which Boethius uses for his explanation of the three "genera of melodies", viz. diatonic, chromatic and enharmonic, are also found in the Anonymous commentary — the lay-out is identical (see Plate 3).[51] The same is true for diagrams for the visualization of intervals and their proportions, *i.e.* the fourth with its sesquiterce ratio (4:3), the fifth with its sesquialter ratio (3:2), the octave with its duple ratio (2:1), and the tone with its sesquioctave ratio (9:8). These figures are both used by Boethius as illustrations for book I.16,[52] and also in several *De nuptiis* manuscripts with the Anonymous commentary, here added to section 11 of *De nuptiis*

---

[50] There are also a few Carolingian manuscripts in which Martianus' ninth book is copied alongside with *De institutione musica*: the ninth-century manuscript Paris, BnF, lat. 7200 contains Boethius' *De institutione musica* (with diagrams), Augustinus, *De musica* and an unglossed copy of Martianus' ninth book; the tenth-century manuscript Bamberg, Staatl. Bibl., Class. 9 contains the ninth-century treatise *Musica enchiriadis*, Boethius' *De institutione musica* and a mutilated copy of Martianus Capella's book IX; and the eleventh-century manuscript Ivrea, Biblioteca Capitolare, Ms. 84 contains a part of book IX and of Boethius' *De institutione musica*; Paris, Bibl.Nat., Lat. 7200 . (Cf. C. LEONARDI, *I codici* II, nos. 71, 7 and 155 respectively.)

[51] G. FRIEDLEIN, p. 213; C.M. BOWER, *Fundamentals of Music*, Fig. A.17 (p. 41).

[52] G. FRIEDLEIN, p. 201; but in C.M. BOWER, *Fundamentals of Music*, Fig. A.1-4 (pp. 23-24) the manner in which the diagrams are represented is much closer to the original representation in the manuscripts.

on the music of the Apollonian woods. (see Plate 4-5-6)

On the other hand, there are also examples of Martianian material in the *Glossa maior* on *De institutione musica.*[53] Obvious examples are, for example, places where Martianus was cited or referred to by name by the authors of the *Glossa maior*, such as:

> [I,8,32] INTERVALLUM VERO EST ...
> Felix Capella: "Tonus est spatium cum legitima quantitate, qui diversis inter se sonis conficitur."[54]
> (Felix Capella: "A tone is a distance with a legitimate quantity, which is accomplished between different pitches.)

Or:

> [I,2,108] MUNDI MUSICA
> [1]*Musicam mundi* alii dicunt graviorem in cursu Saturni, acutiorem in cursu Lunae ob id, quod gravitas in longiori spatio corde consistit, acumen vero in breviore. [2]Alii autem gravitatem in cursu Lunae intelligentes, acumen dicunt in cursu Saturni, *sicut Martianus per arborem intelligi voluit.* [3]Ea videlicet, quae terrae proprius erant, rauciori sono, quae sperae caelesti proprius, acutiori sono deputans, musicam caelestem designavit.
> ([1]*Cosmic music*, some say that the lower sound is in the orbit of Saturn, the higher in the circle of the Moon, because lowness is located in a string with a longer length, highness, on the other hand, in a shorter (string). [2]Others, however, perceiving that lowness is found in the orbit of the Moon, say that highness (is found) in the orbit of Saturn, *just as Martianus wanted to show by means of a tree.* [3]For he explained the music of the spheres by assigning those (branches) which were close to the ground a harsh and low sound, and those which were close to the heavens a higher sound.)

Thus, the controversy over highness and lowness in the music of

---

[53] Since the series of publication is not yet complete, and there is not yet an index to facilitate searches in the corpus of glosses, it is not possible to present here a complete overview of places in which echoes of Martianus can be heared. I limit myself to a few examples, found by leafing through the three published volumes of the *Glossa maior*.

[54] *Glossa maior*, I, p. 167. See also 202,26, 57: Est enim sesquioctava proportio tonus, i. legitimum spatium inter chordam et chordam. (*Ibid.*, p. 233.)

the planets, which received many glosses in Martianus Capella commentaries, is here referred to when Boethius speaks of *musica mundana*. The gloss in question refers to Martianus Capella I.11-12, where the music of the spheres is illustrated with the allegory of the Apollonian woods, in which the trees with their high, middle and low branches bring forth a whole range of tones and a miraculous harmony.[55]

Furthermore, the commentaries on *De institutione musica* and on *De nuptiis* both seem to rely on a common corpus of antique or late-antique texts for their explanations and elaborations. Etymologies and myths, for example, are often taken from Isidore's *Etymologiae*, from Pliny, or Servius' Vergil commentaries, while comments on the music of the cosmos often refer to Macrobius' Commentary on *Scipio's Dream*. These sources seem to form, both for the Martianus Capella commentator and for the Boethius commentator, a common base of knowledge, from which they could draw their explanatory material. Aside from these identifiable sources, the glosses sometimes show a remarkable concurrence with each other. Thus, for example, the names of Orpheus[56] and Aristoxenos received similar etymologies in the *Glossa maior* and in the Anonymous commentary:

Anonymous commentary	*Glossa maior*
345.19-21 ORPHEUS orios phone, pulchra vox *Leb* Orpheus dicitur quasi orios phau id est pulcra vox *Par*	I,20,8a-d ORPHEUM 8a pulchra vox *M*^{2} 8b orea pulchra ΦONH VOX *M*^{2} 8c [1]OPHA grece pulchra latine, ΦONE vox. [2]Inde Orpheus pulchra vox dicitur. *McMf*^{2}*MhMkMmPeShSo*

[55] This allegory will be discussed in the section on the harmony of the spheres later in this chapter.

[56] The source for this etymology seems to be Fulgentius, *Mythologiarum libri tres*, III.10: "Orpheus dicitur oreafone, id est optima vox, ..."

AutHwgLryPts
8d Oreaphone pulchra vox; inde Orphaeus pulchre canens dicitur. *Wq*

354.9-10
ARISTOXENUS Virtutis acumen vel xenos hospitium, unde xenodochium domus hospitum *Leb*
Si ab eo quod est aris et oxos virtutis acumen, si ab aris et xenos, virtutis hospicium interpretatur *Par*

II,31,98
ARISTOXENUS
98 virtutis hospitium [interpretatur] *OSOlPe*^{5Wq}
(see also III,1,1 and III,1,14)

Also, the words *chromaticus*, *diatonicus* and *hypaton* received similar etymological explanations:

Anonymous commentary

357.7-8
Cromaticon gen<u>s dicitur colorabile. Croma enim Grece Latine color dicitur. Sicut enim color dicitur, quicquid est medium inter album et nigrum, sic cromaticum modulandi gen<u>s in medio inter enarmonion et diatonon positum est *La*
Cromaticus in tertia parte, tonos in dimidia. Vocantur a proprietate et qualitate, quod non habet nec grave nec gracile i. mixtum de duobus i. de duabus extremitatibus medium, ut aliquis color inter album et nigrum *LebPar*, cromaticus in - et qualitate] colorabile *Par*
aliquis] aliqua *Par*

Glossa maior

I,15,16a
[1]*Diatonum* et diatonicum unum est; et inde dicitur, quod tonis abundat. [2]*Chroma* et chromaticum unum est, i. colorabile, quia ab utrisque accipit colorem. ...
M^{2}*OLaMcMf*^{2}*MhMkMm*^{1}*Pe*$^{3+5}$*R1***S hWqAutVrf*

357.15-7 Diatonus dicitur, quia abundat tonis *Leb*	
362.14 ypatos, consul vel rector *Leb* Pato calco. Quia sicut dux calcat omnes sub se, ita ista omnes cordas praemit *Leb* Unde ypate dicitur: ypatos enim consol vel rector interpretatur *Par*	I,20,32 [1]Pato calco, ypato subcalco, quasi ypopato. [2]Inde ypates dicitur consul, qui sub se calcat omnes, sicut ypate reliquas chordas. $M^2Mf^2MhMkMm^1Pe^5ShSo^*Wq$ I,20,33 Hypate supercalcans, pato calco; unde hypate quasi aliis praecellens ut consul. *CgMcPn* (see also I,20,155)

4. *Conclusions*

The manuscript evidence shows that Boethius' *De institutione musica* was read, studied and copied on a more limited scale than *De nuptiis* in the ninth century. From the ninth century we have roughly 50 manuscripts of *De nuptiis* (almost all glossed), against only eleven of *De institutione musica* (six of which were glossed in the ninth century).[57] This reception on a smaller scale is not surprising: Boethius' treatise on the *ars musica* was elaborate, technical and mathematical; characteristics which made it a fairly inaccessible work. The numerics, diagrams and schemes which illustrate Boethius' treatise made it not only hard to read, but also hard to copy.

On the basis of this manuscript evidence, Duchez reconstructed a chronological path of the reception of the two music theoretical works. She argued that the study of *De nuptiis* preceded the study of *De institutione musica*. The study of *De*

---

[57] Cf. C.M. BOWER, *Boethius' De institutione musica*; C.M. BOWER, *Die Wechselwirkung*, pp. 164-168.

nuptiis, in which all the basic concepts of Greek musical theory were introduced (like *sonus, modus, diastema, symphonia*, etc.), paved the way for a fuller grasp of *De institutione musica*. But — as I have shown above — since there are clear traces of Boethius in the earliest collections of glosses on *De nuptiis*, we can no longer uphold such a chronological reconstruction.

Manuscript evidence further shows that the earliest manuscripts of and earliest glosses on Boethius' *De institutione musica* are to be located in the very same region of the Carolingian empire in which collections of glosses to Martianus Capella were composed, and that the active study of the two works can be traced back to the same period of time: the reigns of Louis the Pious and Charles the Bald. Despite the differences of the texts they comment upon, the two gloss collections show similar characteristics, use of the same sources of antique knowledge, and even similarities in choice of words and linguistic usage. Rather than speaking of a phased reception, where the study of Martianus Capella influenced the reception of Boethius, we must conclude that there was a mutual influence: there is Boethian material in glosses on Martianus and Martianian material in the glosses on Boethius. The study of the two musical treatises was evidently stimulated by comparing the two. There are bits of evidence which justify these conclusions in John the Scot's commentary, but the lion's share of evidence is found in the Anonymous commentary on *De nuptiis*, and the *Glossa maior* on *De institutione musica*.

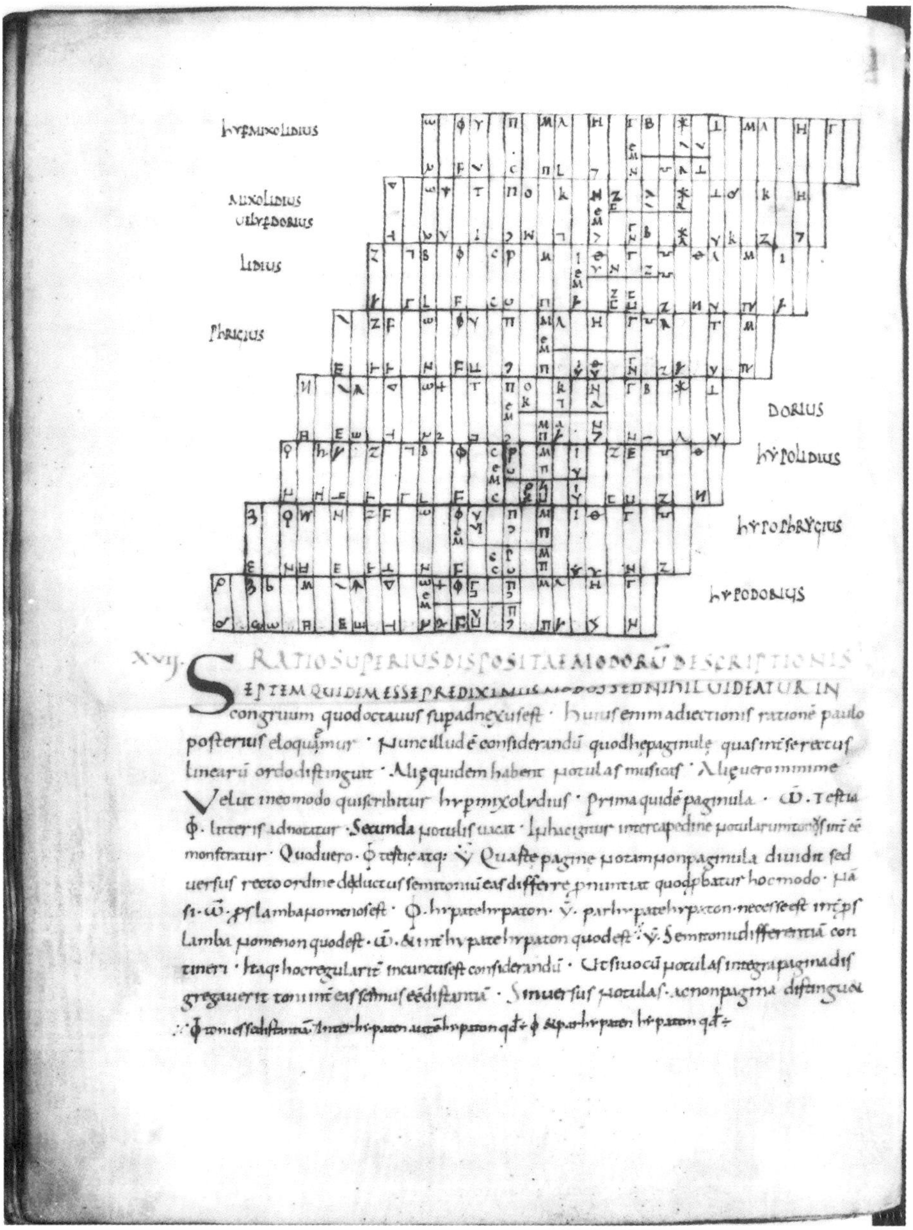

Plate 1 Diagram showing the relationships between the modes in a ninth-century Boethius manuscript, *Paris, BnF, lat. 13020,* fol. 51v.

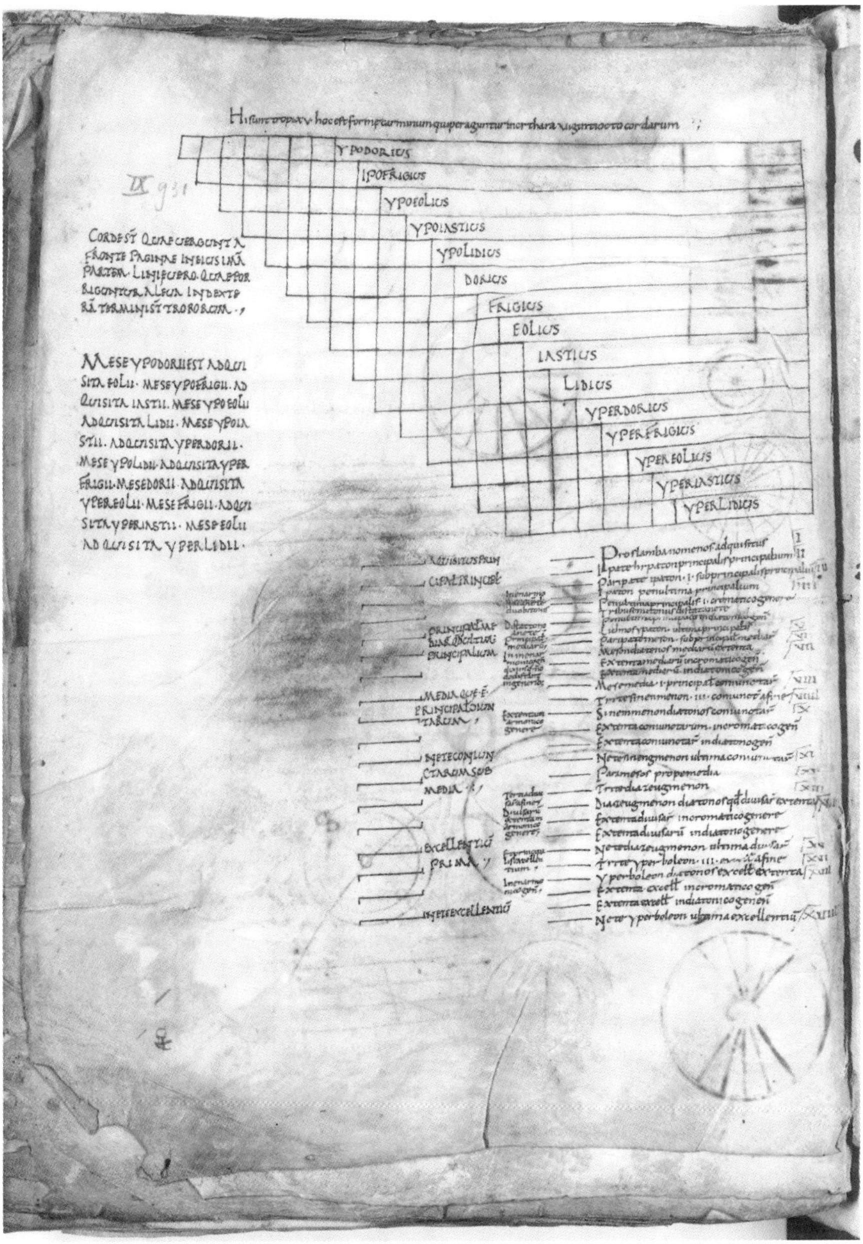

Plate 2 A similar diagram as found in a ninth-century Martianus Capella manuscript, *Leiden, UB, BPL 36*, fol. 129v.

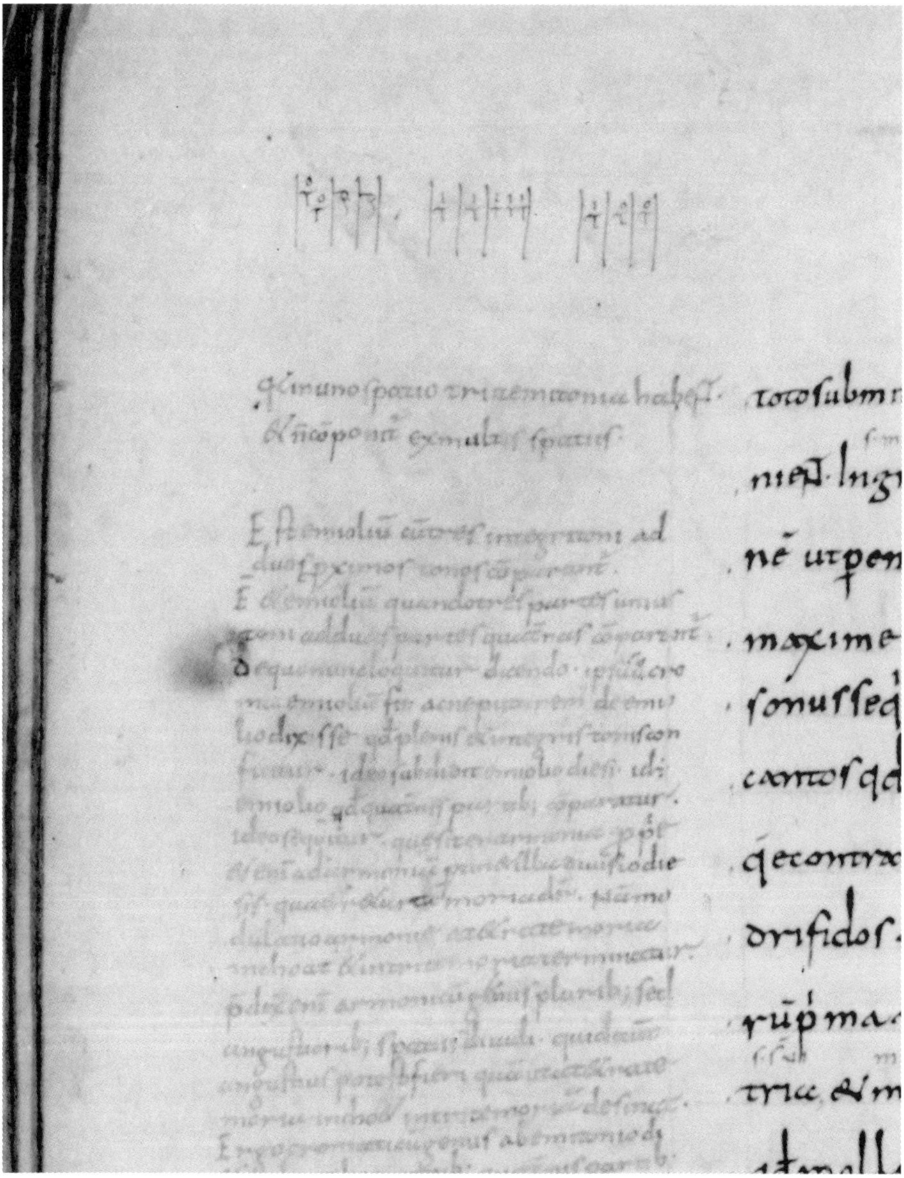

Plate 3 A visual representation of the tone steps in a tetrachord according to the three different *genera* (diatonic, chromatic and enharmonic), as found in *Leiden, UB, Voss. Lat.Fol. 48,* in the margin of fol. 88v.

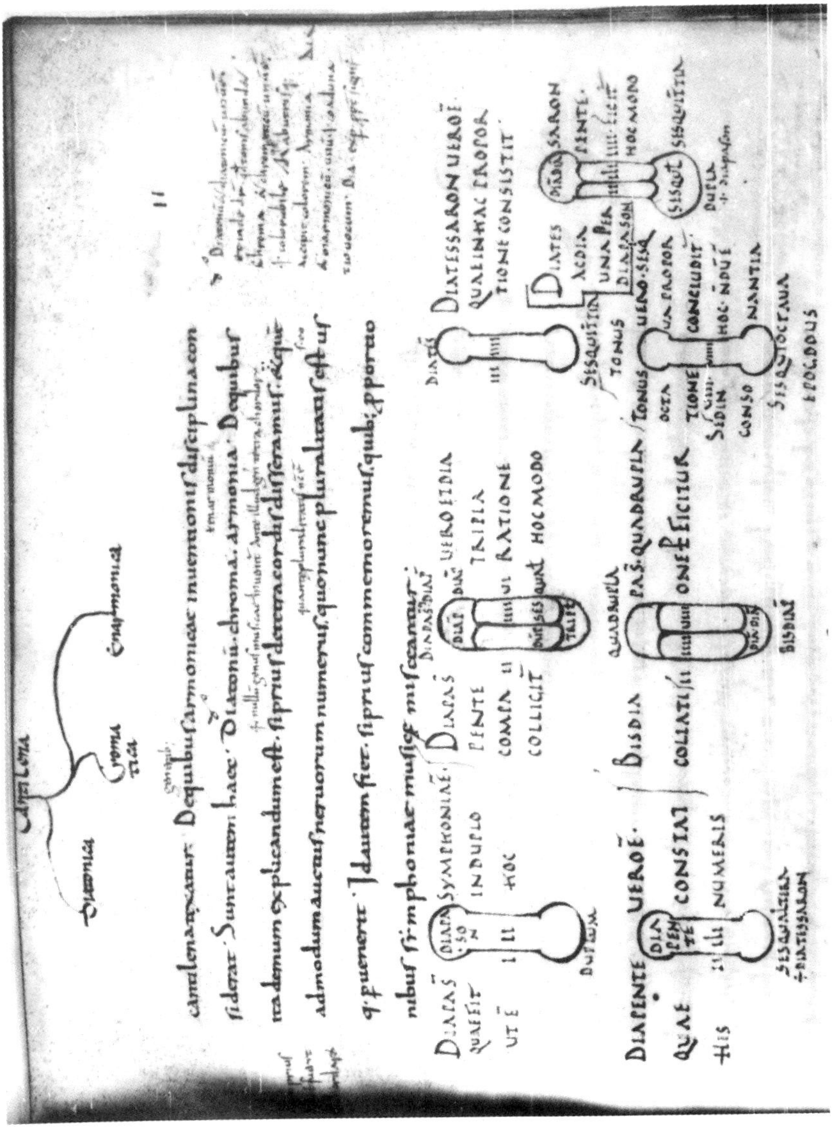

Plate 4 Visual representations of the consonant intervals and their ratios, as found in a ninth-century Boethius manuscript, *Paris, BnF, lat. 7200*, fol. 11r.

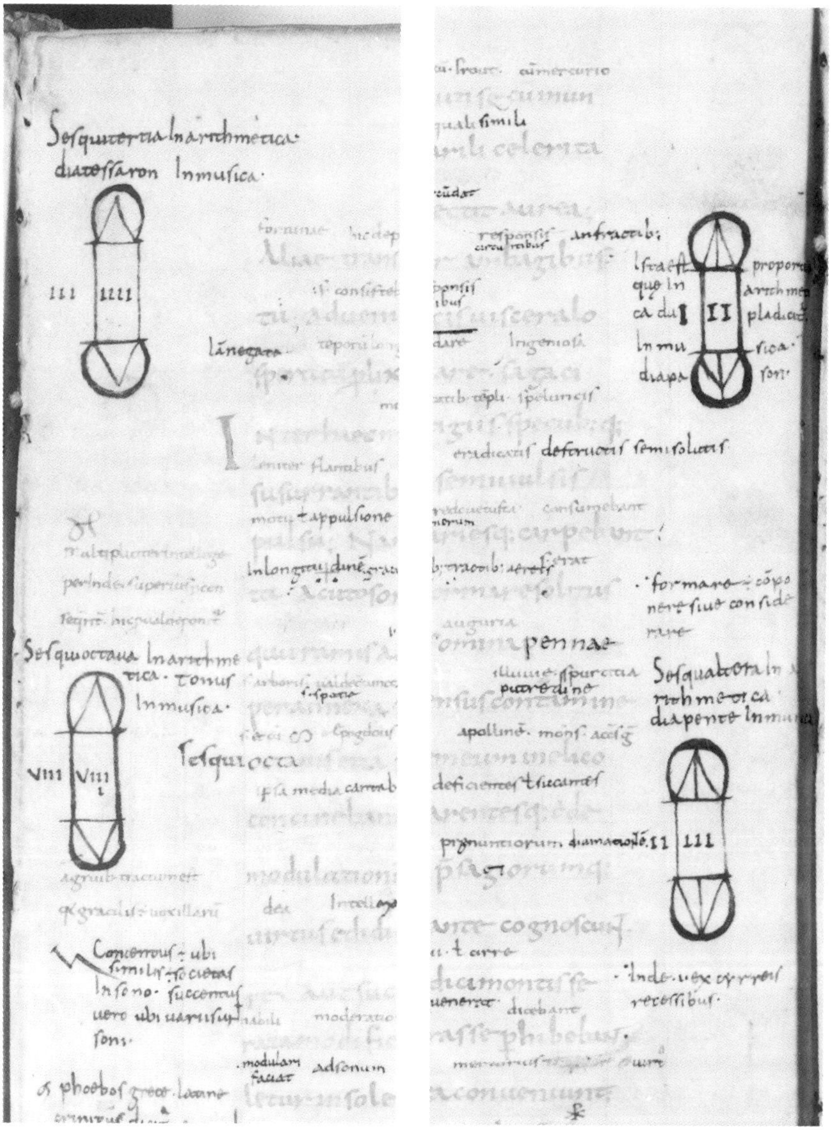

Plate 5 Similar representations as found in the Martianus Capella manuscript *Leiden, UB, BPL 88,* in the margins of fol. 3v-4r.

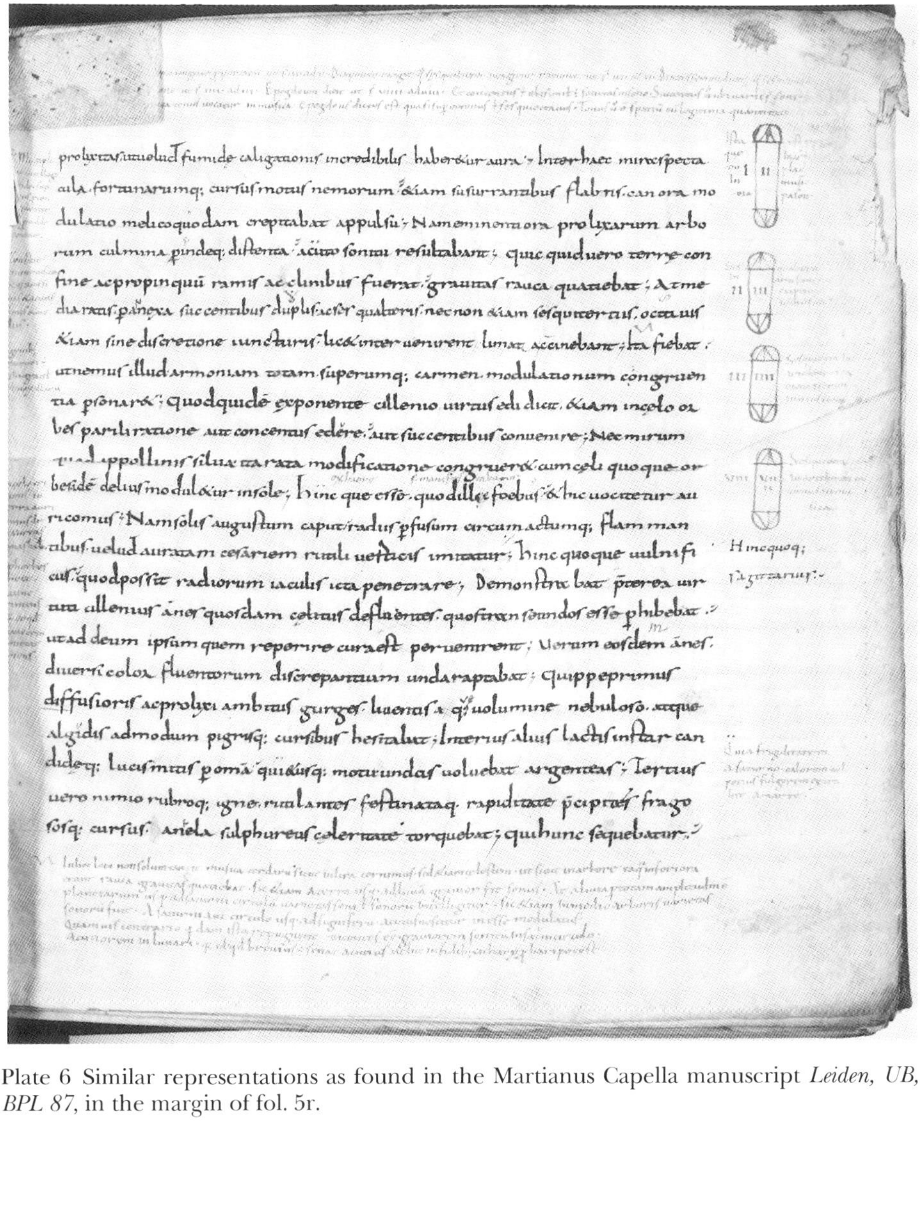

Plate 6 Similar representations as found in the Martianus Capella manuscript *Leiden, UB, BPL 87*, in the margin of fol. 5r.

Harmony of the spheres: philosophical controversies

1. *Introduction*

In ancient and medieval speculation on the order of the cosmos, the idea of a sounding harmony, produced by the heavenly spheres, is very prominent.[58] It is usually found side by side with a world view in which the earth stands still in the centre of the cosmos. The eight planets revolve around it, each in its own orbit, and because of their movements they bring forth tones. Their individual pitches are defined by the sizes of their orbits and their velocities. Astronomers complicated this view of the cosmos in order to find explanations for the phenomena they observed while watching the stars. They developed systems with elliptic and epicyclic orbits, orbits in orbits, and retrograde movements. The image of the sounding cosmos, however, remained a simple one: a speculative system that — once understood — could help mortals to grasp the divine order of creation.

According to Joscelyn Godwin, who wrote the classic *Harmonies on Heaven and Earth: The Spiritual Dimension of Music*

---

[58] The idea of cosmic harmony in this sense is treated by L. SPITZER in his elaborate article in two parts for *Traditio*: *Classical and Christian Ideas of World Harmony*. This article, which rather bears the character of a very lengthy lecture on harmony in general and celestial harmony in particular, ranges from Antiquity to Modern Time, from Archytas to Milton. Although it is quite difficult to use because of its associative or "stream of consciousness"-like character, it is a true *Fundgrube* for the intellectual history of the concept in Antiquity, Middle Ages and Renaissance. The rather old but still valuable study of H. ABERT, *Die Musikanschauung des Mittelalters*, likewise has much to offer on the subject (esp. Kap. 3C, *Die auf dieser Grundlage aufgebaute Symbolik der Theoretiker*, pp. 168-193). Many of the texts that are relevant for the subject occur in the "sourcebook" on Music and Esotericism compiled by J. GODWIN, *Music, Mysticism and Magic*, who also wrote the monograph: *Harmonies of Heaven and Earth*. A book which evidently appeals to a broader public is J. JAMES, *The Music of the Spheres*. The subject is also discussed in S. GERSH's recent study: *Concord in Discourse*, esp. 5.5 and 5.6, pp. 255-268. For the development of ideas on cosmic harmony in the later Middle Ages, see also the thorough and lucid study by E. VETTER, *Concentrische cirkels*, esp. the second chapter, "Muziek en sferen", pp. 61-103.

from Antiquity to the Avant-Garde, three types can be distinguished among the systems of planetary music.[59] In the first two types the music of the cosmos consists of a scale, ascending or descending step-wise from planet to planet, in the third type the musical distances between the planets are determined by numerical ratios and form intervals of fifths or octaves. As an example of the first type, one can take Plato's *Myth of Er* (*Republic* X.614B-619B): the orbits of the eight planets circle around the earth and on top of each orbit a Siren is depicted who brings forth a certain tone:

> *ἐπὶ δὲ τῶν κύκλων αὐτοῦ ἄνωθεν ἐφ'ἑκάστου βεβηκέναι Σειρῆνα συμπεριφερομένην, φωνὴν μίαν ἱεῖσαν, ἕνα τόνον: ἐκ πασῶν δὲ ὀκτὼ οὐσῶν μίαν ἁρμονίαν συμφωνεῖν.*
> (and up above on each of the rims of the circles a Siren stood, borne around in its revolution and uttering one sound, one pitch, and from all the eight there was the concord of a single harmony.)[60]

Traditionally it is Pythagoras who "invented" the distance between earth and moon: 126.000 *stadia*, which equals the interval of a major second or a full tone.[61] The rest of the scale of the planets proceeds in whole tones, semitones and one-and-a-half tones. According to Pliny, the planets thus form a chromatic Dorian scale, built out of two disjunct tetrachords plus an added tone as *proslambanomenos*. The entire distance measures a major ninth:

> Sed Pythagoras interdum ex musica ratione appellat tonum quantum absit a terra luna, ab ea ad Mercurium dimidium eius spatii, et ab eo ad Venerem tantundem, a qua ad solem sescuplum, a sole ad Martem tonum, id est quantum ad lunam a terra, ab eo ad Iovem dimidium,

---

[59] London 1986; pp. 124-127, 130-131, 138-143.

[60] PLATO, *De Republica* X, 617B. Transl. P. SHOREY, Loeb Classical Library, pp. 502-505.

[61] A *stadium* (Gr. *στάδιον*) is, in Astronomy, the distance of half a degree of the equator (cf. OLD, 2b). See, for example, PLINY, *Naturalis historia*, II, 19-20: "Intervalla quoque siderum a terra multi indagare temptarunt, et solem abesse a luna undeviginti partes quantam lunam ipsam a terra prodiderunt. Pythagoras vero, vir sagacis animi, a terra ad lunam cxxvi milia stadiorum esse collegit (...). Sed Pythagoras interdum ex musica ratione appellat tonum quantum absit a terra luna (...)."

et ab eo ad Saturnum dimidium, et inde sescuplum ad signiferum; (But occasionally Pythagoras draws on the theory of music, and designates the distance between the earth and the moon as a whole tone, that between the moon and Mercury a semitone, between Mercury and Venus the same, between her and the sun a tone and a half, between the sun and Mars a tone (the same as the distance between the earth and the moon), between Mars and Jupiter half a tone, between Jupiter and Saturn half a tone, between Saturn and the zodiac a tone and a half.)[62]

Pliny, after: J. GODWIN, *Harmonies of Heaven and Earth*, p. 126

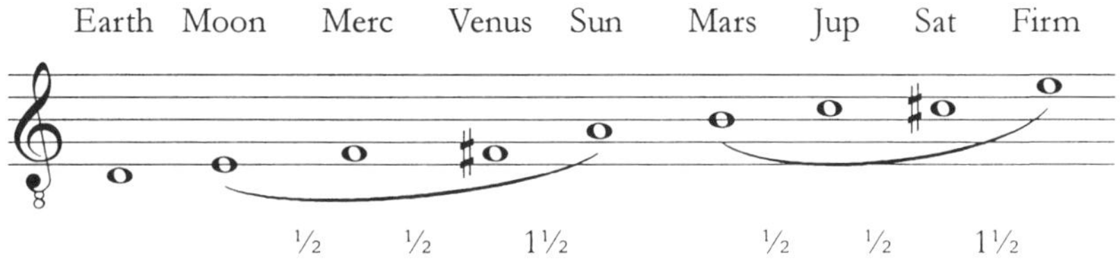

Censorinus and Theon of Smyrna lower the pitch of the Firmament to a d, in order to create the distance of one octave, the ratio 2:1, between earth and Firmament. Since the basic idea of this system of celestial harmony is that it fundamentally reflects the nine-stringed Greek lyre, it changes from age to age and from source to source, in concurrence with the changes in the musical system.

In the second type of musical interpretation of the universe, the pitches of the planets and their relations to each other are not determined by their distances from each other, but by their movements. The pitch of a certain planet is relative to its velocity and the size of its orbit around the earth. It is this system which is found, for example, in Cicero's *Dream of Scipio*:

Quam ob causam summus ille caeli stellifer cursus, cuius conversio est concitatior, acuto et excitato movetur sono, gravissimo autem hic lunaris atque infimus. Nam terra, nona, immobilis manens, una sede

[62] PLINY, *Naturalis historia*, II.20. Transl. H. RACKHAM, Loeb Classical Library, pp. 226-229.

> semper haeret complexa medium mundi locum. Illi autem octo cursus, in quibus eadem vis est duorum, septem efficiunt distinctos intervallos, qui numerus rerum omnium fere nodus est.
> (Consequently the outermost sphere, the star-bearer, with its swifter motion gives forth a higher-pitched sound, whereas the lunar sphere, the lowest, has the deepest sound. Of course the earth, the ninth and stationary sphere, always clings to the same position in the middle of the universe. The other eight spheres, two of which move at the same speed, produce seven different intervals, this number being, one might almost say, the key to the universe.)[63]

In interpretations of this type, the earth, standing still at the centre of the universe, is the only planet which does not bring forth a tone. Boethius suggested tones for Cicero's interpretation of the music of the spheres in *De institutione musica* I.27:

> Sed Marcus Tullius contrarium ordinem facit. Nam in sexto libro de re publica sic ait: Et natura fert, ut extrema ex altera parte graviter, ex altera autem acute sonent. Quam ob causam summus ille caeli stellifer cursus, cuius conversio est concitatior, acuto et excitato movetur sono, gravissimo autem hic lunaris atque infimus. Nam terra nona immobilis manens, una sede semper haeret. Hic igitur Tullius Terram quasi silentium ponit, scilicet immobilem. Post hanc qui proximus a silentio est, dat Lunae gravissimum sonum, ut sit Luna proslambanomenos, Mercurius hypate hypaton. Venus parhypate hypaton, Sol lichanos hypaton, Mars hypate meson, Juppiter parhypate meson, Saturnus lichanos meson, Caelum ultimum mese.
> (Marcus Tullius draws up a different order, for in the sixth book of *De re publica* he asserts: "Nature is so disposed that low sound emanates from its one extreme part, whereas high sound emanates from its other. Therefore that high celestial orbit, that of the stars, the revolution of which is faster, moves with a high and shrill sound, whereas the weak orbit of the moon moves with a very low sound. The earth, in ninth place, remaining immobile, is alone always fixed in place." Tullius thus regards the earth as silent — that is, immobile. Next after the earth he assigns the lowest sound to the moon, which is closest to silence, so that the moon is the proslambanomenos, Mercury the hypate hypaton, Venus parhypate hypaton, the sun the lichanos hypaton, Mars the hypate meson,

---

[63] CICERO, *Somnium Scipionis*, V,2. Transl. W.H. STAHL, *Commentary on the Dream of Scipio by Macrobius*, pp. 73-74.

Jupiter the parhypate meson, Saturn the lichanos meson, and the highest heaven the mese.)[64]

Boethius' interpretation of Cicero,
after: J. GODWIN, *Harmonies of Heaven and Earth*, p. 131

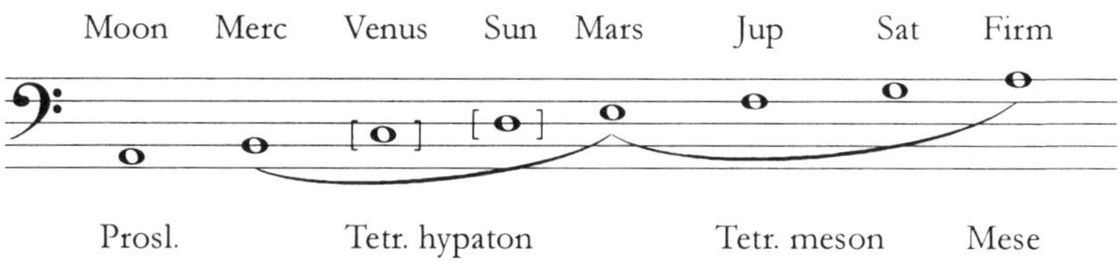

In the same chapter, however, Boethius first presents an interpretation probably taken from Nicomachus. In this interpretation the music of the planets also has the shape of a scale, this time two conjunct tetrachords from the moon to Saturn. But it does not ascend from moon to Saturn, but rather descends:[65]

> Illud tantum interim de superioribus tetrachordis addendum videtur, quod ab hypate meson usque ad neten quasi quoddam ordinis distinctionisque caelestis exemplar est. Namque hypate meson Saturno est adtributa, parhypate vero Ioviali circulo consimilis est. Lichanon meson Marti tradidere. Sol mesen obtinuit. Triten synemmenon Venus habet, paraneten synemmenon Mercurius regit. Nete autem lunaris circuli tenet exemplum.
> (At this point it would seem proper to add concerning the above tetrachords that the disposition from the hypate meson to the nete synemmenon is, as it were, a kind of exemplar of the celestial order and specification. The hypate meson is assigned to Saturn, whereas the parhypate is like the orbit of Jupiter. The lichanos meson is entrusted to Mars. The sun governs the mese. Venus holds the trite synemmenon. Mercury rules the paranete synemmenon. The nete is analogous to the orbit of the moon.)[66]

---

[64] Transl. C.M. BOWER, *Fundamentals of music*, pp. 46-47.

[65] Cf. C.M. BOWER, *Fundamentals of Music*, p. 46 n. 127: "The comparison of the strings to the disposition of the heavenly spheres agrees with that found in Nicomachus *Excerpta* 3 [...] and *Enchiridion* 3." See for Boethius' interpretation of Nicomachus on this point: R. BRAGARD, *L'harmonie des sphères selon Boèce*.

[66] BOETHIUS, *De institutione musica*, I.27; transl. C.M. BOWER, p. 46.

Boethius' interpretation of Nicomachus,
after: J. GODWIN, *Harmonies of Heaven and Earth*, p. 131

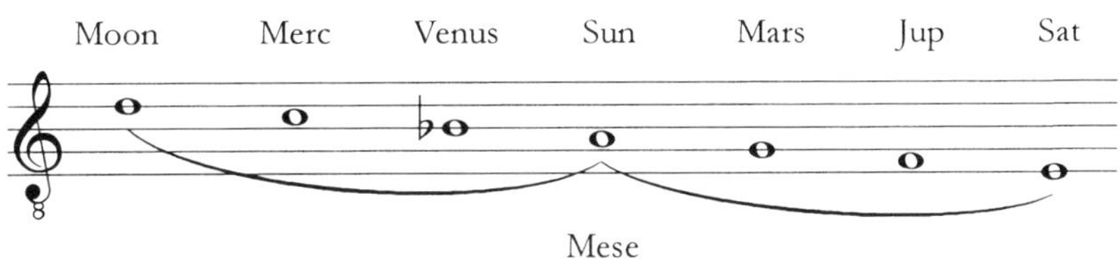

Martianus Capella, whose system is to be categorized in the A-type category, presents his scale-like interpretation of cosmic music in the second book of *De nuptiis* (II.169-199).[67] Philology embarks upon a heavenly journey; for this is the way she, who is mortal and who is to be wedded to the god Mercury, can reach the status of immortality. She is given an egg-shaped cup of immortality potion by Athanasia,[68] regurgitates all her earthly knowledge in the form of books and scrolls flowing from her mouth and is then thin and light enough to begin her celestial journey. For a part of it, she is carried aloft in a palanquin; Labor, Amor, Epimelia and Agrypnia (Work, Love, Attention and Wakefulness) are her four bearers. Partly, however, she herself struggles over the long and difficult path. She climbs from planet to planet and the distance between the planets is measured in tones. The picture of a scale arises: the distance between earth and moon is a full tone, that between moon and Mercury a semitone, between Mercury and Venus a semitone, between Venus and the sun one and a half tones, between the sun and Mars a semitone, between Mars and Jupiter a semitone, between

---

[67] The theme of the music of the spheres is not treated by Harmonia in the ninth book of *De nuptiis*. On book IX, there is one Carolingian gloss, which, in fact, states that the subject of heavenly music is "too profound" to be treated by her: 353.21 IPSA RAPIDITAS ET CONCINAT ET AGNOSCAT Nullus enim rationem debet dare de armonia caelesti, quia nimiae profunditatis est, quia ibi non est vox humana nisi tantummodo sonus *BeLa*. The commentator thus says that although Harmonia brings harmony to the spheres, she is ignorant of the how and why of its functioning, because it has too much profundity — no human sound or human voices are found there, just "abstract sound".

[68] On the myth of the cosmic egg see P. DRONKE, *Fabula*, pp. 79-99.

Jupiter and Saturn a semitone, and, finally, between Saturn and the Firmament one and a half tones. Philology covers the distance of six and a half tones, exceeding the distance of one octave with a semitone:[69]

Martianus Capella

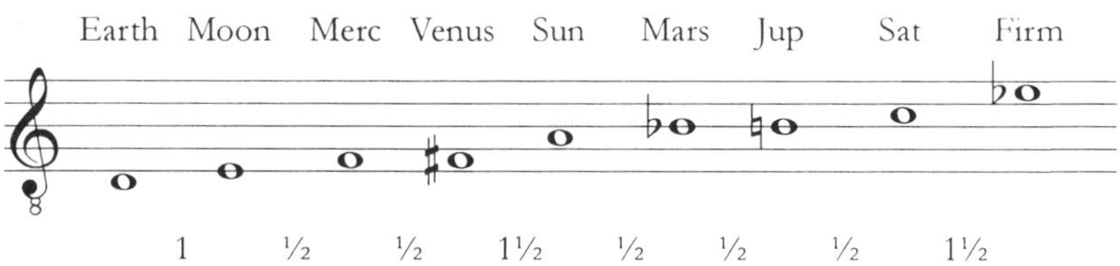

At the end of her climb, however, Martianus says that she, exhausted by her climb over six full tones, gains new strength from the perfect harmony of the octave:

> [199] sicque sex tonorum conscensionibus <et> stadiorum defecta lassitudine fatigati, cum diapason symphoniam quicquid emensi erant adverterent consonare perfectione absolutae modulationis, post labores maximos recreati paululum conquierunt.
> (Thus they were worn out with complete exhaustion after so many stades — an ascent of six tones. When they noticed that all they had traversed gave forth a concord of the whole octave in full and perfect harmony, they were refreshed, and rested a little after their mighty efforts.)[70]

It seems clear that Martianus actually meant to make her travel the distance of one octave, but by mistake made her travel a

[69] Note that the scale Godwin presents in *Harmonies of Heaven and Earth*, p. 126, is not correct. Godwin suggests that Martianus' scale is the same as that of Pliny. In fact, the system Martianus presents could very well be a faulty reflection of Pliny's scale. The fourth between moon and sun, and that between Mars and the firmament could be interpreted as two disjunct tetrachords in the chromatic genus (their distances being ½ - ½ - 1½). In this interpretation, however, the distance between sun and Mars is a problem: for the distance between the two disjunct tetrachords (the tetrachords meson and diezeugmenon, or, in other words, the distance between *mese* and *paramese*) is never a semitone, but always a full tone.

[70] Transl. W.H. STAHL, *Martianus Capella*, II, p. 60.

semitone too far.[71] The matter is further complicated by a corruption in the text, which makes the proper interpretation of the passage difficult. On the distance between Mars and Jupiter, the text, in its edition, tells us that Jupiter resounds in the Phrygian mode, *pthongo Phrygio*, and that the distance between Mars and Jupiter measures a semitone.[72] In the manuscripts, however, the words *pthongo Phrygio* are never found; instead, they are contracted to all kinds of combinations: *ptongio, phtongio, tpongio, tphongio* and even *emipthongio*.[73] To complicate matters further, Martianus does not mention the next distance, between Jupiter and Saturn, explicitly, but refers to the previous distance and says that they have the same value (*parili interiectione*).

Before turning to the medieval reception of these problems, however, there is a third type of interpretation of the music of the universe which needs introduction. In this type (C in Godwin's typology) the universe does not resemble a scale, but consists of larger intervals, which, like the "fixed" notes of the Greek Perfect System, form the skeleton of scales. In the Perfect System the bounding notes of the five tetrachords are fixed and make intervals of fourths, but the two inner notes of each tetrachord vary according to *genus* (diatonic, chromatic, enharmonic). The music of the planets is, according to some of the ancient interpreters of cosmic harmony, the principle on

---

[71] In fact, if the distance between sun and Mars is actually to be interpreted as a full tone, the distance between moon and firmament would add up to one octave. Because the earth stands still and brings forth no sound in many of these interpretations of cosmic music, the distance between earth and moon could be discarded.

[72] J. WILLIS, *Martianus Capella*, p. 53: quo transgresso (neque enim labor fuerat hemitonii interiecta transcurrere) in Iovialis sideris pervenere fulgores, cuius circulus pthongo <Phrygio> personabat. This emendation was already suggested by Kopp in his edition (1836), and accepted by all later editors. It is based on PLINY, *Nat.hist.*, 2,84: Saturnum Dorio moveri pthongo, Iovem Phrygio.

[73] See J. WILLIS, *apparatus* ad l. 16, p. 53. See also L. LENAZ, *Martiani Capellae De nuptiis Philologiae et Mercurii Liber secundus*, p. 222-223.

which the Perfect System was modelled.[74] It is assumed that the source for this kind of interpretation is found with the Pythagoreans or Neo-Pythagoreans. In Plato's *Timaeus* the basis on which the World Soul is created is formed by a series of numbers, the inner proportion of which can also be interpreted musically:

> *ἤρχετο δὲ διαιρεῖν ὧδε· μίαν ἀφεῖλε τὸ πρῶτον ἀπὸ παντὸς μοῖραν, μετὰ δὲ ταύτην ἀφῄρει διπλασίαν ταύτης, τὴν δ'αὖ τρίτην ἡμιολίαν μὲν τῆς δευτέρας, τριπλασίαν δὲ τῆς πρώτης, τετάρτην δὲ τῆς δευτέρας διπλῆν, πέμπτην δὲ τριπλῆν τῆς τρίτης, τὴν δ'ἕκτην τῆς πρώτης ὀκταπλασίαν, ἑβδόμην δὲ ἑπτακαιεικοσαπλασίαν τῆς πρώτης.*
>
> (And He (*i.e. the Creator*) began making the division thus:
> First He took one portion from the whole [1];
> then He took a portion double of this [2];
> then a third portion, half as much again as the second portion, that is, three times as much as the first [3];
> the fourth portion he took was twice as much as the second [4];
> the fifth three times as much as the third [9];
> the sixth eight times as much as the first [8];
> the seventh twenty-seven times as much as the first [27].)[75]

These series of numbers were interpreted as corresponding with the relations between the planets and were given a musical interpretation. There is much deviation in the Type C scales —

[74] L. SPITZER made the astute observation: "In spite of the fact that the simile: world harmony—musical harmony was derived (historically speaking) from a human instrument, Pythagoreans inverted the order by admitting that the human lute (as imagined in the hands of the god Apollo) was an imitation of the music of the stars; human activities had to be patterned on godly activities, i.e. on the processes in nature: human art, especially, had to be an imitation of the gods, i.e. of reasonable Nature. Thus we will witness a continuous flow of metaphors from the human (and divine) sphere to Nature and back again to human activities which are considered as imitating the artistic orderliness and harmony of Nature." (*Classical and Christian Ideas of World Harmony* I, p. 414.) The same can be said, in this case, of the theoretical construct of the Perfect System: although it is clearly designed by humans, its origin is placed in a divinely inspired model: the music of the planets.

[75] PLATO, *Timaeus*, 35B-C. Transl. R.G. BURY, Loeb Classical Library, pp. 66-67.

Godwin shows interpretations by, among others, Plutarch, Ptolemy, and Plato —, but the overriding symbolic intention is that the music of the heavens should not correspond to a single octave species, but rather to the entire range of possible tones in music: "these", Godwin says, "are the bare bones of music, to be clothed in multifarious ways."[76]

In Calcidius' translation and interpretation of Plato's *Timaeus*, the proportional series results in the distance of an octave between moon and sun, a fifth between sun and Venus, a fourth between Venus and Mercury, an octave between Mercury and Mars, a tone between Mars and Jupiter, and, finally, an octave + fifth between Jupiter and Saturn.[77] The total distance from moon to Saturn measures 4 octaves, a fifth and a tone.[78] In the interpretation of Macrobius, on the other hand, it actually measures 14 octaves, 2 fifths and 2 major seconds.[79] From his translation of the relevant passage in the *Timaeus*, it becomes clear how Macrobius arrived at this much bigger *ambitus*. Just as in the *Timaeus* the calculation of the intervals between the planets is based on the proportional series 1:2:3:4:9:8:27. He, however, takes the distance between the earth and the previous planet as the point of reference, instead of the distance between the two previous planets, and thus the musical intervals determined by the proportional numbers are accumulated instead of succeeding each other:

---

[76] J. GODWIN, *Harmonies of Heaven and Earth*, pp. 139-140.

[77] Note that the order of the planets in the texts of Plato, Calcidius and Macrobius is different from that of, for example, Boethius, Cicero, Pliny and Martianus Capella. Instead of using the so called "Chaldean" order of the planets, with the sun in the centre of the series of planets, they use the "Egyptian" order, with the sun next to the moon.

[78] Cf. figure 2 below.

[79] See on this J. FLAMANT, *Macrobe et le Néo-Platonisme latin*, pp. 351-381, and esp. 396-371. In his commentary on *Scipio's Dream*, Macrobius tries to arrive at some kind of synthesis between Plato's *Timaeus* (in Calcidius' translation) and Cicero's *Somnium*. The planetary system used by Cicero, however, is different from that of Plato. Macrobius decided to use the system of Plato/Calcidius for his interpretation of heavenly music. See also J. FLAMANT, "Annexe 8", for a comparison between relevant passages in the texts of Plato, Calcidius and Macrobius, p. 706.

et statuerunt hoc esse credendum, ut quantum est a terra usque ad lunam, duplum sit a terra usque ad solem, quantumque est a terra usque ad solem, triplum sit a terra usque ad Venerem, quantumque est a terra usque ad Venerem, quater tantum sit a terra usque ad Mercurii stellam, quantumque est ad Mercurium a terra, novies tantum sit a terra usque ad Martem, et quantum a terra usque ad Martem est octies tantum sit a terra usque ad Jovem, quantumque est a terra usque ad Jovem, septies et vicies tantum sit a terra usque ad Saturni orbem.
(And they decided that there could be but one opinion, that the distance from the earth to the sun was twice as great as from the earth to the moon, that the distance from the earth to Venus was thrice as great as from the earth to the sun, that the distance from the earth to Mercury was four times as great as from the earth to Venus, that the distance from the earth to Mars was nine times as great as the distance from the earth to Mercury, that the distance from the earth to Jupiter was eight times as great as from the earth to Mars, and that the distance from the earth to Saturn was twenty-seven times as great as from the earth to Jupiter.)[80]

Calcidius:		Macrobius:
	Moon = 1	
1 octave		octave
	Sun = 2	
1 fifth		octave + fifth
	Venus = 3	
1 fourth		double octave
	Mercury = 4	
1 octave		3 octaves + tone
	Mars = 8	
1 tone		3 octaves
	Jupiter = 9	
1 octave + fifth		4 octaves + fifth + tone
	Saturn = 27	

Calcidius and Macrobius, after: J. FLAMANT, *Macrobe et le Néoplatonisme*, pp. 372 and 371.

---

[80] MACROBIUS, *In Somnium Scipionis*, II,iii,14; transl. W.H. STAHL, *Commentary on the Dream of Scipio by Macrobius*, p. 196.

	Boethius I	Boethius II	Pliny	Martianus
Earth-Moon	-	-	1	1
Moon-Merc.	1	1	½	½
Merc.-Venus	½	1	½	½
Venus-Sun	1	½	1½	1½
Sun-Mars	1	1	1	½
Mars-Jup.	½	1	½	½
Jup.-Sat.	1	½	½	½
Sat.-Firm.	1	-	1½	1½
Total	6	5	7	6½
Direction	up	up	Down	up

Figure 1, Planetary scales in different sources
Boethius I = Boethius, *De institutione musica* I.27, interpretation of Cicero
Boethius II = Boethius, *De institutione musica* I.27, interpretation of Nicomachos
Pliny = Pliny, *Natural history* II.20
Martianus = Martianus Capella, *De nuptiis*, book II, § 169-199

	Calcidius	Macrobius
Moon-Sun	octave	Octave
Sun-Venus	fifth	octave + fifth
Venus-Mercury	fourth	double octave
Mercury-Mars	octave	3 octaves + tone
Mars-Jupiter	tone	3 octaves
Jupiter-Saturn	octave + fifth	4 octaves + fifth + tone
Total	4 octaves + fifth + tone	15 octaves + 3 tones
Direction	up	Up

Figure 2, Planetary music in Calcidius and Macrobius
Calcidius, *Timaeus a Calcidio translatus commentarioque instructus* (ed. Waszink), c. 96
Macrobius, *Commentarius in Somnium Scipionis* (ed. Willis), II.3.14

	Martianus	Anonymous	John the Scot	Remigius
Earth-Moon	1	1	1	1
Moon-Merc.	½	½	½	½
Merc.-Venus	½	½	½	½
Venus-Sun	1½	1½	1½	1½
Sun-Mars	½	½	½	½
Mars-Jup	½	1	½	1
Jup.-Sat.	½	1	1	½
Sat.-Firm	1½	1½	1½	1½
Total	6½	7½	7	7

Figure 3, Planetary scales in Martianus Capella commentaries
Martianus Capella, *De nuptiis*, book II, § 169-199
Anonymous commentary, *ad locum*
John the Scot Eriugene, *ad locum*
Remigius of Auxerre, *ad locum*

These, then, are the main antique sources from which medieval scholars could learn about the music of the spheres: Plato's *Timaeus* in the translation by Calcidius, Cicero's *Dream of Scipio*, translated and commented upon by Macrobius, Pliny's *Natural History*, Boethius' *De institutione musica* and Martianus Capella's *De nuptiis*. Their planetary scales and musical systems are represented in combination in Figures 1 and 2 above.

2. *The music of the spheres in the Anonymous commentary*

In the Anonymous commentary the theme of the Harmony of the Spheres receives remarkably lengthy glosses; the chorus of the planets sounded in the inner ears of the Carolingian scholars with much resonance.

For the (musical) distances between the planets and the related problems in *De nuptiis*, the commentator found the following solution. The distance between Mars and Jupiter is interpreted as a whole tone, instead of a semitone, and the distance between Jupiter and Saturn is likewise measured as a whole tone.[81] In sum, the commentator thus makes Philology travel a distance of seven and a half tones instead of Martianus' six and a half; two semitone distances are interpreted as whole tone distances.[82] Subsequently, the section on the total distance of one octave for the entire cosmos is clarified as follows:

> [54.15-6] SICQUE SEX TONORUM [1]A terra secundum istum usque Saturnum sex toni sunt, i. a terra usque ad lunam tonus, a luna usque Mercurium hemitonium, ab illo usque Venerem similiter, a Venere usque ad solem tonum et semissem, ab isto Marte tenus hemitonium, a Marte usque Iovem tonum et hinc usque ad Saturnum tonum. [2]Fiunt simul sex toni. [3]Alii doctores aliter dicunt. *BeLa*
> (SIX TONES [1]From the earth to Saturn there are, according to him (i.e. Martianus Capella), six tones, that is from the earth to the moon one tone, from the moon to Mercury a semitone, from him to Venus the same, from Venus to the sun one and a half tones, from him to Mars a semitone, from Mars to Jupiter one tone, and from

[81] See also the glosses ad 50.3-4; 51.7-8; 51.7-12; 53.14-16; 54.5-7 and 54.13 on the individual distances.

[82] See figure 3 above.

there to Saturn one tone; in sum six tones. [2]Other scholars say otherwise.)

The commentator simply skips the distance between Saturn and the Firmament — a distance of one and a half tones — and only counts the remaining distance of 6 tones between earth and Saturn to reach the correct distance of the octave. This is consistent with some of the ancient sources: Calcidius and Macrobius also assign the last interval in the universe to the distance between Jupiter and Saturn, and so does Boethius when he presents Nicomachean material. The commentator also notes, however, that other scholars have a different opinion.

The late antique sources the commentator could have consulted indeed have different solutions as to how the cosmos actually sounds. Furthermore, other interpretations of the planetary scale are found in the other ninth-century commentaries: their interpretations are combined in Figure 3 above. From this figure we can read that John the Scot counts a total of 7 tones. He does not even try to explain the difference between this sum and the "perfect harmony of the octave" reached at the end of Philology's journey. For his views on celestial music are not expressed in comments upon Martianus' second book, but upon a passage in the first book; they will be treated extensively below.[83] Remigius, who borrowed heavily from both the Anonymous glosses and John the Scot's *Annotationes* for his own commentary on *De nuptiis*, also counts a total of 7 tones. He then argues that the distance of one tone between earth and moon should be discarded, because the earth stands still and is therefore silent.[84] In this way he indeed reaches the harmony of one octave, this time between the moon

[83] See the next section: "Music of the Spheres in the commentary of John the Scot."

[84] C.E. LUTZ, *Remigius Autissiodorensis*, I, 70.8 (p. 191): "quia cum omnia elementa mundi in motu sint et musicam de se efficiant, sola terra, quae in statu est, nullum de se sonum emittit, utitur tamen proximo sono lunari." This is consistent with, for example, both interpretations Boethius presents in *De institutione musica* I.27. It is also consistent with Macrobius and Calcidius, whose first distance is that between the moon and the sun. Cf. figures 1 and 2 in this section.

and the Firmament.

In a gloss added to a different section in book II, the Anonymous commentator envisions a scale divided into two conjunct tetrachords, the sun forming the tone which both ends the first and begins the second tetrachord. This becomes clear in a gloss ad § 187, in which Philology prays to Apollo the sun, and says:

> nonne a principio geminum tu das tetracordon?
> (Do you not bestow upon the two twin tetrachords a beginning?)[85]

The commentator explains:

> [52.13] [1]Inter caelum et terram gemina sunt tetracorda i. duo. [2]Septenarius enim planetarum numerus duo tetracorda efficit. [3]Prima corda lunaris circulus, secunda Mercurialis, tertia Veneria, quarta solaris, quae etiam finem prioris tetracordi et initium sequentis obtinet, sicut mese in musica. [4]Est igitur secundum tetracordum, cuius quasi prima corda solaris circulus, secunda Martis, tertia Iovis, quarta Saturni. *BeLaLbVaLeaOxf*, 2 Septenarius] septinarius *Va*; 3 Veneria] venerialis *Oxf* prioris] priuris *BeLb* primi *Oxf* obtinet] facit *Oxf*; 4 corda] pars *Oxf*[1] corda *Oxf*[2] tertia] tercia *LaVa*
> (Between heaven (*that is, Firmament; M.T.*) and earth there is a double tetrachord, that is two tetrachords. For the sevenfold number of the planets realizes two tetrachords. The first string is that of the circle of the moon, the second of Mercury, the third of Venus, and the fourth of the sun, which holds the end of the first tetrachord and the beginning of the next, just as the *mese* in music. Then there is a second tetrachord, of which the first string is the circle of the sun, the second of Mars, the third of Jupiter and the fourth of Saturn.)

This interpretation of the planetary scale as two conjunct tetrachords is in fact in contradiction with the image that arises from Philology's journey through the planetary spheres. It does concur with Boethius' interpretation of the musical scale of the planets, when he presents Nicomachean material (cf. Figure 1 above), but according to Nicomachus, the scale from moon to Saturn is an ascending one. The commentator does not explain

[85] Transl. W.H. STAHL, *Martianus Capella*, II, p. 58.

the discrepancy. The central position of the sun remains unrelated to the position of the sun in Philology's heavenly journey.

Ideas on the harmony of the spheres are, however, not only expressed in the glosses on Philology's journey. Throughout book I we find brief excursions on the theme, especially in sections 11-12, where Martianus speaks of the secret and holy Apollinian Grove. Its trees are depicted as resounding in perfect harmony: the vibrating trunks and branches sing in harmonies of the duple ratio (octaves), the hemiolic ratio (fifths), and the sesquitertian ratio (fourths). They resound harmoniously even in consonances of tones and semitones. The whole passage is told in a particularly ornate and expressive language, and appealed very much to the imagination of Carolingian students of the text:

> [11-12] Tandem Fama nuntiante cognoscunt quod Phoebo gaudet Parnasia rupes. licet inde quoque ad Indici montis secretum obumbratumque scopulum nube perpetua posterius migrasse perhibebant, tamen Cirrhaeos tunc recessus et sacrati specus loquacia antra conveniunt. illic autem circumstabat in ordinem quicquid imminet saeculorum, Fortunae urbium nationumque, omnium regum ac totius populi. videbantur aliae transacti cursus emensa fugientes; consistebant aliae sub conspectu, adveniebantque quamplures, atque ita nonnullis eminus vanescebat disparata prolixitas, ut velut fumidae caligationis incredibilis haberetur aura. inter haec mira spectacula Fortunarumque cursus [motus] nemorum etiam susurrantibus flabris canora modulatio melico quodam crepitabat appulsu. nam eminentiora prolixarum arborum culmina perindeque distenta acuto sonitu resultabant; quicquid vero terrae confine ac propinquum ramis acclinibus fuerat, gravitas rauca quatiebat. at media ratis per annexa succentibus duplis ac sesqualteris nec non etiam sesquitertiis, <sesqui>octavis etiam sine discretione iuncturis, licet intervenirent limmata, concinebant. ita fiebat, ut nemus illud harmoniam totam superumque carmen modulationum congruentia personaret. quod quidem exponente Cyllenio Virtus edidicit etiam in caelo orbes parili ratione aut concentus edere aut succentibus convenire. nec mirum Apollinis silva ita rata modificatione congrueret, cum caeli quoque orbes idem Delius moduletur in Sole.
>
> (Thanks to Rumour's report, they eventually learn that the rock of Parnassus is graced by the presence of Phoebus. Although one also

heard that he had since departed for the peak of an Indian mountain hidden and shaded by perpetual cloud, they nevertheless repair to his Cirrhaean abode and the talkative recesses of his sacred cave. All around there stood in due order of events which time will bring to pass: the fortunes of cities and nations, of all kings and every people. Some of these appeared to flee on completing their measured course; others stood in full view; many were approaching. Some vanished afar, with so great a distance between them that a mysterious gust of dark vapour seemed to seize them. Amid these wondrous sights and the vicissitudes of Fortune, a sweet harmony arose from the movement of the trees as the breezes whispered through them with a certain melodious effect. For the crests of the largest trees, being more elevated and therefore extended, resonated with the highest pitch. The roughness of the lowest pitch shook whatever was close or near to the ground in drooping boughs. But the crests of the median trees in the fixed intervals of their mutual accompaniments sounded duple (2:1), sesquialter (3:2), sesquiterce (4:3), and even indivisible sesquioctave (9:8) intervals not without intervening limmata. So it happened that this grove resounded with the whole of harmony and the divine song in the consonance of its modulation. As the Cyllenian explained these things, Virtue also learned that the heavenly spheres emit melodies or contribute to their accompaniments in a similar manner. And it is not surprising that Apollo's grove is concordant in such a fixed harmony, since the same Delian god in the shape of the sun also modulates the heavenly spheres.)[86]

In all ninth-century commentaries this allegorical allusion to the music of the cosmos, governed by Apollo, the Sun, is further explained. Several long glosses are attached to this passage in the Anonymous commentary:

[6.17-9] ILLIC AUTEM CIRCUMSTABAT IN ORDINEM QUICQUID IMMINET SAECULORUM, FORTUNAE URBIUM NATIONUMQUE, OMNIUM REGUM AC TOTIUS POPULI
(All around there stood in due order the events which time will bring to pass: the fortunes of cities and nations, of all kings and every people.)

[1]Quod ait in specu Apollinis omnium esse Fortunas quicquid

[86] Transl. S. GERSH, *Concord in Discourse*, pp. 1-2.

imminet seculorum sub specie fabulae figuratae loquens. [2]Hoc videtur intimasse, ut omnis seu eventus prosper sive sinister, quae ab illis Fortuna dicebatur, Apollinem de exactis, de instantibus et de futuris consultum sibi credebant provenire. [3]Nec non etiam quod nemus Apollinis succentibus duplis, sesqualteris ac sesquitertiis reliquisque acutas ac graves dicitur personasse modulationes. [4]Hoc indicare videtur, quod circulus solaris circulis trium planetarum superpositarum et totidem suppositarum comparatur; musica videtur ratione provenire. [5]Quia sicut mese in lira medio posita tetracordum facit, sic sol in medio planetarum. *BeLaLbVaLea*[87]
([1]And it (i.e. the text) says that the Fortunes of all times, whatever is about to happen, are in the cave of Apollo, speaking in the manner of a fictitious tale. [2] This seems to have been meant: that every event is either favourable or unfavourable, which is called Fortuna by them: they believed that Apollo, consulted about the past, the present and the future, emerged for them (from his cave). [3]The Apollinian grove is also said to sing loudly high and low melodies with octaves (ratio 2:1), fifths (ratio 3:2) and fourths (ratio 4:3) and other intervals. [4]This seems to indicate that the orbit of the sun is compared to the orbits of the three lower and just as many higher positioned planets; music seems to originate from reason. [5]For just as the *mese*, positioned in the middle on a lyre, makes the tetrachord, so does the sun in the middle of the planets.)

[7.1-2] QUICQUID VERO TERRAE CONFINE AC PROPINQUUM RAMIS ACCLINIBUS FUERAT, GRAVITAS RAUCA QUATIEBAT
(The roughness of the lowest pitch shook whatever was close or near to the ground in drooping boughs.)

[1]In hoc loco non solum tangit musicam cordarum, sed etiam celestem, ut sicut in arbore ea quae inferiora erant rauca gravitas quatiebat, sic etiam a terra usque ad lunam gravior sit sonus. [2]At a luna per totam amplitudinem planetarum usque ad Saturni circulum varietas sonorum intellegitur. [3]Sic etiam in medio arboris varietas sonorum fuit. [4]A Saturni autem circulo usque ad signiferum acutus noscitur inesse modulatus. [5]Quamvis contrario quidam ista repugnent, dicentes graviorem esse sonitum in Saturni circulo, acutiorem in lunari, quia quod brevius est, acutius sonat.

[87] For the *apparatus* see the edition.

BeLaLbVaLea[88]
[1]Nec non etiam quod nemus Apollinis succentibus duplis sive sesqualteris ac sesquitertiis reliquisque acutas ac graves dicitur personasse modulationes. [2]Hoc indicare videtur quod circulus solaris circulis planetarum trium super se positarum ac totidem suppositarum comparatur; musica videtur ratio<ne> provenire. [3]Sicut enim corda, quae in medio est, principalis est, ita sol medium tenet planetarum. *BeLa*[89]
([1]At this point he touches not only upon the music of strings, but also upon celestial music, for just as in that tree a hoarse lowness shook (the branches) that were low, so is the lowest pitch from the earth to the moon. [2]But from the moon through the entire extent of the planets up to Saturn a variety of pitches is understood. [3]So too a variety of pitches was in the middle range of the tree. [4]From the orbit of Saturn to the star-bearing (Firmament) a high pitch is observed to be brought forth. [5]However, some people argue against this, saying that the lowest pitch is in the orbit of Saturn, the highest in the moon, because that which is shorter sounds higher.
[1]The Apollinian grove is also said to sing loudly high and low melodies with octaves (ratio 2:1), fifths (ratio 3:2) and fourths (ratio 4:3) and other (intervals/ratios). [2]This seems to mean that the orbit of the sun is compared to the orbits of the three lower and three higher positioned planets; music seems to originate from reason.
[3]For as the string which is in the middle (i.e., on a lyre) is the first (the most important), so the sun holds the central position among the planets.)

Concerning these two glosses, which partly overlap, two striking observations can be made. First of all, comments are made on the central and guiding position of Apollo, the sun. Secondly, there is a controversy as to which of the planets has the highest, and which the lowest pitch of the universe. In other words: is the scale of the planets from the earth to Saturn or the Firmament an ascending or a descending one?

In *De nuptiis*, and consequently also in the glosses, there are several passages which allude to the sun as the leader of the

[88] For the *apparatus* see the edition.
[89] For the *apparatus* see the edition.

universal harmony.[90] In a gloss added to section 19, for example, where a lyre-playing Apollo is evoked, we read:

> [9.24] FIDIBUS PERSONANTI [1]Sol cum armonia movetur, et in hoc loco informitatem vocum designat. [2]Fidis enim tensa intellegibilem vocem profert, ita ut aliquod nomen usurpet sibi. [3]Laxa enim melodiam cum informitate sonus emittit. [4]Informitates enim vocum ponit, quia dum sit semper stabilis et nullam vocem proferre videatur, tamen nulla vox fieri potest, nisi ex informitate illius formetur. [5]Verbi causa: cum prima corda figitur in cithara, laxissimum atque gravissimum sonum sine ulla pulchritudine emittit. [6]Si tamen extenta fuerit, aliquam formam assumit, quae in numero troporum hypodorius vel sub dorio positus vocatur. [7]Sol enim quamvis magnum faciat sonum, nos tamen expertes illius modulationis sumus. *BeLaLbVaLea*[91]
> (PLAYING ON STRINGS [1]The sun is moved with harmony, and indicates at this point the "formlessness of sounds". [2]For a tense string brings forth a perceptible sound, so that it assumes a certain name for itself. [3]Loosened, however, it utters a melody with formlessness of sound. [4]He puts "the formlessness of sounds", because while it is always steady and seems to bring forth no sound, yet no sound can be created unless it is formed out of its formlessness. [5]For example: when the first string is tied on the cithara, it produces a very loose and very low sound, without any beauty. [6]If this string, however, has been stretched, it assumes a certain form, which is called "hypodorius" in the number of modes, or "positioned under the doric mode." [7]For even though the sun produces a big sound, we remain unaware of its melody.)

This idea of the sun bringing forth a massive, formless sound, which we are unable to perceive because of its formlessness, is an idea which, as far as I can find, does not appear in any other source, at least not any earlier one.[92] It appears, however, for a

[90] See also the gloss ad 7.8-9 NEC MIRUM Non mirum si in terris talia agit, qui in caelo etiam armoniam regit. *BeLa* (It is not strange that he (i.e. Apollo) who rules the harmony even in the heavens, does these things on earth.).

[91] For the *apparatus* see the edition.

[92] The verb "formare" and the adverb "informiter" are, however, used for sound by Augustine in one of his confessions: AUGUSTINUS, *Confessiones* XII,29 (40), (CCSL 27, p. 239): "At in cantu non ita est. Cum enim cantatur, auditur sonus eius, non prius informiter sonat et deinde formatur in cantum. Quod

second time in a long gloss attached to section 27, in which the entire universe joins in a happy song and the Muses each look for their own planetary sphere with matching musical tone. The gloss explains the meaning of the different names of the Sun, and the allegorical meaning of the Muses. Sentences 10-12 express the idea of the formless sound of the universe:

> [12.20] ET APOLLINE CONSPICATO AERIA TEMPERIES SUDIS TRACTIBUS RENIDEBAT
> (the region of Air shone without clouds in the sight of Apollo)[93]

> [1]Sol diversis utitur nominibus et pro variis significationibus accipitur. [2]Vocatur sol, eo quod prae ceteris elementis plurimum solus luceat vel quod cuncta illuminet. [3]Dicitur Delius, declaratio. [4]Dicitur Apollo id est exterminans, eo quod suo calore omnes humores arborum pellere videatur. [5]Dicitur Phithius id est fidem auferens. [6]Pro diversis etiam ponitur significationibus: nunc pro ipso sole, nunc pro deo, nunc pro divino, nunc pro mundo, nunc pro significatione humanae vocis sicut hic. [7]Cui bene novem musae applicantur, quia humanae vocis officia novem sunt. [8]Omnis enim sermo his novem rebus formatur: primo appulsu quattuor dentium, repercussione duorum labrorum, plectro linguae, cavo gutturis, adiutorio pulmonum. [9]Si enim ex his quid defuerit, vox perfecta non

---

enim primo utcumque sonuerit, praeterit, nec ex eo quidquam reperies, quod resumptum arte componas: et ideo cantus in sono suo vertitur, qui sonus eius materies eius est. Idem quippe formatur, ut cantus sit. Et ideo, ut dicebam, prior materies sonandi quam forma cantandi: non per faciendi potentiam prior; neque enim sonus est cantandi artifex, sed cantanti animae subiacet ex corpore, de quo cantum faciat; nec tempore prior: simul enim cum cantu editur; nec prior electione: non enim potior sonus quam cantus, quando quidem cantus est non tantum sonus verum etiam speciosus sonus. Sed prior est origine, quia non cantus formatur, ut sonus sit, sed sonus formatur, ut cantus sit. Hoc exemplo qui potest intellegat materiam rerum primo factam et apellatam caelum et terram, quia inde facta sunt caelum et terra, nec tempore primo factam, quia formae rerum exerunt tempora, illa autem erat informis iamque in temporibus simul animadvertitur, nec tamen de illa narrari aliquid potest, nisi velut tempore prior sit, cum pendatur extremior, quia profecto meliora sunt formata quam informia, et praecedatur aeternitate creatoris, ut esset de nihilo, unde aliquid fieret."

[93] Transl. W.H. STAHL, *Martianus Capella*, II, p. 16.

erit.[94]
[10]Et phisica naturarum ratione Apollini musae applicantur. [11]Sicut enim terrestris musicae MHCH id est media principatum in omni tropo tenet, ita sol armoniae caelestis, quae viiii ordines habere dinoscitur scilicet propter vii planetas octavamque spheram caelestem atque ipsam terram. [12]Quae totius consonantiae informitatis vocumque materiei proportionem possidet.
[13]Iuxta vero leges allegoriae apertissimo musarum usus est ordine. [14]Urania igitur primo ponitur id est humanae intellegentiae sublimitas. [15]Post hanc Polimnia id est capacitas memoriae. [16]Tertia hinc ponitur Euterpe, delectatio voluntatis, quia intellegentia memoriae iungitur et voluntati. [17]Inde Erato. [18]Nulla enim re plus voluntas delectatur quam similium collatione. [19]Hoc autem sine profunda cogitatione fieri non potest. [20]Ideo subsequitur Melpomene. [21]Huic Terpsicore quasi artium delectatio sociatur. [22]Cogitationum enim perfectio sine disciplinarum exercitatione fieri non potest. [23]His omnibus omnis humanae locutionis honestas gignitur, quae Calliope signatur. [24]Inde Clio hoc est bona fama nascitur. [25]Sub qua Talia quasi in ultimo loco ponitur, hoc est positio vel germinatio virtutum. [26]Omnis enim virtus propter pulchram appetitur famam. [27]Ideoque albo alite id est cigno sustollitur stagnisque deponitur, quia bonae famae semina per terram spargi monstrantur.
[28]Totius mundi armoniam viii ptongis constare manifestum est: vii quippe planetarum motibus atque ipsa sphera ultima, quae semper celerrimo volvitur motu. [29]Terra vero in proportione i. in ratione. [30]Sicut sol in armonia et mese in lyra tetracordum faciunt, ita lingua in voce. *BeLaLbVaLea, lin. 10-27 Oxf et lin. 20-29 BeLa: ad* 12.23 QUIPPE MUSAS.

([1]The sun uses various names, and takes on different meanings. [2]He is called "sol", because he alone (solus) brings most light, more than the other elements, or because he enlightens all of them. [3]He is called "Delius", "declaration". [4]He is called "Apollo", that is "the destroyer", because with his heat he seems to drive away all moisture from the trees. [5]He is called "Pythius", that is, "he who removes faith". [6]For he symbolizes different things; the sun itself, god, the divine, the cosmos, a symbol of the human voice, as is the case here.

---

[94] This is a quote from ISIDORUS, *Etym.*, III.20.14. The context is in fact quite different: "Perfecta autem vox est alta, suavis et clara. alta, ut in sublime sufficiat; clara, ut aures adinpleat; suavis, ut animos audientium blandiat. *Si ex his aliquid defuerit, vox perfecta non est.*" (the italics are mine)

[7]And the nine Muses are rightly joined with him, for the instruments of the human voice are nine. [8]For every act of speaking is formed out of these nine things: first the coming together of the four teeth, the repercussion of the two lips, the striking of the tongue, the cavity of the throat, the help of the lungs. [9]For if one of these were absent, the voice would not have been perfect.
[10]The Muses are also joined with Apollo for a physical reason, (pertaining to) things of nature. [11]For just as in earthly music the *mese*, that is *media*, occupies the first place in each mode, so does the sun in heavenly harmony, which is distinguished as having nine successive steps, that is because of the seven planets, the eighth celestial sphere (the Firmament) and the earth itself.
[12]And this (natural philosophical reason) possesses the proportion of the consonance of the entire formlessness and the matter of sound.
[13]He used a very clear order of the Muses according to the rules of allegory. [14]Thus, Urania comes first, that is the loftiness of human intelligence. [15]After her <comes> Polymnia, that is the capacity of memory. [16]The third is Euterpe, the charm of the will, because intelligence is joined with memory and will. [17]Next is Erato. [18]For there is nothing in which the will finds more delight than in the comparison of similar things. [19]This, however, is not possible without profound thought. [20]Therefore Melpomene follows immediately. [21]With her Terpsichore is associated, as the delight in the arts. [22]For the perfection of thoughts is not possible without excercise in the disciplines. [23]Out of all this the entire honesty of human speech is brought forth, which is signified by Calliope. [24]And then Clio, that is good fame, is born. [25]And after her Talia is posed, as in the last place, that is the planting or sprouting of the virtues. [26]For every virtue is sought after because of good fame. [27]And therefore it is lifted up by a white bird, that is a swan, and put down in lakes, because the seeds of good fame are shown to be spread over the earth.
[28]It is clear that the harmony of the entire cosmos consists of seven "ptongoi", that is of the motions of the seven planets and the outer sphere itself, which is always turned round at highest speed. [29]The earth (belongs to this) proportion, that is in this reason. [30]Just as the sun in the (heavenly) harmony and the mese in the lyre produce the tetrachord, so does the tongue in the voice.)

A similar allegory of Apollo and the nine Muses as representatives of the human voice is found in Fulgentius, *Mythologiae I.15*. Fulgentius, however, describes 10 elements of the human voice: 9 for the Muses, and the 10th for Apollo himself. His order of the Muses is also different. In his myth Clio comes first, as the will to be taught. Euterpe is the second Muse, as the delight in that which is wanted. Third comes Melpomene, the pursuit of that in which delight is found. Talia is next, as the ability to grasp what was pursued. Polymnia is put in fifth place, and she symbolizes the ability to remember what was grasped. Erato is the act of finding similarities to that which is remembered. Terpsichore is the ability to judge that which was found, Urania to select that which was judged and, finally, Calliope is the eloquent presentation of that which was selected.[95]

---

[95] FULGENTIUS, *Myt. I.15*, "Fabula de novem Musis". "Huic etiam Apollini novem deputant Musas ipsumque decimum Musis adiciunt illa videlicet causa, quod humanae vocis decem sint modulamina; unde et cum decacorda Apollo pingitur cithara. Sed et lex divina decacordum dicit psalterium. Fit ergo vox quattuor dentibus, id est e contra positis, ad quos lingua percutit et quibus si unus minus fuerit sibilum potius quam vocem reddat necesse est. Duo labia velut cimbala verborum commoda modulantia, lingua ut plectrum quae curvamine quodam vocalem format spiritum, palatum cuius concavitas profert sonum, gutturis fistula quae tereti meatum spiritalem praebet excursu et pulmo qui velut aerius follis concepta reddit ac revocat. Habes ergo novem Musarum vel Apollinis ipsius redditam rationem, sicut in libris suis Anasimander Lamsaenus et Zenopanes Eracleopolites exponunt; quod et alii firmant ut Pisander fisicus et Euximenes in libro teologumenon. Nos vero novem Musas doctrinae atque scientiae dicimus modos, hoc est: prima Clio quasi cogitatio prima discendi — cleos enim Grece fama dicitur, unde et Homerus: *κλέος οἶον ἀκούσαμεν*, id est: famam solam audivimus, et alio in loco: *πεύθετο γὰρ Κύπρονδε μέγα κλέος*, et quoniam nullus scientiam quaerit nisi in qua famae suae protelet dignitatem, ob hanc rem prima Clio appellata est, id est cogitatio quaerendae scientiae —, secunda Euterpe quod nos Grece bene delectans dicimus, quod primum sit scientiam quaerere, secundum sit delectari quod quaeras, tertia Melpomene quasi melenpieomene, id est meditationem faciens permanere, ut sit primum vele, sucundum desiderare quod velis, tertium instare meditando ad id quod desideras, quarta Talia, id est capacitas velut si dicatur tithonlia, id est ponens germina, unde et Epicarmus comicus in Difolo comedia ait: *λήια μὴ ἰδὼν λιμόν τις ἀρτύνει*, id est: germina dum non viderit, famem consumit, quinta Polymnia quasi polymnemen, id est multam memoriam faciens dicimus, quia post capacitatem est memoria necessaria, sexta Erato, id est euonchomoeon, quod nos Latine inveniens simile dicimus, quia post scientiam et memoriam iustum est ut aliquid simile et de suo inveniat, septima Terpsicore ed est delectans instructionem, unde et Hermes in Opimandrae libro ait: *ἐκ*

Of course, the reason for this different order is clear — the writer of the gloss is commenting upon Martianus Capella, and should follow his order, not that of Fulgentius. But in this gloss, we see how the allegory of Fulgentius is kept intact — the Muses being symbols of nine consecutive steps in the process of learning —, while on the other hand it is cleverly altered — the commentator changes the symbols slightly to make them fit Martianus' order of Muses. Furthermore, it is this adaptation of the myth, not Fulgentius' original, which has had a "Nachleben": Remigius copied it almost word for word in his commentary on *De nuptiis*,[96] and the allegory is found again in the *Epistola de harmonica institutione* of Regino of Prüm.[97]

The second aspect which surfaced in the reading of the glosses to sections 11-12, the controversy over the direction of the planetary scale, whether the moon produced the lowest pitch or quite the contrary, is found in all ninth-century gloss collections. It is perhaps implicitly present in the first gloss: *Alii doctores aliter dicunt.*[98] The confusion can also be seen in one of the glosses on section 197 — the distance between Jupiter and Saturn in Philology's heavenly journey: some manuscripts[99] comment

---

κόρου τροφῆς ἢ ἐκ κούφου σώματος, id est: absque instructione excae et vacuo corpore, ergo post inventionem oportet te etiam discernere ac diiudicare, quod invenies, Urania octava, id est caelestis — post enim diiudicationem eligis quid dicas, quid despuas; eligere enim utile caducumque despuere caeleste ingenium est —, nona Calliope, id est optimae vocis, unde et Homerus ait: θεᾶς ὄπα φωνησάσης, id est deae vocem clamantis. Ergo hic erit ordo: primum est velle doctrinam, secundum est delectari quod velis, tertium est instare ad id quod delectatus es, quartum est capere ad quod instas, quintum est memorari quod capis, sextum est invenire de tuo simile ad quod memineris, septimum iudicare quod invenias, octavum est eligere de quo iudicas, nonum bene proferre quod elegeris."

[96] C.E. LUTZ, *Remigius Autissiodorensis*, I, 19.11 (pp. 101-102).

[97] ed. M. GERBERT, *Scriptores*, I, p. 245. See also M. BERNHARD, *Studien zur Epistola de armonica institutione des Regino von Prüm*, p. 42.

[98] Cf. 54.15-6 SICQUE SEX TONORUM.

[99] For example, *Paris, BnF, lat. 12960*, the manuscript Lutz used for her edition of "Dunchad", ad 75.17 "MELO DORIO id est modulatione Dorica, id est acuto sono" (C.E. LUTZ, *Dunchad Glossae in Martianum*, p. 10), and *Leiden, UB, BPL 36*, fol. 21r.

upon the *melo Dorio* of Saturn with: "that is a high tone" (*acuto sono*). All other manuscripts of the Anonymous commentary, however, have: "the lowest tone" (*sono gravissimo*). This difference in interpretation is, again, understandable: after all, of the five main Greek modes, the Doric mode was known to be the lowest one. On the other hand, Saturn was, in many of the ancient interpretations of the music of the cosmos, supposed to resound with the highest pitch. The controversy is treated most explicitly in the gloss to 7.1-2, quoted above, in sentences 1-5:

> [7.2] [1]At this point he touches not only upon the music of strings, but also upon celestial music, for just as in that tree a hoarse lowness shook (the branches) that were low, so is the lowest pitch from the earth to the moon. [2]But from the moon through the entire extent of the planets up to Saturn a variety of pitches is understood. [3]So too a variety of pitches was in the middle range of the tree. [4]From Saturn to the star-bearing (Firmament) a high pitch is observed to be brought forth. [5]However, some people argue against this, saying that the lowest pitch is in the orbit of Saturn, the highest in the moon, because that which is shorter sounds higher.

And although the commentator here still seems to be keeping an open mind on the subject, saying *quamvis contrario quidam ista repugnent*, his opinion is completely clear in yet another gloss, attached to section 196, on Philology's journey from Mars to Jupiter:

> [53.14-6] QUO TRANSGRESSO (NEQUE ENIM LABOR FUERAT HEMITONII INTERIECTA TRANSCURRERE), IN IOVIALIS SIDERIS PERVENERE FULGORES, CUIUS CIRCULUS PTHONGO <PHRYGIO> PERSONABAT
> (When she had crossed this (for it was no effort to cross the space of a half tone), she came to the brightness of Jove's planet, whose circle rang with Phrygian sound.)[100]
>
> In hoc differt musica caelestis a terrestri: quia in quantum sunt breviores fistulae sive cordae in terrestri, tantum acutiorem sonitum reddunt, et quantum sunt longiores, tantum graviorem; in caelesti vero musica quantum sunt angustiores circuli planetarum, tantum

[100] Transl. W.H. STAHL, *Martianus Capella*, II, p. 59.

> sunt e contrario graviores, ut est in circulo lunae, quantum autem longiores vel vastiores, tantum acutiores, ut in circulo Saturni. *LbVaLea*[101]
>
> (In this aspect the music of the heavens differs from that of the earth. Because in earthly music the shorter pipes or strings are, so much higher is the pitch they produce, and the longer they are so much lower; in heavenly music, on the other hand, the narrower the planetary orbits, so much lower on the contrary (is their pitch) — as in the orbit of the moon —, and the longer or wider (their orbit), so much higher — as in the orbit of Saturn.)

A source for this confusion could, of course, be Boethius' Nicomachean material on heavenly harmony, which does indeed present a descending scale from moon to Saturn. I would argue, however, that this is not very likely, since Boethius does not really go further into the matter; he merely offers two opposing views, one with an ascending, one with a descending scale, and does not defend or reject either. The source of the controversy must be sought elsewhere, and can perhaps be found in the commentary of John the Scot.

3. *Music of the spheres in the commentary of John the Scot*

John the Scot's commentary is known in two versions, which are believed to reflect different stages in his teaching career.[102] In both versions he argues that the pitch of a certain planet is related to the length of its orbit, and to its speed. From Saturn to the moon he envisions a rising scale — just the opposite of the

---

[101] See for the *apparatus* the edition.

[102] John's commentary survived in its entirety in two manuscripts: *Paris, BnF, lat. 12960* and *Oxford, Bodl.Lib., Auct. T.2.19.* The manuscript that is now kept in Paris is a late ninth-century miscellany of glosses in a running form on Martianus Capella, written in the monastery St. Pierre in Corbie. A few years after Lutz published the edition of John's commentary based on this source, Labowsky discovered a second version of his teachings in a late ninth- or early tenth-century manuscript from the Bodleian Library in Oxford, originally probably from St. Vincent in Metz. As the differences in the two versions are mainly found in the glosses added to the first book of the encyclopedia, Jeauneau edited this part of the commentary according to the Metz source. See for the *status quaestionis* on this the relevant section in Chapter 1.

scale presented in Martianus Capella and in the Anonymous commentary. In the version which is chronologically believed to be the first, John the Scot merely presents this view on cosmic harmony. In the second version, however, we can see him arguing actively against those who believe in an ascending scale from the moon to Saturn. His image of the machine of the Heavenly Spheres, their movements and their music is here expressed in an extremely long gloss, which can almost be read as a separate treatise. It expresses a unique and idiosyncratic idea of celestial music and the physical laws that rule cosmic movements:[103]

> DE ARMONIA CAELESTIUM MOTUUM SIDERUMQUE SONIS
> (On the harmony of heavenly movements and the sounds of the stars)
>
> Octo sunt soni, planetarum videlicet VII et spherae unus. Quorum gravissimus Saturni est, acutissimus vero spericus. Proinde spherae sonus sono Saturni ratione quadrupli consonat, ac bis diapason efficiunt, ut in organo vel fidibus principalis principalium et ultima excellentium. Sonus vero Solis est inter Saturnum et spheram sicut MECH inter duas praedictas cordas: superat enim Saturnum duplo eique consonat diapason, superatur autem a sphera duplo et aliam diapason componunt. Ubi mirabilis naturae virtus admiranda est. Nam quod in tetrachordis quinis conficitur, hoc in octo sonis caelestibus completur. Sed qua ratione peragitur, diligenti indagatione querendum est.
>
> (There are eight pitches, seven of the planets and one of the sphere.[104] Of these the lowest one is the pitch of Saturn, the highest that of the sphere. Likewise the pitch of the sphere concords with the pitch of Saturn in a fourfold ratio, and they make a double

---

[103] The gloss is edited by E. JEAUNEAU, in: *Quatre thèmes érigéniens*, pp. 123-130. I suggested some emendations in *Harmonie der hemelsferen*. The gloss is added to book I, sections 11-12 of *De nuptiis*, the sections on the secret and holy Apollinian Grove. For the sake of easy comparison between text and translation, I have chosen to divide the text of the gloss in sections and present the translation section per section.

[104] *Sphera* stands for the highest sphere, the sphere of the fixed stars, the Firmament.

octave, just like on an organ or on strings (the concord of) *principalis principalium* and *ultima excellentium*.[105] Well then, the pitch of the sun is located between Saturn and the sphere, just like the *mese* between the above mentioned strings: for it surpasses Saturn twice and concords with it in the ratio of the octave, and it (i.e. the sun) is surpassed twice by the sphere and they bring forth a second octave. Here the power of miraculous nature is to be admired. Because what is accomplished with five tetrachords together is completed with the eight tones of the heavenly spheres. But according to which order (*qua ratione*) this is achieved must be studied with meticulous examination.)

Primo igitur intellige tres planetas quae supra Solem locorum situ sunt graviorum sonituum esse. Nec immerito, quoniam et in amplioribus mundi spatiis moventur et nimia caeleritate spherae, cui contrarium cursum peragunt, ne tantae velocitatis sint, impediuntur. Quae autem sub Sole localiter, quoniam et longius a spherica velocitate distant et in brevioribus mundi spatiis discurrunt, acutiores sonos extendunt. Ac per hoc, non locorum positio sed proportionis sonorum ratio caelestem efficit armoniam, presertim cum non sinat ratio sursum et deorsum localiter in universo. In qualitate vero sonorum gravitas et acumen mediaeque varietates succinentes diversas efficiunt symphonias.

(First, you should understand that the three planets placed above the sun exhibit the lower sounds. And this is not without cause, because first of all they move along longer orbits in the cosmos, and secondly they are impeded from having such a high speed, because of the high speed of the sphere, compared to which they move in the opposite direction. The planets, however, which are under the sun in terms of their location stretch toward the higher sounds, because they a) are further removed from the fast moving sphere and b) revolve along the smaller orbits in the cosmos. This then is the reason that not the positioning of the planets, but the arithmetical ratio of the pitches produces the heavenly harmony, especially since this ratio does not permit ascent and descent in terms of their location. Lowness, highness and variants in between bring forth different consonances within the quality of the tones.)

---

[105] The interval between these two is not a double octave. Instead of *principalis principalium* John should have named its neighbour, *adquisitus* (or *proslambanomenos*).

Fiat igitur diatonicum genus pro exemplo. Sol ad Saturnum diapason dupla ratione, sphera ad Solem similiter diapason alteram reddit, ac per hoc, sphera Saturno in quadrupli ratione consonat bis diapason. Et notandum quod omnis diapason octo sonis, septem spatiis, sex tonis consistat. Est ergo primus in gravioribus Saturnus, cui proximus Iovis tono coniungitur, Iovi similiter Mars tono, Sol Marti hemitonio et diatessaron in sesquitercia proportione Sol ad Saturnum consistat. In eisdem quoque sonis Sol ad Saturnum sesqualteram consonat simphoniam sic: a Saturno ad Iovem tonus, ab Iove ad Martem tonus, a Marte ad Solem tonus, et habes diapente simul et diatessaron inter Solem et Saturnum. Et ne mireris Solem caeteris planetis multiplici proportione convenire. Diximus enim eum tribus modis concinere Saturno, in dupla videlicet et sesquitercia et sesqualtera copulatione. Cum videas non eisdem intervallis semper soni approquinquare sed secundum absidarum altitudinem, quid ergo mirum si Sol Saturno diapason in duplo concinat dum in longissimis ab eo distantiis currit; ubi vero ceperit ei appropinquare, diapente in sesqualtera; at si ei proxime accesserit, diatessaron in sesquitercia sonabit.

(Let us take the diatonic genus as an example. The sun and Saturn bring forth an octave in a duple ratio, the sphere and the sun similarly a second octave, and that is why the sphere and Saturn resound in the fourfold ratio of the double octave. It should be noted that every octave consists of eight pitches, seven steps and six whole tones. Well then, the first of the lower sounds is Saturn, and the nearest (planet) Jupiter is connected to it by a whole tone, similarly Mars to Jupiter by a whole tone, the sun to Mars by a half tone, and (from these) originates a fourth in the ratio of 4:3. Between these same tones of sun and Saturn the ratio of 3:2 (the fifth) resounds as well, viz.: from Saturn to Jupiter a tone, from Jupiter to Mars a tone, from Mars to the sun a tone, and you have a fifth and a fourth simultaneously between the sun and Saturn. Do not be surprised that the sun joins the other planets in manifold ratios. For we have said that it resounds with Saturn in three different manners, viz. in a ratio of 2:1, 4:3 and 3:2. When you see that (the planets) do not always approach each other with the same intervals of sound, but according to the heights of their orbits, why then should you be surprised that the sun resounds in a duple ratio with Saturn when it revolves around the sun in the furthest position; when it begins to approach the sun, it will resound in the ratio 3:2, with a fifth; and when it has approached the sun closest, it will resound in the ratio

4:3, with a fourth.)

Hac autem ratione considerata, non te movebit, ut opinor, quod diximus etiam de Marte, tono videlicet aliquotiens a Sole distare, aliquotiens emitonio. Quod enim valet in cordis extentio et remissio, longitudo et brevitas, sicut in fistulis organi, in quibus spatia longitudinis vocum facit distantiam, hoc idem in planetis absidarum altitudo et a Sole elongantia aut ei propinquitas. Et quod de Sole diximus, hoc ipsum de omnibus planetis inter se invicem intelligendum est. Non enim eisdem intervallis semper aut a se invicem distant aut sibi invicem appropinquant pro conditione absidarum, ac per hoc, in octo caelestibus sonis omnes musicas consonantias fieri posse credendum est, non tantum per tria genera, diatonicum, dico, chrommaticum, enarmonicum, verum etiam in aliis ultra omnium mortalium ratiocinationem. Supra vero Solem altitudinem acuminum, infra vero positionem locorum ut sensibus mortalium videtur.[106] Ita diapason potest inveniri. Sol ad spheram diapason co<n>iungit, et primo habet diatessaron ad Lunam, et primum ei Venus tono consonat, Veneri Mercurius tono altero, Luna Mercurio hemitonio. Diapente vero in eisdem spatiis consonat, dum Venus Soli tono, Mercurius Veneri tono, Luna Mercurio emitonio, spera Lunae tono respondeant.

(Now that we have considered this rule, what we have said about Mars also, i.e. that it sometimes stands at the distance of a tone from the sun, sometimes half a tone, will not bother you, I think. For what is made possible by the tension and release of strings, by length and width — as in organ pipes, where the difference in length is responsible for the difference in pitch — the same is accomplished in the planets by the lengths of their orbits and their withdrawal from or approach to the sun. And what we said about the sun, the same must be understood for all the planets among one another. For they are of course not always apart from each other or close to each other at one and the same distance, because of the nature of their orbits, and that is why one should believe that with the eight sounds of the heavenly spheres all possible musical consonances can be made, not only through the three genera — to sum up: diatonic, chromatic and enharmonic — but also in other genera, which are beyond human

---

[106] At this point I suggest a punctuation which differs from that of Jeauneau. He has in his edition: "ut sensibus mortalium videtur, ita diapason potest inveniri."

reasoning. As it appears to the human senses, the sharpness of the high sounds is above the sun, below the position of places. This is how the octave can be found. An octave joins the sun and the sphere, and at first it (i.e. the sun) has a fourth to the moon, and at first Venus resounds (with the sun) with a whole tone, Mercury with Venus with a second whole tone, the moon with Mercury with a half tone. Along these same distances, however, a fifth resounds when Venus corresponds to the sun by a tone, Mercury to Venus by a half tone, the moon to Mercury by a half tone, the sphere to the moon with a tone.)

Et notandum quod illi toni qui a terra computantur ad spheram, verbi gratia, a terra ad Lunam tonus, non sint in proportionibus vocum, sed in intervallis locorum. Tonorum enim multae species sunt. Siquidem toni sunt intervalla siderum, hoc est, quantum distat unumquodque ab alio quantumque Luna elongatur a terra: qui toni pro diversitate absidarum et circulorum variantur. Quam speciem tonorum MARTIANUS diffinit dicens: "Tonus est spatium cum legitima quantitate". Quae species in musica diastema vocatur. Sunt toni temporum, in longitudine et brevitate constituti. Sunt toni spirituum in spisitudine et exilitate vocum. Sunt toni armonici, de quibus nunc agitur, in gravitate et altitudine sonorum, ex quibus omnis proportionalitas simfoniarum constituitur. Itaque, quemadmodum in organo non consideratur in quo loco sit fistula, sed qualis vox ipsius est, et quot et quibus coniungitur et quales proportiones efficit — unaqueque fistula diversis copulata diversas efficit simfonias — sic non locus siderum sed sonus caelestem componit armoniam. Quod verum moveantur voces siderum iuxta spatia absidarum, nec mirum dum et colores eadem causa mutent et cordam in breviori et in longiori spatio positam aut extentam et remissam, cum sit eadem, non eandem vocem reddere videmus.

(Further it should be noted that these pitches, which are calculated from the earth up to the sphere (e.g. from earth to moon one tone), are not found in the ratios of the pitches, but in the distances between their positions. For there are many kinds of tones. In this case then tones are distances between the planets, i.e. how much one is apart from the other and how far the moon is removed from the earth. These pitches differ according to the diversities of their orbits and circles. And it is this kind of tone that MARTIANUS defines when he says: "A tone is a distance with a measure, determined by rule." And this kind is called 'interval' (*diastema*) in music. There are sounds of time, defined in long or short duration. There are sounds of breath, defined in density or sparseness of sounds. There are harmonic sounds, which is what we are discussing at present, defined in lowness and highness of sound, with which the entire proportional system of consonances is composed. And thus, just as on an organ it does not matter where the pipe is placed but which pitch it has and with how many and which (other pipes) it is joined and which ratios it makes — every pipe makes different harmonies in conjunction with other (pipes) —, so it is not the position of the stars but the sound that composes heavenly harmony. For, because the pitches of the planets are moved according to the lengths of their orbits, it is not strange that they also change colours for this same reason and that we do not hear a string, stretched over a shorter or longer distance or — keeping the same length — tightened or released, bring forth the same pitch.)

Tonus autem ex tendendo dicitur, et est grecum, et dirivatur a verbo TEINO, hoc est extendo. Quod vero in octavis proprie ponitur, usus musicorum fecit, quoniam omnium proportionum communis mensura est. Et ne mireris quia diximus a Luna ad Solem tonum. Non enim locorum tonos, hoc est spatia, hic consideramus, sed vocum consonantias. Dum enim rationabiliter ex gravissimo omnium sonorum, Saturno videlicet, inchoavimus, et proportionali ascensione ad solaris soni medietatem ascendimus, indeque superius rationem extendentes, ad acutissimum planetarum omnium, lunarem sonitum — nec immerito, quoniam angustissimum circulorum omnium meatum optinet — pervenimus, quo altius ascendere non valentes, ad caelestis spherae acutissimum omnium sonorum motum ratio nos perduxit, ideoque acutissimum omnium planetarum sonorum acutissimo extimoque totius mundi motui tonica proportione copulavimus.

(*Tonus* then is derived from *tendere*, and it is Greek, and it derives from the verb TEINO, which means to stretch. What one in fact means by 'octave' is the custom of musicians, because (the octave) is the common measure of all ratios. Do not wonder that we said that the distance between moon and sun is a tone. For here we are studying not the tones of position, i.e. the distances, but the consonances of their pitches. And when, according to reason, we start with the lowest of all pitches, i.e. Saturn, and then in a proportionate climb ascend to the middle, the pitch of the sun, and from there — following the rule — reach the highest pitch of all planets, that of the moon — not without reason (is the pitch of the moon the highest), because it has the shortest orbit of all —, and from this point we cannot ascend any further, then the rule has brought us to the highest movement of all pitches, the one of the heavenly sphere, and thus we have attached the highest movement of all planetary sounds to the highest and utmost movement of the entire universe in a sounding (*tonica*) proportion.)

Causa igitur erroris multis fit ignorantia tonorum, existimantes tonum illum, quo Luna distat a terra, ad proportiones vocum caelestium pertinere, non animadvertentes primum quidem quod tonus musicus nonnisi inter duos sonos constituitur, terra autem quia in statu est nullum efficit sonum, inter terram igitur et Lunam musicus tonus non est; deinde quod nunquam musica intervalla stadiorum numero mensurata sunt, sed sola rationabili extentionum ascensione secundum regulas numerorum. Aliud est enim centum viginti sex milia stadiorum mensurare a terra ad Lunam, aliud inter numerum centum nonaginta duo et ducentos XVI: XXIIII^or^ unitates. Ibi tonus est CXXVI stadiorum; hic octava pars minoris numeri tonus est XXIIII.

(The cause then for the error of many is ignorance of the tones, because they believe that the tone with which the moon is separated from the earth refers to the ratios of the heavenly (sounding) voices, and because they do not pay attention to the fact that a) (what we define as) a musical tone only exists between two pitches and that the earth — because it stands still — does not bring forth a sound; b) (they do not pay attention to the fact) that musical intervals are never measured in *stadia*, but only in rational ascension in steps according to the rules of numbers. The measuring of 126,000 stades between earth and sun is of course something else than the difference between the numbers 192 and 216 = 24 units. In the one case a tone is 126 (thousand) stades; in the other case it is 24, the eighth part of the smaller number.)[107]

Utamur ergo exemplo quodam quo manifestius apareat quod conamur asserere. In choro, ubi multi simul cantantes consonant, non locus in quo unusquisque constituitur sed proportio suae vocis

---

[107] 216:192 = 9:8, the ratio that determines the tone. In his *De institutione musica* Boethius uses the same numbers to describe the ratio of the tone, e.g. in I.17.

consideratur. In quocunque enim loco fuerit qui gravissimam vocem emittit, necesse est gravissimam vocum omnium proportionem optineat. Eadem ratione, ubicunque in choro sit qui acutissimam vocem profert, necessario sonorum omnium acumen tenebit. De succinentibus similiter intelligendum, quorum non localis possitio sed proportionalis vociferatio in universitate modulaminis diiudicatur. Frustra igitur localium intervallorum rationibus caelestem musicam coartari arbitrantur. In quo nihil aliud conspicitur nisi gravitatis et acuminis ascensus atque descensus. Ut enim ascendit gravitas in altitudinem decrescendo donec in acumine finem inponat, sic altitudo descendit similiter <crescendo>[108] quatenus in gravitatem terminum suum constituat.

(Let us, however, present an example to show more clearly what we are trying to establish. In a choir, where many singers sound together simultaneously, one does not mind the place where each of the singers is arranged, but the pitch of his voice. Naturally, wherever the person who sings the lowest pitch stands, it is necessary that he has the lowest ratio of all pitches. Following the same reasoning: wherever in the choir the person is who sings the highest pitch, inevitably he holds the top of all sounding pitches.[109] Consonances should be understood similarly; of these also one does not distinguish the position of the placements, but the sound of the ratio in the universal melody. In vain therefore one believes the heavenly harmony to be limited by the relations of the distances according to their placement. Because in this we can see nothing but ascent and descent in lowness and highness. Now, just as the graveness (of pitch) rises by decreasing in height (of placement) until it stops on the highest point (of pitch), so the sharpness (of pitch) decreases similarly by increasing (in height) until it reaches its limit in the graveness (of pitch).)

Saturn, in John's world view, holds the lowest pitch of the cosmos, the Firmament (*sphera*) the highest. The sun is located exactly halfway between these two and the entire cosmos stretches over (at most) two octaves: one from Saturn to the sun, the other from the sun to the Firmament. The earth is silent,

---

[108] This emendation will be explained in the following.

[109] I shall go further into this part of *De harmonia*, which links music theory with music practice, in the last section of this Chapter, *Theory and Practice*.

because while movement produces sound the earth stands still. John builds a system in which the speed of a planet in combination with the size of its orbit determines the pitch of that planet. He distinguishes three planets above the sun (in ascending order Mars, Jupiter and Saturn) and three below (Venus, Mercury and the moon in a descending sequence). The planets above the sun bring forth the lowest pitches, because of their long orbits and low speeds,[110] those under the sun produce the highest pitches, due to the shortness of their paths and their higher velocities. In addition to intervals of the octave between Saturn - sun - Firmament, other intervals can occur as the distances between the wandering planets change. An example of a possible interpretation, elaborated in the diatonic genus (i.e. by means of whole and half tones), produces the following distances:

Sat. -1- Jup. -1- Mars -½- sun	(Sat.-sun = fourth)
Sat. -1- Jup. -1- Mars -1- sun	(Sat.-sun = fifth)[111]

Similarly, according to John, the interval between moon/Firmament and the sun can differ:

sun -1- Venus -1- Merc. -½- moon	(sun-moon = fourth)
sun -1- Venus -1- Merc. -½- moon -1- Firm.	(sun-Firm. = fifth)[112]

On closer examination several problems can be spotted in the text. The distance between Saturn and the sun in the first example does not in fact differ between fourth and fifth, but between fourth and tritonus. In the second example, John has not presented a single, varying interval, but two intervals, that is,

---

[110] Compared to the planets, the Firmament moves in the opposite direction. Because the latter spins round with high speed, it slows down the movement of the planets. The closer they are to the Firmament, the more counterforce they experience. Because of this they revolve more slowly and bring forth lower tones.

[111] As a matter of fact the distance described is not a fifth, but a tritonus. A fifth encompasses three and a half tones.

[112] Although John in this passage tries to illustrate that the planets through their changing distances bring forth ever changing intervals, it is in fact an example of different intervals between different endings.

a fourth between sun and moon, and a fifth between sun and Firmament. It seems clear, however, that he means to depict a framework of three fixed pitches (the lowest, middle and highest sounds in the cosmos) and that the other planets do not assume one fixed pitch in this framework, but wander, so to speak, between them. The intervals they produce are thus ever changing and not bound by any modal system or set of fixed ratios.[113]

In what follows John explains that when Martianus speaks of whole or half tones between the planets — as he does when describing Philology's heavenly journey as a climb from planet to planet in the second book (II.169-199) —, he is using the term *tonus* in a different meaning. In this passage Martianus does not mean to indicate a certain pitch with the word *tonus*, but to define a distance with a fixed measure. This definition is present in book IX (*De Harmonia*, IX.930): *verum tonus est spatium cum legitima quantitate* (a tone then is a distance with a measure, determined by rule). By redefining the word *tonus* as a certain 'distance' in Philology's scale, John solves the contradiction between his own description of celestial harmony and Martianus's. Furthermore he breaks the connection between the position of a planet and the sound it produces, which sets him free to fill in the sounding cosmos according to an entirely new scheme. He is now able to abandon the principle of an ascending or descending scale from earth to Firmament. Thus, in his world view, the Firmament brings forth the highest pitch, the planet that revolves in the orbit next to the Firmament, Saturn, the lowest. Similarly, the orbits of moon and Firmament, while physically furthest removed from each other, are musically closest to each other. The order of the pitches of the planets in ascending order is now interpreted as shown below.

---

[113] John's view on the harmony of the Heavens thus concurs remarkably well with Godwin's C-type of cosmic music. Godwin does indeed discuss John's ideas, using the little known gloss: cf. J. GODWIN, *Harmonies of Heaven and Earth*, pp. 140-142.

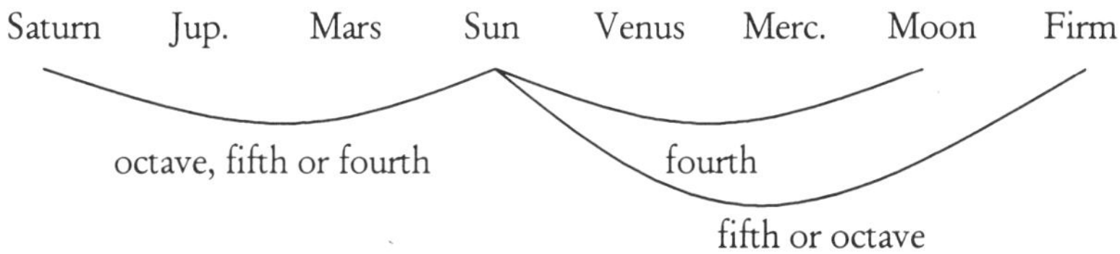

In the last words of *De armonia*, in which the principle of the release of position and pitch is once more formulated in a brief and to the point manner, yet another problem arises. The sentence reads as follows:

> Now, just as the graveness (of pitch) rises by decreasing in height (of placement) until it stops on the highest point (of pitch), so the sharpness (of pitch) decreases similarly by decreasing (in height) till it reaches its limit in the graveness (of pitch).[114]

To my mind, it argues that placement and pitch are inversely proportional (cf. Figure 4): the higher the placement of the planet (in relation to the earth, the centre of the cosmos), the lower its pitch. Conversely, the closer the planet is placed to the earth, the higher its pitch. In other words, John proposes a system wherein the moon is the planet that produces the highest pitch and Saturn the lowest, although others argue the opposite.[115] The very last part of the reasoning as cited above, however, seems then to be false. The pitch of a planet does not decrease by *decreasing* (*decrescendo*) in height in the universe, but, on the opposite, by *increasing* (*crescendo*) in height. My proposal is therefore to emend the sentence as follows:

> ... sic altitudo descendit similiter *crescendo* quatenus in gravitatem terminum suum constituat.

---

[114] É. JEAUNEAU, *Quatre thèmes Érigéniens*, 129-130: "Ut enim ascendit gravitas in altitudine decrescendo donec in acumine finem inponat, sic altitudo descendit similiter decrescendo quatenus in gravitatem terminum suum constituat."

[115] This remark, then, brings us back to the Anonymous commentary, as I shall argue below. But first another aspect of *De harmonia* will be discussed: its supposed heliocentric view on the cosmos.

Pitch:		Placement:
Low	Firmament	High
↓	Saturn	↑
	Jupiter	
	Mars	
	Sun	
	Venus	
	Mercury	
	Moon	
High	(Earth)	Low

Figure 4

De armonia sheds new light on another recent point of debate: the supposed heliocentrism in John's works. Despite the fact that John in his commentary stresses the independence of position and pitch of the planets in the cosmos, his new and idiosyncratic interpretation of the music of the heavens has always been explained as a natural consequence of his idiosyncratic view of the physics of the universe. It has been assumed that John depicted a cosmos with a larger heliocentric element than any astronomer before Copernicus dared to believe. Four planets, Venus, Mercury, Jupiter and Mars, were understood to circle not around the earth, but around the sun in his cosmos. In a series of papers on heliocentrism before Copernicus, Bruce Eastwood has overthrown this image, which he thinks is a misunderstanding and false interpretation of John's words.[116]

The hypothesis that John presents a cosmos where the sun is the centre for the orbits of a number of planets is based on two texts: a passage in his commentary on Martianus Capella (I.14-16) and one in *Periphyseon* (III.27). Eastwood argues convincingly that

[116] B.S. EASTWOOD, *Heraclides and Heliocentrism*; *Plato and Circumsolar Planetary Motion*; and particularly in *The astronomies of Pliny, Martianus Capella and Isidore of Seville*, pp. 171-173. See also Eastwood's latest article: *Astronomical Images and Planetary Theory in Carolingian Studies of Martianus Capella*. Some 50 years earlier E. VON ERHARDT-SIEBOLD and R. VON ERHARDT published two books specifically on John's astronomy, where his heliocentrism is also denied: *The astronomy of Johannes Scotus Eriugena*, and *Cosmology in the "Annotationes in Marcianum"*.

both cases are concerned, not with a description of planets orbiting the sun, but with a description of the influence of the sun on the planets in their orbits around the earth. John divides the cosmos into three fundamentally different parts: a. the outer part, which is influenced by the stars in the sky (the outer sphere and Saturn), b. the middle part, which is affected by the sun (the sun itself and the four planets Jupiter, Mars, Venus and Mercury), and c. the inner part (earth and moon).[117]

Turning back to *De armonia* the following conclusion comes to mind. Heliocentrism is not needed as an explanation of John's exceptional image of cosmic harmony. More than that, it is even unlikely. He derives the pitch of a planet from its speed and the length of its orbit. In a (partly) heliocentric system, the planets that revolve around the sun (Venus, Mercury, Jupiter and Mars) would probably have smaller orbits than the sun itself, while two of these planets (Jupiter and Mars) in John's system are attributed lower pitches than the sun, based on the lengths of their orbits. Furthermore, he divides the pitches of the planets into two zones: below the sun are the high voices, above are the low voices. This suggests that the orbits of these planets are either below or above the sun *in their entireties*. In a heliocentric cosmos John would have had to account for orbits that are partly below, and partly above the sun.

4. *An (implicit) relation between the two commentaries?*

Returning to the controversy on the direction of the planetary scale in the Anonymous commentary, the following observations can be made. John expressed a unique and idiosyncratic idea of the mechanics that rule the music of the cosmos, in which the sun — as a fixed, central point — occupies the most important place. The range of planetary pitches stretches from Saturn, on the one side, to the moon/the Firmament on the other. For the other planets John — in contradistinction to other theorists — does not posit a set combination of the place of a planet in the universe and the pitch it produces. He conceives of a celestial

[117] B.S. EASTWOOD, *The astronomies of Pliny, Martianus Capella and Isidore of Seville*, p. 173.

harmony in which the planets move freely between three fixed points (Saturn - sun - moon/Firmament), and change pitch through their movements to and from the sun. In the Anonymous commentary, on the other hand, the scholar seems to have felt much less freedom with regard to the text: he largely follows Martianus' presentation of the music of the cosmos as it arises from Philology's heavenly journey, and tries subtly to explain away its problematic passages, using antique sources wherever useful. The sun occupies an important place in the music of the heavens, but only in an allegorical way; its importance is not reflected in the musical interpretation of the cosmos.

It is important that two versions of John the Scot's commentary exist: in the first he merely states his views on the harmony of the spheres and in the second he explicitly argues against scholars who give interpretations which contradict these views. In the Anonymous commentary, the writer of the glosses clearly states that he is aware of different and contradictory interpretations of celestial music, and declares himself openly in favour of one of them. Consequently, we could imagine that the Anonymous commentary reacts to John the Scot's first teachings, and that John's second version might in its turn be considered a reaction to the Anonymous commentary. In any case, it seems safe to conclude that the controversy on celestial harmony was a lively one, and engaged scholars who were contemporaries. New arguments in favour of the preferred view and against the view of the other were brought forward. The controversy over the direction of the heavenly scale seems to suggest that the Anonymous commentary and the — as always — differing views of John the Scot in some way formed the context for each other. In both commentaries, the theme of the harmony of the spheres was treated with much originality. Looking at the sources used by the Anonymous commentator, we can conclude that he was very knowledgeable, and that he enjoyed bringing the sources together and comparing them. The glosses enable us to cast a glance into the mind of the commentator, who, like a Carolingian "chef de cuisine", skilfully mixes ingredients from Martianus Capella, Macrobius, Isidore of Seville, Boethius and Fulgentius together into a tasty dish.

Sounds, pitches and notes:
their number, their Greek and Latin names, their characteristics

1. *Martianus Capella*

After a lengthy introduction in which the story of Mercury's and Philology's wedding is continued and the power of music is assessed, Harmony opens the technical discussion of her own art with the subject of *soni* or *toni*, that is, sounds or pitches (§ 930-931).[118] She gives a general definition of the word *tonus* in the meaning of interval of a major second, and a brief excursion on intervals smaller than a second. In 931 she says that the terms *tonus* and *sonus* are interchangeable. She now uses the term *tonus/sonus* for a certain point in a scale, a certain note. She gives a list of the Greek names with their Latin translation of the 18 notes of the Perfect System, canonical in ancient Greek music theory, in ascending order.[119] (See Figure 5 below)

She concludes by saying that a melody is made of high and low sounds, and goes on to the subject of consonant intervals (*symphoniae*).

Martianus lets her return to the subject of notes in § 939-947. Here the term *tonus* is no longer used, and instead the terms *sonus* and *phthongus* are presented as interchangeable terms. She

---

[118] The terminology of these concepts, which includes *sonus, sonitus, tonus, vox, phtongos* and *corda*, is extremely ambiguous. The words available in the English language, however, are often equally vague and equivocal. In my translations (or, rather, interpretations) of the Anonymous glosses, I tried to obtain clarity by using a consistent set of terms. The term "sound" is used for the most general category of musical utterance, an unspecified sound. The term "tone" is, in general, avoided; it is only used in the meaning of "the interval of a tone", that is a major second. The term "pitch" is used when a certain frequency of sound is implied, and the term "note" is used when the author is referring to a specific point in the scale of the Greek system, such as the *proslambanomenos* or the *mese*, or the sign used to represent these points in a notational system.

[119] There is, of course, no absolute pitch attached to the Greek and Latin note names. The pitches suggested here are just there to help us visualize the intervals which use the Greek or Latin note names. See M.L. WEST, *Ancient Greek Music*, p. 222; L.P. GRIJP & P. SCHEEPERS (eds.), *Van Aristoxenos tot Stockhausen* 1, p. 416.

goes into the subject of the movement of pitches, highness and lowness, and then mentions that there are 28 legitimate notes. This is in apparent contradiction with her earlier statement about the 18 notes in § 931. In fact, Martianus, who was earlier probably rephrasing Varro's introduction to the subject of music, is now translating and compiling from a different source: Aristides Quintilianus' *De musica libri tres*. In the first source, the notes of the diatonic *genus* are mentioned, in the second, the notes of the chromatic and enharmonic *genera* are counted as well, which accounts for the difference. Harmony briefly touches upon the existence of the three *genera* (diatonic, chromatic and enharmonic), and then goes on to technical Greek terms which describe certain characteristics of sounds: *βαρύπυκνος*, *ὀξύπυκνος*, *ἄπυκνος*, *παρυπατοειδες*, *λιχανοειδες*, *σύμφωνος*, *διάφωνος*, *ὁμόφωνος*.

	Greek name	Latin name	Pitch
1	*προσλαμβανόμενος*	adquisitus	A
2	*ὑπάτη ὑπάτων*	principalis principalium	B
3	*παρυπάτη ὑπάτων*	subprincipalis principalium	c
4	*ὑπάτων διάτονος*	principalium extenta	d
5	*ὑπάτων μέσων* (*λίχανος*)	principalis mediarum	e
6	*παρυπάτη μέσων*	subprincipalis mediarum	f
7	*μέσων διάτονος*	mediarum extenta	g
8	*μέση*	media	a
9	*τρίτη συνημμένων*	tertia coniunctarum	[bes]
10	*συνημμένων διάτονος*	coniunctarum extenta	[c']
11	*νήτη συνημμένων*	ultima coniunctarum	[d']
12	*παράμεσος*	prope media	b
13	*τρίτη διεζευγμένων*	tertia divisarum	c'
14	*διεζευγμένων διάτονος*	divisarum extenta	d'
15	*νήτη διεζευγμένων*	ultima divisarum	e'
16	*τρίτη ὑπερβολαίων*	tertia excellentium	f'
17	*ὑπερβολαίων διάτονος*	excellentium extenta	g'
18	*νήτη ὑπερβολαίων*	ultima excellentium	a'

2. *Glosses*

2.1 *Definition and terminology*

Martianus' own definition of a tone, *verum tonus est spatium cum legitima quantitate* (a tone is a distance with a legitimate quantity, IX.930), is cited several times in all commentaries on Martianus Capella. In the Anonymous commentary, for example, it occurs in the glosses to book I, in glosses added to section 11, where the miraculous music brought forth by the trees of the Apollonian grove is described:

> [7.2-4] AT MEDIA RATIS PER ANNEXA SUCCENTIBUS DUPLIS AC SESQUALTERIS NEC NON ETIAM SEQUITERTIIS, <SESQUI>OCTAVIS ETIAM SINE DISCRETIONE IUNCTURIS, LICET INTERVENIRENT LIMMATA, CONCINEBANT
> (But the crests of the median trees in the fixed intervals of their mutual accompaniments sounded duple (2:1), sesquialter (3:2), sesquiterce (4:3) and even indivisible sesquioctave (9:8) intervals not without intervening limmata.)[120]

The glosses elaborate upon <SESQUI>OCTAVIS:

> Epogdous in arithmetica, tonus vocatur in musica. Epogdous dictus est quasi superoctonus vel sesquioctavus. Tonus vero est spatium cum legitima quantitate. *LbVaLea*
> (What is called *epogdous* in arithmetics, is called *tonus* in music. It is called *epogdous* as *superoctonus* or *sesquioctavus*, one and one eighth times as great. A tone then is a distance with a legitimate quantity.)

> [...] Epogdoum dicit ut sunt viiii ad viii. Epogdous autem vocatur in arithmetica, tonus in musica. Epogdous vero dictus quasi superoctonus vel sesquioctavus tonus. Tonus vero est spatium cum legitima qualitate, emitonium semis tonus *BeLa*
> ([...] He calls the ratio 9:8 *epogdoum*. What is called *epogdous* in arithmetics, is called *tonus* in music. It is called *epogdous* as a *superoctonus* or *sesquioctavus* tone, one and one eighth times as great. A tone then is a distance with a legitimate quality.)

Be and *La* have the curious corruption *qualitate* instead of *quantitate*: a strange variant, especially since the citation occurs in a context of intervals explained by their numerical ratios. I would

[120] The translation is by S. GERSH, *Concord in Discourse*, p. 2. Although this translation does not stay close to the original Latin and is in fact rather ornate, it does provide the most satisfactory solutions to the difficult passages in this section, such as the meaning of *per annexa*, *succentibus* and *sine discretione*.

argue, therefore, that the variant should be read as a corruption, and that the correct version is preserved in *Lb, Va* and *Lea*
The quotation was not used by the other Carolingian commentators, John the Scot and Remigius of Auxerre.

Martianus' definition of a tone does occur, however, in the oldest glosses on Boethius' *De institutione musica* (I,8):

> [195,6] INTERVALLUM VERO EST ...
> [32] Felix Capella: "Tonus est spatium cum legitima quantitate, qui diversis inter se sonis conficitur."[121]
> (Felix Capella: "A tone is a distance with a legitimate quantity, which is accomplished between different pitches.")

And again, though not *verbatim*, in the glosses to I,16 of *De institutione musica*:

> [202,26] RETINENT SESQUIOCTAVAM PROPORTIONEM
> [57] Est enim sesquioctava proportio tonus, i. legitimum spatium inter chordam et chordam.[122]
> (This is the ratio "sequioctava" (9:8), that is the legitimate distance from string to string.)

It can also be recognised in the definition given for *tonus* in the ninth-century treatise *Musica enchiriadis*:

> Tonus est spacii legitima magnitudo a sono in sonum. Hocque spacium musicorum sonorum, quia in sesquioctava proportione est, Greco nomine dicitur epogdous.[123]
> (A tone is a legitimate dimension of a distance from sound to sound. This distance of musical sounds is called *epogdous* with a Greek name, because it is in the sesquioctave ratio.)

The terms *tonus, sonus* and *phthongus*, which are regardlessly used as interchangeable by Martianus, are used with considerably more care by the commentator. The glossator uses *sonus* in the meaning of "a certain pitch", and *tonus* in the meaning of

---

[121] M. BERNHARD & C.M. BOWER, *Glossa maior* I, p. 167.

[122] *Ibid.*, p. 233

[123] Ed. H. SCHMID, *Musica et scolica enchiriadis*, p. 21; cited from N. PHILLIPS, *"Musica" and "Scolica enchiriadis"*, p. 124.

"interval of a major second" with some consistency. There are two glosses which confirm this distinction:

> [357.11] VERUM SONI SUNT PER SINGULOS QUOSQUE AC PER OMNES TROPOS NUMERO XVIII
> [...] <I>nter sonum et tonum dif<f>ert: sonus est <proprie> in voce, <t>onus in spatio <v>oce (*sic, lege* vocis), scilicet inter diversas cordas. *La*
> (There is a difference between *sonus* and *tonus*: *sonus* is the proper term for pitch (*vox*), *tonus* for tone distance, that is between different strings.)

> [361.11] PHTHONGOS SONOS DICIMUS
> PTONGOS Inter sonum et tonum differt: sonus est proprie in voce, tonus in spatio vocis inter cordas et cordas *BeLa*
> (There is a difference between *sonus* and *tonus*: *sonus* is the proper term for pitch (*vox*), *tonus* for tone distance between these strings and those.)

In *Scolica enchiriadis*, the echo of Martianus' definition is also found, but even more an echo of the glosses just mentioned:[124]

D<iscipulus>:	Tonus quid est?
M<agister>:	Legitimum acuminis vel gravitatis spacium inter sonum ac sonum veluti inter cordam et cordam.
(Student:	What is a tone?
Master:	It is a legitimate distance of highness or lowness between sound and sound, just as between string and string.)

The tone as a distance with a certain quantity is also apparent when the subject occurs in a different context: the harmony of the heavenly spheres. The distance of a whole tone is there found between planets, for example earth and moon, and is measured in *stadia* (*De nuptiis*, II.169):

> [169] Tunc portitores divae correpta lectica magno eam molimine subvexere. sed postquam centum viginti sex milia stadiorum aeria subvecti levitate conscenderant ac tonum primum ex pthongis

[124] Ed. H. SCHMID, *Musica et scolica enchiriadis*, p. 65; cited from N. PHILLIPS, *"Musica" and "Scolica enchiriadis"*, p. 125.

complevere caelestibus, lunarem ingressa circulum virgo divae congruis nidoribus supplicabunda de proximo conspicatur globosum quoddam tenerumque corpus e superni roris levitate compactum instar speculi praenitentis adiaculati fulgoris radios revibrare.
(Then the bearers picked up the goddess's palanquin and bore it aloft with great effort. And when they had ascended 126,000 *stadia* supported by the lightness of the air and completed the first tone among the celestial notes, the maiden, who had entered the moon's circle while making burnt offerings suitable for a goddess, saw from close by a round and soft body composed from the lightness of heavenly dew, reflecting, like a gleaming mirror, the rays of light that fell upon it.)

The number of *stadia* is repeated in, *e.g.*, glosses added to 54.15-6 (*La*) and 356.23 (*Par*):

[54.15-6] SICQUE SEX TONORUM
Unus tonus habet cxxvi milia stadiorum. <...> *La*
(One tone has 126,000 stades.)

[356.23-1] VERUM TONUS EST SPATIUM CUM LEGITIMA QUANTITATE
cxxvi milia *Par*
(126,000 stades)

2.2 The number of notes

On the number of notes, Martianus' information is, as noted above, not very clear. There are 18 notes in the Greek system, if one counts the notes of the diatonic *genus*. But if the notes of the chromatic and enharmonic *genera* are counted as well, their number is 28. The glossator understands the mistake and provides an explanation:

[362.7-8] SED SPECIALITER PER SINGULOS TROPOS VIGINTI OCTO TANTUM POTERUNT CONVENIRE
XXVIII Supra xviii posuit, nunc xxviii. Si volueris diatonicum tantum commemorare, tunc x et viii habebis. Si autem cromaticum vel enarmonium intromiseris, x supra habebis et erunt xxviii. Sub uno tamen nomine v cromatici et v enarmonici adduntur i. pro una corda. Sed subtiliter et obscure iste dixit. Boetius vero enucleatius patefecit. *BeLa*
(28: Above he presented 18 (notes) now 28. If you would want to enumerate just the diatonic (genus), you would have 18 (notes). But if you would put in the chromatics and enharmonics, you would

1	proslambanomenos	1
2	hypate hypaton	2
3	parhypate hypaton	3
4	diatonic lichanos hypaton	4
	chromatic lichanos hypaton	5
	enharmonic lichanos hypaton	6
5	hypate meson	7
6	parhypate meson	8
7	diatonic lichanos meson	9
	chromatic lichanos meson	10
	enharmonic lichanos meson	11
8	mese	12
9	trite synemmenon	13
10	diatonic paranete synemmenon	14
	chromatic paranete synemmenon	15
	enharmonic paranete synemmenon	16
11	nete synemmenon	17
12	paramese	18
13	trite diezeugmenon	19
14	diatonic paranete diezeugmenon	20
	chromatic paranete diezeugmenon	21
	enharmonic paranete diezeugmenon	22
15	nete diezeugmenon	23
16	trite hyperbolion	24
17	diatonic paranete hyperboleon	25
	chromatic paranete hyperboleon	26
	enharmonic paranete hyperboleon	27
18	nete hyperboleon	28

Figure 6

> have 10 (notes) extra, and there will be 28. Yet under one name 5 chromatics and 5 enharmonics are added, that is for one string. But he said this in a subtle and obscure manner. Boethius however made this clear in a more precise manner.)

The glossator refers here to Boethius' *De institutione musica* I.22, where the 28 tones are enumerated and laid out in a diagram (ed. G. FRIEDLEIN, pp. 215-216, see also C.M. BOWER, *Fundamentals of Music*, Fig. A.18, p. 44). In this chapter, Boethius offers a set of names which is designed to show how the three *genera* relate to each other, and presents the chromatic and enharmonic *genera* as variations on the diatonic *genus* (see Figure 6 above).[125]

2.3 *Greek and Latin names of notes (i.e. strings)*

In many glosses Martianus' Latin note names are identified with their Greek equivalents, or vice versa. Examples of this are found in the glosses added to sections 941-944 (362.9-364.5), where the contents of the Greek system with its 28 notes and its 5 tetrachords is discussed, and in the glosses on the eight *systemata* (octaves), which Martianus discusses in section 954 (367.13-368.6). Some examples of this kind of glosses are:

> [362.9] ADQUISITUS i. proslambanomenos *BeLa*
> (*ADQUISITUS*, that is *προσλαμβανόμενος*)
>
> [362.11] MEDIAE i. mese *Be*
> (TO THE *MEDIA*, that is *μέση*)
>
> [362.12] A PRINCIPALI PRINCIPALIUM i. ab *υπατε ιπατον BeLa*
> (FROM THE *PRINCIPALIS PRINCIPALIUM*, that is from the *ὑπάτη ὑπάτων*)
>
> [362.22] PRINCIPALIS AUTEM MEDIARUM i. ypate meson *BePar*, i.] haec dicitur *Par*
> (THE *PRINCIPALIS MEDIARUM*, that is the *ὑπάτη μέσων*)
>
> lychanos vel ipate meson *Leb*
> (*λίχανος* or *ὑπάτη μέσων*)

---

[125] In doing so he distorts his antique sources. See Bower's footnotes 120-122 about the inconsistencies in Boethius' chapter on the names of the 28 tone names of the Greek system: "This, however, is a highly idiosyncratic set of names [...]. The emphasis on the diatonic terminology in all genera may reflect a conception of the cromatic and enharmonic genera as alterations or shades of the diatonic [...]." (C.M. BOWER, *Fundamentals of Music*, p. 42, n.120)

[362.22-1] SUBPRINCIPALIS MEDIARUM paripate meson *Be*
pariπατε μεσον *La*
(*SUBPRINCIPALIS MEDIARUM* παρυπάτη μέσων)

[363.16] ΠΑΡΑΜΕCOC prope mediam *La*
iuxta mese *LebPar*
(ΠΑΡΑΜΕΣΟΣ next to the *media*/*μέση*)

In two cases, the identifying gloss was accidentally attached to the wrong lemma. In *La* the gloss ad 362.12 PRINCIPALIS PRINCIPALIUM *πηριπατη ιπατον* (*παρυπάτη ὑπάτων*), in fact belongs to the lemma 362.14 SUBPRINCIPALIS PRINCIPALIUM, and is found there in manuscript *Be*. Also in *La*, at 367.17 ΠΡΟΣΛΑΜΒΑΝΟΜΕΝΟΝ the gloss *qui est pro<pe> mediam* is placed, which belongs exactly one line below, at the lemma 367.19 ΠΑΡΑΜΕΣΟΝ, and is in fact placed there in *Be* and in *La* for the second time.

Many of these glosses also comprise some kind of indication of location in the Greek tetrachord system or tonal system. Of the note names *tertia coniunctarum* (*τρίτη συνημμένων*), *tertia divisarum* (*τρίτη διεζευγμένων*) and *tertia excellentium* (*τρίτη ὑπερβολαίων*) for example, the "three"-part is explained each time. This is due to Martianus' presentation of the note names in ascending order, which in a way contradicts their original conception in a descending tetrachord system: the *tertia coniunctarum*, for example, is the *third* pitch in the conjunct tetrachord in descending order, the second, however, in ascending order:

[d']	1	ultima coniunctarum	4
[c']	2	coniunctarum extenta	3
[bes]	3	tertia coniunctarum	2
[a]	4	media	1

Figure 7

[358.1] NONUS ΤΡΙΤΗ ΣΥΝΗΜΜΕΝΩΝ, ID EST TERTIA CONIUNCTARUM
(THE NINTH NOTE IS THE ΤΡΙΤΗ ΣΥΝΗΜΜΕΝΩΝ, THAT IS THE *TERTIA CONIUNCTARUM*)
NONUS TRITE CYNENMENON Secunda est a media, tertia ab ultima. *BeLa*
(It is the second note from the *media*, the third from the last.)
TERTIA CONIUNCTARUM a fine *Leb*
(from the end)

[363.7-8] VERUM POST MEDIAM EXTENDENTI HEMITONIUM CONIUNCTARUM ERIT TERTIA
(ONE SEMITONE STEP BEYOND THE *MEDIA* IS THE *TERTIA CONIUNCTARUM*)
i. trite senemmene i. secunda a prima, tercia ab ultima. *BeLa*
(the *τρίτη συνημμένων*, that is the second from the first, the third from the last.)
Est enim emitonio in mediam et tertiam coniunctarum a fine. *La*
(For it is a semitone from the *media* and the third of the conjunct notes from the end.)
TERCIA a fine *LebPar*
(THE THIRD from the end)

[363.19-1] UNDE DIVISARUM TERTIA PERHIBETUR
(HENCE IT IS CALLED THE *TERTIA DIVISARUM*)
†CONIUNCTARUM TERTIA i. ab ultima *Be*
(† THE *TERTIA CONIUNCTARUM* that is, (the third) from the last)
i. a fine tertia et a prima secunda. *BeLa*
(the third from the end, and the second from the first.)

The *extenta*-part of the note names *principalium extenta, mediarum extenta* and *coniunctarum extenta* is explained with *penultima* — for these notes are the penultimate ones (in ascending order) in each of the tetrachords (see figure 7 above):

[357.16-7] QUARTUS ΥΠΑΤΩΝ ΔΙΑΤΟΝΟΣ, ID EST PRINCIPALIUM EXTENTA
(THE FOURTH IS THE ΥΠΑΤΩΝ ΔΙΑΤΟΝΟΣ, THAT IS *PRINCIPALIUM EXTENTA*)
Ipsa est penultima, quia acutiorem sonum habet i. penultimam. *Leb*
(That is the penultimate note, because it has the higher sound, that is the penultimate note.)
Quia acutiorem sonum habet i. penultima. *Par*

(Because it has the higher sound, that is the penultimate note.)

[357.19] SEPTIMUS ΜΕΣΩΝ ΔΙΑΤΟΝΟΣ, ID EST MEDIARUM EXTENTA
(THE SEVENTH IS THE ΜΕΣΩΝ ΔΙΑΤΟΝΟΣ, THAT IS THE *MEDIARUM EXTENTA*)
i. penultima *Leb*
(that is the penultimate note)

[358.2] CONIUNCTARUM EXTENTA i. penultima *Leb*
(THE *CONIUNCTARUM EXTENTA* that is the penultimate note)

In the section on the eight octave species (*systemata*, § 954),[126] the places of the first and last notes of each of the octaves are identified in the glosses:

[367.16-8] ΠΡΟCΛΑΜΒΑΝΟΜΕΝΟΝ i. primo *BeLa*
(ΠΡΟΣΛΑΜΒΑΝΟΜΕΝΟV that is from the first)
MESEN i. octavam *BeLa*
(ΜΕΣΗΝ that is the eighth)

[367.18-9] A PRINCIPALI PRINCIPALIUM ypate ipaton. secundo *BeLa*, ipate *La*

---

[126] The section runs as follows: [954] Nunc quid sit systema perhibendum [...] sunt autem absoluta et perfecta systemata numero octo: et primum est, quod ab adquisito, quem προσλαμβανόμενον dicimus, ad mediam, quam μέσην diximus, omne conficitur; secundum, quod a principali principalium in παράμεσον usque tenditur; tertium, quod a subprincipali principalium in divisarum tertiam iungitur; quartum, quod ab extenta principalium usque in divisarum diatonum profertur; sextum, quod a subprincipali mediarum in neten divisarum usque progreditur; sextum, quod a subprincipali mediarum in tertiam excellentium trahitur; septimum, quod a mediarum extenta in excellentium diatonon excurrit; octavum, quod a media excellentium in ultimam ducitur.
(Now we must explain what a system is. [...] There are eight absolute and perfect systems: and the first begins at the *adquisitus*, which we call προσλαμβανόμενος, and is completed at the *media*, which we have called μέση; the second extends from the *principalis principalium* to the παράμεσος; the third, from the *subprincipalis principalium* to the *tertia divisarum*; the fourth, from the *extenta principalium* to the *divisarum diatonus*; the fifth, from the *principalis mediarum* to the *nete divisarum*; the sixth, from the *subprincipalis mediarum* to the *tertia excellentium*; the seventh, from the *mediarum extenta* to the *excellentium diatonus*; and the eighth, from the *media* to the *ultima excellentium*.)

(FROM THE PRINCIPALIS PRINCIPALIUM ὑπάτη ὑπάτων. from the second)
IN ΠΑΡΑΜΕCON i. duodecimum, hoc est prope mediam *BeLa*
(TO THE ΠΑΡΑΜΕΣΟΣ that is the twelfth, the *prope media*)

[367.19-1] IN DIVISARUM †TERTIUM i. in trite diezimenon i. tertium decimum *BeLa*
(TO THE TERTIA DIVISARUM that is the τρίτη διεζευγμένων, the thirteenth)

[368.1-2] EXTENTA PRINCIPALIUM i. ipaton diatonos. quarto *BeLa*
(FROM THE PRINCIPALIUM EXTENTA that is ὑπάτων διάτονος. from the fourth)
IN DIVISARUM DIATONUM i. in quartum decimum *BeLa*, in *om. Be*
(TO THE DIVISARUM EXTENTA that is the fourteenth)

[368.2-3] A PRINCIPALI MEDIARUM ipate meson. hoc est quinto *BeLa*
(FROM THE PRINCIPALIS MEDIARUM ὑπάτη μέσων. that is from the fifth)
IN NETEM DIVISARUM i. usque in ultimam divisarum *BeLa*
unicam (*sic, lege forte* xv-mum (quintum decimum)?) *La*
(TO THE ULTIMA DIVISARUM (ΝΗΤΗ ΔΙΕΖΕΥΓΜΕΝΩΝ) that is to the *ultima divisarum*. the fifteenth (?))

[368.3-4] A SUBPRINCIPALI MEDIARUM i. paripate meson. sexto *BeLa*, sexto] quinto *Be*
(FROM THE SUBPRINCIPALIS MEDIARUM that is the παρυπάτη μέσων. from the sixth)
IN TERTIAM EXCELLENTIUM i. in trite iperboleon. sexto decimo *BeLa*
triten hiperboleon *LebPar*, yperboleon *Par*
(TO THE TERTIA EXCELLENTIUM that is the τρίτη ὑπερβολαίων. the sixteenth)

[368.4-5] A MEDIARUM EXTENTA i. meson diatonon *BeLa*
(FROM THE MEDIARUM EXTENTA that is the μέσων διάτονος)
IN EXCELLENTIUM DIATONON i. septimum decimum *BeLa*
(TO THE EXCELLENTIUM EXTENTA (ΥΠΕΡΒΟΛΑΙΩΝ ΔΙΑΤΟΝΟΣ) that is the seventeenth)

[368.5-6] A MEDIA a mese *LebPar*[App], a *om. Par*[App]
(FROM THE MEDIA from the μέση)
IN †ULTIMUM i. in nete iperboleon *BeLa*

(TO THE ULTIMA <EXCELLENTIUM> to the *νήτη ὑπερβολαίων*)

A different type of glosses on names of notes is the one in which some kind of explanation of the name is provided. The *media/μέση* for example, is explained many times as the note which is found in the middle of the system — the note which stands at the end of the first octave between the *adquisitus/προσλαμβανόμενος* and the *media/μέση*, and which is the first note of the second octave between *media/μέση* and *ultima excellentium/νήτη ὑπερβολαίων*. Because of its central pitch position, the *media* is credited with a special kind of harmonious or binding quality within the Greek tonal system.

[362.14-5] Prima dicitur mese, quia finis est graviorum et initium acutorum, vel quia quantum superat graviorem (i. adquisitus), tantum superatur ab acutiore (i. nete yperboleo, quae est ultima excellentium) i. proportione diapason. *BeLa*
(The *μέση* is called the first (note), because it is the end of the low pitches and the beginning of the high pitches, or because as much as it surpasses the lowest (that is the *adquisitus*), that much is it surpassed by the highest (that is the *νήτη ὑπερβολαίων* which is the *ultima excellentium*), viz. with the ratio of the octave.)

[363.4] MEDIA Sicut mese est finis gravitatis, sic etiam est initium acuminis. Nam graves cordae praecedunt usque ad mesen, acutae subsequuntur ab illa. *LebPar*
(*MEDIA* Just as the *μέση* is the end of the lowness, so it is also the beginning of altitude. For the low strings run up to the *μέση*, and the high ones follow upon it.)[127]

[362.10-2] SED EXTRINSECUS VELUT ADQUISITUS ACCEDIT PROPTER CONSORTIUM MEDIAE, CUI CONCINERE CONSUEVIT
(BUT IT (the *adquisitus*) JOINS (the other notes) AS AN ADDED NOTE FROM OUTSIDE BECAUSE OF ITS RELATIONSHIP WITH THE *MEDIA*,

---

[127] See also the similar glosses added to 357.19 and 363.16: [357.19] OCTAVUS MECH Usque mece sunt grave<s> soni, a mece autem acuti sunt soni. *La* (THE EIGHTH TONE IS THE ΜΕΣΗ Up to the *μέση* are the low tones, from the *μέση* are the high tones); [363.16] Mese ideo media dicitur, quia ibi finitur omnis gravitas et incipit omnium cordarum acutio. *Leb* (The *μέση* is therefore called *media* (the middle one), because there ends all lowness and begins the highness of all strings).

WITH WHICH IT IS USED TO SOUND IN HARMONY)
MEDIAE Id est mese, quae omnibus accedit. *LebPar*
(WITH THE *MEDIA* That is the μέση, which agrees with all (pitches).)
CONCINERE Duplo enim superatur a media et efficit proportionem diapason. *BeLa*
(TO SOUND IN HARMONY It (the *adquisitus*) is surpassed by the *media* by the factor two and makes the harmony of the octave.)

The first pitch of the Greek system, the προσλαμβανόμενος or *adquisitus*, is also called *prosmelodos* by Boethius (*De institutione musica*, I.20 and 22). It is special in a different way: it does not belong to any of the five tetrachords, but is added in order to complete the range of two octaves. There are two glosses which comment upon its place outside the tetrachords, and in these the *adquisitus*, just as the *media*, seems to be credited with a special quality which extends over all tetrachords in the system:

[357.12-3] QUORUM PRIMUS DICITUR APUD GRAECOS ΠΡΟΣΛΑΜΒΑΝΟΜΕΝΟΣ
(OF WHICH THE FIRST IS CALLED ΠΡΟΣΛΑΜΒΑΝΟΜΕΝΟΣ BY THE GREEKS)
Secundum Boetium prosmelodos dicitur proslambanomenos. *BeLa*
(According to Boethius, the *proslambanomenos* is called *prosmelodos*.)
Iste est in omnibus. *LebPar*
(This note is "in all" (tetrachords or modes?))

[357.14] QUIA EADEM VOCE NOS UTI SUMMUS IUPPITER VETUIT, IDEM DICITUR ADQUISITUS
(BECAUSE JUPITER FORBID US TO USE THE SAME NAME (that is προσλαμβανόμενος), THIS SAME ONE IS CALLED *ADQUISITUS*
Quia conveniunt illi omnes soni. *BeLa*
(Because with it (the *adquisitus*) all pitches agree.)
Quia adquisitur ad omnes tropos. *LebPar*
(Because it is added to all tetrachords (modes).)

Of the first note of the first tetrachord, the ὑπάτη ὑπάτων or *principalis principalium*, the Greek word ὕπατη is explained in several glosses:[128]

---

[128] See also a similar explanation in the *Glossa maior* on Boethius' *De institutione musica*: I.20,32 and I,20,33 (M. BERNHARD & C.M. BOWER, *Glossa maior* I, p. 258).

[362.13-4] QUAE PRINCIPALIS PRINCIPALIUM, QUIA PRIMA IN TETRACHORDO COLLOCATUR, QUASI CUIUSDAM RECTORIS NOMEN ACCEPIT
ypatos, consul vel rector *Leb*
(ὕπατος, consul or ruler)
Unde ypate dicitur: ypatos enim consol vel rector interpretatur. *Par*
(Why is it called ὕπατη? Because ὕπατος means consul or ruler.)
Pato calco. Quia sicut dux calcat omnes sub se, ita ista omnes cordas premit. *Leb*
(πατέω means "to trample on". Because just as a leader tramples on everyone under his authority, so this string takes over all the others.)

In manuscript *Par* a strange association came to the mind of the annotator at the note name ὑπάτη ὑπάτων: he wrote in the margin the name of a well-known *proconsul*, viz. "Macrovius Theodosius" (ad 362.13-4, i.e. Ambrosius Macrobius Theodosius, author of the commentary on *Scipio's Dream*).

Of the note name ὑπάτων διάτονος (*principalium extenta*) the διάτονος-part is explained in one gloss by referring to the diatonic *genus*, which consists of tones and semitones:

[357.16] Diatonus dicitur, quia abundat tonis. *Leb*
(It is called "diatonus", because it consists mostly of tones.)

And, finally, the note name παράμεσος (*prope media*) is explained in two glosses, which explain that it is the note next to the μέση:

[358.4] HOC EST PROPE MEDIA iuxta MHCHN *La*
(THAT IS THE *PROPE MEDIA* next to the μέση)

[363.16] TONUS ΠΑΡΑΜΕΣΟΣ
Quasi ad mese, quia iungitur spatio toni ad mesen. *BeLa*
(As if "to the μέση", because it is joined to the μέση with the distance of one tone)
iuxta mese *LebPar*
(next to the μέση)

2.4 *Characteristics of sounds — more (Greek) terminology*

In § 945-947 Harmony introduces a whole list of Latin and Greek terms which refer to certain characteristics of sounds. Some, she says, are fixed (*consistere et perseverare*), others are moveable (*vagus*). The glosses elaborate on this, and provide us with examples of "fixed" versus "moveable" sounds. In so doing, they show a fairly good understanding of the Greek system, in which the outer notes of each of the five tetrachords are fixed, but the inner ones vary according to the *genus* used.

> [945/364.6-7] HORUM IGITUR SONORUM, ID EST ΦΘΟΓΓΩΝ, SUNT ALII, QUOS CONSISTERE ET PERSERVERARE NECESSE EST
> (Of these sounds or *phthongi*, some are necessarily firmly fixed and continuous)
> SUNT ALII sicut proslambanomenos *BeLa*
> (such as the προσλαμβανόμενος)
> CONSISTERE ET PERSEVERARE ut mese et nete excellentium *Be*
> (such as the μέση and the νήτη (ὑπερβολαίων, that is *ultima) excellentium*)
> Sine ulla mutatione eum sustinent. *La*
> (They sustain it without any alteration.)

> [364.7] ALII VERO SUNT VAGI
> (others are moveable)
> i. mobiles sicut paramese. *BeLa*, mobile *La*
> (That is: moveable such as the παράμεσος.)
> Varii ut penultimae, quia una semper modulatio esset, nisi ipsae variarentur. *LebPar*, modulatio esset] esset modulatio *Par*
> (Varying, like the penultimate notes (that is the *principalium extenta, mediarum extenta, coniunctarum extenta, divisarum extenta* and *excellentium extenta*, the penultimate notes of each of the tetrachords), because except if these were changed, there would always be one *modulatio*.)

In the manuscripts *Leb* and *Par*, however, the example given for fixed notes is not correct:

> principalis et subprincipalis et nete *LebPar*, et nete *om. Par*
> (the *principalis (principalium)*, the *subprincipalis (principalium)* and the νήτη (ὑπερβολαίων))

The *prinicipalis principalium* and the νήτη ὑπερβολαίων are indeed outer notes of tetrachords, but the *subprincipalis principalium* is not.

Harmony then introduces four terms which define the density (*spissum*) of a composition of three pitches (*compositivus* — a *hapax* of Martianus): *βαρύπυκνοι*, *μεσόπυκνοι*, *ὀξύπυκνοι*, and *ἄπυκνοι*.[129] Fixed pitches are called *βαρύπυκνοι* or *ἄπυκνοι*, moveable ones are called *μεσόπυκνοι* or *ὀξύπυκνοι*. The pitches that are densely spaced in the lowest region of the melody are called *βαρύπυκνοι*, the pitches that are densely spaced in the middle region *μεσόπυκνοι*, and the pitches that are densely spaced in the highest region *ὀξύπυκνοι*. When there is no rule or formula underlying the composition of the three pitches, they are called *ἄπυκνοι*. The long and intricate sentences of this section are difficult to understand, and three different categories of qualification for pitches — the extent to which sounds are moveable or fixed, the degree to which sounds are densely or widely spaced, and the position of sounds in the high, low or middle region of the pitch range — seem to be confused. The confusion is reflected in the glosses: in the manuscript pair *Be-La*, the Greek term *πυκνός* (close, compact) is translated with the Latin *spissus* (dense, compact), in *Leb* and *Par* on the other hand, it is translated with *fixus* (fixed).

[364.7-8] DENIQUE ALII ΒΑΡΥΠΥΚΝΟΙ NOMINANTUR, ALII ΜΕΣΟΠΥΚΝΟΙ, <ALII ΟΞΥΠΥΚΝΟΙ, ALII ΑΠΥΚΝΟΙ> PERHIBENTUR
(Some, then, are called *βαργπυκνοι*, others *μεσοπυκνοι*, others *οξγπυκνοι*, others are named *απυκνοι*)

ΒΑΡΥΠΙΓΝΟΥ i. primi spissi *BeLa*
(the first "densely spaced" pitches)
Graviter fixi; semper enim stant. *LebPar*, stant] sunt *Leb*
(Fixed in the low region; because they are always fixed in the same position.)

†ΜΕCΟΠΙΓΝΟS i. medii spissi *BeLa*
(the middle "densely spaced" pitches)
Medii fixi; per vices gravantur et acuuntur. *LebPar*, fixi *om.* *Leb*
(Fixed in the middle region; they are by turns moved downwards and upwards.)

PERHIBENTUR i. ultimi spissi vel alii okyπικνοι. *BeLa*, difficile lectu in *La*

[129] This passage is corrupt in all manuscripts: they have only two of these four terms, either *βαρύπυκνοι* and *μεσόπυκνοι*, or *βαρύπυκνοι* and *ὀξύπυκνοι*. See J. WILLIS, *Martianus Capella*, 364.8 and L. CRISTANTE, *Martiani Capellae Liber IX*, pp. 142 and 308-310. In her explanation of these four terms which follows directly below, however, Harmony confusingly treats all four of them.

(THEY ARE NAMED, that is, the last "densely spaced" pitches or "others [are named] ὀξύπυκνοι".)

[364.10-1] ΒΑΡΥΠΥΚΝΟΙ AUTEM SUNT, QUI VELUT REGIONES PRIMAS SPISSI RETENTANT, ΜΕΣΟΠΥΚΝΟΙ VERO, QUI MEDIA POSSIDENT
(The *βαργπυκνοι* are those densely spaced (pitches), which occupy as it were the lowest regions, the *μεσοπυκνοι*, on the other hand, those which occupy the middle range)
ΒΑΡΥΠΙΓΝΟΥ i. spissi in gravioribus cordis i. in tetracordo principalium. *BeLa*
(ΒΑΡΥΠΥΚΝΟΙ that is the densely spaced (pitches) in the lower strings, that is in the first tetrachord (the *tetrachordum principalium*).)
†MECOΠΙΓΝOS Tetracorda continet mediarum, coniunctarum, disiunctarum. *BeLa*
(ΜΕΣΟΠΥΚΝΟΙ It encompasses the middle, conjunct and disjunct tetrachords.)
MEDIA Nec tam graviter nec acute sonant. *LebPar*, sonant] sonat *Par*
(THE MIDDLE RANGE They sound neither very low, nor high.)

[364.11-3] ΟΞΥΠΥΚΝΟΙ, QUI ULTIMA TENENT, ΑΠΥΚΝΟΙ QUI IN POSITIONE SONORUM TRIUM, QUI SUNT SPISSI, NULLO GENERE AUT LEGE IUNGUNTUR
(the *οξγπυκνοι* (are those), which encompass the highest range, the *ἄπυκνοι* (are those) which in their position of the three pitches, which are densely spaced, are joined by no *genus* or rule)
ΟΚΥΠΙΓΝΟΥ Ultimi spissi s. in tetracordo excellentium. *BeLa*
  (the highest densely spaced (pitches), that is in the last tetrachord (the *tetrachordum excellentium*).)
  acute fixi *LebPar*
  (the (pitches) fixed in the high range)
ΑΠΙΓΝΟΙ inspissi *BeLa*
  (the "not densely spaced" (pitches))
  non fixi *LebPar*, fixi *om. Leb*
  (the "not fixed" (pitches))

[364.13-4] STANTES AUTEM PERSEVERANTESQUE DICUNTUR VEL ΑΠΥΚΝΟΙ VEL ΒΑΡΥΠΥΚΝΟΙ
(the (pitches) which are fixed and continuous are called the *απυκνοι* or *βαργπυκνοι*)
VEL ΑΠΙΓΝΟΥ i. non spissi *BeLa*
(ΑΠΥΚΝΟΙ that is not densely spaced)
VEL BARYΠΙΓΝΟΙ i. spissi *BeLa*

(ΒΑΡΥΠΥΚΝΟΙ that is densely spaced)

In her final remarks on the subject of sounds, Harmony mentions five more Greek terms which define some aspect of a certain sound. The terms *παρυπατοειδεῖς* (the notes that are next to the lowest) and *λιχανοειδεῖς* (the notes which are made by the index finger) are introduced as names for moveable or wandering notes in the Greek system. The terms *σύμφωνος*, *διάφωνος* and *ὁμόφωνος* are introduced for, respectively, consonant sounds, dissonant sounds, and sounds which have the same pitch, but a different function in the Greek system.[130] The translation and commentary upon these terms in the glosses is very straightforward. With these terminologically coloured remarks Harmony concludes her remarks on the subject of sounds, notes and pitches.

[364.17-8] SED HI ALII ΠΑΡΥΠΑΤΟΕΙΔΕΙΣ, ALII ΛΙΧΑΝΟΕΙΔΕΙΣ VOCANTUR
(but of these, some are called *παρυπατοειδεις*, others *λιχανοειδεις*)
ΠΑΡΙΠΑΤΟΥΔΙC i. subprincipales *BeLa*
(that is the *subprincipales*)
iuxta principales vel subprincipales *LebPar*
(next to the *principales* or *subprincipales*)
ALII †ΛΥΚΑΟΝΙΔΙC i. dactilici *BeLa*
(that is "finger notes")
vigilantes *LebPar*
("watchful notes")

[364.19-2] ΛΙΧΑΝΟΕΙΔΕΙΣ DICUNTUR A DIGITO, QUI MINISTER ARTIS <EST> ET CANTUS SINGULOS MOVET, EST AUTEM PRIMUS A POLLICE
(The *λιχανοειδεις* get their name from the finger, which is the servant of the art and moves each melody — it is, however, the finger next to the thumb)
†ΛΥΚΑΟΝΙΔΙC digitales *Leb*
("finger notes")
PRIMUS A POLLICE Primo pollice movetur corda. *BeLa*
(The string is moved by the first (finger) next to the thumb.)

[130] These are treated in the next section of this chapter, on intervals.

2.5 *Conclusion*

In her treatment of the subject of sounds and pitches, Harmony speaks of its definition, its terminology (*tonus, sonus, phthongus*), of the number of notes (18 versus 28), of their Greek and Latin names and of their characteristics (*πυκνός* or *spissus* — densely spaced or not; moveable or fixed; consonant, dissonant or homophonous, etc.) In conclusion it can be said that the Anonymous commentator was well-informed on the Greek Perfect System and its division into tetrachords, which is reflected in the names of the notes and in the distinction of fixed and moveable notes. Greek names are usually correctly identified by their Latin counterparts and vice versa. Fixed and moveable notes are — most of the time — illustrated with correct examples.

A concept which did cause confusion, however, is *πυκνός* (or *spissus*), which in one manuscript tradition is interpreted as "densely spaced or not" (*Be* and *La*), in the other (*Lb* and *Va*) as "fixed versus moveable". In defence of the commentators it can be argued that the term is introduced in a section in which Martianus himself was apparently unable to grasp the complexity of the theory. Its language is woolly and unclear. Moreover, *πυκνός* is a term that does not appear in any other antique source on music theory. *Spissus*, on the other hand, is introduced again by Martianus Capella in section 950, where he uses the term to characterize intervals as narrow (*spissus*) or wide (*rarior*). It is also found in Boethius' *De institutione musica* in yet another meaning: he distinguishes *motus spissus*, fast or frequent motion that brings forth high sounds, and *motus rarior*, slow or less frequent motion that produces low sounds.[131]

[131] Cf. BOETHIUS, *De institutione musica*, I.3, 189.28-190.2. See also my treatment of *spissus* in this context in "4. *More Greek and Latin terminology*" in the next section.

(Consonant and dissonant) intervals and harmonies

1. *Martianus Capella*

As we have seen in the case of Martianus' treatment of the subject of sounds and pitches, their number and their names, Martianus treats the technical aspects of intervals at several instances in his book on music. The terms he uses for intervals are on the one hand *symphonia* for two sounds sounding (harmoniously) at the same time, and on the other hand *spatium* and its Greek equivalent *diastema* — two neutral terms for a "distance" between two pitches, which sound consecutively.
At the very beginning of Harmony's introduction to the *ars musica* (section 930), she tells us that the art of melody is a matter of right proportions, and that these proportions are found in the legitimate pitch distances of major second, semitone and different sizes of microtones (*dieseis*): the *tetartemoria* or enharmonic diesis (a quartertone), the *tritartemoria* or chromatic diesis (⅓ of a tone), and the *hemiolia* or enharmonic diesis (⅜ of a tone).

In section 933-934, where Martianus speaks on the subject of *symphoniae* (two sounds sounding at the same time), he defines three consonant intervals: *diatessaron, diapente* and *diapason* (fourth, fifth and octave). He gives their Latin names, the number of tones (*soni*), of steps (*spatia*), their distance in whole tones (*productiones*), in semitones (*hemitonii*) and in quartertones (*dieseis*), and he defines their numerical ratio (cf. figure 8).

In section 951, here speaking on the subject of *diastemata* or *spatia*[132] (two pitches sounding consecutively) he enumerates not three, but six consonant intervals: fourth, fifth and octave, and the octave and fourth (*ex omnibus et ex quattuor*), the octave and fifth (*ex omnibus et ex quinque*) and the double octave (*bis ex*

[132] Note that the term *spatium* is used in two different meanings: interval in general, that is, a space between two pitches, and the interval of a major second, that is the space between two consecutive pitches.

omnibus).[133] He defines them in the same way (cf. figure 9).

In section 948, Martianus gives a definition of the Greek term for interval, *diastema*:

> diastema est vocis spatium, quo acuta et gravior includitur.
> (A *diastema* is an interval of pitch, in which a high pitch and a lower one are involved.)

In sections 948-950, Martianus treats all kinds of twin concepts, which relate to intervals. Thus, there are small intervals (the smallest being the enharmonic diesis) and large ones (the largest being the double octave). There are composite and non-composite intervals (*spatia composita* versus *spatia incomposita, disiuncta* or *asyntheta*) — that is, there are intervals which are built out of several tone steps, and those which run consecutively. There are rational intervals (*logica* or *rationabilia*), which can be represented by a numerical ratio, and irrational intervals (*aloga* or *irrationabilia*). There are intervals which are consonant (*convenientia*) and those which are dissonant (*discrepantia*);[134] there are enharmonic, chromatic and diatonic intervals. Some intervals are even (*artia* or *aequalia*) and can be divided into two equal parts, others are odd (*perissa*). Some intervals, which are held together by small intervals like dieseis, are crowded or dense (*spissa*), some are extended or roomy (*rariora*), and are held together by tones (major seconds).

---

[133] The canonisation of these six intervals as consonant intervals is not universal. In Macrobius' commentary on *Scipio's Dream*, for example, five intervals (*symphoniae*) are called consonant: fourth, fifth, octave, octave plus fifth and double octave. (MACROBIUS, *Commentarium in Somnium Scipionis*, 2.1,24.)

[134] It is not clear how this pair of concepts differs from the pair consonant-dissonant, which is introduced later. I would suggest a meaning of "in tune" and "out of tune", which can apply to either consonant intervals or dissonant ones: a consonant interval such as the octave can be in tune or out of tune, but this is also true of the interval of a third or sixth, which are not consonant.

Table of section 933-934:

Greek name:	διὰ τεσσάρων	διὰ πέντε	διὰ πασῶν
Latin name:	ex quattuor	quinaria (ex quinque)	ex omnibus
number of *soni*:	4	5	8
number of *spatia*:	3	4	7
number of *productiones*:	2½	3½	6
number of *hemitonii*:	5	7	12
number of *dieseis*:	10	14	24
numerical ratio:	*epitrita*, 4:3	*hemiolia*, 3:2	*diplasia*, 2:1

Figure 8

Table of section 951

Greek name:	-	-	δὶς διὰ πασῶν
Latin name:	ex omnibus et ex quattuor	ex omnibus et ex quinque	bis ex omnibus
number of *soni*:	11	12	15
number of *spatia*:	10	11	14
number of *productiones*:	8½	9½	12
number of *hemitonii*:	17	19	24
number of *dieseis*:	34	38	48
numerical ratio:	*diplasiepidimoera*, 8:3	*triplasia*, 12:4	*tetraplasia*, 12:3

Figure 9

Apart from this technical treatment in Book IX, the basic intervals, which are defined by relatively simple numerical ratios — the octave (2:1), the fifth (2:3), the fourth (4:3) and the major second (9:8) —, play an important role in the first two books. The numbers which form the ratios of these intervals were, in antique and medieval speculation on the order of the cosmos, seen as an earthly reflection of the numerical ratios which the divine Creator used for the making of the universe. They not only represent harmonious intervals, but also form the key to the harmony of creation itself.

In sections 11-12 (Book I) for example, where the trees of the Apollonian grove sound together in perfect harmony, the terms *symphonia, concentus* and *succentus* are frequent and caused the commentators to write lengthy glosses. Furthermore, in sections 105-107 (Book II), where Philology ponders the numerical values of Mercury's name and her own, the perfection of the numbers 3 and 4 is also established in the field of music: within the number 4 — that is, with the numbers 1, 2, 3 and 4 —, the ratios of three of the aforementioned intervals (octave, fifth and fourth) are realised.

2. *Glosses*

2.1 *Definitions of intervals and their ratios*

The longest glosses on the subject of intervals, their numerical ratios and their consonance are found in the glosses added to the books I and II. In the section on the sounding trees of the Apollonian grove, for example, we find the following gloss, defining the intervals of octave, fifth, fourth and major second:

> [7.2-4] AT MEDIA RATIS PER ANNEXA SUCCENTIBUS DUPLIS AC SESQUALTERIS NEC NON ETIAM SESQUITERTIIS, <SESQUI>OCTAVIS ETIAM SINE DISCRETIONE IUNCTURIS, LICET INTERVENIRENT LIMMATA, CONCINEBANT
> (But the crests of the median trees in the fixed intervals of their mutual accompaniments sounded duple (2:1), sesquialter (3:2), sesquiterce (4:3) and even indivisible sesquioctave (9:8) intervals not without intervening limmata)[135]

[135] Translation S. GERSH, *Concord in Discourse*, p. 2.

SUCCENTIBUS DUPLIS [1]Diapason dicit quae dupla iungitur proportione, ut sunt duo ad unum. [2]Diapente per sesqualteram iungitur rationem ut sunt iii et ii. [3]Diatesseron quae sesquitertia copulatur ratione ut sunt quatuor ad iii. [4]Epogdoum dicit ut sunt viiii ad viii. [...] *BeLaLbVa*
(IN DUPLE RATIOS A diapason, he says, is bound together by a duple ratio, as are the numbers 2 to 1. A fifth by the hemiolic ratio, just as 3 to 2. A fourth is joined by the ratio *sesquitertia*, just as 4 to 3. He calls the ratio 9:8 *epogdous*. [...])

In the Oxford manuscript we find a different definition of these intervals, echoing Macrobius' description of the ratios of the intervals in his commentary on *Scipio's dream*:[136] the ratios are explained by taking the minor number and comparing it to the larger one, as, for example, the ratio of the fourth, which is explained by the sum 4 = 1⅓ x 3 (or 4 = 3 + ⅓ x 3), and the ratio of the major second by the sum 9 = 1⅛ x 8 (or 9 = 8 + ⅛ x 8):

SUCCENTIBUS DUPLIS Diapason i. ex omnibus in musica, quia per omnia. Maior numerus minorem bis continet. [...] AC SESQUALTERIS iii ii. Sesqualter i. ii ad iii in aritmetica. Idem est et diapente hoc est de quinque in musica [...] SEQUITERTIIS iiii iii. Sesquitertius ut est iii ad iiii in arithmetica. Idem est et diatesseron hoc est de quatuor. Sed inter duas extremas fides peragit, quando sonus alterius fidis sonum alterius tertia minoris parte superat [...] †OCTAVIS s. tonis, ubi octava superat pars. *Oxf*
(IN DUPLE RATIOS Diapason, that is "ex omnibus" in music, because

[136] Cf. MACROBIUS, *Commentarium in Somnium Scipionis*, 2.1,15-20 (ed. J. WILLIS, pp. 97-98): Et est epitritus cum de duobus numeris maior habet totum minorem et insuper eius tertiam partem, ut sunt quattuor ad tria, nam in quattuor sunt tria et tertia pars trium, id est unum, et is numerus vocatur epitritus, deque eo nascitur symphonia quae appellatur διὰ τεσσάρων. hemiolius est cum de duobus numeris maior habet totum minorem et insuper eius medietatem, ut sunt tria ad duo; nam in tribus sunt duo et media pars eorum, id est unum, et ex hoc numero, qui hemiolius dicitur, nascitur symphonia quae appellatur διά πέντε. duplaris numerus est cum de duobus numeris minor bis in maiore numeratur, ut sunt quattuor ad duo; et ex hoc duplari nascitur symphonia, cui nomen est διὰ πασῶν. <...> epogdous numerus qui intra se habet minorem et insuper eius octavam partem, ut novem ad octo, qui in novem et octo sunt et insuper octava pars eorum id est unum. hic numerus sonum parit quem τόνον musici vocaverunt.

(it runs) through all (the notes). The larger number holds the smaller one twice. AND HEMIOLIC 3 to 2. The hemiolic ratio, that is 2:3 in mathematics. This same (ratio) is the fifth, diapente or "de quinque" in music. THE RATIO SESQUITERTIA 4 to 3. The *ratio sesquitertia* is as 3:4 in mathematics. This same (ratio) is the fourth, diatesseron, that is "de quattuor". It is accomplished between two outer strings, when the sound of one string exceeds the sound of the other by the third part of the smallest one (that is, the highest one) THE RATIO <SESQUI>OCTAVIS viz. a tone, where it (that is, the string) exceeds (the other string) by one eighth part.)

We find this same echo in the main manuscripts of the Anonymous corpus of glosses in the commentary added to the beginning of book II, where Philology ponders the nature of her marriage with Mercury by contemplating the numerical symbolism of their names. Her own name can be reduced to the number 4, which comprises in its parts the perfection of the number 10 (since 1+2+3+4=10), which is associated with the four seasons of the year, the four regions of heaven, the four elements of the earth, and — the aspect which concerns us here — within which the ratios of the musical harmonies are produced.

[30.22-4] IGITUR QUATERNARIUS NUMERUS OMNES SYMPHONIAS SUIS PARTIBUS PERFECTUS ABSOLVIT OMNIAQUE MELA HARMONICORUM DISTRIBUTIONE CONQUIRIT
(And so the perfect number four solves all the consonant intervals in their own parts, and by the distribution of the harmonies it comprises all melodies.)

SUIS PARTIBUS [1]Plenitudo totius musicae tribus simphoniis continetur, quae sunt diatesseron, diapente, diapason, quarum rationem inter quaternarium contineri manifestum est. [2]Nam sicut quaternarius ad ternarium, ita est diatessaron i. maior numerus ad minorem et tertiam partem minoris colligit in se: sic extremi soni simphoniae diatessaron. [3]Similiter ut sunt tria ad duo in ratione sesqualtera, sic diapente in extremis suis sonis. [4]Habet enim maior minorem et dimidiam partem minoris intra se. [5]Quemadmodum autem sunt duo ad unum in dupla proportione, sic extremi soni diapason simphoniae. [6]Habet enim maior minorem bis. [7]Hinc igitur inter quaternarium numerum et ternarium plenitudo artis musicae continetur. [8]Nulla enim musicae ratio extra has tres simphonias

protenditur; quicquid enim extra has est, ex his compositum est. *BeLaLbVaOxf*
(IN THEIR OWN PARTS The fullness of all music is contained by three harmonies, which are the fourth, the fifth and the octave, and it is clear that the ratios of these are contained within the number four. Because just as the ratio 4:3 is, so is the fourth: i.e. the larger number holds the smaller one in it plus one third of the smaller one; there you have the outer notes of the harmony of a fourth. Likewise, just as the numbers 3 to 2 are in a hemiolic ratio, so is the fifth in its outer notes. The larger number, after all, contains the smaller one and half of it in itself. Just as the numbers 2 and 1 are in the duple ratio, so are the outer notes of the harmony of the octave. The larger number, after all, has the smaller one twice. Therefore, then, the whole art of music is contained within the numbers four and three. For no musical ratio is reached outside these three harmonies; anything, after all, which is outside these (three harmonies), is made up from them.)

2.2 *Terminology of the interval*

When speaking on the subject of intervals, Martianus uses different terms. He uses the word *symphonia* for two pitches sounding together, he uses *spatium, diastema* and *sistema* for two pitches sounding consecutively. These words are annotated in the commentaries.
Symphonia is a consonant interval,[137] of which there are only three in the whole scala of pitches sounding together, viz. the octave, the fifth and the fourth. These three are the basis of all music, and every other interval can be deduced from these three.

[30.7-8] TRES AUTEM SYMPHONIAS QUIS IGNORAT IN MUSICIS?
(Who does not know that there are three *symphoniae* in music?)

consonantias *La*
(consonant intervals)
diatessaron diapente diapason *LaOxf*, i. *add. Oxf*
(the fourth, the fifth, the octave)
i. diapason diapente diatesseron *LbVaLea*, i. *om. Va*
(that is the octave, the fifth, the fourth)

[137] See 358.15 SYMPHONIAE i. consonantiae s. sunt *BeLa*

[30.22-4] IGITUR QUATERNARIUS NUMERUS OMNES SYMPHONIAS SUIS PARTIBUS PERFECTUS ABSOLVIT OMNIAQUE MELA HARMONICORUM DISTRIBUTIONE CONQUIRIT
(And so the perfect number four solves all the consonant intervals in their own parts and by the distribution of harmonies it brings together all melody.)

OMNES SIMPHONIAS diapason et diapente et diatesseron *BeLaLbVaLea*, et *om. BeLaLea*
(ALL THE CONSONANT INTERVALS the octave and the fifth and the fourth)

diatessaron, diapente, diapason *Oxf*
(the fourth, the fifth, the octave)

SUIS PARTIBUS Plenitudo totius musicae tribus simphoniis continetur, quae sunt diatesseron, diapente, diapason, quarum rationem inter quaternarium contineri manifestum est. <...> Hinc igitur inter quaternarium numerum et ternarium plenitudo artis musicae continetur. Nulla enim musicae ratio extra has tres simphonias protenditur; quicquid enim extra has est, ex his compositum est.[138] *BeLaLbVaOxf*
(IN THEIR OWN PARTS The fullness of all music consists of three harmonies, which are the fourth, the fifth and the octave, and it is clear that the ratio of these are contained within the number four. <...> Therefore, then, the whole art of music is contained within the numbers four and three. For no musical ratio can be beyond these three harmonies; anything, after all, which is beyond these (three harmonies), is made up from them.)

The term *spatium* is mostly used for the space between two consecutive pitches (or two consecutive strings[139]), as a synonym of *tonus* (major second). Sometimes, however, the words *spatium* and *intervallum* are considered synonyms (e.g. gloss 365.7-8 SPATIORUM intervallorum).

The most technical term for interval is *diastema* (treated in § 948ff), which is defined by Martianus himself as *vocis spatium, quo acuta et gravior includitur* (365.12-3) — an interval of pitch in which high and low are included.

---

[138] A parallel for this last sentence is found in John the Scot's *Expositiones in hierarchiam celestem*, 6.187.

[139] Cf. gloss 34.14-5; 356.23-1; 361.11.

[365.12-3] †DIASTEMATA spatium inter cordam et cordam *Be*
(DIASTEMA an interval between a string and a string)

In the glosses the term *diastema* is usually introduced together with the Greek word for octave system, *sistema*. The two definitions are juxtaposed, the one (*diastema*) meaning an undefined interval, the other (*sistema*) a scale with the range of one octave. Nevertheless, the two terms are often mixed up, probably because of their sounding alike. For *diastema* the definition *magnitudo vocis* (range of sound) is used, but exactly the same words are also used to define *sistema*, and *diastema* is sometimes defined as an interval of one octave.

[365.12] NUNC IAM DE DIASTEMATIS EDISSERENDUM
(let us now speak on the subject of intervals)

i. de magnitudine vocis *BeLa*
(that is on the range of sound)
Aliud est sistema, aliud diastema. Sistema magnitudo vocis, diastema spatium vocis hoc est locus inter cordam et cordam. *BeLa*
(A system is one thing, an interval another. A system is a range of sound, an interval is a place between one string and another.)

[365.12-3] DIASTEMA EST VOCIS SPATIUM, QUO ACUTA ET GRAVIOR INCLUDITUR
(An interval is a space of sound, in which high and low are included)

Spatium intervallum, quod fit in viii cordis. *LebPar*
(A space or interval, which is realized in eight strings.)

If the terms used by Martianus Capella, *i.e. symphonia, spatium* and *diastema* are compared to their use by Boethius, several things can be noted. First of all, the term *intervallum*, which is frequently used in *De institutione musica*, is absent from book IX of *De nuptiis*.[140] In the glosses *intervallum* is used as an

---

[140] It does occur once in section 736 of book VII on Arithmetica (265.12): *item naturalia officia, sine quibus esse nihil potest, sunt sex: magnitudo, color, figura, intervallum, status, motus.* (transl. W.H. STAHL, *Martianus Capella* II, p. 280: There are six natural properties without which bodies cannot exist: magnitude, colour, shape, space, rest and motion.) According to Remigius, *intervallum* should here be interpreted as "space" in the sense of a body's size

explanation or further definition of *spatium*. *Spatium*, however, can be interpreted either equally generally or in a more specific meaning than *intervallum*; as shown above, *spatium* is used both for interval in general and for the interval of a major second. In *De institutione musica*, *spatium* is never used in the sense of major second, but always for an interval in general. *Diastema*, the most technical term used by Martianus, does not occur in *De institutione musica*[141] — whence, perhaps, its constant confusion with *sistema*, a term which only resembles *diastema* somewhat in sound, but has an entirely different meaning.

2.3 *Consonance, dissonance and unison*

There are several terms which refer to the consonant or dissonant nature of intervals. The most important pairs of terms introduced in this context are, *concentus* and *succentus*, *consonus* and *dissonus* and *homophonus* and *diaphonus*.

Martianus uses *concentus* and *succentus* as apparent opposites terms in section 12, on the harmony of the trees of the secret Apollonian woods, and, by analogy, the music of the planetary spheres. The phrase in question goes as follows:

> [12] quod quidem exponente Cyllenio Virtus edidicit etiam in caelo orbes parili ratione aut concentus edere aut succentibus convenire.
> (As the Cyllenian (i.e. Mercury) explained this, Virtue learned that in heaven also the spheres either produce harmony or join in the accompaniments in a similar manner.)

Stahl translated *concentus edere* with "to produce harmony", while Gersh in his recent book on *Concord in Discourse* used the phrase

(cf. REMIGIUS, *Commentum*, 370.18).

[141] Boethius does use the adjective from *diastema*, διαστεματική, when he treats the classification of voices (BOETHIUS, *De institutione musica*, 199.04). A voice is, according to Boethius, either continuous (συνεχής) or "interrupted or sustained by intervals" (διαστεματική). The continuous voice is the speaking or reciting voice, which does not distinguish discrete pitches. The "diastematic" voice, on the other hand, is a voice which builds a melody out of defined intervals and discrete pitches. See also C.M. BOWER, *Fundamentals of music*, p. 20 and nn. 76-78. In the ninth-, tenth- and eleventh-century gloss traditions on Boethius, the term is explained with *intervallosa* or *spatiosa*, "rich in intervals" or "spacious", cf. M. BERNHARD & C.M. BOWER, *Glossa maior* I, 199,3 (p. 210).

"to emit melodies", probably to create the opposition implied by Martianus' syntax – the planets either bring forth melodies or join in the accompaniment of those melodies.[142] In fact, however, the terms are not in opposition with each other, but simply repeat each other. *Succentus* as a noun occurs only in *De nuptiis* and is derived from *succinere* (or *succanere*), to sing or play in accompaniment, to "chime in"[143]. *Concentus* is a sounding of several pitches harmoniously joined.[144] Both terms thus refer to the same concept: the spheres sound together in harmony. Like Gersh, however, the medieval commentators also sought to explain some kind of opposition; in the glosses, *concentus* is interpreted as the sounding together of similar (or homophonous) pitches, while *succentus* refers to the joining of different pitches:

> [7.2] Concentus est cum similis est sonus. succentus cum varii sunt soni. *La*
> (*Concentus* is when the sound is the same, *succentus* when there are several different sounds.)
> Et concentus est ubi similis est societas in sono, succentus vero ubi varii sunt soni. *LbVaLea*, similis] similiter *Lea*
> (*Concentus* is when there is a similar union of sound, *succentus* when there are several different sounds.)
>
> [7.6-8] Concentus est sonus consonus vocum, succentus vero ubi aliqua differentia. *Be*
> (*Concentus* is sound joined with similar sound, *succentus* when there is some kind of difference.)
>
> Concentus est ubi simul est societas, succentus vero ubi varii sunt soni. *La*
> (*Concentus* is when there is a joining (of sound) at the same time, *succentus* when there are several sounds.)[145]

---

[142] Cf. W.H. STAHL, *Martianus Capella* II, p. 10; S. GERSH, *Concord in Discourse*, p. 2.

[143] Cf. OLD, *succino*; LEWIS & SHORT, *succentus*.

[144] Cf. OLD.

[145] The word *simul* used here is strange: it implies that the commentator interpreted the term *concentus* as tones sounding together, and *succentus* as tones sounding consecutively – in other words, harmony opposed to melody. The other glosses, however, do not make this distinction, and one could therefore

AUT SUCCENTIBUS Ubi maior minorque vox auditur. *Oxf*
(OR WITH *SUCCENTI* Where a large and a small (i.e. low and high?) pitch are heard.)

When the terms, which are rare anyway, are found separately, the somewhat artificial opposition seems to be forgotten. *Succentus* and *concentus* both mean that there are more pitches than just one, and that they sound in some kind of harmony. They seem to occur mainly in poetic or lyrical passages. *Concentus*, for example, occurs again in section 910, when Harmonia enters the banquet hall, carrying a round shield stringed with circular chords from which a miraculous harmony streams forth.

[347.19] Denique mox ingressa atque eiusdem orbis sonuere concentus: cuncta illa, quae dissona suavitas commendarat, velut mutescentia tacuerunt
(As soon as (Harmony) entered a *concentus* poured forth from this shield: all other music, which was rendered dissonant by the sweetness (of Harmonia's music), fell silent.)

A gloss to this passage defines *concentus* with a phrase which rather resembles definitions of harmony in general:

[347.18-19] CONCENTUS multarum vocum coadunatio *BeLa*
(CONCENTUS the union of many pitches)[146]

In Boethius' *De institutione musica* the terms are absent. Boethius does use *concinentia* in the meaning of "consonance" or "harmony",[147] and he uses *succinere* in section 1 of book I (184.20-185.3), in the introduction of the story of a drunk and raving young man, who is brought to reason by someone playing the right kind of music to him.

---

also assume that the word is simply an error, and that the text should have read: "Concentus est ubi similiter est societas <in sono>."

[146] Cf., *e.g.*, 7.5-6 ARMONIAM adunationem vocis *BeLa*; 353.10 ARMONIAE armonia est adunatio vocum *BeLa*. See also BOETHIUS, *De institutione arithmetica*, 2.32.16: Est enim armonia plurimorum adunatio et dissidentium consensio.

[147] Cf., *e.g.*, I.5.1, *concinentia harmoniae*, translated by Bower as "consonance of harmony" (C.M. BOWER, *Fundamentals of Music*, p. 14).

[I.1] Cui enim est illud ignotum, quod Pythagoras ebrium adulescentem Tauromenitanum sub phrygii modo sono incitatum *spondeo succinente* reddiderit mitiorem et sui compotem?
(Who does not know that Pythagoras, *by mixing in a spondee*, restored a drunk adolescent of Taormina incited by the sound of the Phrygian mode to a calmer and composed state?)[148]

In section 947, Martianus introduces the concepts of consonance, dissonance and unison, and here uses the terms *σύμφωνοι* for consonant, *διάφωνοι* for dissonant intervals, and *ὁμόφωνοι* for unison — that is, for sounds with the same pitch but a different function within the mode. *Σύμφωνοι*, however, does not appear in the manuscripts, but instead *ὁμόφωνοι* appears twice.
Διάφωνοι is translated, by the commentator, with *dissonus* or, curiously, *ἄφωνος* (cf. 365.4). *Ὁμόφωνοι* is translated with *aequisoni* (or †*aequivoci*), and the shift in function is paraphrased:

[365.4-6] ΟΜΟΦΩΝΟΙ, QUI VOCIS QUIDEM ALIAM SIGNIFICATIONEM GERUNT, EUNDEM TAMEN IMPETUM SERVANT
(homophones, that is, sounds which serve a different function, but coincide in pitch)

i. in simili voce vel aequivoci *BeLa*
(that is, with the same pitch or †*aequivoci*)
aequisoni *LebPar*
Id est aliud nomen i. vel mediam, vel gravem, vel acutam. *BeLa*, Id est *om. La*
(That is a (*vox* with a) different name, such as *media*, *gravis* or *acuta*.)
i. non in uno genere intelleguntur. *La*
(They are not interpreted in one *genus*.)

---

[148] Transl. C.M. BOWER, *Fundamentals of Music*, p. 5. Bower, however, translates *spondeo succinente* with "by *performing* a spondee", while I chose to translate "by *mixing in* a spondee", which, in my opinion, justifies the preference for the rare word *succino* better. In one of the glosses added to the passage in Boethius the verb *succino* is understood in this same way: [185,2:468] Spondeus durus et asper, *mixtus et succinens frigio molli*, demit molliciem et infert mansuetudinem. (M. BERNHARD & C.M. BOWER, *Glossa maior* I, p. 56). Note, however, that this gloss and also other glosses added to this passage (466, 467a-b) have the mollifying influence reversed: the harsh spondee is, in their explanations, softened by the Phrygian mode.

The three Greek terms are not found in Boethius' *De institutione musica*, but the term *aequisonus* is, viz. in book V section 11, where Boethius presents Ptolemy's classification of consonances. This classification is based on music practice, and not on the theory of numerical ratios in musical intervals. Ptolemy distinguishes two basic kinds of "consonances", i.e., two pitches sounding together: he distinguishes unisons (*unisoni*) and non-unisons (*non unisoni*). The non-unisons are further divided into five categories:

1. equisons (*aequisoni*) — pitches which are so alike that they sound like unisons, *e.g.* octave and double octave;
2. consonants (*consoni*) — pitches which sound harmoniously together, such as fourth and fifth;
3. melodic consonants (*emmeles*) — pitches which are not, technically spoken, consonants, but which nevertheless sound pleasant within the melody (for example the sounds which lead from consonant to consonant);
4. dissonants (*dissoni*) — pitches which, when joined, do not please the ear;
5. non-melodic consonants (*ekmeles*) — sounds which are not melodically apt.

Thus, we see here that *aequisonus* in Boethius does not have the same meaning as *aequisonus* in the glosses on Martianus Capella. In Boethian terms the correct translation of ὁμόφωνοι would be *unisoni*, and not *aequisoni*. The commentator either misread the Boethian passage in question, or — which seems more likely — did not refer to it here.[149] It is, of course, perfectly possible that *aequisonus* is the commentator's own translation of ὁμόφωνος.

A gloss added to another pair of terms, ἄπυκνος and βαρύπυκνος (that is fixed pitches in the tetrachord in opposition to moveable pitches), should actually have been attached to the lemmata διάφωνοι and ὁμόφωνοι:

> [364.13-14] ΑΠΥΚΝΟΙ VEL ΒΑΡΥΠΥΚΝΟΙ
> A Boetio unisonae et dissonae vocantur. Fortassis Aristoxenum

[149] The passage occurs in book V, which is at any rate the least studied and least copied book of the five, especially in the early Middle Ages.

secutus est. *BeLa*
(By Boethius these are called *unisoni* (unisons) and *dissoni* (dissonants). Perhaps Aristoxenos is followed here.)

It is not very surprising that Aristoxenos turns up in this gloss, for he is often referred to as a source in *De institutione musica*, and is therefore in the eyes of the Carolingian commentators a likely source for Greek terminology not used by Boethius.

There is one more gloss on consonance and dissonance in the Anonymous commentary, not attached to any section in the technical treatment of Harmonia, but added to section 108 in book II, where Philology contemplates the numbers three and four and their relation to music.

[31.7-8] TANTUMQUE PENSAT IN NUMERIS, QUANTUM SYMPHONIA [DIAPASON] IN MELICIS QUAE TONON FACIT, QUI EST CONSONAE UNITATIS CONTINUA MODULATIO
(And it has the same value in number as the consonance (of the octave), which produces the tone, has in music, which (that is the octave) is the sustained tone of consonant unity.)[150]

[1]Sunt enim in melicis <voces> consonae et dissonae. [2]Dissonae vero sunt cum a se separantur aut emitonio aut tritemoria i. tertia parte toni aut tetratemoria i. diesi, quarta scilicet parte toni, quae spatia sunt in diatesseron et diapente. [3]Consonae vero voces sunt, cum uno eodemque spatio toni separantur, ut est in simphonia diapason. *BeLaLbVaOxf*
(In music there are consonant and dissonant sounds. Those that are separated from each other by a semitone, a *tritemoria*, that is one third of a tone, or a *tetartemoria*, that is a *diesis* or a quarter part of a tone, are dissonant; these intervals are found in the fourth and fifth. Consonant sounds, on the other hand, are those which are separated by one and the same space only, as is the case in the harmony of the octave.)

It is obvious that in this gloss the ninth-century commentator struggled with and did not fully grasp the music theory involved. There are, as elsewhere correctly explained, three consonant

[150] Transl. W.H. STAHL, *Martianus Capella* II, p. 37.

intervals (fourth, fifth and octave, and in some lists three more: octave plus fourth, octave plus fifth and double octave, cf. sections 933-934 and 951). In this gloss, however, the fourth and fifth are mentioned as examples of dissonant intervals, as their distances are indeed measured in whole tones *and* a semitone: the distance of the fourth measures 2½ tones, of a fifth 3½ tones. In this line of reasoning the only truly consonant interval of the three is the octave, which measures 6 whole tones.

The confusion is caused by a corruption in Martianus' text — the addition of *diapason* to *symphonia* —; because of this it states, mysteriously, that the concord of the octave (*symphonia diapason*) produces the major second (*tonon*). Is the *diapason* thus taken as the unity of one whole tone, and the smaller intervals, *diapente* and *diatessaron*, accordingly as (dissonant) microtones? The whole passage remains unclear.

2.4 *More Greek and Latin terminology*

In sections 948-950 Martianus briefly introduces opposing pairs of terms which define certain aspects of intervals.[151] The glossing of these sections is in general straightforward. The smallest and largest intervals are identified as the *diesis* (a quarter part of a tone) and the double octave respectively (365.14-16), or, in terms of the note names of the Greek System, the step from the first to the second and from the first to the last note:

> [365.15-16] Exemplum minus in <proslambanomenon et> ipate ipaton, exemplum maius in proslambanomenon et nete excellentium. *BeLa*
>
> (The smallest example (is found) in the (step from the) *proslambanomenos* to the *hypate hypaton*, the largest example, in the

[151] Note that none of these terms is used by Boethius in the same meaning. *Asynthetos*, *logica*, *aloga*, *conhibentia*, *discrepantia*, *archia* (or *artia*) and *perissa* are completely absent from Boethius' text, the other terms are used in different contexts or with different meanings. *Compositus* and *incompositus*, for example, are used for intervals, but in a different meaning: Boethius calls the consonances of fourth, fifth and octave incomposite, while those of octave plus fourth, octave plus fifth and double octave are composite — that is, composed of two consonances. *Disiunctus* only occurs in the setting of conjunct and disjunct tetrachords.

(step from the) *prolambanomenos* to the *nete hyperbolaion* (or *ultima excellentium*).)

Composite intervals (*composita*, *disiuncta* or *asyntheta*; that is intervals which consist of a space larger than a tone) and incomposite intervals (*incomposita*; intervals which run consecutively) are more or less correctly interpreted in the following glosses:

[365.17-18] ET COMPOSITA SUNT i. quando una corda consentit alia. *BeLa*
(COMPOSITE is when one string agrees with the other.)
INCOMPOSITA AUTEM i. quando una corda non consequenti[s][152], <sed> tertiae aut quartae coniungitur in sono. *BeLa*, i. *om. La* consequentis] consentientis *La*
(INCOMPOSITE is when in the melody one string is not joined with the next, but with the third or fourth.)

[365.16-7] Compositum est, quando multa spatia similiter iunguntur, nulla interposita corda. Adiunctum est, quod, quamvis videatur esse compositum, tamen videtur esse incompositum; interponuntur enim cordae. Unaquaque corda, quae cum alia iungitur, loica dicitur; quae non iungitur aloga. *BeLa*
(Composite is when several spaces are joined at once, without (playing) the chords in between. When an (interval) is added, it, although seemingly composite, is seen as incomposite; for strings are played in between. Every string, when joined with another, is called *loica*; when not joined *aloga*.)

This brings us to the next set of terms: rational (*rationabilia* or *logica/loica*) and irrational (*irrationabilia* or *aloga*) intervals, that is, those we can represent as proportional and those which do not have a numerical ratio underlying them. These terms are glossed as follows:

[366.1-2] RATIONABILIA i. possumus dicere, quam proportionem ad

[152] Note that there seems to be a confusion between *consentus* and *consequens* — for the explanation of *compositus* and *incompositus*, the adjective *consequens* is most apt, because it implies a progress from one string to the next in line. In the first gloss and, in manuscript *La*, also in the second, however, *consentus* appears instead of *consequens*.

invicem habeant. *BeLa*
(RATIONAL, that is, we can tell what their mutual proportion is.)
Utrum dupla an sesqualtera vel sesquitertia sit. *LebPar*, vel] an *Par*
(Whether they bear the ratio 2:1, 3:2 or 4:3.)

[366.2-3] IRRATIONABILIA Nesciens, quod facit. *Leb*
(When someone does not know what he is doing.)
Quod nesciens facit. *Par*
(What someone does who is ignorant.)

Some intervals, Martianus continues, are in harmony (*convenientia* according to a conjecture by Grotius, but *conhibentia* in the manuscripts[153] — in tune?), others are discordant (*discrepantia* — out of tune?);[154] some are enharmonic (*enharmonia*), others chromatic (*chromatica*), and still others diatonic (*diatonica*). The commentators elaborate:

[366.3-4] †CONHIBENTIA s. consentanea sibi *BeLa*
(COHERENT(?) viz. mutually agreeing)
coeuntia sibi vel concordantia. *LebPar*
(that is they go well together or are in harmony.)
DISCREPANTIA inconsona *La*
(DISCORDANT not in harmony)

ENARMONIA quarta pars *LebPar*
(ENHARMONIC (characterized by) a fourth part (of a tone))
CROMATICA emitonium *LebPar*
(CHROMATIC (characterized by) a semitone)
DIATONICA toni *LebPar*
(DIATONIC (characterized by) whole tones)

Some intervals are even (*artia*, but *archia* in the manuscripts, glossed with *aequalia*[155]), others are odd (*perissa*, glossed with *inaequalia*[156]): even intervals are those which can be divided by the factor two, such as a tone, which is divided into two semitones; odd intervals, on the other hand, are intervals which are divided by the factor three, such as a tone which is divided

---

[153] Cf. J. WILLIS, *Martianus Capella*, ad 367.3.
[154] Cf. note 134 above.
[155] Cf. 366.4-5
[156] Cf. 366.6-7

into three *dieseis* or microtones. Martianus Capella, however, failed to understand the theory correctly, and says: *perissa autem, quae in tria* hemitonia *discernuntur* (odd intervals, which are divided into three *semitones*). This error is not noticed by the commentators, but is instead applied to the tone step of one and a half tone in the chromatic tetrachord:[157]

> [366.6-7] PERISSA Excurrentia vel discurrentia, quia currunt in tria emitonia, quod fit in cromatico genere. *BeLa*, quia] que *La*
> (ODD running out or through, because they run over three semitones. This happens in the chromatic genus.)
> IN TRIA EMITONIA tonum et emitonium *LebPar*, et *om. Leb*
> (IN THREE SEMITONES a tone and a half)

The last set of terms by which intervals can be determined are *spissa* (narrow or crowded), and *rariora* (extended). The narrow intervals are held together by *dieseis* or microtones, extended intervals by tones. The commentators again recognize a reference to the three genera to divide the tetrachord, viz. enharmonic, chromatic or diatonic:

> [366.7-8] DEINDE ALIA DIASTEMATA SPISSA, ALIA RARIORA
> (Some intervals are furthermore narrow, others extended.)
>
> SPISSA Hoc fit in <en>armonio genere. *BeLa*, genere *om. La*
> (NARROW This happens in the enharmonic genus.)
>
> [366.8] SPISSA SUNT, QUAE PER DIESIS COLLIGUNTUR, RARIORA, QUAE TONIS
> (Narrow are those which are held together by dieseis, more extended (those held together) by tones.)
>
> SPISSA SUNT Quia non potest esse minus spatium quam in illis. *LebPar*
> (THEY ARE NARROW Because there cannot be a smaller space than in those (intervals).)
> QUAE PER DIESIS COLLIGUNTUR Cum enim quarta pars differentiae ponitur spissum, non enim multum separatur. *La*
> (WHICH ARE HELD TOGETHER BY DIESEIS When there is a quarter

[157] See also the section on *genera* in this chapter.

tone of difference (it is) narrow, because it is not widely spaced.)

It is noteworthy that this set of terms (*spissus* - *rarior*) can also be found, as a pair, in Boethius' *De institutione musica*. He used them in book I,3 to explain the relation between motion and pitch: slow motions (*rarior motus*) cause low pitches, and faster motions (*motus spissi*) high pitches (189.28-190.2):

> Et si tardus quidem fuerit ac *rarior* motus, graves necesse est sonos effici ipsa tarditate et raritate pellendi. Sin vero sint motus celeres ac *spissi*, acutos necesse est reddi sonos.
> (If motion is slow and less frequent, low sounds are necessarily produced by the very slowness and infrequency of striking. But if motions are fast and more frequent, high sounds are necessarily produced.)[158]

This connection of *spissus* and highness of sound on the one hand, *rarior* and lowness on the other, though not intended by Martianus Capella in this passage, is still the association which pops up in some of the glosses:

> [366.7-8] SPISSA i. spatia acuta *BeLa*, i. *om. La*
> (NARROW that is high or sharp intervals)
> RARIORA graviora *La*
> (MORE EXTENDED lower (intervals))

2.5. *Glosses on individual intervals*

On occasion the Anonymous commentator wrote glosses on individual intervals in a quite straightforward manner. On the double octave he notes that it is the largest interval (cf. 365.13-4 MAIORA), he recognizes its Greek and Latin names (*bis* or *δὶς διὰ πασῶν* and *bis ex omnibus*, cf. 365.14-5 BIS EX OMNIBUS), and he identifies the tone names which identify the interval: *προσλαμβανόμενος* and *νήτη* (*ὑπερβολαίων* or *ultima*) *excellentium* (cf. 365.15-6).

On the octave the commentator elaborates on its Greek name and its Latin translation:[159]

---

[158] Transl. C.M. BOWER, *Fundamentals of Music*, p. 11.

[159] For the identification of *diapason* as *ex omnibus* and vice versa, see, among others, 7.2-4 (*Oxf*); 30.21-2 (*La*); 54.16-7 (*La*, *Oxf*).

[359.5-6] TERTIA DIAPASON, QUAE EX OMNIBUS DICITUR
(the third (consonant interval) is the octave, which is called *ex omnibus* — of all (tones))

Ideo ex omnibus, quia veteres non <h>abebant nisi viii cordas. *La*
("Of all", because the ancient had no more than eight strings.)

The ratio of the octave, 2:1, frequently comes up in the glosses, for example in 7.3 "maior numerus minorem bis continet" (*Oxf*; the larger number contains the smaller one twice), or in 31.5 "duplam proportionem" (*La*; the duple ratio), or in 362.11:[160]

[362.11] Duplo enim superatur a media et efficit proportionem diapason. *BeLa*
(For (the *adquisitus*) is twice exceeded by the *media* and makes the ratio of an octave.)

In the Greek tonal system, the octave is treated as a unit — the μέση or *media* is both the perfection of the first octave and the beginning of the next one. In the glosses, expressions such as *plenum sistema*, but also *perfectio musicae* or *perfecta modulatio* occur at several instances where Martianus speaks of the interval of one octave. The meaning of *perfectio* is then not only taken literally — the *media* makes the first system complete —, but also metonymically: the octave is presented as the most perfect interval in music, the interval which sounds in the harmony of the spheres and which comprises, in itself, all other intervals. Thus, for example:

[363.14] MEDIAE <...>, QUAE PERFECTAM SYMPHONIAM PRIMA COMPLEVIT,
(the *media*, which for the first time completes the perfect consonance,)

Ac subprincipali principalium (*sic, lege* adquisitus!) facit plenum sistema. *BeLa*
((The *media*) makes a full system with the *subprincipalis principalium*. — *sic*, but read: *adquisitus*)

[54.15-6] Perfectio musicae id est plenitudo diapason. *La*

---

[160] See also 54.16-7 (*La, Lb, Va, Lea*); 366.11-2 (*La*).

(The perfection of music, that is the wholeness of the octave.)

[54.16-7] DIAPASON SYMPHONIAM (THE CONSONANT OF THE OCTAVE)
i. perfectam modulationem musicae sonoritatis. *Oxf*
(that is the perfect sound of musical harmony.)

Other glosses comment that the octave consists of eight steps or strings, and define the strings which form the first and last notes of an octave:

[363.14] Octo cordas habet ante se i. diapason. *LebPar*
(It has eight strings before it, that is the octave.)

Thus, an octave runs from *adquisitus* or *proslambanomenos* to *media* or *mese*, and from the *media* or *mese* to the *ultima excellentium* or *nete hyperbolaion*:[161]

[362.14-5] Prima dicitur mese, quia finis est graviorum et initium acutorum, vel quia quantum superat graviorem (i. adquisitus), tantum superatur ab acutiore (i. nete yperboleo, quae est ultima excellentium) i. proportione diapason. *BeLa*, vel *om. La* quantum] quartum *La* excellentium] excellente *Be*
(The *mese* is called the first (note), because it is the end of the low pitches and the beginning of the high pitches, or because by as much it surpasses the lowest (that is the *adquisitus*), by that much it is exceeded by the highest (that is the *nete hyperbolaion* which is the *ultima excellentium*), viz. the ratio of the octave.)

As to the individual intervals of fifth and fourth, the

[161] See also 362.10-2 (*BeLa*); and — erroneously — ad 363.14, where the octave is defined as the interval between *subprincipalis principalium* and *media* instead of *adquisitus* and *media*. In 363.17-8 (*La*) the distance between *subprincipalis principalium* and *tertia divisarum* is correctly identified as one octave. To the same lemma a gloss is added which does not comment upon the distance of one octave, but instead identifies the tone which is separated eight steps from the *tertia divisarum* in the Greek system, that is, the *principalis mediarum* or λίχανος: Quia a trite divisarum usque ad lychanos principalem mediarum octo cordae sunt (*LebPar*). Because of the overlapping of the middle and the conjunct tetrachord (*mediarum* and *coniunctarum*), however, the distance indeed runs over eight tone names, but does not add up to one octave. See the next section on tetrachords.

commentator explains their Greek names and Latin translations — *διὰ πέντε* versus *ex* or *de quinque* (see for example 7.2-4 *Oxf*), and *διὰ τεσσάρων* versus *ex* or *de quattuor* (see for example 7.2-4 *Oxf*, 358.15 *Leb*). Further, there are glosses on the ratios of these intervals, i.e., on the *ratio hemiolia* or *sesqualtera* of the fifth,[162] and on the *ratio epitrita* or *sesquitertia*[163] of the fourth. Both ratios are illustrated by figures in *Be* and *La* (see Figures 1 and 2 in the edition, ad 358.15 and 358.21-1 respectively). The size of the interval of a fifth is identified as a fourth plus one tone (358.21-1), and the commentator briefly elaborates on the *dia* in diatessaron:

> [30.18-9] *Dia* si per *i* nostrum scribitur, nihil aliud nisi *ex* significat, sed per *y* Grecum, duo ostendit. *BeLa*, nihil] nil *La*
> (When *dia* is written with our *i*, it does not mean anything else but *ex* or "out of", but when it is written with the Greek *y*, it means two.)
> Diatesseron: *di* quando per *i* scribitur, pro *de* vel pro *ex* accipi debet, quando per *y*, duo significat. *LbVaLeaOxf*, *di*] *om. LbVaOxf* *post* quando] vero *add. Oxf* per *om. Va*
> (*Diatessaron*: when *di* is written with an *i*, you must understand it as *de* or *ex* (from or out of), when written with a *y*, it means two.)

For the ratio of the interval of a whole tone, the terms *sesquioctavus*, *superoctonus* and *epogdous* are used, and the numerical ratio of 9:8 (cf. glosses ad 7.4; 31.5; 31.6; 31.7; 356.23-357.1). Furthermore, a whole tone is defined as a space between two strings (ad 35.15 and 361.11).

By error the Greek term λεῖμμα, which literally means "remnant" or "residence" and is in music theory used for the interval 256:243,[164] is interpreted as the distance of a semitone, for which the terms *emitonium, semitonium* or *semisonus* are used (cf. gloss ad 7.4).

---

[162] Cf. among others the glosses 7.3; 30.19-1; 357.9-10; 359.3-4; 366.18. Sometimes *hemitonium* (half a tone) was confused with *hemiolium* (the ratio 3:2), cf. 356.22-3; 357.2.

[163] Explained by the terms *supertrinus* (30.18-9 *LbVa*) or *supertertius* (358.21 *Leb*).

[164] That is, the interval which remains when two whole tones in the ratio of 9:8 are subtracted from the distance of a fourth.

On distances smaller than a semitone the commentator adds almost nothing to Martianus' material. He identifies a *diesis* as a quarter part of a tone (cf 356.23 *Be*; 357.3 *Leb, Par*; 365.13 *Be, La*; 365.14 *Leb, Par*; 366.8 *La*), but he also shows awareness of different sizes of the *diesis*, which are introduced, for example, in section 930 of *De nuptiis*:[165]

> [356.23] DICO QUICQUID RITE SONUERIT AUT TONUM ESSE, AUT HEMITONIUM, AUT QUARTAM <PARTEM> TONI, QUAE DIESIS APPELLATUR
> (I maintain that whatever sounds strike the ear in the right proportion form either a whole tone, a half tone, or a quarter tone, the last being called a *diesis*)[166]
>
> AUT QUARTAM TONI sed non semper quarta pars toni. *La*
> (A QUARTER TONE but not always a quarter part of a tone.)

2.6 *Summary and conclusion*

In Harmonia's treatise of music theory, different terms are used for the concept of the "interval" (*spatium* and *diastema*) and for "consonants" (*symphoniae*). Martianus' material on consonants

---

[165] The *tetartemoria* or quarter part of a tone is actually the smallest size of the *diesis*, and the term *diesis* can also be used for ⅓ part of a tone (*tritemoria*) or ⅜ parts of a tone (*divisio hemiolia*). The relevant section runs as follows:
dieseon vero distantiae tres sunt. nam prima brevior, quae tetartemoria nominatur ex eo, quod quartam partem toni recipiat; enarmonios quoque dicitur propterea, quod enarmonion modulandi genus per hanc maxime dimetimur. secunda ab illa maior est, nam tritemoria nominatur, quoniam habet partem tertiam toni, itemque chromatice appellatur, quod chromaticum modulandi genus per ipsam funditur; tertia vero habet toni quartam partem ac dimidiam quartae, et vocatur hemiolia enarmoniae divisionis, quoniam hemiolii modum complet.
(There are three sizes of the *diesis*. The first is the smallest one, which is called *tetartemoria*, for it consists of a quarter part of a tone. It is also called *enharmonic*, because the enharmonic genus is mostly determined by this (size). The second is larger than the first, and is called *tritemoria*, for it has the third part of a tone. It is also called *chromatic*, because it lies at the basis of the chromatic genus. The third has a quarter part and half of a quarter part (viz. ⅜ parts), and is called the *hemiolia* of the enharmonic division, since it completes the hemiolic mode.)

[166] Transl. W.H. STAHL, *Martianus Capella* II, p. 359.

defines their number as three or six.[167] Their numerical ratios form — in a Neoplatonic view on creation — an important key to the understanding of the order of the universe, and are as such amply treated by both Martianus and his ninth-century commentators.

In the glosses of the Anonymous corpus, material from Macrobius' *Commentary on Scipio's Dream* was used to explain the mathematics of the proportions. The commentators were fully able to understand Martianus' treatment of intervals and their characteristics and aptly translated or paraphrased his Greek terminology, though there are some traces of a struggle with the concepts of consonance and dissonance. The individual intervals of octave, fifth, fourth and tone received glosses which correspond with Martianus' material from passages in book I, II and IX.

When the glosses of the Anonymous corpus are compared to those of Remigius of Auxerre, it can be noted that many glosses appear in both corpora, and that the same material is presented. The most striking difference, however, is that Remigius illustrates the intervals with examples in Gregorian chants. On the consonance of the fourth, for example, Remigius notes:[168]

> [496.18] SYMPHONIAE TRES SUNT, QUARUM PRIMA EST DIATESSARON ut est "Tibi dixit cor meum," ...
> (there are three consonants, of which the first is the fourth as in "Tibi dixit cor meum," ...)
>
> [497.6] SED HAEC id est diatessaron, SYMPHONIA EST id est consonantia vocis, ut est "Deus in adiutorium meum," quod fit in quattuor cordis ...
> (BUT THIS CONSONANCE IS that is the fourth, the harmonious consonance, as in "Deus in adiutorium meum," which happens in

---

[167] This number does not correspond to, e.g., the number of consonants according to Macrobius: in his *Commentary on Scipio's Dream* (2.1,21-23) he mentions not three or six, but five consonants (*symphoniae*): fourth, fifth, octave, octave+fifth and double octave. There is no evidence, however, that the ninth-century scholars noticed this difference or that they found it of importance.

[168] Cited from C.E. LUTZ, *Remigius Autissiodorensis*, II, pp. 332-333. The examples Remigius cites are *introitus*, and are provided with neumes.

four strings ...)

And on the fifth:

> [497.13] HEMIOLII sesquialteri, id est diapente, POSSIDET RATIONEM ut est "Oculi mei, semper," quod fit in quinque chordis ...
> (IT, that is the fifth, POSESSES THE HEMIOLIC RATIO, that is the sesquialter ratio, as in "Oculi mei, semper," which happens in five strings ...)

The Anonymous corpus (and also the commentary of John the Scot) such do not contain any illustrations with Gregorian chants, which relate antique music theory directly to Carolingian music practice.

Tetrachords

1. *Martianus Capella*

The Perfect System in antique music theory consisted of five tetrachords: the tetrachord of the "principal notes", ὑπάτων or *principalium*; the tetrachord of the "middle notes", μέσων or *mediarum*; the so-called "conjunct" tetrachord, συνημμένων or *coniunctarum*; the "disjunct" tetrachord, διεζευγμένων or *divisarum*;[169] and the tetrachord of the "highest notes", ὑπερβολαίων or *excellentium* (see Figure 10).

Martianus does not deal with these tetrachords separately, but different aspects of the tetrachords are discussed when, for example, the 18 or 28 notes of the Perfect System are treated, or when Harmonia comes to speak on the subject of the three *genera*. In § 935, Martianus defines a tetrachord as *affectio quaedam sonorum quattuor per ordinem compositorum, quorum extremi sibi debeant convenire* — "a pleasing relationship of four notes arranged in order, in which the two bounding notes are consonant".[170] In § 940, he explains that the *adquisitus* does not

[169] Note that Boethius here uses the term *disiunctarum*; cf., for example, *De institutione musica* I,26, and see C.M. BOWER, *Fundamentals of music*, p. 46 n. 125.

[170] Transl. W.H. STAHL, *Martianus Capella*, II, p. 362.

	προσλαμβανόμενος	adquisitus
	ὑπάτη ὑπάτων	principalis principalium
ὑπάτων (principalium)	παρυπάτη ὑπάτων	subprincipalis principalium
	ὑπάτων διάτονος	principalium extenta
	ὑπάτων μέσων (λίχανος)	principalis mediarum
μέσων (mediarum)	παρυπάτη μέσων	subprincipalis mediarum
	μέσων διάτονος	mediarum extenta
	μέση	media
συνημμένων (coniunctarum)	τρίτη συνημμένων	tertia coniunctarum
	συνημμένων διάτονος	coniunctarum extenta
	νήτη συνημμένων	ultima coniunctarum
	παράμεσος	prope media
διεζευγμένων (divisarum)	τρίτη διεζευγμένων	tertia divisarum
	διεζευγμένων διάτονος	divisarum extenta
	νήτη διεζευγμένων	ultima divisarum
ὑπερβολαίων (excellentium)	τρίτη ὑπερβολαίων	tertia excellentium
	ὑπερβολαίων διάτονος	excellentium extenta
	νήτη ὑπερβολαίων	ultima excellentium

Figure 10

belong to the first tetrachord, but is added to the whole system of notes to complete the range of two octaves. In § 943-944, he introduces the concepts of the conjunct and disjunct tetrachords. In § 959 ff. he returns to the subject: the concept of *genera*, ways of dividing the tetrachord, is treated. He defines the tetrachord for a second time in § 961: *tetrachordum quippe est quattuor sonorum in ordinem positorum congruens fidaque concordia* — "a tetrachord is a coherent and faithful concord of four sounds arranged in a sequence",[171] and (for the first time since he mentioned them in § 935) enumerates all five of them. The principal tetrachord, he explains, runs from the *principalis principalium* to the *principalis mediarum*, the conjunct tetrachord[172] from the *media* to the *ultima coniunctarum*, the disjunct tetrachord from the *prope media* to the *ultima divisarum*. He does not mention the bounding notes of the middle or of the highest tetrachord.

2. *Glosses*

In several glosses it is determined that the bounding notes of a tetrachord form the consonant interval of a fourth, and the tetrachord is acknowledged as a basic element in music:

> [360.1-3] TETRACHORDUM AUTEM EST AFFECTIO QUAEDAM SONORUM QUATTUOR PER ORDINEM COMPOSITORUM QUORUM EXTREMI SIBI DEBEANT CONVENIRE
> (a tetrachord is a pleasing relationship of four notes arranged in order, in which the two bounding notes must be consonant)
>
> †TETRACORDA i. diatessaron *BeLa*
> (TETRACHORD that is a fourth)
>
> Ideo per diatessaron dividitur, quia elimentum est totius musicae. *Par*
> (It (that is, the Perfect System) is divided by the fourth, because it is a basic element of the whole of music.)

---

[171] *Ibid.*, p. 370.

[172] A lacuna occurs, in which Martianus should have mentioned the bounding tones of the middle tetrachord. Cf. W.H. STAHL, *Martianus Capella*, II, p. 371 n. 88.

Verbi gratia: si quis dixerit mihi quomodo resonare debet ultima corda ad primam, tetracordon dicetur sub epitriti ratione. *BeLa*
(For example: if anyone asked me how the last string should sound in relation to the first, I would answer: the tetrachord bears the ratio of a fourth.)

[370.8] VERUM HORUM EXTREMOS SIBI APTANDOS ESSE NON DUBIUM EST
(there is no doubt that their (that is of the tetrachords) bounding notes must be adjusted to each other)

i. primam cordam et ultimam; necesse est enim, ut in his concordia inveniatur. *BeLa*
(that is the first string and the last; for it is necessary that consonance should be found in these.)[173]

The basic principle of the conjunct tetrachord — that is, the tetrachord which starts with the *media* which is also the final note of the middle tetrachord — is explained in a gloss ad 370.13-4:

[370.13-4] VERUM PRINCIPALIUM ET MEDIARUM CONEXUM, QUOD CONIUNCTARUM VOCATUR
(the (tetrachord of the) principal notes and middle notes are conjunct, hence it is called "of the conjunct (notes)")

CONEXUM Quia unam cordam habent, quae terminat tetracordum praecedens et subsequens <inchoat>. *BeLa*
(CONJUNCT Because they have one string, which concludes the preceding tetrachord and <begins> the next one.)

Quia una corda coniungit. *LebPar*[App]

[173] To this last lemma, however, two manuscripts added a gloss in which the text was evidently misunderstood:
Inter penultimam et ultimam; in omnibus tetracordis duae ultimae <cordae> sesquitertiam proportionem habent *Leb*
(Between the penultimate tone and the last one; in every tetrachord the two last strings have the sesquitertian ratio.)
And:
Inter penultimam et ultimam sesquitertia est *Par*[App]
(Between the penultimate tone and the last one is a sesquitertian (ratio).)

(Because one string conjoins (them).)

But when Martianus earlier explained the difference between the conjunct and disjunct tetrachords in § 943-944, a problem arises in the text. For he wrote:

> [943-4] omne autem tetrachordum coniunctarum ideo dictum, quia mediae ipsi, quae perfectam symphoniam prima complevit, adiungitur. †ac producente media sequentem† tonum, ille qui sequitur tonus παράμεσος nominatur, quod hunc satis proximum sonus adiunctae modulationis offendat. dehinc divisarum tertia, quae plenum systema diapason finit, unde divisarum tertia perhibetur; post hanc soni ceteri consequuntur. genus vero divisarum ideo hoc nomen accepit, quia uno et dimidio tono a mediae fine divellitur.
> (Every conjunct tetrachord is called thus because it is joined to the *media* itself, which first completes the perfect consonance. [Advancing from the *media* to the next] note, that next note which follows is called the *paramese*, because it, close enough to its adjacent modulation, collides unpleasantly with its neighbour. Next comes the *divisarum tertia*, which completes a full system of an octave, hence it is called the third note of the disjunct tetrachord. After this the other notes follow. The *genus divisarum* therefore received this name, because it is separated from the final *media* by a tone and a half.)

In this last sentence Martianus clearly states that the distance between *media* and *paramese* measures one and a half tones, instead of one tone.[174] The commentator saw himself faced with the difficult task of trying to make a correct argument on the basis of a wrong one:

> [364.2] QUIA UNO ET DIMIDIO TONO A MEDIAE FINE DIVELLITUR
> Quod post mesen sequitur tetracordum coniunctarum, ita ut primum locum mese[n] teneat, emitonio superatur ipsa media a sequente i. tertia coniunctarum. Quando vero sequitur tetracordum divisarum ita etiam, ut eadem media principium habeat, tunc a proxima corda hoc est a paramese tono superatur. Quando vero tetracordum divisarum inchoat, ita ut mese principium non habeat, sed ipsa paramese, tunc

---

[174] No satisfactory solutions have been found to solve this error. See for several attempts by editors of the text, L. CRISTANTE, *Martiani Capellae De nuptiis Philologiae et Mercurii Liber IX*, pp. 306-307.

> paramese a media uno tono et emitonio separatur atque ideo divisarum tetracordum appellatur. Et ideo dictum est uno et dimidio tono a fine medie divellitur. *BeLa*
> (Because after the *mese* follows the conjunct tetrachord, in such a way that the *mese* occupies the first place, and that this *media* is surpassed by a semitone by the next (note), that is the *tertia coniunctarum*. However, when the disjunct tetrachord follows in such a way that the *media* occupies the first place, then it is surpassed by the next string, that is by the *paramese*, by a full tone. But when the disjunct tetrachord begins in such a way that it is not the *mese* that occupies first place, but the *paramese* itself, then the *paramese* is separated from the *media* by one and a half tones, and therefore it is called "disjunct tetrachord". And so it is said that it is separated from the final *media* by a tone and a half.)

The commentator in fact construes two disjunct tetrachords, one in which the *mese* and *paramese* are separated by one tone, and one in which they are separated by a tone and a half. His conjunct tetrachord begins from the *mese*, which was also the final note of the middle tetrachord. Its second note is the *tertia coniunctarum*, which is separated from the *mese* by a semitone. So far his explanation concurs with antique music theory. Then, however, he hypothesizes two disjunct tetrachords: his first disjunct tetrachord starts with the *mese*, and its second note is the *paramese*, which is separated from the *mese* by the distance of a full tone. His second disjunct tetrachord begins from a *paramese*, which is separated from the *mese* by a tone and a half.

It seems clear that even if the commentator knew the exact way in which the five tetrachords were positioned in the Perfect System, he gave the text too much authority for him to overrule it and point out Martianus' error. Rather he tries to maintain it — erroneous as it is — and uses subtle artifice to explain the error away.

Genera (enharmonic, chromatic, diatonic)

1. *Martianus Capella*

The fourth topic of Harmonia's discourse is *genus*, the way in which a tetrachord is divided into tone steps (§ 955-959). She defines three *genera*, that is, three kinds of melodic movement (*genera modulandi*): enharmonic, chromatic and diatonic. When playing in the enharmonic *genus* a musician uses the smallest steps (*dieseis*), in the diatonic *genus* the division is at its largest (using semitones and whole tones), and the chromatic *genus* lies in between — just, Harmonia explains, as colour is something which is found in between black and white.

In ascending order the division of the three *genera* of tetrachords is as follows:
- enharmonic: diesis, diesis, incomposite ditone
- chromatic: half tone, half tone, three incomposite halftones
- diatonic: half tone, tone, tone

In descending melodic progression the order is, naturally, reversed. At present, Harmonia adds, the diatonic *genus* is the most common.

There are, she explains in §958, different ways of melodic movement within the three *genera*. There is movement by skipping, which is called *πλοκή*. There is movement by consecutive pitches (*ἀγωγή*), which is called *recta* (or *εὐθῖα*) when the movement is ascending; but *revertens* (or *ἀνακάμπτουσα*) when it is descending; and *circumstans* (or *περιφερής*) when it is both ascending and descending.
In §959 Harmonia enumerates six familiar ways of dividing tetrachords; one in the enharmonic *genus*, three in the chromatic *genus* (soft chromatic, hemiolic chromatic and tonic chromatic) and two in the diatonic *genus* (soft diatonic and shrill diatonic). These six tetrachord divisions are defined in Martianus' source (see Figure 11 below, and cf. Aristides Quintilianus, *De musica libri tres*, I.9 [17][175]); Martianus, however, limits himself to mentioning their names and does not explain their division.

---

[175] Cf. table 8 in W.H. STAHL, *Martianus Capella*, I, p. 216, and T.J. MATHIESEN, *Aristides Quintilianus On Music*, p. 85 and n. 101.

Tetrachord division	Intervals (in descending order)		
divisio enharmonia (enharmonic division)	2	1/4	1/4
divisio chromatis mollis ac soluta (soft and loose chromatic division)	1 5/6	1/3	1/3
divisio chromatis hemiolia (hemiolic chromatic division)	1 3/4	3/8	3/8
divisio chromatis toniaea (tonic chromatic division)	1 1/2	1/2	1/2
divisio diatonica mollis (soft diatonic division)	1 1/4	3/4	1/2
divisio diatonica robusta (firm or sharp diatonic division)	1	1	1/2

Figure 11

2. *Glosses*

In the glosses, the fact that there are three *genera* is repeated several times. In a gloss added to a passage on the legitimate distances in music, for example, where Martianus speaks of different sizes of *dieseis*:

> [357.4-5] ENARMONIOS QUOQUE DICITUR PROPTEREA, QUOD ENARMONION MODULANDI GENUS PER HANC MAXIME DIMETIMUR
> (The enharmonic (diesis) is also called thus because we define the enharmonic kind of melody mostly by that (diesis))
>
> Tria sunt genera musicae: cromaticum, diatonum, enarmonion i. adiunctum *BeLa*, adiunctum] adunatum. *Be*
> (There are three kinds of music: chromatic, diatonic, enharmonic, that is conjoined.)

And, when Martianus speaks of "28 notes in each of the tropi",[176] the commentator explains that this number comprises diatonic, chromatic and enharmonic notes:

> [362.7-8] PER SINGULOS TROPOS In omni tropo tria genera considerantur musicae i. enarmonicum, diatonicum, cromaticum. *BeLa*
> (IN EACH OF THE MODES In each mode three *genera* of music are considered: enharmonic, diatonic and chromatic.)
>
> XXVIII Supra xviii posuit, nunc xxviii. Si volueris diatonicum tantum commemorare, tunc x et viii habebis. Si autem cromaticum vel enarmonium intromiseris, x supra habebis et erunt xxviii. Sub uno tamen nomine v cromatici et v enarmonici adduntur i. pro una corda. Sed subtiliter et obscure iste dixit. Boetius vero enucleatius patefecit. *BeLa*, et erunt xxviii *om. La*
> (28: Above he presented 18 (notes) now 28. If you wanted to enumerate just the diatonic (genus), you would have 18 (notes). But if put in the chromatics and enharmonics, you would have 10 (notes) extra, and there would be 28. Yet under one name 5 chromatics and 5 enharmonics are added, that is for one string. But he said this in a subtle and obscure manner. Boethius however made this clear in a more precise manner.)[177]

In their Greek and Latin names the penultimate notes of each tetrachord which determine the *genus* received additions which identified them as such. The penultimate note of the first tetrachord in the diatonic *genus*, for example, was called the "diatonic lichanos hypaton", while the same note in the chromatic *genus* was called "chromatic lichanos hypaton", and in the enharmonic *genus* "enharmonic lichanos hypaton".[178] This principle is commented upon in glosses added to § 942, on tetrachords:

---

[176] IX.940: sunt igitur innumerabiles soni, sed specialiter per singulos tropos viginti octo tantum poterunt convenire, quorum nomina superius memoravi. (There are an unlimited number of sounds, but in particular only twenty-eight tones can be asserted in each of the modes — I mentioned their names above.)

[177] See also my treatment of this gloss in the section on the related reception of Boethius' *De institutione musica* and Martianus Capella's *De nuptiis.*

[178] Cf. Figure 6 in the section on tones.

[362.20-1] MULTIPLEX QUIPPE TETRACHORDORUM RATIO SIC FORMATUR
(several varieties of tetrachords are thus formed)[179]

i. per additionem istarum penultimarum. *LebPar*
(By the addition of those penultimate (notes).)

i. ex distinctionibus harum cordarum i. cromaticum, enarmonicum, diatonicum. *BeLa*
(That is, out of the distinguishing characteristics of those strings, that is chromatic, enharmonic, and diatonic.)
[362.15-6] Tres penultimae istae movent tria genera modulandi. *LebPar*
(The three penultimate notes (of each tetrachord) decide on the three kinds of melodic movement.)

[363.1] Quia faciunt ipsae tres penultimae distantiam in tribus generibus modulandi *LebPar*, ipsae. *om. Par*
Because these three penultimates make the distance in the three *genera* of melodic movement.)

In book II.120, a verse sung by the muse Polymnia in honour of Philology, the rare word *crusma*[180] occurs:

melos probare ac tonos et crusmata
(<Philology! you are used to> examine melodies, sounds and tunes)

In some of the glosses this word is associated with *chroma*:[181]

[35.14-5] Croma varietas, inde crusmata quasi cromata i. crusmata i. colores varii scilicet motus. *BeLaLbLea*
(*C(h)roma* is variety, thus *crusmata* or *cromata*, different colours or (melodic) motion.)

---

[179] That is, by means of the "three penultimate" tones of each tetrachord; the possible ranges of movement of the penultimate tones were classified in the three *genera*, viz. diatonic, chromatic and enharmonic.

[180] A loan-word from Ancient Greek, κροῦσμα, technically a musical piece or tune played upon a stringed instrument and in a more general sense an air or tune (OLD, LEWIS & SHORT).

[181] In one of the manuscripts, however, viz. *Oxf*, the commentator gives an explanation which is remarkably close to the original technical meaning: CRUSMATA tractus cordarum (the plucking of strings).

> Per cromaticum tria genera musice vult intellegi, quae sunt diatonicum, cromaticum et enarmonium. *BeLa*
> (By *cromaticum* he implies the three kinds of music, which are diatonic, chromatic and enharmonic.)

In one manuscript — *Leiden, UB, Voss.Lat.F. 48* — the three divisions of the tetrachord are visualised in a marginal figure, which can also be found in manuscripts of Boethius' *De institutione musica* (see Plate 3 in the section on the related reception of Boethius and Martianus Capella).

At the end of §957 Martianus states that the diatonic *genus* is the most common one in his time, and upon this the commentator adds:

> [369.4] SED NUNC MAXIME DIATONO UTIMUR i. usui habemus eo, quod pulcrior caeteris sit. *BeLa*, pulchrior *La*
> (BUT NOWADAYS THE DIATONIC MODE IS MOSTLY USED that is, we commonly use it, because it is more beautiful than the others.)

Furthermore the term *diatonicus* is explained by *quia abundat tonis* — because it has an abundance of tones —, not only in the Martianus Capella commentary, but also in the glosses on Boethius:[182]

Anonymous commentary	*Glossa maior*
357.15-6 Diatonus dicitur, quia abundat tonis *Leb*	I.15.16a *Diatonum* ... inde dicitur, quod tonis abundat. ... *M^{2}OLaMcMf^{2}MhMkMm^{1}Pe^{3+5}R 1*S hWqAutVrf*

The diatonic division is also explained in a gloss ad 368.10:

> [368.10-1] DIATONON VERO, QUOD TONIS COPIOSUM
> (diatonic, which has an abuncance of tones)
>
> Quia tonus in primo, tonus in secundo. *BeLa*
> (Because it has a tone as the first (step), and a tone as the second (step).)

[182] Eds. M. BERNHARD & C.M. BOWER, *Glossa maior*, I, p. 224.

tonus tonus emitonium *Par*^{App}
(tone, tone, semitone)

The term *chromaticus* is explained as follows by Martianus himself:[183]

[956] sicut enim quod inter album nigrumque est color dicitur, ita hoc chroma, quod inter utrumque <enharmonia et diatonon> est nominatur.
(For just as that which is in between white and black is called colour, so is that which is in between both <enharmonic and diatonic> called *chroma*.)

This explanation is also found in the glosses:[184]

[357.7-8] Cromaticon gen<u>s dicitur colorabile. Croma enim Grece Latine color dicitur. Sicut enim color dicitur, quicquid est medium inter album et nigrum, sic cromaticum modulandi gen<u>s in medio inter enarmonion et diatonon positum est. *La*
(The chromatic genus is called the colourful one. For χρῶμα in Greek means colour in Latin. For just as something which is in the middle of white and black is called colour, thus the chromatic *genus* of melodic movement is positioned in the middle of the enharmonic and diatonic (*genera*).)

Cromaticus in tertia parte, tonos in dimidia. Vocantur a proprietate et qualitate, quod non habet nec grave nec gracile i. mixtum de duobus i. de duabus extremitatibus medium, ut aliquis color inter album et nigrum. *LebPar*, cromaticus in - et qualitate] colorabile *Par* aliquis] aliqua *Par*
(Chromatic (melodic movement goes) in third parts (of tones), (dia)tonic in half (tones). They are named after their feature and quality, which is neither heavy nor graceful but a mixture of the two; it holds the middle between the two extremes, just like some sort of colour in between white and black.)

---

[183] Cf. ISIDORUS, *Etym.*, 10.45: Chromaticus, quia non confunditur nec colorem mutat. Graece enim χρῶμα color vocatur.

[184] See also the echo in the *Glossa maior* on Boethius' *De institutione musica*: I,15,16a ... *Chroma* et chromaticum unum est, i. colorabile, quia ab utrisque accipit colorem. ... *M*^{2}*OLaMcMfMhMkMm*^{1}*Pe*$^{3+5}$*R1***Sh WqAutVrf.* (Eds. M. BERNHARD & C.M. BOWER, *Glossa maior*, I, p. 224.)

The incomposite distance of one and a half tones of the chromatic *genus* is explained in glosses ad 368.16-7:

> [368.16-7] CHROMA AUTEM HOC MODO <...> MELOS ACCIPIT: PER HEMITONIUM ET HEMITONIUM ET TRIA HEMITONIA, QUAE INCOMPOSITA PROVENIENT
> (The chromatic receives its melody in this way: by semitone, semitone and three semitones, which proceed incompositly)
>
> Quia in uno spatio tria emitonia habent et non componitur ex multis spatiis. *La*
> (Because in one space they have three semitones and (this space) is not composed out of several spaces.)
>
> Quia in uno spatio sunt. *LebPar*^{App}
> (Because they are in one space.)

On the enharmonic *genus* it should be noted that both in Martianus Capella's text and in the glosses *en(h)armonius* and *(h)armonius* are used.[185] In §957, for example, the text reads:

> genera modulandi sunt tria: *ἁρμονία, χρῶμα, διάτονον*. et harmonia quidem dicitur, quod pluribus spatiis et angustioribus separatur; diatonon vero, quod tonis copiosum; chroma, quod de hemitoniis componitur.
> (There are three kinds of melodic movement: [en]harmonic, chromatic, diatonic. That which is seperated by several and smaller steps is called enharmonic; that which is rich in full tones is called diatonic; and that which is composed of semitones is called chromatic.)

In §959, however, where he lists the six familiar divisions of tetrachords, *enharmonius* is used in stead of *harmonius*:

> ... tetrachordorum ... divisiones ... sex sunt notae: enharmonia una; chromatis tres ...; diatonici duae ...
> (There are six familiar ways of dividing the tetrachord: one enharmonic (division); three chromatic (divisions); and two diatonic (divisions) ...)

---

[185] For an example of the usage of the two terms as synonyms, see gloss 369.16-7, which is treated below.

It should further be noted that — because of a corruption of the text — the distances of the enharmonic tetrachord division are wrongly enumerated in § 957:

> [957] et harmonia quidem, cum ab immutabilibus sonis cingitur, modulationi oboediet in acumen per diesin <et diesin> et ditonon incompositum; in gravem vero sonum e contrario modulatio tota submittitur.
> (The enharmonic *genus*, since it is surrounded by fixed tones, will observe in ascending direction the order of diesis—<diesis>—incomposite ditone; in descending order, on the contrary, the entire modulation is reversed.)

The absence of *et diesin* is reflected in several glosses, which have the division as tone—tone—diesis, instead of ditone—diesis—diesis:

> [368.9-10] HARMONIA Propter diesis; sic enim dicitur tonus, tonus, diesis. *Par*[App]
> (ENHARMONIC On behalf of the diesis: for thus is called tone—tone—diesis.)
>
> [368.14] PER DIESIN ET DITONOS per diesin et duos tonos *Leb*
> (BY DIESIS AND DITONE by diesis and two tones)

The incomposite ditone is explained in the following gloss:

> [368.14] INCOMPOSITUM Quia simul sunt duo toni. Si fuerint duo toni in uno spatio, incompositum est. Quicquid vero ex diversis colligitur, compositum est, quicquid per se est in uno spatio, incompositum est. *LebPar*[App], vero *om. Leb* *post* incompositum] est *om. Leb*
> (INCOMPOSITE Because the two tones are in one space. If the two tones were in one space, it would be incomposite. For when something is put together out of different (spaces), it is composite; when something is by itself in one space, it is incomposite.)

The language of § 959 — the last section on the subject of the *genera* diatonic, chromatic and enharmonic — is not very clear. Martianus here abbreviated Aristides' explanation so much that it became difficult to make sense of the text:[186]

> [959] sed cum tetrachordorum, quos quadrifidos appellamus, divisiones innumerabiles sint, sex sunt notae: enharmonia una; chromatis tres, quarum prima quae mollis ac soluta, secunda quae hemiolia est, tertia quae est toniaea; diatonici duae, mollis atque robusta. et modos quidem accipit harmonia *τεταρτημορίᾳ διέσει*, id est toni parte quarta; illud vero, quod mollius diximus, per *τριτημορίαν δίεσιν* copulatur; ipsum vero chroma *ἡμιόλιον* fit *ἡμιολίᾳ διέσει* eius, quae sit enharmoniae.
> (Although the tetrachords, which we call *quadrifidi*, have innumerable divisions, six of these are familiar: one enharmonic (division); three chromatic (divisions), of which the first is soft and loose, the second hemiolic, and the third tonic; two diatonic (divisions), soft and sharp. The enharmonic *genus* receives its modifications from the *tetartemoria diesis*, that is a fourth part of a tone. That (division) which we called "softer" (add: chromatic), however, is bound by the *tritemoria diesis*. The hemiolic chromatic is produced from the *hemiolic diesis*, which belongs to the enharmonic *genus*.)

Thus Martianus explains that the enharmonic division uses *tetartemoriae dieseis* — fourth parts of a tone; that the soft and loose chromatic division uses *tritemoriae dieseis* — third parts of a tone; and that the hemiolic chromatic division uses *hemiolic dieseis* — three eighth parts of a tone. He does not explain the divisions of the third chromatic *genus* (the tonic division), nor those in the two diatonic *genera* (soft and shrill).[187]

A long gloss which appears in two of the main manuscripts of the Anonymous corpus illustrates the confusion caused by this section:

> [369.16-7] [1]Est emiolium, cum tres integri toni ad duos proximos tonos comparantur. [2]Est et emiolium, quando tres partes unius toni ad duas partes quaternas comparantur, de quo nunc loquitur dicendo: *ipsum vero croma emiolium fit.* [3]Ac ne putaremus de emiolio dixisse quod plenis et integris tonis conficitur, ideo subdidit *emiolio diesi* i.

[186] Moreover, due to the Greek terms the transmission of this passage in the manuscripts is rather corrupt. Cf. the critical *apparatus* in J. WILLIS, *Martianus Capella*, p. 369.

[187] Cf. Figure 10 above.

emiolio, quod quaternis partibus comparatur. [4]Ideoque sequitur *quae sit enarmonia*<*e*>. [5]Proprie etenim ad <en>armoniam pertinet illa divisio diesis, quae t[r]etartemoria dicitur. [6]Nam modulatio <en>armoniae a tetratemoria inchoat et in tritemoria terminatur. [7]Praedixit enim armonicum genus pluribus, sed angustioribus spatiis dividi. [8]Quid autem angustius potest fieri quam ut a tetratemoria inchoet et in tritemoriam desinatur? [9]Ergo cromaticum genus ab emitonio (*sic, lege* emiolio) diesi inchoat hoc est a tribus quaternis partibus unius toni. *BeLa*, 4 ideoque] ideo *La*; 8 et *om. La* desinatur] desinat *La*; 9 diesi inchoat] diesinchoat *La*

("Hemiolic" is when three full tones are compared to the two adjacent tones. "Hemiolic" is also when three parts of one tone are compared to two fourth parts, of which he is now speaking with the words: *ipsum vero croma emiolium fit*. And lest we should think that he was speaking of the hemiolic which is composed of full tones, therefore he adds *emiolio diesi* below, that is the hemiolic which is compared to fourth parts. And because of that *quae sit enarmonia* follows. After all that division of the *diesis* that is called *tetartemoria* pertains only to the "enharmonic" (genus). For enharmonic melodic movement begins with *tetartemoria*, and ends in *tritemoria*. Above he said, after all, that the enharmonic *genus* is divided in several but narrower steps. How, then, can something be made narrower than that which begins with *tetartemoria* and ends in *tritemoria*? Thus, the chromatic *genus* begins with the *emiolic diesis*, that is three fourths of a tone.)

The confusion stems from the use of the term *diesis* in Martianus' definition of the division in the chromatic *genus*. *Dieseis*, which are associated with *tetartemoria* (fourth part of a tone) and *tritemoria* (third part of a tone) by the commentator, only occur — to his mind — in the enharmonic *genus*, while in Martianus Capella the *tetartemoria* (fourth part of a tone) is indeed associated with the enharmonic division of the tetrachord, but the *tritemoria* (third part of a tone) belongs to the "soft chromatic division" (*divisio chromatis mollis ac soluta*).

Further confusion is caused by the term *emiolia diesis*, which consists of three eighth parts of a tone, i.e. one and a half times a fourth part. This distance is characteristic for the "hemiolic chromatic division" (*divisio chromatis hemiolia*). The commentator, however, did not understand the term and, at the end of his argument, considers it equal to three fourths of a tone

— a distance which is only used in the "shrill diatonic division", *divisio diatonica robusta*, and which was left out of consideration in §959.

Other glosses added to this passage show the same confusion:

> 369.12-3 CHROMATIS TRES, QUARUM PRIMA QUAE MOLLIS AC SOLUTA, SECUNDA QUAE HEMIOLIA EST, TERTIA QUAE EST TONIAEA (three chromatic (divisions), of which the first is soft and loose, the second hemiolic, and the third tonic;)
>
> MOLLIS propter tria emitonia *LebPar*^{App}
> (SOFT because of the three semitones)[188]
>
> EMIOLIA i. semis totum *BeLa*
> (HEMIOLIC one and a half)
>
> tonum et dimidium, vel tonum et semitonium *Leb*
> (a tone and a half, or a tone and a semitone)
>
> In qua considera<n>tur tria semitonia ad duo. *LebPar*^{App}
> (In which (division) three semitones against two are born in mind.)[189]
>
> †TONIA Quia unus tonus dividitur in ii emitonia et ii spatiis. (*sic, lege* spatia) *Leb*
> (TONIC Because one tone is divided into two semitones and two steps.)
>
> 369.13-4 DIATONICI DUAE, MOLLIS ATQUE ROBUSTA
> (two diatonic (divisions), soft and firm.)
>
> †UNA MOLLIS Una mollis, quia <sonat> molliter ubi tonus, mollius ubi tria emitonia. *LebPar*^{App}, Una mollis, quia *om. Par*^{App} emitonia] semitonia *Par*^{App}

---

[188] The *divisio chromatis mollis ac soluta*, the "soft and loose chromatic division" is built out of the following tone steps: $1\frac{5}{6}$ — $\frac{1}{3}$ — $\frac{1}{3}$. It does not contain semitones.

[189] In this gloss the same distinction is emphasized as in the longer gloss treated above: the hemiolic ratio can be applied to whole tones, but also to semitones — thus the *hemiolic diesis* consists of 1½ times a half tone, that is, three eighths of a tone.

(ONE SOFT The one is soft, because where there is a tone it sounds reasonably soft, softest, however, where there are three semitones.)[190]

MOLLIS Quae currit per tonum i. gravis. *Leb*
(SOFT (The division) which runs using a tone, that is low.)

959.14-5 ET MODOS QUIDEM ACCIPIT HARMONIA TETAPTHMOPIA ΔΙΕΣΕΙ, ID EST TONI PARTE QUARTA
(The enharmonic *genus* receives its modifications from the *tetartemoria diesis*, that is a fourth part of a tone.)

MODOS differentias *LebPar*[App]
(MODIFICATIONS distinguishing characteristics)

In prima <differentia> spatia, in secunda proportio, in tertia quod tonus dividitur in duo semitonia consideratur. *LebPar*[App]
(In the first modification spaces are examined, in the second proportion, in the third the way in which a tone is divided into two semitones.)

Eo quod unus tonus divid<itur>. *Par*[App]
(The way in which one tone is divided.)

PARTE QUARTA A quarta parte incoat. *La*
(A FOURTH PART It begins with a fourth part.)[191]

369.15-6 ILLUD VERO, QUOD MOLLIUS DIXIMUS, PER TPITHMOPIAN ΔΙΕΣΙΝ COPULATUR
(That (division) which we called "softer" (add: chromatic), however, is bound by the *tritemoria diesis*.)

ILLUD croma *LebPar*[App]
(THAT (DIVISION) the chromatic one)

---

[190] The actual division of the *divisio diatonica mollis* (soft diatonic division) is 1¼ — ¾ — ½. The commentator did not grasp this from Martianus' cryptic description.

[191] When the commentator says "it *begins* with a fourth part", we can conclude that, in this particular instance, he thought of the tetrachord-divisions in ascending order.

ΤΡΙΤΕΜΟRIAN ΔΙΕCIN i. tertiam partem toni *La*
(TRITEMORIA DIESIS a third part of a tone)

COPULATUR s. armonia *La*
(IT IS BOUND viz. the enharmonic (*genus*))[192]

369.16-7 IPSUM VERO CHROMA ΗΜΙΟΛΙΟΝ FIT HEMIOΛIA ΔΙΕΣΕΙ EIUS, QUAE SIT ENHARMONIAE
(The hemiolic chromatic is produced from the *hemiolic diesis*, which belongs to the enharmonic *genus*.)

EMIOLION i. diatonon *LebPar*^{App}, i. *om. Leb*
(HEMIOLIC that is diatonic)[193]

HEMIOΛIA Quia constat sesqualteris. *Par*^{App}
(HEMIOLIC Because it consists of the sesqualter (ratio).)

†ENARMONII Quia inter primam et ultimam sesqualtera est, in<ter> primam et secundam sesquitertia est. *LebPar*^{App}, *difficile lectu in Par*^{App}
(ENHARMONIC Because between the first and the last (string?) there is a sequalter (ratio), between the first and the second a sequitertian (ratio).)[194]

3. *Conclusion*

From the fact that the commentator did not understand the

---

[192] Although the "soft chromatic *genus*" (*divisio chromatis mollis ac soluta*) is implied here, the commentator adds that it concerns the enharmonic *genus*.

[193] Again the commentator does not follow Martianus' explanation: the hemiolic chromatic division is implied, but "diatonic" is added in the gloss.

[194] The meaning of this gloss remains unclear. Does the commentator mean that there is a sesquialter (or hemiolic) ratio between the first and last string of the tetrachord, and that there is a sesquitertian ratio between the first and the second string? This would mean, however, that the distance between first and last string measures a fifth instead of a fourth, and the distance between first and second measures a fourth. One can also imagine that ratios are implied within the distance of a tone, but even then, things do not add up. After all, the distances of the hemiolic chromatic division are 1¾ — ⅜ — ⅜; neither the hemiolic ratio nor the sesquitertian ratio can be recognised in the comparison of these distances with each other.

somewhat cryptic explanation of the six different ways of dividing the tetrachord, one could perhaps assume that he did not have an active knowledge of Boethius' *De institutione musica* V.16, where the same six divisions of the tetrachord according to Aristoxenos are explained and illustrated in a diagram. The more basic information on *genera*, which certainly was understood correctly by the commentator, is also treated by Boethius in I.21-23. This material, however, is so similar to that found in *De nuptiis*, that it is not possible to determine the extent to which the commentator based himself on *De institutione musica* for his explanation of *De nuptiis*. Nevertheless, in one of the ninth-century manuscripts a visual representation of the *genera* is found which corresponds to a figure in *De institutione musica* I.21. And since this manuscript, *Leiden, UB, Voss.Lat.F. 48*, occupies such a central place in the transmission of the Anonymous corpus of glosses, the appearance of this Boethian figure in the margins of *De nuptiis* strongly suggests that here too the commentator used Boethius as a source for his understanding of Martianus Capella. Knowing this, it seems safe to assume that he would have turned to Boethius for understanding the six ways of dividing the tetrachord had he known that Boethius treated them; consequently, the lack of understanding of this piece of antique music theory seems to imply that he did not know book V of *De institutione musica*.

Since neither the system of the five tetrachords nor the three *genera* were particularly important in the practice of Gregorian chant, it is only logical that there is not much evidence of this material in the musical treatises of, for example, Aurelian of Réôme or the Anonymous *Musica enchiriadis*. Hucbald of Saint-Amand treated the subject of tetrachords quite extensively in his *De musica* (§ 29-42), and on the basis of this piece of antique music theory he developed his famous notational system. Hucbald, however, seems to have borrowed his material largely from Boethius, who presented the theoretical fundament of tetrachords in a much clearer way.[195]

---

[195] On this see Y. CHARTIER, *L'oeuvre musicale d'Hucbald de Saint-Amand*, pp. 48-49. Boethius is mentioned twice as a source by Hucbald, for example, in § 29-30 (*ibid.*, p. 168). See also pp. 56, 124-126.

Modi *or* tropi

1. *Martianus Capella*

Martianus Capella treats the concept of the *tropi* or *modi*[196] of antique music theory only briefly. Because of errors on his side or because of an erroneous transmission in the manuscripts, however, what little he has to say on the subject, however, is quite a puzzle; a puzzle which was solved for me by Lucio Cristante in his study on the ninth book of *De nuptiis*,[197] and by Charles Atkinson in a paper he gave at the Phoenix Meeting of the American Musicological Society in 1997.[198] *Modi* are transpositions of (double) octave systems through the gamut of the Perfect System; scales of whole, half and smaller tone distances in a certain order. In §935, Martianus states the following:

> [935] Tropi vero sunt quindecim, sed principales quinque, quibus bini cohaerent: id est Lydius, cui adhaerent Hypolydius et Hyperlydius; secundus Iastius, cui sociatur Hypoiastius et Hyperiastius; item Aeolius cum Hypoaeolio et Hyperaeolio; quartus Phrygius cum duobus Hypophrygio et Hyperphrygio; quintus Dorius cum Hypodorio et Hyperdorio. verum inter hos tropos est quaedam amica concordia, qua sibi invicem germanescunt, ut inter Hypodorium et Hypophrygium, et item inter Hypoiastium et Hypoaeolium; item conveniens aptaque responsio inter Hypophrygium et Hypolydium, qui tamquam duplices copulantur.[199] mediae vero graviorem troporum his, qui acutiores sunt, προσλαμβανόμεναι fiunt.
> (There are fifteen modes; five principal ones and a pair attached to

---

[196] He also uses the term *tonus* in the sense of "mode"; cf. 371.17.

[197] L. CRISTANTE, *Martiani Capellae Liber IX*, p. 136 and pp. 293-295.

[198] I am sincerely grateful to Charles Atkinson for giving me the opportunity to read his paper before publication. In fact, the presentation of the § 935-puzzle and its solution is largely borrowed from his paper. The paper will, in due time, appear in the *Journal of Musicology*.

[199] Willis is the only editor who has *tamquam duplices*; the others have *tantum duplices* — in Stahl's translation: "these are *the only pairs* that are joined together." J. WILLIS, *Martianus Capella*, p. 359 (and *apparatus*); W.H. STAHL, *Martianus Capella*, II, p. 362.

each of them: the Lydian mode, to which Hypolydian and Hyperlydian are attached; the second is the Iastian, with which the Hypoiastian and Hyperiastian are associated; the same for the Aeolian with Hypoaeolian and Hyperaeolian; the fourth is the Phrygian with the pair Hypophrygian and Hyperphrygian; the fifth the Dorian with Hypodorian and Hyperdorian. Among these modes there is a certain friendly kinship, which makes them like brothers and sisters, such as between Hypodorian and Hypophrygian, and the same between Hypoiastian and Hypoaeolian; a similar agreeing and fitting relation exists between Hypophrygian and Hypolydian, which are joined as doubles. For the *mediae* of the lower modes become the *proslambanomenai* in the higher ones.)

The problem here lies in the "friendly pairs" Martianus mentions as examples. When we compare the examples mentioned here with Boethius' material on modes and relations between them, they seem not to fit the profile in any way. The *media* of the Hypodorian mode does not equal the *proslambanomenos* of the Hypophrygian mode. The comparison falls short in more respects, because Boethius mentions only seven modes, viz. Hypodorian, Hypophrygian, Hypolydian, Dorian, Phrygian, Lydian and Mixolydian, to which later the eighth mode, Hypermixolydian, is added to complete the consonance of the double octave.[200] Boethius illustrates the relations between the modes in two diagrams,[201] of which the second is of importance here: it is the wing diagram of which I spoke earlier in the section on the related reception of Boethius and Martianus Capella. I shall return to this wing diagram later.

The source of Martianus' passage is unknown, but for the greater part of book IX Martianus recapitulated the *De musica libri tres* of Aristides Quintilianus, who, in his turn, treated the subject of the modes in chapter 10 of the first book.[202] Aristides states that there are 13 modes, to which the "younger theorists"

[200] Cf. BOETHIUS, IV, 15-17.

[201] BOETHIUS, *De institutione musica*, IV, 15-16, "descriptiones I & II pagina 343 addendae"; C.M. BOWER, *Fundamentals of music*, Fig. D20 and D21, pp. 155-156.

[202] Ed. R.P. WINNINGTON-INGRAM, *Aristidis Quintiliani*; transl. Th.J. MATHIESEN, *Aristides Quintilianus*.

have added another two. Their relations to each other are aptly illustrated in Mathiesen's translation in a footnote:

Aristoxenos (Aristides' source)		Proslambano menos	"Younger theorists"
		g	Hyperlydian
		f#	Hyperaeolian
Hyperphrygian (or Hypermixolydian)		f	
High and low Mixolydian:	Hyperiastian	e	
	Hyperdorian	e♭	
High and low Lydian:	Lydian	d	
	Aeolian	c#	
High and low Phrygian:	Phrygian	c	
	Iastian	B	
	Dorian	B♭	
High and low Hypolydian:	Hypolydian	A	
	Hypoaeolian	G#	
High and low Hypophrygian:	Hypophrygian	G	
	Hypoiastian	F#	
	Hypodorian	F	

Figure 12[203]

[203] Reproduction of Mathiesen's diagram in n. 109, p. 87, based on Aristides Quintilianus' text in I,10 and tables of Alypius' *Eisagoge music*e.

Aristides describes the relations of the modes in terms of pitch as follows:

> Each of the tonoi (*i.e. modes*) will exceed the former by a semitone if we choose to begin from the lowest pitch, or will be lower by a semitone if we make a beginning from the highest pitch. Their proslambanomenoi comprise, as I said, an octave and a tone.[204]

Somewhat further in his text he speaks of a diagram illustrating these relations:

> Having made a beginning of the deepest of the modes, we ascend again by a semitone and prescribe the proslambanomenos of the next mode; and by placing together the sequential tones by the same excesses, we fill out the number of the fifteen modes. <...> The diagram of the modes is akin to a wing, explaining the excesses which the tonoi have to one another.[205]

Unfortunately the diagram does not seem to have been a part of Aristides' book, but the "wing diagram" can be construed from Alypius' *Eisagoge music*e, which is what Mathiesen did in his introduction to *De musica libri tres*; on p. 23 he presents a "Diagram of the Modes". A simpler version of it is here represented as Figure 13. In this diagram the *proslambanomenos* and *mese* of each mode are given.

Now it becomes clear what Martianus' "friendly pairs" should have been: the *mese* of the Hypodorian mode corresponds to the *proslambanomenos* of the *Hyper*phrygian mode, not the *Hypo*phrygian; likewise, the *mese* of the Hypoiastan mode corresponds to the *proslambanomenos* of the *Hyper*aeolian mode, not the *Hypo*aeolian; and, finally, the *mese* of the Hypophrygian mode corresponds to the *proslambanomenos* of the *Hyper*lydian mode, not the *Hypo*lydian.[206]

---

[204] *Ibid.*, p. 87.

[205] *Ibid.*, p. 89.

[206] Although I do not suggest a direct relation, it cannot be a coincidence, to my mind, that in the corrupted version of the text the prefix *hyper* disappeared and was replaced by *hypo*, and that this same elimination took place when the eight church modes received their Greek names in medieval music theory.

	Proslambanomenos	Mese
Hypodorian	F	f
Hypoiastian	F#	f#
Hypophrygian	G	g
Hypoaeolian	G#	g#
Hypolydian	A	a
Dorian	A#	a#
Iastian	B	b
Phrygian	c	c'
Aeolian	c#	c#'
Lydian	d	d'
Hyperdorian	d#	d#'
Hyperiastian	e	e'
Hyperphrygian	f	f'
Hyperaeolian	f#	f#'
Hyperlydian	g	g'

Figure 13

The correct text, which is actually found in some of the many Martianus Capella manuscripts, and which was also already emended by Lucio Cristante in his study on the ninth book of *De nuptiis*, should have read:

> Verum inter hos tropos est quaedam amica concordia, qua sibi invicem germani sunt, ut inter *ὑποδώριον* et *ὑπερφρύγιον*, et item inter *ὑποιάστιον* et *ὑπεραιόλιον*; item conveniens aptaque responsio inter *ὑποφρύγιον* et *ὑπερλύδιον*, qui tamquam duplices

copulantur:[207] mediae vero graviorem troporum his, qui acutiores sunt, προσλαμβανόμεναι fiunt.[208]

In our four most important manuscripts for book IX, *Besançon 594, Leiden Voss.Lat.F. 48, Leiden BPL 36* and *Paris lat. 8671*, the text was transmitted as shown below. A distant echo of the correct text still shines through in the first pair of modes, which read *hypodorium* and *hypephrygium* in the manuscripts *Besançon 594* and *Leiden Voss.Lat.F. 48*, but most of the Greek terms are corrupt. From these four examples it can also be seen that the Carolingian copyists considerable trouble with the Greek characters which must have been used in the original.

In his paper Atkinson mentions only one (un-glossed) Carolingian manuscript in which the scribe got the text right: *Bamberg, Staatsbibliothek, Misc.Class. 39* (a ninth- or tenth-century manuscript of French origin).[209]

---

[207] In the manuscripts the text here shows two variants: *qui tantum duplices copulantur* and *qui tamquam duplices copulantur*, the first meaning, in essence, "these are the only pairs thus joined", the second, "which are joined as in a duple ratio". Nearly all the editors of *De nuptiis*, including L. Cristante, chose the first reading, *qui tantum duplices copulantur*, only Willis chose the second. Both readings are possible: there are indeed only three pairs of modes which are related in the way Martianus describes (the *media* of the lower mode can be equalled to the *proslambanomenos* of the higher one); *and* the ratio which arises between the two modes of each pair is indeed the duple ratio. However, I would prefer *qui tamquam duplices* to *qui tantum duplices*, since in this context it makes more sense that the ratio between the two modes of each pair is emphasized. Moreover, the glosses seem to react to this reading, even if their text in fact has the other, as we will see in *Leb* and *Par*. (Cf. the table on p. 309, showing the glosses ad 359.17 QUI TANTUM DUPLICES.)

[208] L. CRISTANTE, *Martiani Capellae Liber IX*, p. 136. In the "commento" we read that the same emendation was first suggested by F.A. GEVAERT in *Histoire et théorie de la musique dans l'antiquité*, I-II, Gand 1875-81, I, p. 225; and repeated by G. WILLE in *Musica Romana. Die Bedeutung der Musik im Leben der Römer*, Amsterdam 1967, p. 645 n. 460. It was thus available to nearly all the modern editors of the text: Dick, Préaux and Willis.

[209] According to him 22 manuscripts of the 241 have the passage copied correctly; none of the glossed manuscripts he saw had the correct pairs.

Emended text:	ὑποδώριον et ὑπερφρύγιον	ὑποιάστιον et ὑπεραιόλιον	ὑποφρύγιον et ὑπερλύδιον
Bes. 594 (f. 78r)	ΥΠΙΟΔΟΡΙΥΜ et ΥΠΕΦΡΙΓΙΥΜ	ΥΡΟΓΑCTIEΥΜ et ΙΠΟΕΟΛΙΕΥΜ	ΥΠΟΦΡΙΓIUM et ΥΠΟΛΥΔΙΟΥΜ
Leiden VLF 48 (f. 86v)	ΥΠΟΔΟΡΙΥΜ et ΥΠΕΦΡΙΓΙΥΜ	ΥΠΟΓΑCTEΥΜ et ΥΠΟΕΟΛΙΕΥΜ	ΥΠΟΦΡΙΓIUM et ΥΡΟΛΙΔΙΟΥΜ
Leiden BPL 36 (f. 121r)	ΥΠΟΔΟΡΙΟΥΜ et ΥΠΟΦΡΙΓΙΟΥΜ	ΥΠΟΙΑCTΙΟΥΜ et ΥΠΟΛΙΔΙΟΝ	ΥΠΟΦΡΙΓΙΟΥΜ et ΥΠΟΕΟΛΙΟΝ
Paris lat. 8671 (f. 79r)	ΥΠΟΔΟΡΙΟΥΜ et ypofrigium	ΥΠΟΙΑCTΙΟΥΜ et ΥΠΟΛΙΔΙΩΜ	ΥΠΟΦΡΙΓΙΟΥΜ et ypolidium [eolium]

2. *Glosses*

In the Anonymous corpus there are not many glosses added to § 935 on the disposition of the modes:

[359.9] TROPI VERO SUNT QUINDECIM, SED PRINCIPALES QUINQUE, QUIBUS BINI COHAERENT
(There are fifteen modes; five principal ones and a pair attached to each of them:)

TROPI i. modi *BeLa*
(TROPI i.e. modes)
PRINCIPALES sunt v *Be*
(THE PRINCIPAL ONES are five)
BINI COHERENT i. superior et inferior *BeLa*, i. *om. La*

Primo <facis> materiam in animo simul cum gravitate aut etiam cum acumine. Ergo si libet tibi, ut ex gravioribus tropis alterum formes, praeparandum est opus aut fistula aut fidibus et caetera, aut etiam voce, similiter <facere potes> fide acutis. *BeLa*, *difficile lectu in La*

(THERE IS A PAIR ATTACHED TO THEM a higher one and a lower one.)

(First <you compose> the (melodic) material in your mind, at the same time for the high region and for the low region. Thus, if it pleases you to form a different (melody) out of the lower modes, — the work has to be adapted for a flute or stringed instrument etcetera, or even for the human voice —, then <you can make> the same (melody) for a high string.)

BINI secundum numerum Grecae linguae *LebPar*, Grecae linguae] grecorum linguarum *Par*
(BINI the second number in Greek)
ascensiones et descensiones *LebPar*
(rising and descending melodic movements)

[359.10] ID EST LYDIUS, CUI ADHAERENT HYPOLYDIUS ET HYPERLYDIUS
(the Lydian mode, to which Hypolydian and Hyperlydian are attached)

ID EST LIDIUS [Qui gravius sonat. *La*]
(LYDIAN [which sounds lower])
in Lydia inventus. *LebPar*
(invented in Lydia)
YPO pro sub *LaLeb*, pro *om. Leb*
i. sub, quia gravius sonat. *Be*
in gravitate *LebPar*
(YPO that is "sub", below
below, because it sounds lower
in the low region)

YPER super *BeLeb*
(YPER that is "super", above)

[359.10-1] SECUNDUS IASTIUS, CUI SOCIATUR HYPOIASTIUS ET HYPERIASTIUS
(the second is the Iastian, with which the Hypoiastian and Hyperiastian are associated)

IACTIUC Iones eo utuntur. *Leb*
(IASTIAN the Ionians used this mode.)
YIIOIACTIOC nominativus singularis *BeLa*
(HYPOIASTIAN singular nominative)

[359.12-13] QUARTUS PHRYGIUS CUM DUOBUS HYPOPHRYGIO ET HYPERPHRYGIO
(the fourth is the Phrygian with the pair Hypophrygian and

Hyperphrygian)

quintus Dorius, cum *υποδοριο* et *υπερδοριο* *BeLa*
(the fifth is the Dorian with Hypodorian and Hyperdorian;)
FRIGIUS a Frigia *Par*
(PHRYGIAN from Phrygia)

[359.14] VERUM INTER HOS TROPOS EST QUAEDAM AMICA CONCORDIA
(Among these modes there is a certain friendly kinship)

VERUM sed TROPOS modos *BeLa*
(VERUM but TROPOS modes)

[359.17] ITEM CONVENIENS APTAQUE RESPONSIO INTER ΥΠΟΦΡΥΓΙΟΝ ET ΥΠΕΡΛΥΔΙΟΝ, QUI TAMQUAM DUPLICES COPULANTUR:
(a similar congruent and fitting relation exists between Hypophrygian and Hypolydian, which are joined as duples.)

ΥΠΟΛΙΔΙΟΥΜ i. ypofrigius *BeLa*, i. *om. La* iprofigius *La*
(HYPOLYDIAN that is (is joined with) Hypophrygian)
QUI †TANTUM DUPLICES COPULANTUR Subaudis: in gravioribus tonis vel tonus ad emitonium. *Leb*
(†ONLY JOINED PAIRS[210] <”in a duple ratio”> Understand: to the lower notes or a tone to a semitone.)
Germani, ut tonus ad emitonium in duplici proportione. *LebPar*
(They are like brothers and sisters, just as a tone and a semitone in the duple ratio.)

[359.17-1] MEDIAE VERO GRAVIORUM TROPORUM HIS, QUI ACUTIORES SUNT, ΠΡΟΣΛΑΜΒΑΝΟΜΕΝΑΙ FIUNT
(For the *mediae* of the lower modes become the *proslambanomenai* in the higher ones.)

MEDIAE cordae *Leb*
(THE MIDDLE strings)
tonae *Par*

---

[210] The glosses, however, seem to react to the variant *tamquam duplices*, “as doubles” or, “as joined in a duple ratio”.

(notes)
HIS tropis *La*
(IN THESE modes)
ΠΡΟCΛΑΜΒΑΝΟΜΕΝΑΥ i. in gravissimis *BeLa*
(*PROSLAMBANOMENAI* that is in the lowest modes)
Non solum proslambanomenai fiunt in gravissimis, sed etiam in accutis tropis graves sunt, sed tamen non sic graves in acutis sicut in gravissimis. *BeLa*
(The *proslambanomenai* exist not only in the lowest modes, but they are also low in high modes; they are just not as low in the high modes as in the low modes.)
adquisiti *Leb*
(*adquisiti*)

In these glosses the names of the modes are explained. To each of the five principal modes a pair of related modes is attached: a lower one, in which name the prefix *hypo* or *sub* occurs, and a higher one, with the prefix *hyper* or *super*. The commentator also construes how these triplets of related modes should work: a melody composed in one of them can easily be transposed to a higher or lower range of pitches by transposing them to the related modes. A work composed for flute can thus be adapted for strings, or even for voice.[211] The names of the five principal modes are also explained: the Lydian mode is so called because it was invented in Lydia; the Iastian (or Ionian) mode was used by the Ionians; and the Phrygian mode originated in Phrygia. Similar explanations are found elsewhere in the Anonymous corpus, for example:

[354.12-3] TIBIAE PER TRITONIDAM NOSTRI COMITEM MARSYAMQUE LYDIUM SONUERUNT
(flutes were blown by my companion the Tritonian and by the Lydian Marsyas)

LIDIUM Quia in Lidia repertus est tonus. *BeLa*, tonus] sonus *La*
(LYDIAN Because in Lydia a mode is found.)
tropus Lidium *La*
(The Lydian mode)

[211] This interesting gloss will also be treated in the last section of this chapter, *Theory and Practice.*

Quintum tonum principalem *Leb*
(The fifth principal mode)[212]

And they are also found in the *Glossa maior* on Boethius' *De institutione musica*, ad 180,16:

[I,1,232] LYDIUS MODUS (THE LYDIAN MODE)
a Lydia provincia (From the province of Lydia)

[I,1,236] PHRYGIUS (PHRYGIAN)
a Frigia vel Frigibus (From Phrygia or the Phrygians)

Furthermore, ad book II.197, on the sound brought forth by Saturn in the harmony of the cosmos, it is said that Saturn resounds in the Dorian mode:

[54.5-8] hinc etiam praetergressa circum ac parili interiectione sublimis deorum rigidissimum creatorem in algido haerentem pruinisque nivalibus conspicata; verum idem, quem circumire nitebatur, orbis melo Dorio tinniebat.
(When she went an equal distance beyond this circle, from her lofty position she saw the creator of the gods, (viz. Saturn), stiff with cold and with chill hoarfrost clinging to him; she tried to circle round the planet, which rang with Dorian sound.)[213]

In all the manuscripts with glosses from the Anonymous corpus it is added that the Dorian mode is the lowest one, information which is not found in *De nuptiis*, but in Cassiodorus' *Institutiones*:[214]

[54.8] MELO DORIO sono gravissimo *BeLaLbVaLeaOxf*, de *add. Be*
gravissimo sono *LaOxf*
(DORIAN SOUND the lowest sound)
Dorus enim trophos in musica graviter sonat. *Oxf*
(Because the Dorian mode sounds low in music.)

---

[212] Note that the order of the modes is apparently reversed here: the Dorian mode, the lowest one, is viewed as the first, and the Lydian one, the highest, as the last. Martianus enumerates them in reverse order in § 935.

[213] Transl. W.H. STAHL, *Martianus Capella*, II, p. 60.

[214] That is, the Dorian mode is the lowest of the five principal modes. Of the fifteen modes the Hypodorian is the lowest. Cf. CASSIODORUS, *Institutiones*, II, v.

On the relationship of the three (false) pairs of modes — Hypodorian and Hypophrygian, Hypoiastian and Hypoaeolian, Hypophrygian and Hypolydian —, which relate to each other in the duple ratio of the octave, the commentators naturally remained in the dark. Their task was doubly difficult: they had to make sense of nonsense. In several manuscripts of the Anonymous corpus, the glosses explain that the distance between Hypodorian and Hypophrygian, Hypoiastian and Hypoaeolian and Hypophrygian and Hypolydian was that of a whole tone. In itself, this is correct — cf. Figure 13 in this section —, but it does not explain the relation between *media* of the one, and *proslambanomenos* of the other mode. It is then argued that this whole tone in itself holds some kind of duple ratio: a whole tone contains two semitones — the basic unit with which the modes are separated from each other (cf. glosses ad 359.17 in *Leb* and *Par*). This same solution appears in a manuscript which contains John the Scot's commentary on book IX, and in Remigius of Auxerre's later commentary:

John the Scot's commentary in *Leiden, BPL 88* (fol. 174)	Remigius of Auxerre's commentary[215]
[359.17] DUPLICES acutior enim duplo superat graviorem. tono (*sic, lege* tonus) quippe ad emitonium sonat. (DUPLE the higher surpasses the lower by two. For a tone sounds in relation to a semitone.) [359.18] MEDIAE VERO GRAVIORUM TROPORUM HIS, QUI ACUTIORES SUNT, ΠΡΟΣΛΑΜΒΑΝΟΜΕΝΑΙ FIUNT vicem proslambanomenes cordae obtinent. (THE *MEDIAE* OF THE LOWER MODES BECOME THE *PROSLAMBANOMENAI* OF THE ONES WHICH ARE HIGHER The strings occupy the function of the *proslambanomenai*.)	[498.9] QUI TANTUM DUPLICES COPULANTUR Acutior enim duplo superat graviorem. Tonus quippe ad hemitonium sonat. Est autem tonus duplum hemitonii. (The higher surpasses the lower by two. A tone sounds in relation to a semitone. A tone is two semitones.)

[215] C.E. LUTZ, *Remigius Autissiodorensis*, p. 334.

In quite a few Carolingian manuscripts a diagram appears at the end of book IX, in which the pitch relations of the modes are visualized in a Boethian-like wing diagram (see also the treatment of this diagram in the section on the related reception of *De institutione musica* and *De nuptiis*). Christian Meyer composed a list of the manuscripts in which the diagram is copied:[216]

1. *Cambridge, Corpus Christi College, Parker Library, Ms. 153*
   (9th century, English origin, eclectic corpus of *collectae glossae*, diagram on f. 85v)
2. *Cambridge, Corpus Christi College, Parker Library, Ms. 330*
   (10th century, English origin, the same corpus as *Ms. 153*, diagram on f. 126r)
3. *Köln, Diözesan- und Dombibliothek, 193*
   (9th century, origin: Cologne, a mixed corpus of glosses from the Anonymous corpus and from Remigius of Auxerre's commentary, diagram on f. 200v-201r)
4. *München, Bayerische Staatsbibliothek, Clm 14729*
   (9th century, origin: St. Gall, a mixed corpus of glosses, diagram on f. 220r)
5. *Wien, Österreichische Nationalbibliothek, cpv 266*
   (10th/11th century, origin: Würzburg?, mixed corpus of glosses related to that of John the Scot Eriugene, diagram on fol. 150v)
6. *Paris, Bibliothèque Nationale de France, Lat. 8671*
   (9th century, origin: region Bodensee?, Lorsch?, mixed corpus of glosses from the Anonymous corpus and from the commentary of John the Scot Eriugene, the key to the diagram is copied on f. 84v, but the diagram itself is lacking)

To Meyer's collection we can add one more manuscript:

7. *Leiden, Universiteitsbibliotheek, BPL 36*
   (9th century, origin: Lorsch?, mixed corpus of glosses, related to those in *Paris, BN, lat. 8671*, diagram on f. 129v)

The accompanying text of the diagram can be edited as follows:

> Hi sunt tropi xv, hoc est formae carminum, qui peraguntur in cithara viginti octo[217] cordarum.

---

[216] C. MEYER, *Métaphore instrumentale et presentation du système acoustique*.

[217] The Wien manuscript has "x et viii" instead of "xxviii" or "viginti octo". On this difference in the number of strings see the section on tones above.

(These are the fifteen modes, that is, the shapes of songs which are performed on a cithara of twenty-eight strings.)

Cordae sunt quae vergunt a fronte paginae in eius imam partem. Lineae vero quae porriguntur a leva in dexteram termini sunt troporum.
(The strings are the (lines) which run from the head of the page to its bottom. The lines, on the other hand, which stretch from the left to the right, determine the limits of the modes.)

Thus a diagram is drawn in which the vertical lines correspond to the 28 or 18 strings of the Perfect System, and the horizontal lines mark the fifteen modes mentioned in § 935 of *De nuptiis*. The "friendly kinship" between some of the modes — i.e., the fact that the *media* of one mode becomes the *proslambanomenos* of the other — is illustrated in this diagram, and explained in an accompanying text:

Mese ypodorii est adquisita eolii.
Mese ypofrigii, adquisita iastii.
Mese ypoeolii, adquisita lidii.
Mese ypoiastii, adquisita yperdorii.
Mese ypolidii, adquisita yperfrigii.
Mese dorii, adquisita ypereolii.
Mese frigii, adquisita yperiastii.
Mese eolii, adquisita yperlidii.
(The *mese* of the Hypodorian mode is the *adquisitus* of the Aeolian mode. The *mese* of the Hypophrygian mode is the *adquisitus* of the Iastian mode. The *mese* of the Hypoaeolian mode is the *adquisitus* of the Lydian mode. The *mese* of the Hypoiastian mode is the *adquisitus* of the Hyperdorian mode. The *mese* of the Hypolydian mode is the *adquisitus* of the Hyperphrygian mode. The *mese* of the Dorian mode is the *adquisitus* of the Hyperaeolian mode. The *mese* of the Phrygian mode is the *adquisitus* of the Hyperiastian mode. The *mese* of the Aeolian mode is the *adquisitus* of the Hyperlydian mode.)

The transmission of the diagram is not, however, very consistent. In *Cambridge 153*, for example, six vertical lines are drawn per mode, i.e., six strings instead of 18 or even 28.[218] In *Cambridge*

---

[218] Cf. C. MEYER, *Métaphore instrumentale et presentation du système acoustique*, Figure 1.

330 the diagram was never finished: for the mode at the top four vertical lines were drawn, but from the fifth mode onwards, no lines are drawn.[219] In *Wien cpv 266*, the top modes have eight vertical lines, but from the Hyperphrygian mode onwards this number gradually decreases, and the final Hyperlydian mode has only four vertical lines.[220]

Correct diagrams are found in three manuscripts: *Leiden BPL 36, München 14729* and *Köln 193*, the Leiden manuscript being the oldest (reproduced in Plate 2 in the section on the related reception of *De institutione musica* and *De nuptiis*). They present eight vertical lines per mode, that is from *proslambanomenos* to *mese* in a diatonic scale; exactly the right number of strings, in other words, to show the relationship at hand. Following the last vertical line of the Hypodorian mode downwards until it becomes the first line of another mode, one can easily conclude that this is the Aeolian mode: the *mese* of the Hypodorian mode corresponds to the *proslambanomenos* of the Aeolian mode. These octave relations are shown for the other seven modes according to the same principle.

This diagram thus attempts to illustrate the "friendly kinship" of which Martianus speaks in § 935. Its actual conclusions, however, are completely different. Martianus mentioned three pairs: Hypodorian and Hypophrygian, Hypoiastian and Hypoaeolian, Hypophrygian and Hypolydian. These pairs do not emerge as octave-related couples from the diagram, not even if we emend the text by replacing the "Hypo-" of the second mode of each pair by "Hyper-". In the diagram Hypodorian and Hypophrygian are neighbours, separated from each other by a whole tone; Hypoiastian and Hypoaeolian are also neighbours, but in a descending direction; Hypophrygian and Hypolydian are separated by four strings. Moreover, the diagram shows that there are not just three pairs of related modes, but eight.

If we compare the diagram to the one in Boethius' *De institutione musica* (ad IV,16),[221] on the other hand, the

[219] *Ibid.*, Figure 2.

[220] *Ibid.*, Figure 5.

[221] See Plate 1 in the same section and C.M. BOWER, *Fundamentals of Music*, Diagram D.21, p. 156.

conclusion that the lay-out of the diagram was at least inspired by him is obvious. It is also obvious, however, that it is not copied exactly from Boethius: in Boethius' diagram 26 vertical lines are drawn for a complete scale of a double octave in the diatonic mode; these vertical lines represent distances of a semitone; the cells (*paginulae*) created by the vertical and horizontal lines contain the Greek notational symbols; both sides, left and right, are arranged in a steplike way, and the lowest mode is presented at the bottom of the page instead of at the top; and finally, only eight modes are presented: Hypodorian, Hypophrygian, Hypolydian, Dorian, Phrygian, Lydian, Mixolydian and Hypermixolydian. His list of modes, furthermore, is quite different from Martianus' list.

In the diagram added to Martianus Capella, a Carolingian scholar cleverly adapted a lay-out found in *De institutione musica* to material found in *De nuptiis*, arriving at a conclusion which is found in neither of the two treatises. It is difficult to decide whether the diagram was part of the Anonymous commentary. It appears only in manuscripts with mixed corpora of glosses, partly the Anonymous commentary, partly that of John the Scot, partly unidentified glosses. Apart from the appearance of the diagram in the two closely related English manuscripts, its transmission seems to have taken place mainly in the German part of the Carolingian empire — Lorsch, Würzburg, Köln, Bodensee. Since the origin of most manuscripts is uncertain, however, this conclusion can only be hypothetical. Nevertheless, it is striking, to my mind, that the diagram appears in none of the manuscripts which are placed with certainty in the French part of the Carolingian empire (Corbie, Auxerre, Laon). It is also remarkable that the glosses of half of these manuscripts are in some way linked to the commentary of John the Scot — the same person who linked a passage from book II of *De nuptiis* to Boethius' wing diagram of modal transpositions, which formed the basic inspiration for our diagram. We should perhaps conclude that the diagram was more part of his gloss tradition than of the Anonymous corpus. More importantly, it also emphasizes the suggestion that the two traditions are inseparably entangled in the manuscripts that survive today.

Theory and Practice

> Drittens: Die karolingische Rezeption der "Institutio musica" liefert keinen einzigen Hinweis auf irgendeinen Aspekt der praktischen Musik. Vergeblich sucht man nach einem Beispiel aus einer Antiphon oder einem anderen liturgischen Gesang bei der Erklärung eines Intervalls. Vergeblich sucht man nach einem Verweis auf die Oktaven, die entstehen, wenn Knaben und Männer zusammen singen. Vergeblich sucht man nach irgendeiner Form der praktischen Anwendung des Tonsystems, das im Haupttext und im Kommentar entwickelt wird. Man sucht sogar vergeblich nach der berühmten Unterscheidung zwischen *musicus* und *cantor*. Obwohl die Musik vage Grenzen mit der Mathematik und der Philosophie teilte, scheint die musikalische Praxis ausgegrenzt zu sein.[222]

This observation of Calvin M. Bower's on the nature of the ninth-century glosses on Boethius' *De institutione musica* is certainly also true for the Anonymous corpus of glosses on Martianus Capella's *De nuptiis*.[223] In general, references to the living musical tradition by the commentator himself are few, and often ambiguous at that. It is only in the commentary of Remigius, for example, that references to antiphons (with neumes even) are found to illustrate intervals,[224] and it is in vain that one searches for attempts to link the Greek modal system to the Gregorian church tones. In itself, this should not surprise us. Martianus transmits classical Greek music theory, which was based on instrumental music; the strings of the cithara are the basis for the Perfect System, the note names, the tetrachords and

[222] C.M. BOWER, *Die Wechselwirkung*, pp. 182-183.

[223] See also the remarks of Nancy Phillips based on her knowledge of the corpora of glosses of Remigius of Auxerre and John the Scot: "A sense of historicity is quite evident in some glosses; there is no confusion with the contemporary musical practice. When, exceptionally, some aspect of the contemporary practice is mentioned, it usually illustrates a fact of harmonics; the pitch content of a ninth-century chant, for example, may be used as the known, the given, to demonstrate the consonant intervals. Thus these glosses are largely designed to explicate the text at hand with no overt concern for the contemporary musical practice." N. PHILLIPS, *Classical and Late Latin Sources*, pp. 101-102.

[224] Cf. C.E. LUTZ, *Remigius Autissiodorensis*, II: 332.16, 332.29, and 333.5.

their genera of division, etc. At the centre of ninth-century musical practice, on the other hand, stood the human voice. If one can interpret the often ambiguous Latin word *vox* here as "human voice," then one of the glosses of the Anonymous corpus explicitly states it: *Mater musicae vox est*, "the human voice is the mother of music."[225] In the centuries of oral transmission of chant repertoire there had been no room for instrumental concepts such as *diastema*, *systema* or *genera* of tetrachord division. It was only in the ninth century, when Charlemagne undertook his famous attempts at unifying liturgy and chant repertoire, that the classical sources with their mathematical approach to the discipline of music were re-studied and, in the writings of the first medieval theorists such as Aurelian of Réôme and Regino of Prüm, integrated into the process of describing and organizing the Gregorian repertoire.[226]

Nevertheless, even in the earliest corpora of glosses references to contemporary musical practice are not entirely non-existent, and they deserve our attention precisely because of their scarcity. For example, in one of the glosses of the Anonymous commentary which we have already discussed in the section on modes, it is strongly suggested that the commentator thought of a practical example to illustrate the theory at hand:

> [359.9] Primo <facis> materiam in animo simul cum gravitate aut etiam cum acumine. Ergo si libet tibi, ut ex gravioribus tropis alterum formes, praeparandum est opus aut fistula aut fidibus et caetera, aut etiam voce, similiter <opus facere potes> fide acutis. *BeLa, difficile lectu in La*
> (First <you compose> the (melodic) material in your mind, at the same time for the high region and for the low region. Thus, if it pleases you to form a different (melody) out of the lower modes, — the work has to be adapted for a flute or stringed instrument etcetera, or even for the human voice —, then <you can make> the same (melody) for a high string.)

Or, when talking on the subject of the *genera* of tetrachord

[225] 360.16 OMNIS VOX mater musicae vox est *BeLa*.

[226] See on this especially M.-E. DUCHEZ, *Description grammaticale et description arithmétique des phénomènes musicaux*.

division, it is striking that the commentator speaks in the first person when he says that the diatonic *genus* is preferably used:

> [369.4] SED NUNC MAXIME DIATONO UTIMUR i. usui habemus eo, quod pulcrior caeteris sit. *BeLa*, pulchrior *La*
> (BUT NOWADAYS THE DIATONIC MODE IS MOSTLY USED that is, *we* commonly use it, because it is more beautiful than others.)

And in the margins of one of the central manuscripts of the Anonymous tradition (*Leiden, Voss.Lat. F. 48*) a gloss is found which can be read as a kind of stage direction for the performance of the song of the Muses; the refrain was sung by all nine muses together, while the stanzas, one can imagine, were sung solo:

> [34.1-2] Generaliter omnes muse hoc cantabunt i. scande caeli.
> (All muses together will sing this, that is (the refrain) *Scande caeli.*)

In this section, three examples of implied contact between theory and practice will be discussed. First of all, in several manuscripts some of the poems of *De nuptiis* are provided with neumes, just as in several Carolingian manuscripts of the poems of Horace, Virgil, Statius, Terence, Lucan or Boethius. Secondly, there is a reference to the sequence as textless melody in one of the glosses of the Anonymous corpus, which carries implications for the early history of the sequence. Thirdly, in the commentary of John the Scot the harmony of the spheres is likened to a choir, singing at different pitches at the same time — a reference to actual polyphony to be dated before the theoretical description of the *organum*.

1. *Neumes in Martianus Capella manuscripts*

In 1975 Solange Corbin published a short article on neumes in Martianus Capella manuscripts in the Festschrift for Sir Jack Westrup.[227] Assisted by an international team she carried out a systematic search for neumes in all of the manuscripts of Leonardi's census which were dated before the twelfth century, about 160 in total, and the article was written when about sixty

---

[227] S. CORBIN, *The Neumes of the Martianus Capella Manuscripts*.

manuscripts remained to be examined. Of these, however, only ten contain all nine books of *De nuptiis*, about fifteen of them contain no more than short fragments or brief quotations, and the rest contain either book VIII on Astronomy, in which no musical notation has been found up to now, or books I and II. On the whole, she argued that there was little likelihood of discovering more notated sources, since only four of the manuscripts studied by her and her co-workers actually contained musical notation:

1. *Oxford, Bodl.Lib., Laud.lat. 118*;
2. *Paris, BnF, Lat. 8670*;
3. *Vatican, BAV, Reg.lat. 1987*;
4. *Leiden, UB, BPL 88.*

Two of these, *Oxford, Laud.lat. 118* and *Paris, lat. 8670*, she studied herself, and the other two were known to her through the publications of others.[228] In my own search for the best sources of the Anonymous commentary, I studied 21 manuscripts of *De nuptiis*, dating from the Carolingian period, in which all nine books of the encyclopedia were present and which contained glosses on the three books most relevant to my subject, books I, II and IX. I thus saw (partly on microfilm and partly in autopsy) seven more manuscripts of the whole of *De nuptiis*, which were not included in her study,[229] and, indeed, found no more notated sources. On the basis of her experience with this kind of manuscripts, Corbin furthermore concludes that it is fruitless to look for musical notation in manuscripts dated later than the twelfth century.[230] Consequently, it seems safe to assume that there are only four manuscripts with neumes for some of Martianus' poetic passages, which — according to Corbin — is considerably less than one would expect on the basis of a comparison with the musical tradition of the Classical poets. In a

---

[228] Viz. J. PRÉAUX, *Deux manuscrits gantois*; and, on *Vatican, Reg.lat. 1987*, H.M. BANNISTER, *Monumenti Vaticani di paleografia musicale latina*, col. 2.

[229] Bruxelles, BR, 9565-9566; Firenze, BML, San Marco 190; Leiden, UB, BPL 36, BPL 87 and Voss.Lat.F. 48; München, BSb, lat. 14729; and Vatican, BAV, Reg.lat. 1535.

[230] S. CORBIN, *The Neumes of the Martianus Capella Manuscripts*, p. 3.

long article on notated manuscripts of Classical texts, studied from the angle of their provenance, Yves-François Riou perhaps provides a reason for this discrepancy.[231] On the basis of data from hundreds of manuscripts, he concludes that the period in which most of the poems were notated were the tenth and eleventh centuries. Since the height of the production of Martianus Capella manuscripts has to be dated in the second half of the ninth century rather than the tenth or eleventh centuries, this could perhaps explain why there are so few manuscripts of *De nuptiis* with notated poems. In fact, the two manuscripts which Corbin did not consult herself do not contain neumes in the main text, but they were copied on a fly-leaf by eleventh-century hands. It appears that there was no wide-spread melodic tradition for Martianus Capella's poems: we have only scattered pieces with notation in two manuscripts from the ninth century, and two witnesses of one and the same melody, which can be dated in the eleventh century.

Meanwhile, in the four manuscripts listed above, several sections are notated. Neumes occur in all four manuscripts for a refrain in which the nine Muses welcome Philology at the gates of heaven. They each sing a stanza, followed by this refrain:[232]

> [117] scande caeli templa, virgo, digna tanto foedere,
> te socer subire celsa poscit astra Iuppiter.
> (Ascend into the temples of heaven, maiden, deserving of such a marriage,
> your father-in-law Jupiter asks you to rise to the lofty stars.)[233]

In both *Leiden, BPL 88* and *Vatican, Reg.lat. 1987*, this refrain is copied on one of the fly-leaves (fol. 2v and 1v respectively), in

---

[231] Y.-F. Riou, *Chronologie et provenance des manuscrits classiques latins neumés*. In this article he focuses on notated manuscripts from the ninth to the twelfth century of Horace (*Carmina, Carmen saeculare, Epodon liber*); Lucan (*Pharsalia*); Statius (*Thebais*); Terence (*Andria, Eunuchus*); and Virgil (*Eclogae, Georgica, Aeneis*). For previous scholarship on neumed passages in secular works, he gives ample bibliographic references in footnote 15, pp. 79-80.

[232] J. Willis, *Martianus Capella*, § 117-126, pp. 34-40.

[233] Transl. W.H. Stahl, *Martianus Capella*, II, p. 40.

the eleventh century in the monastery of St. Pierre in Ghent.[234] In *Oxford, Laud.lat. 118* the refrain with neumes occurs on fol. 11v, and in *Paris, lat. 8670* on fol. 14v.

Furthermore, the third stanza of the nine, sung by Polymnia, is provided with neumes in *Oxford, Laud.lat. 118* (fol. 12r). One should note that each stanza has a different metre, and that the melody composed for each cannot be used for the others. The text runs as follows:

> [120] tandem laboris fructus, aethram fulgidam
> divumque sedes ac Iovis consortia,
> provecta carpis inditoque numine,
> cruenta dudum, quae iugare rhythmica
> ac dissipare mixta sueta regula,
> mox quid iacente, quid iugata linea
> trigonus recurvet circulusque torqueat,
> melos probare ac tonos at crusmata
> artesque cunctas solita, quaeque caelitum
> possunt parare mente adacta culmina.
> (At last you, exalted to the shining sky, reap the fruits of effort,
> the divine abode and the company of Jove,
> now that you have been endowed with divine power,
> once of flesh and blood. You are used to joining
> and dispersing a mixture (of words) according to the rhythmical rules,
> to assess now which syllable, marked with the macron, is pronounced with circumflexion,
> then which syllable with the mark of brevity the micron curves;
> you are used to assessing melodies, sounds and plucked tunes
> and to all such disciplines, which can make us grasp
> the heights of heaven, when the mind is set to it.)[235]

In two manuscripts a gloss is added: *Nunc laudat Philologiam [a] musica* (Now she praises Philology because of her music) — the reason, perhaps, why this stanza is notated, while the others are

---

[234] According to A. Verhulst, the hand shows pre-Gothic characteristics typical of the writing-style of abbot Wichard († 1058). Both manuscripts carry an annotation of ownership by the monastery Saint-Pierre-au-Mont-Blandin in Ghent. A. VERHULST, *L'activité et la calligraphie du scriptorium de Saint-Pierre-au-Mont-Blandin de Gand*; J. PRÉAUX, *Deux manuscrits gantois*.

[235] Transl. cf. W.H. STAHL, *Martianus Capella*, II, pp. 41-42.

not.[236]

In the manuscript *Paris, lat. 8670*, finally, several poems are provided with neumes. First of all, the opening lines of the whole work received neumes for a line and a half (fol. 1r):

> [1] Tu quem psallentem thalamis, quem matre Camena
> progenitum perhibent, copula sacra deum,
> (You, of whom it is said that you grace weddings with your song, and that a Muse
> was your mother, sacred principle of the unity amongst gods, (on you I call);)[237]

Secondly, three lines of Apollo's speech in book I are notated (fol. 2r):

> [21] anxia cum trepidis nutat sententia rebus,
> fluctuat incertis aut sors ignota futuris,
> consultet mortale genus (quippe indiga veri
> cura facit dubium vel spes incerta fatigat);
> (When their minds tremble with apprehension in perilous times,
> or their destiny is unknown and unsettled with the future insecure,
> let the race of men consult the gods (because anxiety without knowledge of the truth
> makes them hesitant, uncertain prospects weary them);[238]

Thirdly, five lines of Apollo's speech to Juno, protectress of Hymen, are provided with neumes (fol. 5v-6r):

> [31] possem minore ambigens fiducia
> solum Tonantem pignoris pro foedere
> bupaeda vixum vel paterna contremens
> praecepta adire, ni iugata caelitum
> omen pararent prosperum consortia,
> tabensque divum nunc moneret nexio.
> (I would only have been able to approach the Thunderer
> for the marriage of heart with little confidence,

---

[236] 35.10-11 IUGARE RYTHMICA Nunc laudat Philologiam [a] musica *LaOxf*, Filologiam *Oxf.* One should note, on the other hand, that to the stanza sung by Calliope (119) a similar gloss is added in *Lb* (=*Leiden, BPL 88*): 34.16 TUNC CALLIOPE laus Philologiae de musica.

[237] Transl. cf. W.H. STAHL, *Martianus Capella*, II, p. 3.

[238] Transl. W.H. STAHL, *Martianus Capella*, II, p. 13.

—I, an overgrown youth trembling at my father's commands—,
where it not that marriages made in heaven seem a prosperous omen,
and the melting union of the gods now promises (a favourable outcome).)[239]

At the end of book I, where Jupiter makes a final speech in favour of the marriage between Mercury and Philology for the assembly of the gods, the fourth notated passage is found (fol. 10r):

[91] Ni nostra, astrigeri, nota benignitas
conferre arbitrium cogeret intimum
et quicquid tacito velle fuit satis
id ferre in medium collibitum foret,
possem certa meis promere ductibus,
nec quisquam illicitis tollere nisibus
concertans cuperet iussa deum patris.
(People of the stars, my well-known benificence
compels me to bring before you my most secret decision,
and it suits me to make public a matter
that required only a silent act of will;
otherwise, I would be able to lay before you accomplished facts
carried out at my commands,
and no rival would desire by unlawful endeavor
to overthrow the commands of the father of the gods.)[240]

The fifth notated passage is found at the beginning of book II, where the opening poem is provided with neumes for three lines (11v):

[98] Sed purum astrificis caelum scandebat habenis
Nox revocans merso fulgentia sidera Phoebo.
ardua tunc senior succendit plaustra Bootes,
et spiris torvo nituerunt astra Draconi;
(Night, with her starry reigns, began to climb the limpid heavens,
when Phoebus had set.
Then old Boötes kindled the lofty wagon [the Great Bear],
and the stars of Draco fierce with coils shone out;)[241]

---

[239] Transl. cf. W.H. STAHL, *Martianus Capella*, II, p. 17.
[240] Transl. W.H. STAHL, *Martianus Capella*, II, p. 31.
[241] Transl. cf. W.H. STAHL, *Martianus Capella*, II, p. 34.

Sixthly, of the above mentioned song of the Muses, not only the refrain (*Scande caeli templa*) is notated, but also the first four lines of Erato's stanza (14v):

[123] caput artibus, inclita virgo,
cui panditur aula Tonantis,
merito tibi subditur orbis
rationibus ante repertus.
(Head of the disciplines, famous maiden,
to whom the palace of the Thunderer is open,
rightly subject to you is the world,
since long apprehended by your rational principles.)[242]

The seventh and last notated passage is found in book IX, in a wedding song Hymen sings to welcome Harmonia into their company. The first 10 lines of this poem are notated (fol. 109r):

[902] Aurea flammigerum cum Luna subegerit orbem,
rosis iugabo lilia.
virgo deusque sacro sociabunt foedera lecto:
fulcris parate cinnama.
Hesperus intactam servet licet usque puellam,
nuptam videbis, Phosphore.
nec matris lacrimae, pressi nec crinibus ungues,
nexus valebunt rumpere.
ne thalamos metuas; eris hoc, quod Iuno Tonanti est,
quae nunc sorore dulcior.
(When golden Moon has raised her blazing orb,
I shall mingle lilies with roses.
The maiden and the god will consummate their vows on the sacred couch:
bring fragrant garlands for their bedposts.
Though Evening Star may keep the maiden pure until then,
you, Morning Star, will look upon her wedded.
A mother's tears, nor fingers clawing in hair,
will not avail to break the bonds of wedlock.
Fear not the marriage chamber: you too will be what Juno is to the Thunderer:
she is now dearer (to him) than a sister.)[243]

---

[242] Transl. cf. W.H. STAHL, *Martianus Capella*, II, p. 43.
[243] Transl. cf. W.H. STAHL, *Martianus Capella*, II, pp. 349-350.

The nature of the notation in these manuscripts has been described by Solange Corbin.[244]

The neumes in *Oxford, Laud.lat. 118* (Plates 7 and 8) are Aquitanian, a type of musical notation common in Southern France and Northern Spain.[245] Especially the resemblances between the forms of *quilisma* and *scandicus* in this manuscript and in a tenth/eleventh-century troparium of Saint-Pierre in Apt are remarkable.[246] The neumes are contemporary with the manuscript, and can be dated in the second half of the ninth century. Corbin furthermore observes that they belong to a region in which the *punctum* has an ornamental, elongated shape, instead of the classic square form. This form is found in two regions: Clermont-Ferrand and the area to the West of the Massif Central (Limoges-Angoulême). The second seems to her the most probable place of origin. On the basis of the type of script used in this manuscript, however, it has been tentatively located in Auxerre, together with, for example, *Leiden, Voss.Lat.F. 48* and *Vaticano, Reg.lat. 1535* — much further to the North, in other words, than the analysis of the musical notation suggests.[247] Nevertheless, the neumes are written in exactly the same ink as is used by one of the two contemporary glossators and are therefore supposed to be contemporary with the layers of glosses. In fact, the neumes seem to have been written before the glosses, since these are mainly located in the margin or wherever the neumes did not fill interlinear space. Did a wandering scholar or monk travel to the North and take residence in Auxerre? Did he study this manuscript of *De nuptiis* and add, in addition to the glosses he was to copy into this manuscript, a melody for the song of the Muses?

It should further be noted that for the melody of Polymnia's stanza the same formula is reproduced for each line (cf. Plate 8): a

---

[244] *The Neumes of the Martianus Capella Manuscripts*, pp. 5-6.

[245] This type of notation is described in detail by S. CORBIN in her comprehensive study, *Die Neumen*, pp. 3.94-3.99.

[246] I thank my colleague, Ike de Loos, for pointing out the resemblance to me. See f. 88r of a tenth/eleventh-century troparium from Saint-Pierre in Apt, reproduced in *Paleographie Musicale* I (Plate 83).

[247] See the description in Chapter 2, section 3.

sequence of *punctum, pes, punctum, pes, clivis, punctum, pes, clivis, clivis, punctum, pes, punctum* is repeated ten times, the only variation being that in the third, sixth, ninth and tenth line, the *clivis* on the fifth syllable is replaced by a *climacus* with an *oriscus*.

The neumes in *Paris, lat. 8670* concur with the proposed origin in Corbie for the manuscript. They are similar to the ones Corbin describes as accent neumes from the Northern region of Central France.[248] They too, are contemporary with the manuscript. I reproduce the notated passage on fol. 109r in Plate 9. Just as in *Oxford, Laud.lat. 118* the melody of the longest notated stanza on fol. 109v has a repetitive character: as far as we can see from the neumes which seem to have been scribbled down in a rather careless manner, a sequence of neumes for two lines is repeated five times, that is, for the ten lines of the poem.

The eleventh-century neumes on the two fly-leaves of *Leiden, BPL 88* (Plate 10) and *Vatican, Reg.lat. 1987* (Plate 11) are of the same date and origin, but not by the same hand. The writing style, degree of angularity and boldness of the notational signs are considerably different. The notation is Lorrainian, as is consistent with the fact that they were written for the monastery of St. Pierre in Gent.[249] Both fly-leaves show the same annotations: part of the *accessus ad auctores* belonging to the Anonymous glosses (*Iste Martianus genere Kartaginiensis fuit ... sed incertum manet*), and the refrain *Scande caeli templa virgo digna tanto foedere* with neumes. A Greek alphabet with the names of the letters and their numerical values, and the annotation of ownership of the church of Saint Pierre in Ghent (*Liber s. petri gandensis ecclesiae. Servanti benedictio. Tollenti maledictio. Qui folium ex eo tulerit. vel curtaverit. anathema sit.*) are also found in both manuscripts, but in different places: in *Leiden, BPL 88* they are copied on fol. 2v, in *Vatican, Reg.lat. 1987* on fol. 146r-v. The proposed melody for the refrain is the same in the two manuscripts.

For the comparison of the melodies, only the refrain *Scande*

---

[248] S. CORBIN, *Die Neumen*, pp. 3.100-3.131.

[249] On the neumes of this region see Dom HOURLIER, *Le domaine de la notation messine*.

caeli templa is available, for it is the only piece of verse which occurs in more than one manuscript. The two eleventh-century melodies from Ghent are consistent with each other — they were clearly written at the same time, transmitting the same melody and using largely the same notational signs. The two refrains in the Oxford and Paris manuscripts, which are to be dated in the second half of the ninth century, have few points of similarity, as can be seen in the table below (figure 14).[250]

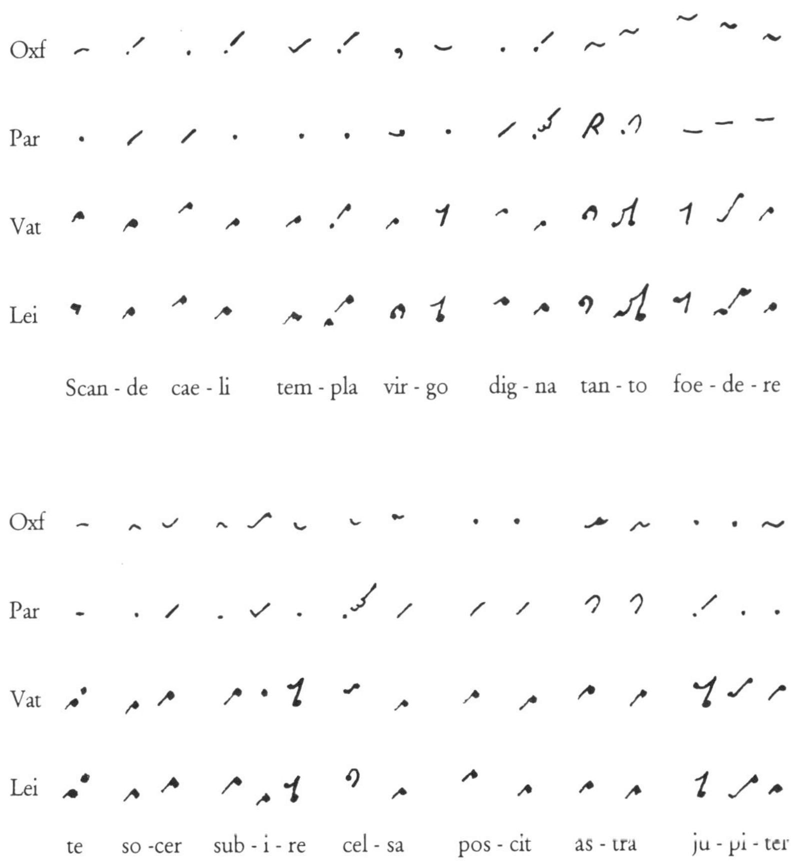

Figure 14

[250] See also S. CORBIN, *The Neumes in the Martianus Capella Manuscripts*, p. 7.

2. *"Ut sunt sequentiae apud cantores"*

In the two most important manuscripts of the Anonymous corpus of glosses on book IX of *De nuptiis*, *Leiden, UB, Voss.Lat.F. 48* and *Besançon, BM, Ms. 594*, a long gloss is added to § 936. After a long introduction on the magical powers of music, *Harmonia* has come to the core of her business: the explanation of the technical principles of her *ars*. She treats the basic terminology of music (*tonus*, *spatium*, *modulatio*, *tropus*, *symphonia*, *diastema*, *sistema* etc.) and in section 936, she discusses the ways in which music can be classified:

> [936] Priusquam me Lasus, ex urbe Hermionea vir, mortalibus divulgaret, tria tantum mei genera putabantur; *ὑλικόν*, *ἀπεργαστικόν*, *ἐξαγγελτικόν*, quod etiam *ἑρμηνευτικόν* dicitur. et *ὑλικόν* est, quod ex perseverantibus et similibus consonabat, id est sono, numeris atque verbis. sed quae ex his ad melos pertinet, harmonica, quae ad numeros, rhythmica, quae ad verba, metrica dicuntur. *ἀπεργαστικόν* est quidam materiae tractatus efficiens exercitium eius, cuius tres itidem partes, id est *μελοποιία*, *λῆψις*, *πλοκή*. *ἐξαγγελτικόν* autem ad expositionem pertinere videtur et habet partes tres: *ὀργανικόν*, *ᾠδικόν*, *ὑποκριτικόν*, quae inferius rerum ordo disponet.
>
> (When Lasus, a man from the city of Hermione, first taught the principles of harmony to mortal men, only three aspects were recognized: *hylikon* [subject matter], *apergastikon* [practice], and *exangeltikon* [exposition], also called *hermeneutikon*. *Hylikon* refers to things that sound together in a continuous and similar manner — melody, measures, and words: those that pertain to melody are called harmonics; those that pertain to measures, rhythmics; and those that pertain to words, metrics. *Apergastikon* has to do with the handling and exercise of the materials of the above subjects: it likewise is subdivided into three parts: *melopoeia* [musical composition], *lêpsis* [choice of pitch], and *plokê* [relation of pitches]. *Exangeltikon* is seen to refer to exposition, and it has three divisions: *organikon* [instrumental music], *ôdikon* [vocal music], and *hypokritikon* [recitation]. Each of these will be explained later in its appropriate place.)[246]

The long gloss from the Anonymous commentary concerns the

[246] Transl. W.H. STAHL, *Martianus Capella*, II, pp. 362-363.

term ὑλικόν, matter, and its division into three elements: *harmonica/soni* (melody), *rhythmica/numeri* (rhythm) and *metrica/verba* (metre, words). In the manuscripts the transmission of ὑλικόν is corrupt, and it is transformed into *ydikon* — a transformation which is easily explained from the similarity of capital lamda to capital delta (Λ - Δ):

[360.8] YΔICON
[1]Primum genus est idicon, quo genere forme inveniuntur.
[2]Formarum autem tres sunt species, quarum prima in sonis, secunda in numeris, tertia in verbis.
[3]Illa autem prima i. quae in sonis invenitur, ad armoniam pertinet.
[4]Non etenim in ea verba aliquid significantia, sed soni tantum sibimet ipsis aliqua ratione coniuncti queruntur, ut sunt sequentiae apud cantores.
[5]Secunda autem que ad numeros pertinet nichil aliud requirit nisi tantum convenientem numerum sonorum, nichil significantium [aut nichil] aut aliquid significantium absque ullo certo termino, ut est scansio versuum.
[6]Tertia vero, quae ad verba pertinet et vocatur metrica, non solum sonos certis temporibus terminatos inquirit, sed etiam cogit, ut aliquid significent et certum numerum terminumque non excedant.
[7]Quorum exempla sunt versus et metrum, in quibus duobus certus terminus numerusque pedum ponitur, <in> versu[i] quidem in senario, in metro vero in octonario.
[8]Sed notandum, quod hae tres formae in humanis vocibus inveniri possunt.
[9]In armonia vero, i. in his vocibus quae carent sensu, non est nisi una i. prima.
[10]In ea enim sonus tantum invenitur et non sine ratione numerorum.
[11]In hac vero i. humana voce aliquando sonus sine sensu, non tamen sine numero invenitur, si aliqua proportione copuletur; aliquando sonus cum sensu et numero, quae verborum nomine meruit appelari.
BeLa, 2 in numeris] in nu in numeris *Be* in verbis] in menbris *Be*; 3 *post* invenitur] et *add. La*; 4 etenim] enim *Be* queruntur] quierunt *Be*[1] querunt *Be*[2]; 5 aut nichil] aut nihil *La*; 7 certus terminus] certis terminis *Be*; 10 et] tamen *Be*

([1]The first aspect is *idicon*, in which aspect (the different kinds of) forms are found.
[2]There are, however, three kinds of forms, of which the first is found in sounds, the second in numbers, the third in words.

> [3]The first, that is the one found in sounds, pertains to *harmonia*.
> [4]For in this (form) one seeks not words with a certain meaning, but only sounds, which are joined with each other according to some rational basis, just as sequences are among singers.
> [5]The second (form), which pertains to numbers, seeks nothing but a fitting rhythm of sounds; without meaning, or with some kind of meaning without a certain measure, just as in the scanning of verses.
> [6]The third (form) then, which pertains to words and is called *metrica*, seeks not only sounds measured in certain time units, but also compels that they mean something and that they do not exceed a certain rhythm or a certain measure.
> [7]Examples of this (third form) are verse and metre, in both of which a certain measure and number of feet are posited; in verse in six (a *hexameter*, for example?), in metre in eight (an *octameter*?).
> [8]Let it be noted, however, that these three forms can be found in human utterances.
> [9]In harmony, that is in those utterances that do not bear meaning, just one can be found, that is the first (form).
> [10]In this, (that is harmony) only sound is found, though not without the logic of numbers.
> [11]In this, (that is in human utterance) sometimes sound is found without meaning, not, however, without number, if it is joined together according to some (rhythmic) proportion; sometimes sound (is found) with meaning and number, which deserves to be named "words".)

In the sentences 3-4, the commentator illustrates the first aspect of ὑλικόν: *harmonica* or *soni*, that is, things that pertain to sound only. He illustrates this with an example from his own experience: the singing of sequences. In sentence 5, he examines the second aspect: *rhythmica* or *numeri*, that is, things that pertain to measure, rhythm. Again, this abstract concept is related to a practice from his own experience: the scanning of verses. In the sentences 6-7 he goes into the third aspect, *metrica* or *verba*, that is, things that pertain to metre. Then he explains how these three aspects can be found in vocal music: sentences 9-10 treat the first aspect (sound without rhythm and without meaning), the first half of 11 the second aspect (sound without meaning but with rhythm), and the second half of 11 the third aspect (sound with rhythm *and* meaning, viz. text sung to melody). The sequence, mentioned in sentence 4, *ut sunt sequentiae apud cantores*, is thus used as an illustration of the first aspect of *hylikon* — sound without meaning, that is, without words.[247]

The early history of the sequence is a much disputed issue in musicological scholarship. Richard Crocker, who wrote, among others, the influential article on the sequence for the New Grove's dictionary and for the New Oxford History of Music, sketched a historical picture in which, from their earliest existence, sequences are structured, new compositions, in which both melody and text are original.[248] The relationship between text and melody is the crux of the matter; scholars such as Bruno Stäblein had always argued that there was a historical development from textless melismas to texted chant.[249] David Hiley and Lori Kruckenberg recently brought the discussion to a synthesis: they argued that from its emergence in the ninth century the sequence existed both as a texted and as a textless chant.[250] To tackle the complex problem of the early history of the sequence, scholars concentrated either on the liturgical sources, or on non-musical documents. The earliest liturgical sources, on the one hand, are:

- The *Antiphonale Blandiniense* (Bruxelles, Bibliothèque Royale, 10127-10144), dated ca. 800, probably written in Northern France, in which to six of the 25 Alleluias for the Sundays after Pentecost the rubric *cum sequentia* is added;

---

[247] In a manuscript of Augustine's *De musica*, dated at the end of the eighth century or the beginning of the ninth, Nancy Phillips noticed a gloss with the same implication. The manuscript, *Tours, BM, Ms. 286*, is written in the scriptorium of Saint-Martin of Tours, and contains glosses written before the death of Charlemagne in 814. On f. 62v the gloss "sine verbis sicut sequentiae sunt" is found. N. PHILLIPS, *Classical and Late Antique Sources*, p. 120 and n. 46.

[248] "Sequence", in: *New Grove*; *Medieval Chant*, in: *New Oxford History of Music* Vol. II, pp. 256-264. See also, of course, his influential study on the early sequence, *The Early Medieval Sequence*.

[249] See, for example, *Zur Frühgeschichte der Sequenz*; "Sequenz", in: *MGG*, Band 12, cols. 522-549.

[250] See, for example, D. HILEY, *Western Plainchant*, pp. 172-195; L. KRUCKENBERG, *The Sequence from 1050-1150*, pp. 42-85; and her article for the new edition of the *MGG*. In Kruckenberg's dissertation, a full bibliography on the subject is found.

- late ninth-/early tenth-century copies of sequences: melodies, texts, or texts with music notation.

On the other hand, the non-musical documents adduced are:[251]
- A passage in the *Liber officialis* of Amalarius of Metz, dated in 823, in which he speaks of a *iubilatio, quam cantores sequentiam vocant*, in which it becomes clear that this *iubilatio* is linked with an Alleluia and is explicitly *without* words;[252]
- A canon from the Synod of Meaux, held in 845, in which the practice of adding *compositiones, quas prosas vocant* or other novelties to the sequences that follow Alleluia's is forbidden in strong terms;[253]
- Notker's famous preface to his *Liber Ymnorum* (ca. 880), in which he dedicates his collection of newly composed poems for sequences to Bishop Liutward of Vercelli. In the preface, he recollects how, as a small boy, he had difficulties in remembering the *melodiae longissimae*, and searched for a way to make it easier to learn them by heart. A monk, who came from Jumièges and sought refuge in St. Gall, carried with him a book of chant, in which texts were written which fitted the

---

[251] See on these A. HAUG, *Ein neues Textdokument zur Entstehungsgeschichte der Sequenz.*

[252] I cite here the complete passage from Haug's article (p. 13): "Versus alleluia tangit cantorem interius, ut cogitet in quo debeat laudare Dominum, aut in quo laetari. Haec iubilatio, quam cantores sequentiam vocant, illum statum ad mentem nostram ducit, quando non erit necessaria locutio verborum, sed sola cogitatione mens menti monstrabit quod retinet in se. (Lib. III, cap. sci, 3; HANSSENS ed. III, p. 304. This passage is also cited in the fifth *Ordo Romanus* (OR V 30/31), ed. Andrieu II, p. 215) and by Pseudo-Alcuin in *De divinis officiis liber* cap. 39 (PL 101, col. 1245) (cf. A. HAUG, *Ein neues Textdokument*, p. 9).

[253] Haug discusses this passage in great detail (*Ein neues Textdokument*, pp. 10-12). The complete text runs as follows: "Propter inprobitatem quorundam omnino dampnabilem, qui novitatibus delectati puritatem antiquitatis suis adinventionibus interpolare non metuunt, statuimus, ut nullus clericorum, nullus monachorum in ymno angelico, id est Gloria in excelsis deo, et in sequentiis, quae in alleluia sollempniter decantari solent, quaslibet compositiones, quas prosas vocant, vel ullas fictiones addere, interponere, recitare, submurmurare aut decantare presumant. Quodsi fecerit, deponatur." (MGH Concilia 3, p. 129).

> sequence melodies (*versus ad sequentias erant modulati*). Notker thus found a solution to his problems, and started to compose texts for sequences himself.

Crocker mainly investigated the (relatively late) musical material: in *The Early Medieval Sequence* he analysed the West-Frankish repertoire, and reached the conclusion that both text and melody were original in the compositions he studied.[254] In his article for the New Grove's dictionary, he makes a distinction (which seems rather unnatural) between the wordless, structure-less "sequence" (*prosa*), and the "*sequentia*", a composition in which text and melody form a unity. He argues that there is not enough evidence to suggest an historical development from a troped melisma to the new compostion form found in his West-Frankish sources.

Nevertheless, the way in which the word *sequentia* is used in our gloss seems to fit with the meaning of the same term in the above-mentioned, older text documents for the earliest history of the sequence. Like Amalarius of Metz, the Anonymous commentator uses the word without any doubt to describe a type of chant without words — as a way of illustrating to the aspect *sound* (without rhythm and without meaning) in a theoretical classification of the *ars musica*. Moreover, the fact that the term is introduced to clarify and illustrate suggests that the term was *generally* understood in this meaning — melody without text.
Moreover, Remigius of Auxerre, who copies the gloss into his commentary almost word for word, deliberately left out the illustrating sentence *ut sunt sequentiae apud cantores*. One could assume that, in Remigius' time, the illustration no longer unequivocally corresponded to contemporary liturgical practice:

> ... Formarum autem tres sunt species, quarum prima in sonis, secunda in numeris, tertia in verbis. Illa autem prima, id est quae in sonis invenitur, ad armoniam pertinet. Non enim in ea verba aliquid significantia sed soni tantum sibimet ipsis aliqua ratione coniuncti quaeruntur. Secunda autem, quae ad numeros pertinet ...[255]

---

[254] R. CROCKER, *The Early Medieval Sequence*.

[255] C.E. LUTZ, *Remigius Autissiodorensis*, II, 499.7 (pp. 334-335).

(... There are, however, three kinds of forms, of which the first is found in sound/pitches, the second in numbers, the third in words. The first, that is the one found in pitches, pertains to *harmonia*. For in this (form) it is not words with some kind of meaning which one seeks, but only sounds, which are related with each other according to some rational basis. The second, which pertains to numbers ...)

This is, of course, an argument *ex silentio*, which does not, in itself, have evidential value. It is, however, imaginable, or even probable, that the term *sequentia* was changing in meaning and that it had come to signify a different kind of chant in the time of Remigius (viz. the end of the ninth century or beginning of the tenth) than it did in the time of the Anonymous commentary. The specific example — *sequentiae* — is no longer illustrative, and is therefore omitted by Remigius. The new meaning of *sequentia* could, then, easily correspond to the meaning espoused by Crocker and acknowledged by Hiley and Kruckenberg: a type of chant in which melody and text are both an integral part of the composition.

3. *Polyphony in John the Scot's commentary on the harmony of the spheres*

In the long gloss discussed earlier in my section on the harmony of the spheres, John the Scot introduced actual music to illustrate his ideas on the music of the cosmos in a remarkable manner. He uses the example of a choir to explain the independence of position and pitch of the planets. For just as the place of the singer in the choir does not influence the pitch he is singing, so the physical order of the planets is (relatively) irrelevant for their individual pitches. It is the sound of a singer's voice in relation to the other voices that makes him sound high or low, not the place where he is standing in the choir. The example undoubtedly suggests a polyphonic musical practice:[256]

Utamur ergo exemplo quodam quo manifestius apareat quod conamur asserere. In choro, ubi multi simul cantantes consonant,

[256] Cited from E. JEAUNEAU, in: *Quatre thèmes érigéniens*, pp. 129-130. See also my article *Harmonie der hemelsferen*.

> non locus in quo unusquisque constituitur sed proportio suae vocis consideratur. In quocunque enim loco fuerit qui gravissimam vocem emittit, necesse est gravissimam vocum omnium proportionem optineat. Eadem ratione, ubicunque in choro sit qui acutissimam vocem profert, necessario sonorum omnium acumen tenebit. De succinentibus similiter intelligendum, quorum non localis possitio sed proportionalis vociferatio in universitate modulaminis diiudicatur. Frustra igitur localium intervallorum rationibus caelestem musicam coartari arbitrantur. In quo nihil aliud conspicitur nisi gravitatis et acuminis ascensus atque descensus. Ut enim ascendit gravitas in altitudinem decrescendo donec in acumine finem inponat, sic altitudo descendit similiter <crescendo> quatenus in gravitatem terminum suum constituat.
>
> (Let us, however, present an example to show more clearly what we are trying to establish. In a choir, where many singers sound together simultaneously, one does not consider the place where each of the singers is arranged, but the pitch of his voice. Naturally, wherever the person who sings the lowest pitch stands, it is necessary that he has the lowest ratio of all pitches. Following the same reasoning: wherever in the choir the person is who sings the highest pitch, inevitably he holds the top of all sounding pitches. Consonances should be understood similarly; of these also one does not distinguish the position of the placements, but the sound of the ratio in the universal melody. In vain therefore one believes in linking the heavenly harmony to the relations of the distances according to their placement. Because in this we can see nothing but ascent and descent in lowness and highness. Now, just as the graveness (of pitch) rises by decreasing in height (of placement) until it stops on the highest point (of pitch), so the sharpness (of pitch) decreases similarly by increasing (in height) until it reaches its limit in the graveness (of pitch).)

The question of polyphony in John's texts is, again, a topic which has been discussed vehemently among musicologists for many years. The discussion has been based on a passage from John's main philosophical work, *Periphyseon* (*De divisione naturae*), from which a lively discussion was spun on the existence or non-existence of evidence for polyphonic practice:

> Ut enim organicum melos ex diversis vocum qualitatibus et quantitatibus conficitur dum viritim separatimque sentiuntur longe a se discrepantibus intentionis et remissionis proportionibus segregatae,

> dum vero sibi invicem coaptantur secundum certas rationabilesque artis musicae regulas per singulos tropos naturalem quandam dulcedinem reddentibus, ita universitatis concordia ex diversis naturae unius subdivisionibus a se invicem dum singulariter inspiciuntur dissonantibus iuxta conditoris uniformem voluntatem coadunata est.[257]
> (Now, since the *organicum melos* is composed of different kinds and numbers of sounds, which are perceived (one by one and apart from each other) to be separated by proportions of tension and release that differ greatly from each other (i.e. the rising and descending of a melody), but that — as soon as they are joined together according to certain rational rules of the *ars musica* — bring forth a certain natural sweetness, such then is the harmony (*concordia*) of the universe out of different parts of one nature, in disharmony when they are perceived one by one, united in accordance with the uniform will of the Creator.)

The main problem of this passage is the expression *organicum melos*, which was interpreted by several musicologists as referring to the music-technical term *organum* — the earliest form of polyphony according to a certain technique, described for the first time in the treatise *Musica enchiriadis* (3rd/4th quarter of the ninth century). Handschin, who initiated the discussion, interpreted the passage as a "gewaltiges Zeugnis für die mehrstimmige Kunst der Zeit".[258] Others, however, have pointed out that the passage should be understood in its context of a philosophical idea of a harmonious cosmos, interpreting "harmony" (*concordia, harmonia*) not in its musicological sense of "a concord of several pitches which pleases the ear", but rather in its philosophical sense of the divine and perfect order of creation.[259] It was argued that John was not referring to musical

---

[257] JOHANNES SCOTUS, *Periphyseon*, III.6, ed. I.P. SHELDON-WILLIAMS, p. 68.

[258] J. HANDSCHIN, *Die Musikanschauung des Johannes Scotus (Eriugena)*. Handschin concludes from the use of the words *organicum melos* that John must have known the treatise *Musica enchiriadis* and the polyphonic practice of *organum* while writing *Periphyseon*. This provides him with a reason to date *Musica enchiriadis* before 867, the date of origin of *Periphyseon*.

[259] On this aspect of John's philosophy see B. MÜNXELHAUS, *Pythagoras musicus*, pp. 199ff; S. GERSH, *Concord in Discourse*, (passages throughout the book); W. BEIERWALTES, *Eriugena. Grundzüge seines Denkens*, pp. 159-179; and

practice of his own time, but to the classical concept of harmony as a well proportioned equilibrium. *Organicum melos* was to be read in a classical context, where the word refers to an instrumental practice, which could never have been used to explain a contemporary chant practice, and the entire passage had to be seen from the perspective of a Pythagorean world view of numerical order in the cosmos.[260]

The passage cited above, however, seems to put things in a different light.[261] For here the same abstract idea — harmony of the cosmos — is without doubt illustrated with a practical example. Furthermore, because of the didactic character of the text, which *explains* Martianus Capella's *De nuptiis*, we must assume that the example was intended to be recognisable to the audience of the text. The passage in which John illustrates the harmony of the spheres with a choir singing different pitches simultaneously, *in choro, ubi multi simul cantantes*, is evidence for the existence of a polyphonic singing practice in the second half of the ninth century. John knew, from his own experience, of the practice of polyphonic singing. This does not, however, answer the question of whether, here or in *Periphyseon*, he refers to the specific technique of *organum*. There is, to my mind, no proof of the specific compositional technique in either of the two passages.

K.W. NIEMÖLLER, *Die Musik im Weltbild des Johannes Scotus Eriugena.*

[260] See, among others, W. WIORA, *Das vermeintliche Zeugnis des Johannes Eriugena*, pp. 39-40: "Bei aller Verehrung für Handschin [...] scheint es mir vielmehr, daß sich bei Eriugena kein Satz oder Terminus eindeutig auf Mehrstimmigkeit oder gar eine neue Art der Mehrstimmigkeit bezieht."; and E.L. WAELTNER, *Organicum melos. Zur Musikanschauung des Iohannes Scottus (Eriugena).*

[261] In his article *Der Harmonie-Gedanke im frühen Mittelalter* (reprinted in *Eriugena. Grundzüge seines Denkens*, pp. 159-179) W. BEIERWALTES already opened the door again for a concrete musical interpretation of the passage in *Periphyseon.*

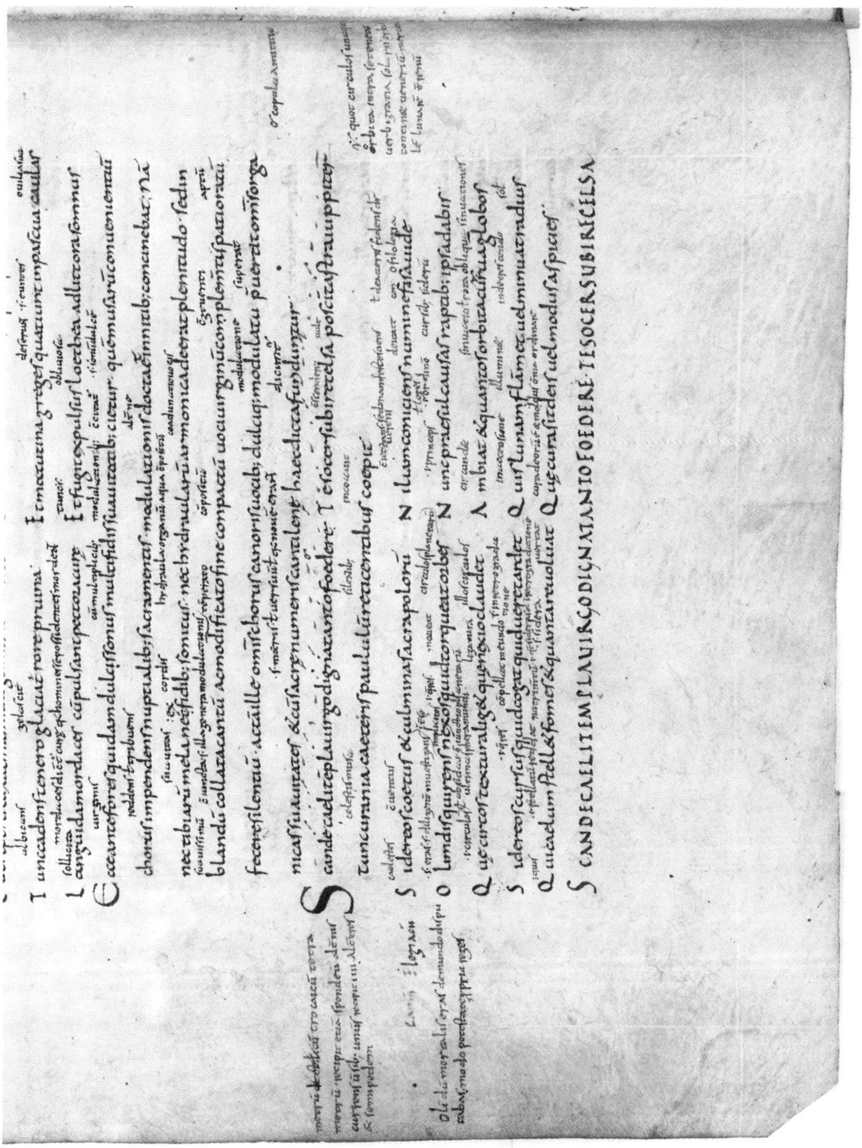

Plate 7 *Oxford, BL, Laud.lat. 118*, fol. 11v: Ninth-century Aquitanian point neumes.

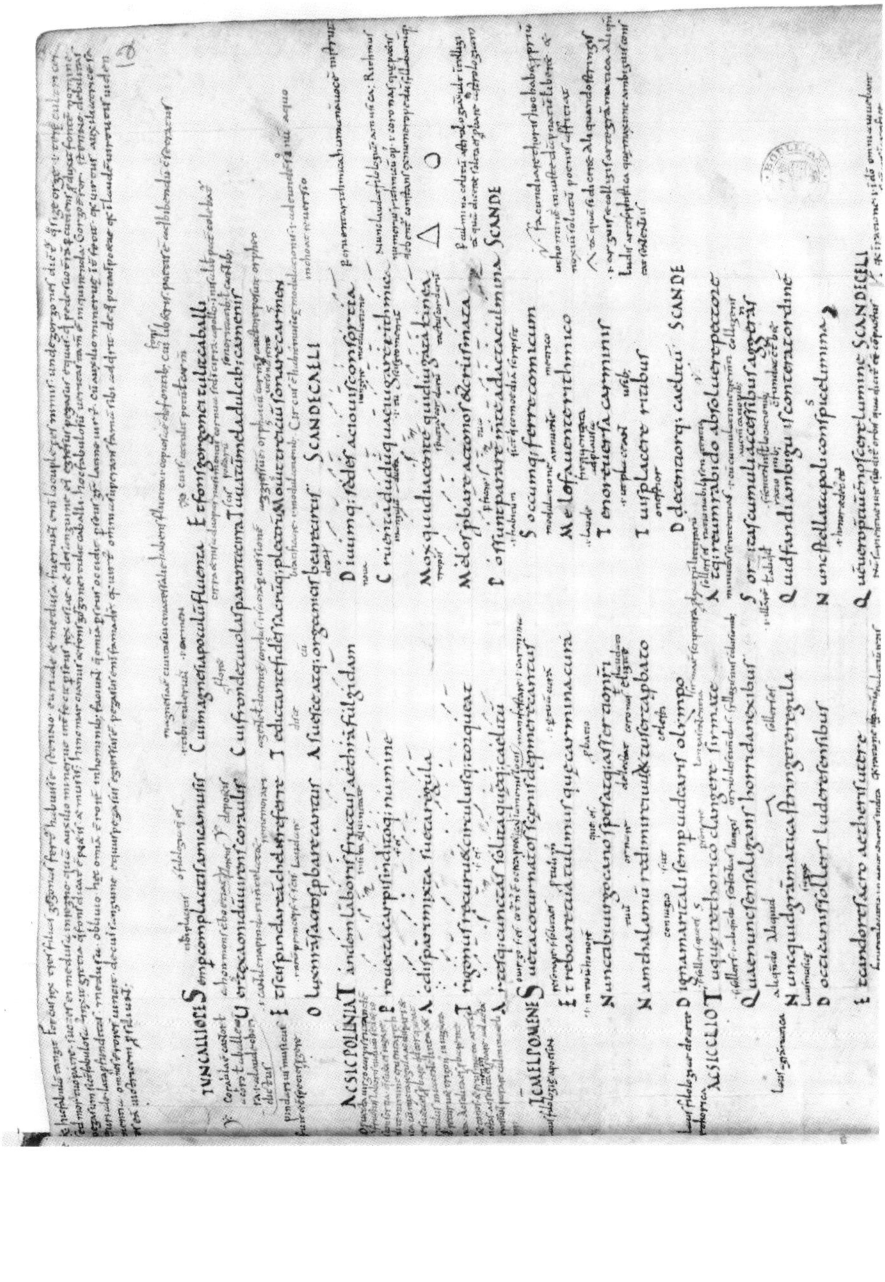

Plate 8 *Oxford, BL, Laud.lat. 118*, fol. 12r: Ninth-century Aquitanian point neumes.

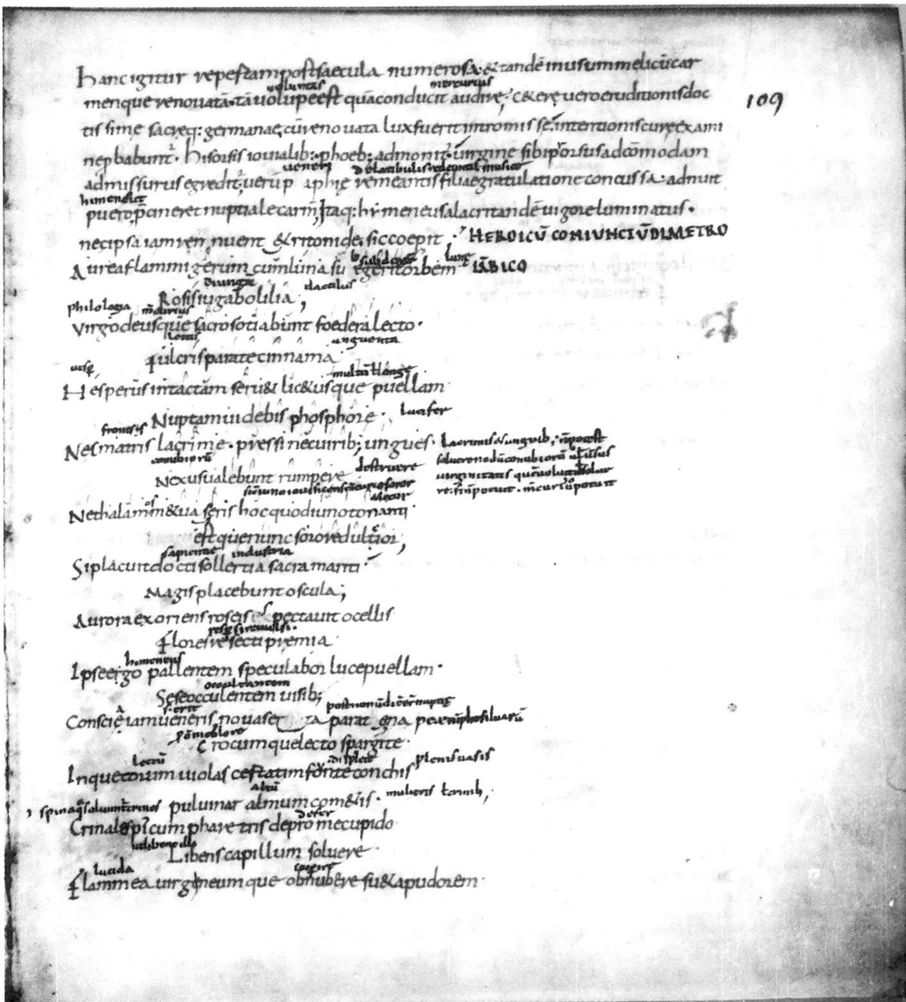

Plate 9 *Paris, BnF, lat. 8670,* fol. 109r: Ninth-century accent neumes from Central France.

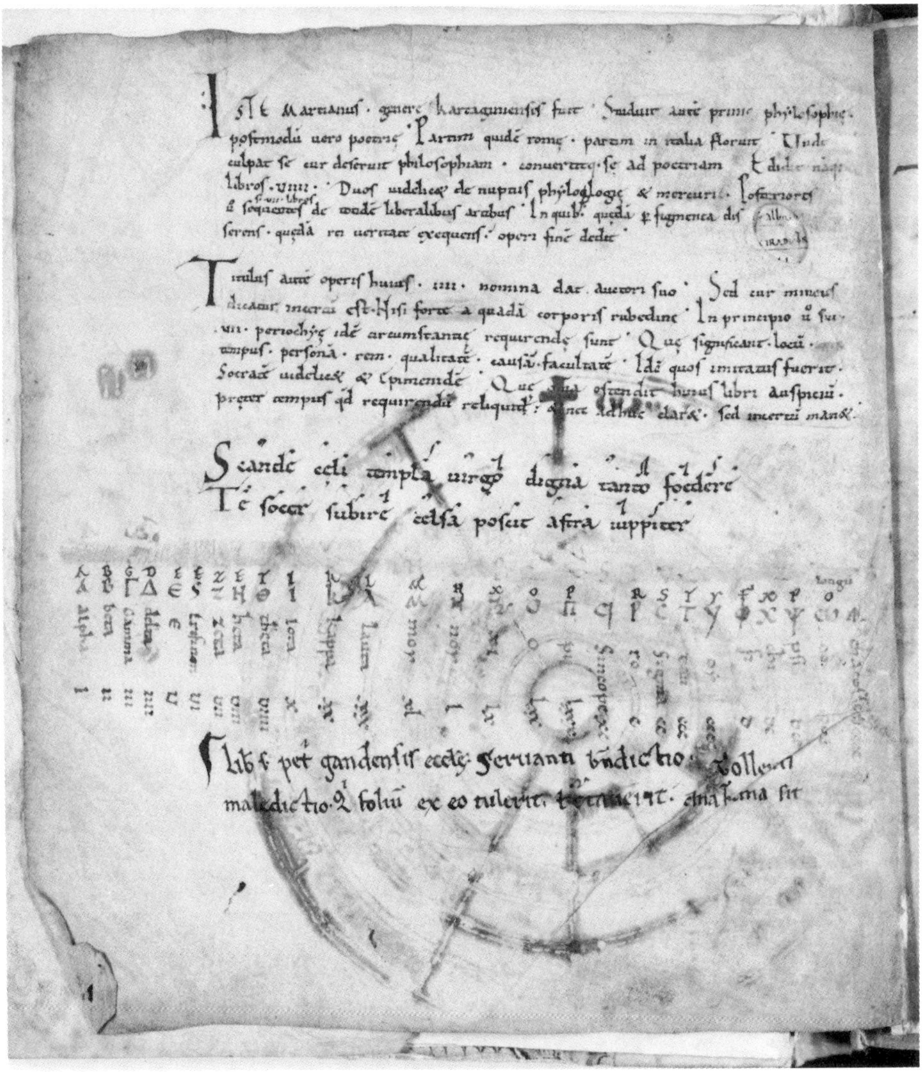

Plate 10 *Leiden, UB, BPL 88*, fol. 2v: Eleventh-century Lorrainian notation on a fly-leaf.

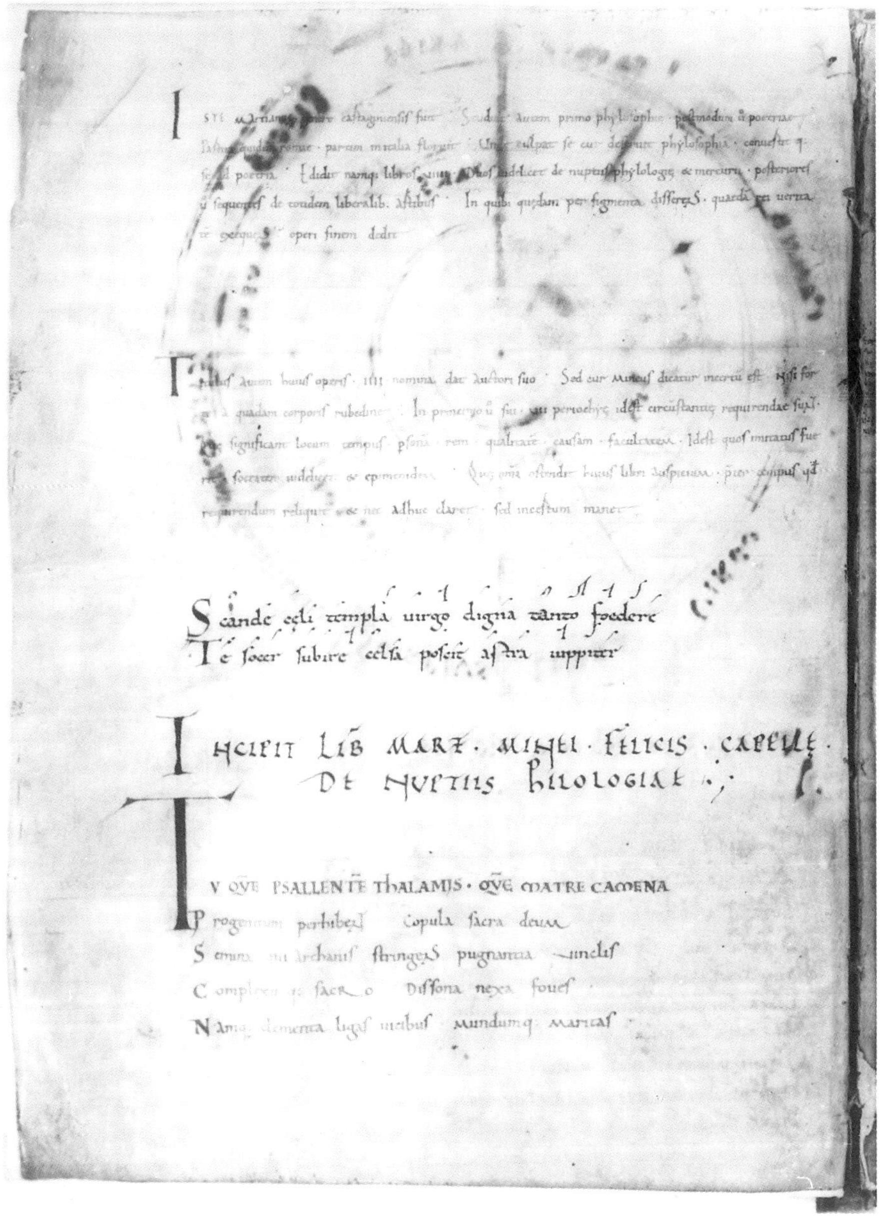

Plate 11 *Vaticano, BAV; Reg.lat. 1987*, fol. 1v: The same notation, also on a fly-leaf.

SUMMARY AND CONCLUSION

This study focuses on the ninth-century reception of Martianus Capella's musical material in his *De nuptiis Philologiae et Mercurii*. At the centre stands, more specifically, an Anonymous commentary, which can be dated before the middle of the ninth century, and which may be placed in the intellectual circle of the Carolingian courts of Louis the Pious and Charles the Bald. Those parts of this commentary that pertain to music are edited in full in the second part of this study. The first part has been divided into three sections: a first chapter on Martianus Capella himself and the history of the reception of his work; a second chapter on the manuscripts and the selection of the manuscripts for the present edition, and a third and final chapter on the contents of the musical glosses themselves.

The first chapter aims to show that the history of *De nuptiis* is by no means clear. The text, which must have been written in the fifth century in the capital of the province Africa of the Roman empire (Carthage), turns up in Rome in the sixth century, as is shown by a subscription of its corrector, dated in 534. Its fate, however, is shrouded in darkness from the sixth to eighth centuries. The earliest manuscripts — which contain, however, only fragments of the text — date from the late eighth century. Nevertheless, it was shown by Préaux that there must have been a sixth-century Roman manuscript in *scriptio continua*, which functioned as the oldest exemplum in the Carolingian period, and left its mark on a group of ninth-century manuscripts. This manuscript, then, was first copied in the heart of the Carolingian empire, the region Corbie - Laon - Auxerre, and from there the text was disseminated to the other parts of the empire.

Apart from the manuscripts themselves, which are remarkably numerous (there remain today nearly fifty ninth-century manuscripts with the entire text or part of it), several gloss traditions or commentaries testify to the popularity of the text in the Carolingian period. Roughly speaking, there are three, closely related, gloss traditions in the ninth and early tenth centuries, which are, in chronological order:

1. An Anonymous commentary, earlier attributed to Dunchad and to Martin of Laon. The date of this commentary is unknown, but it must have been written before the middle of the century.
2. A commentary or collection of annotations by John the Scot Eriugene. The date of the commentary, again, is uncertain. Moreover, the text exists in two different versions, which makes the dating problem even more complex. It can, however, be dated roughly around the middle of the century.
3. A commentary by Remigius of Auxerre, which has been dated around the turn of the ninth century and uses the two earlier gloss collections.

It is the Anonymous commentary that is at the centre of attention in this study. For, while there are decent editions of the other two commentaries, only a small part of the Anonymous commentary has been made available in a modern edition. The rest remains hidden in the margins of circa fifteen manuscripts. It is also the earliest and — at my estimate — largest commentary. Both John the Scot and Remigius, furthermore, are related to it, albeit in different ways: while the commentary of John the Scot may be read as a deliberately different opinion on the interpretation of the text, or as an addition to the Anonymous glosses, Remigius by and large incorporated the Anonymous commentary into his own. Both commentaries, therefore, will perhaps be easier for modern scholars to understand, once the Anonymous commentary can be consulted in a modern edition.

In the selection of manuscripts, two preliminary criteria were set:

1. a dating in the Carolingian period;
2. the presence of glosses in the three books containing musical material, viz. I, II, and IX.

This preliminary selection yielded a list of twenty-one manuscripts, which were then studied on microfilm and from which eight manuscripts were selected, viz. those that looked most promising for an edition of the Anonymous commentary. These eight can further be divided into two groups: four manuscripts of primary importance and four secondary sources. The four primary sources are *Leiden, Voss.Lat.F. 48* (*La*), *Besançon, ms. 594* (*Be*), *Leiden, BPL*

88 (*Lb*) and *Vaticano, Reg.lat. 1987* (*Va*). The four secondary sources are *Leiden, BPL 87* (*Lea*), *Leiden, BPL 36* (*Leb*), *Oxford, Laud.lat. 118* (*Oxf*), and *Paris, lat. 8671* (*Par*).

Only two of these eight sources transmit a more-or-less complete version of the Anonymous commentary on book nine, viz. *La* and *Be*. The manuscript tradition of the Anonymous glossator is much stronger for the first two books of *De nuptiis*.

The conclusions of the codicological and paleographical analysis of the manuscripts, standing on the shoulders of those scholars who had already studied the history of the text, *e.g.* Préaux and Leonardi, made it possible to take the discussion on authorship, dating and intellectual context of the commentary one step further. The codicological data of the manuscripts suggest that from its earliest dissemination in the first half of the century, the glosses were considered part of the text, although they had less authority than the text itself. For the greater part the manuscripts contain glosses, and most of them are especially laid out to contain glosses; for example, they were pricked and ruled to contain an extra, narrow column for glosses next to the writing space for the text. Secondly, the paleographical evidence shows that there is not one specific place that can be recognized as 'the centre' for the dissemination of this gloss tradition. The earliest witnesses (which are all copies and not originals) are located throughout Northern or Northeastern France, *e.g.* Auxerre, Corbie and Reims.

Thirdly, although the Anonymous commentary is often quoted or referred to by John the Scot and Remigius of Auxerre in their own commentaries on *De nuptiis*, the author of the Anonymous commentary is never mentioned by name. This may imply that the commentary was composed by a group of people rather than by a single scholar, and that it was, perhaps, one of the texts that were brought into circulation on the initiative of scholars who resided at the courts of Louis the Pious or Charles the Bald. The character of the glosses comfortably fits this hypothesis. They do not seem to be designed for the class room, but rather to attempt, first, to establish an authoritative reading of the text, and, secondly, to gather as much ancient knowledge as possible on a wide range of subjects.

In the third chapter the musical content of the glosses has been analysed and contextualised. In the introduction to chapter three, the ancient sources that were available to the Carolingian scholars are described in short: Plato's *Timaeus*, transmitted to the early Middle Ages by the translation and commentary of Calcidius; Cicero's *Somnium Scipionis*, known through the lengthy commentary of Macrobius; Augustine's *De musica*; Martianus Capella's *De nuptiis*; Boethius' *De institutione musica*; and the works of the early medieval encyclopedists: Cassiodorus' *Institutiones* and Isidore of Seville's *Etymologiae*.

The earliest sources of Carolingian musical thinking are in fact the corpora of glosses on these ancient texts, especially those on Boethius' *De institutione musica* (viz. the earliest layer of the so-called *Glossa maior*) and Martianus Capella's *De nuptiis* (viz. the Anonymous commentary). Furthermore, the first treatises on music theory from the later ninth and early tenth centuries are introduced: Aurelianus' *Musica disciplina*; the Anonymous treatises *Musica enchiriadis* and *Scolica enchiriadis*; Hucbald of Saint-Amand's *Musica*; and Regino of Prüm's *Epistola de armonica institutione*.

While the earliest gloss traditions were, clearly, mainly concerned with the *understanding* of the ancient knowledge, these ninth- and tenth-century treatises tried to incorporate it into their own musical practice: they *transformed* the ancient knowledge in their attempt to reflect systematically on their own chant repertoire.

In the first section of the third chapter, the relation between the oldest gloss traditions on Boethius' *De institutione musica* and Martianus Capella's *De nuptiis* has been analysed in detail. It is shown — by means of textual parallels; the usage of common ancient sources; common diagrams and linguistic similarities — that it is very likely that active study of these two works took place in the same period, and in the same intellectual context. The old hypothesis of Marie-Elisabeth Duchez, that the study of *De nuptiis* paved the way, so to speak, for a fuller grasp of Boethius' *De institutione musica*, can no longer be upheld. Rather than a phased reception of first Martianus' and then Boethius' work, the study of the two musical treatises was evidently simultaneous: the influence is mutual.

The second section introduces the theme of the harmony of the spheres, which plays a major role in *De nuptiis*, and, consequently, in the glosses. From the point of view of the history of ideas it is interesting to see how the Pythagorean tradition of a cosmos created according to numerical ratios found a warm welcome in the minds of the Carolingian scholars. This theme, which occurs only in the first two books of *De nuptiis*, received exceptionally long glosses in the Anonymous commentary. In these glosses, the commentator brings together all kinds of antique and late antique sources, such as texts by Macrobius, Isidore of Seville, Boethius and Fulgentius, as if to form a compendium of knowledge on the subject.

There may also have been some kind of interaction between the Anonymous commentary and John the Scot's idiosyncratic view of the harmony of the cosmos. From the Anonymous commentary, it is clear that there were contradictory interpretations of celestial music: was the scale of the planets from earth to firmament a rising one, or did it fall? The Anonymous glossator openly declares himself in favour of the first option — which concurs, by the way, with Martianus' presentation of the music of the spheres —; John the Scot, on the other hand, defends exactly the opposite, especially in what has been acknowledged as a later version of his commentary. Without suggesting that the two texts are directly dependent on each other, one can conclude that the controversy over the direction of the heavenly scale was a lively one. Even though the outcome of the discussion is completely different for the two authors, they both depart from a very similar intellectual context.

After these two studies of the Anonymous gloss material of a more thematic nature, five sections follow which rather bear the character of inventories. Five collections of gloss material have been made on five subjects within Martianus' classification of the *ars musica*: notes, intervals, tetrachords, *genera* (enharmonic, chromatic, diatonic) and modes (Dorian, Lydian, Phrygian, etc.).

In general, we conclude that the commentator was well-informed on the Greek Perfect System and its division in tetrachords, on the Greek names of the notes and their Latin translations. He used Boethius' *De institutione musica* frequently, especially the first book of the five, to clarify Martianus' more difficult passages.

Secondly, the glosses are a particularly revealing source when it comes to the vocabulary of the *ars musica*, the technical terminology of music. In the Anonymous glosses on *De nuptiis*, the student (or students) clearly struggled with terms such as *tonus, sonus, diastema, spatium, consonantia, dissonantia,* etc., and it is in this area that the main interest of these texts lies. The struggle with vocabulary and technical terms was continued by the first medieval authors of music-theoretical treatises, who sought to incorporate the ancient vocabulary into their own systematic reflection on chant practice. Some of the definitions phrased by the Anonymous commentator found their way into the *Glossa maior* on Boethius' *De institutione musica*, into the commentary on *De nuptiis* of Remigius of Auxerre, and from there into later treatises on music theory.

The last section of the third chapter concerns the aspect of theory versus practice. One of the main conclusions of Michael Bernhard and Calvin M. Bower in their study of the *Glossa maior* is that in the earliest glosses the author does not refer to contemporary chant practice. There is, in other words, a remarkable sense of historicity, and ancient and contemporary means of organising music are not confused. This also holds true, by and large, for the Anonymous commentary on Martianus Capella's *De nuptiis*, but there are some exceptions. Precisely because of their rarity, they deserve our attention. Firstly, there are several manuscripts that contain neumes for a few passages in verse. Some of these are contemporary with the manuscripts, and can be dated in the third quarter of the ninth century. Secondly, there is one long gloss on Martianus' classification of the *ars musica*, in which the word *sequentia* is explicitly used to refer to melodies without words. This is one of the earliest unequivocal witnesses of the use of the word *sequentia* for a type of chant without text. We must add it to the few sources of the — much disputed — earliest history of the sequence. Thirdly, there is a long gloss in John the Scot's commentary in which the music of the spheres is illustrated by a remarkably early reference to polyphonic singing.

To conclude, I would like to emphasize that this study has left open as many questions as it tried to answer. Many issues remain to be pursued further than has been possible in this study. The last word has definitely not been said, for example, on the authorship of the

Anonymous commentary, and my suggestions remain to be carefully verified or proven false. Furthermore, the focus of this study has mainly been on the glosses of the Anonymous commentary themselves, and relationships with other texts have not been explored in full depth. This too, then, remains a task for the future.

In the course of the present research, moreover, the importance of the Anonymous commentary has become increasingly clear, and a complete edition seems to me a necessity. Since the edition of even such a small part of the Anonymous tradition as I have completed here already yields so many insights into the reception of classical knowledge and its transformation in the Carolingian renaissance, an edition of the complete commentary would undoubtedly solve many more puzzles and shed more light on the intellectual networks of the time.

PART TWO

EDITORIAL PRINCIPLES

For the present edition, the editorial principles of Michael Bernhard and Calvin M. Bower, as laid out in their edition of the *Glossa maior in institutionem musicam Boethii*, have by and large been adopted.[1] The material they present is highly comparable with the glosses presented here, and their solutions to the many problems editing such complex texts as gloss-collections proved both useful and easily adoptable. In the following mainly the differences are noted.

The glosses are keyed to the latest edition of Martianus Capella's *De nuptiis* by James Willis, Leipzig (*Teubner*) 1983. In an appendix a concordance is given allowing readers to find the relevant passages in the earlier edition of A. Dick. This also greatly facilitates comparison with the corpora of glosses edited by Cora E. Lutz, which are keyed to this earlier edition. Each gloss will be treated as a separate entity, so that the variations in each collection of glosses can be taken into account. The same four sections as found in the *Glossa maior* can be found here: *lemma*; *gloss(es)*; *list of manuscripts*; and *critical apparatus*.

The choice of lemma is not, as in the *Glossa maior* edition, determined according to the principle of clarity, but rather a faithful rendering of the situation in the manuscripts themselves. Most glosses are tied to their lemmata by means of certain signs, and it is felt that these are part of the ninth- or tenth-century commentaries and should not be discarded. The text of the lemma is printed in small capitals, preceded by page- and line number of the Willis edition. The numbers of the different sections of *De nuptiis* are also given. Generally, a somewhat longer passage from *De nuptiis* is given as it appears in the Willis edition, and is followed by a list of glosses on that passage. Where the text of the ninth-century manuscripts deviates from Willis' reconstructed text, the *lemma* in question is marked with

[1] *Glossa maior* I, pp. lxiii-lxxi.

a dagger (†). This method of representation has been chosen in order to avoid the main difficulty editions of glosses in general face, viz. that it results in a largely incomprehensible text. It was felt that, by giving the reader somewhat longer passages, the context of the glosses becomes clearer.

As stressed before, this edition does not contain all the glosses added to *De nuptiis*, but only those added to passages on music — that is, a selection of sections from the first two books of the encyclopedia and the entire ninth book. From these sections of book I and II, however, all the glosses are edited, even though many of them are in fact not relevant for the study of music or musical theory. This was felt necessary in order to have a clear picture of the relations between the manuscripts and their gloss-collections.

Those glosses pertaining to the unusual location of a gloss (e.g. *Require in precedente pagina*, *Va* fol. 6v), those clarifying the syntaxis of a sentence by means of signs (e.g. in the first books of ms. *Be*) are not recorded in the edition, and neither are variant readings added by contemporary or later scribes. In *La* Ratherius of Verona marked unusual words in the text by repeating them in capitals in the margins, presumably to enrich his vocabulary. Since these annotations do not comment on the primary text, they are not recorded in the edition.

The *sigla* of the manuscripts are not composed according to the same principle as used in the *Glossa maior* edition, that is revealing their age by the number of letters in the *siglum*. In the present study all manuscripts used date from approximately the same period: shortly before or after the middle of the ninth century. Therefore, I chose to compose *sigla* of a minimum of two letters, which reveal the present location of the manuscripts (e.g. *Va* for Vatican City, *Be* for Besançon). Two letter *sigla* are used for the four most important sources, and for the four sources of secondary importance three letter *sigla* are used (cf. Chapter 2).

Two or more hands copying the same gloss tradition are not differentiated; they are only differentiated when they add a different tradition to the already existing one. Raised numbers following the *sigla* indicate layers of text, and — following Bernhard and Bower — a superscript "t" is used to indicate an addition made by the hand that wrote the principal text.

The typical introductory formula used in most glosses, *i.* for "id est", or *s.* for "scilicet", are not solved in the edition. Variants which are merely orthographical in nature are not mentioned in the *apparatus*; the spelling used in most manuscripts or in the most reliable manuscripts has been adopted in the edition. The punctuation and the use of upper and lower case letters are from the editor according to modern practice, and strive to make the meaning of the text as clear as possible.

Cross references to other glosses or passages in *De nuptiis* are given after the list of *sigla*. References to other sources or parallel passages in contemporary or later writings are given in footnotes, marked with ** to avoid confusion with superscript numbers added to the text for other reasons.[2] These footnotes are also used for comments on the specific problems of individual glosses, such as variant readings in the lemmata.

Text-critical signs used in the edition are:

<...>	addition of the editor, which is not given in the source
[...]	corrupt addition in the source, which obviously does not belong to the original text
(...)	comment of the editor for a better understanding of the text (such as *sic*, *lege* ...: suggests an alternative for a corrupt reading)

Abbreviations used in the edition are:

add.	addidit
cancell.	cancellavit
cf.	confer
lin.	linea
om.	omisit
superscr.	superscripsit

---

[2] It should be noted that these references are far from complete. The time-consuming task of identifying sources and parallels has not been one of the major goals of my research, and remains to be carried out more thoroughly in a future study.

List of sources referred to in the footnotes, alphabetically ordered:

Ars Laur. Exp.Don.mai.	*Ars Laureshamensis: expositio in Donatum maiorem*
Aug. Civ.Dei	Augustinus, *De civitate dei*
Aug. Quaest.ad Simpl.	Augustinus, *De diversis quaestionibus ad Simplicianum*
Beda Ars met.	Beda, *De arte metrica et de schematibus et tropis*
Beda Luc.evang.exp.	Beda, *In Lucae evangelium expositio*
Beda Orth.	Beda, *De orthographia*
Ps-Boeth. Geom.	Ps-Boethius, *Geometriae*
Boeth. Inst.Ar.	Boethius, *De institutione arithmetica*
Boeth. Inst.Mus.	Boethius, *De institutione musica*
Bon. Ars gram.	Bonifatius, *Ars grammatica*
Don. Ars gram.	Donatus, *Ars grammatica*
Don.Ort.	Donatus Ortigraphus, *Ars grammatica*
Fulg. Myt.	Fulgentius, *Mythologiarum libri tres*
Hier. Hier.proph.	Hieronymus, *In Hieremiam prophetam*
Hier. Trac.Marc.	Hieronymus, *Tractatus sive homiliae in Marci evangelium*
Joh.Scot.Er. Exp.	Johannes Scotus Eriugena, *Expositiones in hierarchiam celestem*
Isid. Etym.	Isidorus, *Etymologiarum sive originum libri XX*
Ps-Juv. Praef.	Ps-Juvencus, *Praefatio operi Iuvenci postea anteposita*
Lact. Theb.	Lactantius Placidus, *In Statii Thebaida commentum*
Mar.Vict. Ars gram.	Marius Victorinus, *Ars grammatica*
Mart. Cap.	Martianus Capella, *De nuptiis Philologiae et Mercurii*
Mur. Don.art.mai.	Murethach, *In Donati artem maiorem*
Myth.Vat.I	Mythographus Vaticanus I
Myth.Vat.II	Mythographus Vaticanus II
Myth.Vat.III	Mythographus Vaticanus III
Plin. Nat.Hist.	Plinius, *Naturalis Historia*
Prisc.	Priscianus, *Institutiones grammaticarum*
Rem.	Remigius Autissiodorensis, *Commentum in Martianum Capellam*
Sed.Scot. Col.misc.	Sedulius Scottus, *Collectaneum miscellaneum*

Sed.Scot. Don.art.mai.	Sedulius Scottus, *In Donati artem maiorem*
Serv. Com.Verg.A.	Servius, *Commentarius in Vergilium, Aeneas*
Serv. Com.Verg.E.	Servius, *Commentarius in Vergilium, Eclogae*
Serv. Com.Verg.G.	Servius, *Commentarius in Vergilium, Georgica*
Smar. Lib.part.Don	Smaragdus, *Liber in partibus Donati*
Ps-Tert. Adv.Marc.	Ps-Tertullianus, *Carmen adversus Marcionem*
Verg. Aen.	Vergilius, *Aeneas*
Verg. Ecl.	Vergilius, *Eclogae*

LIST OF MANUSCRIPTS

Be	Besançon, Bibliothèque Municipale, Ms. 594 (book I, II, IX)
La	Leiden, Universiteitsbibliotheek, Voss.Lat.F. 48 (book I, II, IX)
Lb	Leiden, Universiteitsbibliotheek, BPL 88 (book I, II)
Va	Città del Vaticano, Biblioteca Apostolica Vaticana, Reg.lat. 1987 (book I, II)
Lea	Leiden, Universiteitsbibliotheek, BPL 87 (book I, II)
Leb	Leiden, Universiteitsbibliotheek, BPL 36 (book IX)
Oxf	Oxford, Bodleian Library, Laud.lat. 118 (book I, II)
Par	Paris, Bibliothèque Nationale de France, lat. 8671 (book IX)
Par^{App}	Paris, Bibilothèque Nationale de France, lat. 8671, fol. 89r (book IX)

ACCESSUS

[1]Iste Martianus genere Cartaginensis fuit. [2]Studuit autem primo phylosophiae, postmodum autem poetriae. [3]Partim quidem Romae, partim in Italia floruit, unde culpat se cur deseruerit phylosophyam converteritque se ad poetriam. [4]Edidit namque novem libros: duos primos de nuptiis Phylologiae et Mercurii, septem sequentes de totidem liberalibus artibus. [5]In quibus quaedam per figmenta disserens, quaedam rei veritatem exsequens operi finem dedit. [6]Interpretatur autem Philologia amor rationis, Mercurius quasi mercatorum chirios i. dominus vel medius currens vel sermo discurrens, quia inter mercatores sermo discurrit, quia ermes Grece Latine sermo dicitur. [7]Cui etiam a pictoribus caduceum datur in manibus, geminis serpentibus innexum. [8]Quia proprium est sermonis iurgia et lites et rixas mercatorum sedare et sedando separare.

[9]Titulus autem huius operis quattuor dat nomina auctori suo. [10]Sed cur Mineus dicatur, non claret, nisi forte a quodam rubore. [11]Periochia i. circumstantia, quae significat personam, locum, tempus, rem, qualitatem, causam et facultatem. [12]Id est quos imitatus fuerit Socratem videlicet et Epimenidem. [13]Quae omnia ostendit libri huius initium, praeter tempus quod reliquit inquirendum, quod adhuc incertum est.

[14]Titulantur hi duo priores libri "De nuptiis Mercurii et Phylologiae", et non absque ratione. [15]Cum enim Phylologia amor rationis interpretetur, Mercurius vero in significatione sermonis ponatur, patet perfecto, quia quibus non est ratio ac vigor, sapientiae quae propalatur per sermonis acumina in ipsis quasi fiant competentes copulae et nuptiae Mercurii et Philologiae. [16]Multos enim novimus habere quasi philologiam i. amorem sapientiae et rationis, carere tamen Mercurio i. eloquenti luculentoque sermone; e contrario multos facundiam copiamque sermonum amittere, tamen fundamenta perfectionemque rationum <habere>. [17]Desiderat enim Mercurius copulari Philologiae itemque Phylologia adnecti Mercurio, quia et sermo ex plenitudine rationis subsistere, et ratio per hostia sermonis

desiderat propagari. *BeLaLb2Va2Lea*^{t}, 1 Cartaginensis] Kartaginensis *Lb2* Cartaginiensis *Va2*; 2 phylosphiae] sophiae *Be* poetriae] poetricae *Be*; 3 floruit *om. LaLea*^{t} converteritque] convertitque *Lb2Va2*; 4 novem libros] libros viiii *BeLb2Va2* primos] scilicet *Be* videlicet *Lb2Va2* et Mercurii *om. LaLea*^{t} septem sequentes] posterimus vero vii *Be* posteriores vero sequentes *Lb2Va2* s. vii libros *add. Lb2* artibus *om. LaLea*^{t}; 6 Philologia amor rationis *om. LaLea*^{t} chirios] chirius *Be* kyrios *Lea*^{t} quia ermes] qui hermes *Lea*^{t}; 6-8 Interpretatur autem - et sedando separare *om. Lb2Va2*; 9 huius operis] operis huius *Lb2*; 10 sed cur] cur autem *Be* non claret] non certum est *Be* incertum est *Lb2Va2* forte *om. Be* quodam rubore] quadam corporis rubedine *BeLb2Va2*; 11 Periochia - facultatem] In principio vii sunt periochiae i. circumstatiae inquirendae sunt, quae significant locum, tempus, personam, rem, qualitatem, causam et facultatem. *Be* In principio vero sui vii periochyae id est circumstantiae requirendae sunt, quae significant locum, tempus, personam, rem, qualitatem, causam, facultatem. *Lb2Va2* In principio vero sui vii] *add. Lea* *post* id est circumstantiae] sunt requirendae *add. Lea*; 12 fuerit Socratem videlicet] est scilicet Socratem *Be* vidilicet *La*; 13 libri huius initium] huius libri auspicium *Lb2Va2Lea* reliquit inquirendum] requirendum reliquit *Lb2Va2Lea* quod adhuc incertum est] et nec adhuc claret, sed incertum manet *Lb2Va2Lea*; 1-13 Iste martianus - sed incertum manet *post* 14-18 Titulantur - desiderat propagari *Lea*; 14-18 Titulantur hi duo - desiderat propagari *om. Lb2Va2*; 14 libri *om. Lea*^{t}; 15 interpretur] interpraetetur *Lea*^{t} et *add. Lea*^{t} perfecto] profecto *Lea*^{t} fiant] fiunt *Lea*^{t}; 16 carere] carent *Lea*^{t}; 18 hostia] ostia *LaBeLea*^{t} propagari] propulari *Be*

GLOSSES ADDED TO SECTIONS OF BOOK ONE

11/

6.14-5 TANDEM FAMA NUNTIANTE COGNOSCUNT QUOD PHOEBO GAUDET PARNASIA RUPES

TANDEM ad extremum *BeLaLbVaLea*

FAMA illa dea *BeLaLbVaLea*

filia terre *Oxf*

FOEBO sole *LbVaLea*

GAUDET PARNASIA RUPES Mons Archadie. Quod gaudet Parnasia rupes de Phebo i. de presentia illius vel quod ipse gaudet stare in Parnasia rupe *La*

mons Archadie. Parnassus *LbVaLea*, Parnassus] parnasus *Lea*

a Parnaso monte *Oxf*

6.15-6 LICET INDE QUOQUE AD INDICI MONTIS SECRETUM

OBUMBRATUMQUE SCOPULUM NUBE PERPETUA POSTERIUS MIGRASSE PERHIBEBANT

INDE inde i. ex Cirreis recessibus *LbVaLea*, Cirreis] cyrreis *Lb*

AD INDICI propinqui Parnaso i. Cirram *BeLa*, Cirram] circa *Be*
propinqui. vel Cirre *LbVaLea*
i. Cirrei *Oxf*

SECRETUM misticum *LbVaLea*

OBUMBRATUMQUE tenebrosum *BeLaLbVaLea*

SCOPULUM montem *LbVaLea*

NUBE caligine *LbVaLea*

PERPETUA continua *LaLbVaLea*

POSTERIUS ad ultimum *La*
Postea postquam illuc venerat *LbVaLea*

MIGRASSE <migrav>isse *LbLea*, <migrav>esse *Va*

PERHIBEBANT dicebant *LbVaLea*

6-7 TAMEN CIRRHAEOS TUNC RECESSUS ET SACRATI SPECUS LOQUACIA ANTRA CONVENIUNT

TAMEN s. ad *BeLa*, s. *om. Be*

CHIRREOS Mons est. Cirra et Nisa montes sunt Indie vel cornua Parnasi montis. In Cirra Apollo, in Nisa Liber pater colitur* *La*, *Be ad* 6.16-7 LOQUACIA. *Cf. LbVaLeaOxf ad* 34.21-2 CYRRA.
Cyrra mons est in Archadia *LbVaLea*
Cirra et Nisa duo capita Parnasi montis sunt. In Cirra Apollo, in Nisa Liber colebatur* *Oxf*

RECESSUS id est secreta *La*
secreta loca *LbVaLea*

SACRATI Apollini *Be*
dicati *La*
Apollini dicati *LbVaLea*

LOQUACIA responsa dantia *BeLaLbVaLea*
In tantum ibi venerabatur Apollo, ut quicquid excultores ibi audissent, pro certo provenire sibi sperarent *Be*

CONVENIUNT scilicet simul *La*
Mercurius, Virtus *LbVaLeaOxf*, *post* Mercurius] et *add. LeaOxf*

7-9 ILLIC AUTEM CIRCUMSTABAT IN ORDINEM QUICQUID IMMINET

---

* Cf. Isid. Etym. 14.8.11

SAECULORUM, FORTUNAE URBIUM NATIONUMQUE, OMNIUM REGUM AC TOTIUS POPULI

ILLIC AUTEM [1]Quod ait in specu Apollinis omnium esse Fortunas quicquid imminet seculorum sub specie fabulae figuratae loquens. [2]Hoc videtur intimasse, ut omnis seu eventus prosper sive sinister, quae ab illis Fortuna dicebatur, Apollinem de exactis, de instantibus et de futuris consultum sibi credebant provenire. [3]Nec non etiam quod nemus Apollinis succentibus duplis, sesqualteris ac sesquitertiis reliquisque (*sic, lege* reliquasque) acutas ac graves dicitur personasse modulationes. [4]Hoc indicare videtur, quod circulus solaris circulis trium planetarum superpositarum et totidem suppositarum comparatur; musica videtur ratione provenire. [5]Quia sicut mese in lira medio posita tetracordum facit, sic sol in medio planetarum. *BeLaLbVaLea*, *LbVaLea ad* 6.17-9, FORTUNAE. 1 quicquid] quicquit *La* imminet] inminet *Be* seculorum] seclorum *LbVaLea*; 2 intimasse] intimare *Be* seu *om. LaLbVaLea* prosper] sive prosper *LbVaLea* quae] quod *LbVaLea* de *om. LbVaLea* consultum] consultus *Be*; 3 duplis - sesquitertiis] duplis ac sesqualteris, nec non sequitertiis *Lea* sesquitertiis] *La*[2] sesquitiis *Be* acutas ac graves] acutus vel gravis *LbVaLea* personasse modulationes *om. LbVaLea*; 4 Hoc indicare videtur *om. LbVaLea* circulis] circulus *Be, om. Lea* planetarum] panetarum *Be* et totidem suppositarum comparatur] comparatus *LbVa* et totidem suppositarum *om. Lea* suppositarum] superpositarum *Be*; 5 Quia sicut mese - planetarum. *om. LbVaLea*

ILLIC in specu Apollinis *LbVaLea*

in illo loco *LbVa*

CIRCUMSTABAT in circuitu *LbVaLea*

IN †ORDINE per ordinem *LbVaLea*

QUICQUID INMINET Venturum est vel instat *La*

Omne venturum est *LbVaLea*

i. quicquid venturum esse saeculis futuris *Oxf*

FORTUNAE eventus causa *La*

Causam occultam, quam nos praescientiam vel providentiam dicimus *BeLa*, causam occultam] causa occulta *Be* praescientiam] presentiam *La*

Fortuna dicitur eventus, quae et in sinistra parte et prospera intellegi potest seu causa occulta, quam nos praescientiam extentam vel providentiam dicimus *LbVaLea*

fortunium et infortunium *Oxf*

URBIUM circumstabant *LbVaLea*

NATIONUMQUE regionum fortunae *LbVaLea*

9-0 VIDEBANTUR ALIAE TRANSACTI CURSUS EMENSA FUGIENTES
VIDEBANTUR s. fortunae *LbVaLea*
ALIAE s. fortunae *LaLbVaLea*, s. *om. LbVa*
hic de praeterito *LbVaLea*
s. fortunarum. Aliae enim erant prae<teri>torum temporum, aliae futurorum *Oxf*
CURSUS genitivus *BeLaLbVa*
†EMENSO mensurato *LbVaLea*
s. spatio *Oxf*
FUGIENTES praetereuntes *La*
praeter *Be*
Iam spatio vitae praetereuntes et velociter recedentes *LbVaLea*

0-1 CONSISTEBANT ALIAE SUB CONSPECTU, ADVENIEBANTQUE QUAMPLURES
ALIAE Quia in memoria tenebantur *LbVaLea*
de presentibus *La*
Quia in memeria (*sic, lege* memoria) retinentur. Capta enim fuit Ierusalem, et licet transactum sit, adhuc in memoria retinetur *Oxf*
SUB CONSPECTU s. in presenti consistebant *BeLa*, s. in *om. Be*
consistebant] s. consistebant *Be*
de presenti s. consistebant *LbVaLea*
i. praesentes *Oxf*
ADVENIEBANTQUE de futuro* *LaLbVaLea*
futurum *Be*
QUAMPLURES s. fortunae *BeLa*
multae fortunae *LbVaLea*

1 ATQUE ITA NONNULLIS EMINUS VANESCEBAT DISPARATA PROLIXITAS
NONNULLIS fortunis *La*
in s. fortunis *LbVaLea*, in *om. Lea*
Illorum figurae qui cito moriuntur *Oxf*
EMINUS longe vel continuo *La*
a longe *LbVaLea*
VANESCEBAT Citius transiebat *LbVaLea*

---

* Tironian notes are added to this gloss in *La*?

Negata temporum longitudo, quia non equaliter cum omnibus perseverabat illa fortuna. Alios ad caelum usque ducebat, alios ad terminum vitae, alios cito deserebat *BeLa*

DISPERATA segregata. non sperata *La*

DISPARATA iam negata *VaLea*

desperata *Lea*

DESPERATA iam negata *Lb*

PROLIXITAS temporum longitudo et vivendi *LbVa*

longevitas *Oxf*

6.22 UT VELUT FUMIDAE CALIGATIONIS INCREDIBILIS HABERETUR AURA

FUMIDAE sicut caligo fumi i. tenebrosae obumbrationis *LbVa*

INCREDIBILIS HABERETUR AURA i. ut esset *BeLa*

Esset ipsa vita *LbVa*, vita *om. Va*

6.22-4 INTER HAEC MIRA SPECTACULA FORTUNARUMQUE CURSUS [MOTUS] NEMORUM ETIAM SUSURRANTIBUS FLABRIS CANORA MODULATIO MELICO QUODAM CREPITABAT APPULSU

MIRA miranda *La*

mirabilia *Be*

mirabilia praedicta *LbVa*

FORTUNARUMQUE s. et inter *LbVa*

CURSUS eventus *BeLa*

accusativus *LbVa*

MOTUS genitivus *LaLbVa*

accusativus *Oxf*

NEMORUM et inter *Lb*

silvarum *LbVa*

SUSURRANTIBUS leniter flantibus *BeLaLbVa*

murmurantibus *Oxf*

FLABRIS ventis *BeLaLbVaOxf*

CANORA sonora dulcis *BeLa*, dulcis *om. Be*

suavis *LbVa*

MODULATIO cantilena *LbVa*

MELICO musico *BeLaLbVaOxf*

suavi dulci *LbVa*

QUODAM motu *La*

cum *LbVa*

CREPITABAT resonabat *La*
  sonabat *Be*
  sonabat simphonizabat *LbVa*
APPULSU pro apulsione vel motu *La*
  motu *Be*
  repercussione. motu vel appulsione *LbVa*
  Quia omnis musica in pulsu fuit *Oxf*

-1 NAM EMINENTIORA PROLIXARUM ARBORUM CULMINA PERINDEQUE DISTENTA ACUTO SONITU RESULTABANT
EMINENTIORA excelsiora *BeLa*
  excellentiora altiora *LbVa*
PROLIXARUM altarum *BeLa*
  procerarum magnarum *LbVa*
PERINDEQUE i. valde vel consequenter *La*
  i. valde *Be*
  valde *LbVa*
  Multipliciter intellege perinde; superius pro consequenter, hic pro valde ponitur *LbVaLea*
  i. crebro vel valde *Oxf*
DISTENTA diffusa *BeLa*
  extenta in longitudinem *LbVa*
  i. altum subrepta *Oxf*
ACUTO SONITU maiore *BeLa*
  gracili *LbVa*
  Constat enim musica ex gravissimis et acutissimis et mediis *LbVaLea*
RESULTABANT resonabant musicam *La**
  recinebant vel resonabant *LbVa*
  resonabant *Oxf*

2 QUICQUID VERO TERRAE CONFINE AC PROPINQUUM RAMIS ACCLINIBUS FUERAT, GRAVITAS RAUCA QUATIEBAT
TERRAE dativus *LbVa*
CONFINE proximum *LaLbVa*
RAMIS ACLINIBUS pronis *BeLa*
  i. curvis *LbVa*

---

* Again: Tironian notes are added to this gloss?

incurvatis *Oxf*

GRAVITAS RAUCA sonus dulcis *BeLa*

sonus raucus. ypallage est *LbVa*

ipallage. gravis raucitas *Oxf*

QUATIEBAT concutiebat personabat *BeLa*, personabat *om. Be*

movebat concutiebat *LbVa*

[1]In hoc loco non solum tangit musicam cordarum, sed etiam celestem, ut sicut in arbore ea quae inferiora erant rauca gravitas quatiebat, sic etiam a terra usque ad lunam gravior sit sonus. [2]At a luna per totam amplitudinem planetarum usque ad Saturni circulum varietas sonorum intellegitur. [3]Sic etiam in medio arboris varietas sonorum fuit. [4]A Saturni autem circulo usque ad signiferum acutus noscitur inesse modulatus. [5]Quamvis contrario quidam ista repugnent, dicentes graviorem esse sonitum in Saturni circulo, acutiorem in lunari, quia quod brevius est, acutius sonat. *BeLaLbVaLea*, *La ad* 7.2-4 NEC NON ETIAM, *LbVaLea ad* 7.5-6 ITA FIEBAT. 1 *post* musicam cordarum] sicut in lira cernimus *add. LbVaLea*; 2 intellegitur] intelligitur *BeLea*; 3 sic etiam] sicut *Be* varietas sonorum] varietas soni vel sonorum *LbVaLea*; 4 noscitur] nascitur *Va*; 5 Quamvis] e *superscr. La* ista repugnent] stare pugnent *Va* graviorem esse] esse graviorem *LbVaLea* acutius sonat] sonat acutius *LbVaLea* *post* acutius sonat] velut in fidibus cithare probare potest *add. LbVaLea*

[1]Nec non etiam quod nemus Apollinis succentibus duplis sive sesqualteris ac sesquitertiis reliquisque (*sic, lege* reliquasque) acutas ac graves dicitur personasse modulationes. [2]Hoc indicare videtur quod circulus solaris circulis planetarum trium super se positarum ac totidem suppositarum comparatur; musica videtur ratio<ne> provenire. [3]Sicut enim corda, quae in medio est, principalis est, ita sol medium tenet planetarum. *BeLa, Be: cf.* 6.17-9 ILLIC AUTEM

7.2-4 AT MEDIA RATIS PER ANNEXA SUCCENTIBUS DUPLIS AC SESQUALTERIS NEC NON ETIAM SEQUITERTIIS, <SESQUI>OCTAVIS ETIAM SINE DISCRETIONE IUNCTURIS, LICET INTERVENIRENT LIMMATA, CONCINEBANT

AT MEDIA RATIS s. loca arboris vel nemoris. per efficientem id quod efficitur *BeLa*

s. loca silvae s. arboris *LbVa*

s. culmina i. silvae vel rationabilibus *Oxf*

PERANNEXA valde coniuncta *BeLaLbVa*

s. spatia *Lb*

Constat enim musica ex acutissimis et gravissimis et mediis *La*, *cf. LbVa* 6.25-1, ACUTO SONITU

SUCCENTIBUS DUPLIS melodiis *La*

i. organis *LbVa*

[1]Diapason dicit quae dupla iungitur proportione, ut sunt duo ad unum. [2]Diapente per sesqualteram iungitur rationem ut sunt iii et ii. [3]Diatesseron quae sesquitertia copulatur ratione ut sunt quatuor ad iii. [4]Epogdoum dicit ut sunt viiii ad viii. [5]Epogdous autem vocatur in arithmetica, tonus in musica. [6]Epogdous vero dictus quasi superoctonus vel sesquioctavus tonus. [7]Tonus vero est spatium cum legitima qualitate, emitonium semis tonus.* *BeLaLbVa*, 1 dicit] dicte *Be*; 2 per sesqualteram iungitur rationem] tangit quae sesqualtera iungitur ratione *BeLbVaLea*; 3 *post* Diatesseron] dicit *add. BeLea* sesquitertia] sestertia *LaLbVaLea* copulatur ratione] concordat modulatione *BeLbVaLea*; 5 Epogdous autem vocatur in arithmetica, tonus in musica] Epogdous enim in aritmetica, tonus vocatur in musica *Be*; 6 Epogdous vero] Vel epogdous enim *Be*; 7 Tonus - semis tonus *La ad* 7.2-4 LIMMATA tonus] tomus *La*; 5-7 Epogdous autem - semis tonus *om. Lea, cf. ad* 7.2-4 †OCTAVIS; 6-7 Epogdous vero - semis tonus *om. LbVa, cf. ad* 7.2-4 †OCTAVIS

Concentus est cum similis est sonus. succentus cum varii sunt soni *La*

Et concentus est ubi similis est societas in sono, succentus vero ubi varii sunt soni *LbVaLea*, similis] similiter *Lea*

tonis *Oxf*

Diapason i. ex omnibus in musica, quia per omnia. Maior numerus minorem bis continet *Oxf*

AC SESQUALTERIS concentibus *LbVa*

iii ii. Sesqualter i. ii ad iii in aritmetica. Idem est et diapente hoc est de quinque in musica *Oxf*

SEQUITERTIIS iiii iii. Sesquitertius ut est iii ad iiii in arithmetica. Idem est et diatesseron hoc est de quatuor. Sed inter duas extremas fides peragit, quando sonus alterius fidis sonum alterius tertia minoris parte superat *Oxf*

†OCTAVIS epogdous *La*

si et cum *LbVa*

Epogdous in arithmetica, tonus vocatur in musica. Epogdous dictus est quasi superoctonus vel sesquioctavus. Tonus vero est

---

* Linea 7: cf. Mart.Cap. 9.930

spatium cum legitima quantitate* *LbVaLea*
s. tonis, ubi octava superat pars *Oxf*
SINE DISCRETIONE sine differentia vocis *BeLaLaVa*, sine *om. LbVa*
spatio carminis *Oxf*
LIMMATA i. emitonia *BeLa*, i. *om. La*
semitonia, semisoni vel minus sive plus *LbVa*, semitonia] seminonia *Va*
Tomus (*sic, lege* tonus) vero est spatium cum legitima qualitate.* Emitonium semis tonus. *La, cf. Be* 7.2-4 (lin. 7) SUCCENTIBUS DUPLIS
spatia carminis sonabilis *Oxf* emitonia *Oxf2*
CONCINEBANT ipsa media cantabant *LbVa*

7.5-6 ITA FIEBAT, UT NEMUS ILLUD HARMONIAM TOTAM SUPERUMQUE CARMEN MODULATIONUM CONGRUENTIA PERSONARET
ITA FIEBAT s. ille sonus *LbVa*
In hoc loco ... *LbVaLea, cf.* 7.1-2 QUATIEBAT
ARMONIAM adunationem vocis *BeLa*, vocis] vocum *Be*
Armonia dicitur adunatio vocis *LbVa*
diapason, diapente, diatesseron, epiogdous, limmata *Oxf*
SUPERUMQUE divinum *Be*
divinum caeleste *LbVa*
CONGRUENTIA cum: ablativus *LaLbVa*, s. *add. Va*
convenientia *La*
i. convenienti modulatione *Be*
A gruibus tactum est, quia gracilis vox illarum est** *LaLbVa*, est *om. Va* quia gracilis est vox illarum *Lb*
Congruo est tractum a gravibus (*sic, lege* gruibus), eo quod exiles habeant voces et bene convenientes, quia in exilitate vocum est consonitas** *Be*
ablativus; quadam convenientia *Oxf*

12/
7.6-8 QUOD QUIDEM EXPONENTE CYLLENIO VIRTUS EDIDICIT ETIAM IN CAELO ORBES PARILI RATIONE AUT CONCENTUS EDERE AUT SUCCENTIBUS CONVENIRE

---

* Cf. Mart.Cap. 9.930
** Cf. Sed.Scot. Don.art.mai. 154.97

EXPONENTE interpretante *LbVa*
CYLLENIO Mercurio *LbVa*
VIRTUS dea *LbVa*
EDIDICIT intellexit *LbVa*
ORBES PARILI RATIONE s. planetarum. sicut nemus *BeLa*
circulos planetarum aequali *LbVa*
AUT CONCENTUS Concentus est sonus consonus vocum, succentus vero ubi aliqua differentia *Be*
Concentus est ubi simul est societas, succentus vero ubi varii sunt soni *La*
Concentus est ubi similis est societas in sono, succentus vero ubi varii sunt soni *LbVa*
i. pariles sonos *Oxf*
EDERE formare *LbVa*
procreare *Oxf*
AUT SUCCENTIBUS Ubi maior minorque vox auditur *Oxf*
CONVENIRE concordari *Oxf*

9 NEC MIRUM QUOD APOLLINIS SILVA ITA RATA MODIFICATIONE CONGRUERET, CUM CAELI QUOQUE ORBES IDEM DELIUS MODULETUR IN SOLE

NEC MIRUM Non mirum si in terris talia agit, qui in caelo etiam armoniam regit *BeLa*, Non] nec *La* armoniam] armenia *La*
s. erat *LbVa*
SILVA s. in *LbVa*
RATA rationabili *BeLaLbVa*
MODIFICATIONE moderamine *Be*
moderatione *LaLbVa*
modulatione *Oxf*
CONGRUERET personaret. conveniret sibi in consonantiis *LbVa*, consonantiis] consonantes *Va*
conveniret *Oxf*
ORBES circulos *BeLbVa*
circuitus *La*
i. circuitus septem planetarum qui celestem musicam consonant *Oxf*
DELIUS Apollo *LbVa*
MODULETUR moveat. modulari faciat *LbVa*
temperetur *Oxf*

IN SOLE ad sonum *LbVa*
cum sole *Oxf*

7.10 HINCQUE ESSE QUOD ILLIC PHOEBUS ET HIC VOCITETUR AURICOMUS
HINCQUE s. didicit ex <h>ac re, i. crinitus *BeLa*, ac] hac *Be*[2]
ex hac re *LbVaLea*
In caelo crinitus, in terra auricomus dicitur id est aureas comas habens *LbVa*
QUOD s. manifestabatur *LbVaLea*
ILLIC i. in caelo *Be*
PHOEBUS Delius i. crinitus s. in celo *La*
novus divinus *LbVa*, divinus *om. Va*
Phebos Grece Latine crinitus dicitur, eo quod mane crines emittat oriens *LbVa*
HIC i. in terra *BeLa*
i. in sole *LbVa*
AURICOMUS sol *La*
splendens *LbVa*
Quia quando surgit, videtur quosdam crines habere, cum vero alta celorum penetrat, ad nos radios quasi aureos mittit *BeLa*, crines] crinos *La*
Crinitus, eo quod mane crines emittat ovans (*sic, lege* oriens?) *LbVa, cf. LbVa ad* 7.10 PHOEBUS. emittat] emittans *Va*

19/
9.24 FIDIBUS PERSONANTI cordibus modulanti *BeLaLbVa*, cordibus] cordis *LbVa*
[1]Sol cum armonia movetur, et in hoc loco informitatem vocum designat. [2]Fidis enim tensa intellegibilem vocem profert, ita ut aliquod nomen usurpet sibi. [3]Laxa enim melodiam cum informitate sonus emittit. [4]Informitates enim vocum ponit, quia dum sit semper stabilis et nullam vocem proferre videatur, tamen nulla vox fieri potest, nisi ex informitate illius formetur. [5]Verbi causa: cum prima corda figitur in cithara, laxissimum atque gravissimum sonum sine ulla pulchritudine emittit. [6]Si tamen extenta fuerit, aliquam formam assumit, quae in numero troporum hypodorius vel sub dorio positus vocatur. [7]Sol enim quamvis magnum faciat sonum, nos tamen expertes illius modulationis sumus

BeLaLbVaLea, 1 informitatem] infirmitatem *Va*; 2 intelligibilem *Lea*; 4 enim] omnium *La*; 5 Verbi causa *om. Be* verbi gratia *LbVaLea* cithara] cythara *LaLb* atque] adque *La* sonum *post* sine ulla pulcritudine *Be* pulchritudine] pulcritudine *BeLa* emittit *post* sonum *LbVaLea*; 6 extenta] extensa *LbVaLea* hypodorius] ypodorio *La* ypodorium *Be* sub dorio positus] sub dorius *LbVaLea*; 7 Sol enim - sumus *om. BeLa*

3 TUM VERO CONSPICERES TOTIUS MUNDI GAUDIA CONVENIRE

TUM deinde *BeLa*

tunc *LbVa*

VERO CONSPICERES Poteras conspicere *Be*

Conspicere poteras, si adfuisses *La*

Videre posses, si adesses *LbVa*

Si adesses *Lea*

TOTIUS MUNDI GAUDIA CONVENIRE Ideo terra letata est in ascensum Mercurii, quia ipse est deus linguarum et in terra utique lingue <...>* *La*

Bene dicit convenire gaudia, quia sole superiores partes ascendente omnis terrena substantia hilarescit i. splendente sole omnia elementa laetantur *LbVaLea*

CONVENIRE confluere *Lb*

9-0 NAM ET TELLUS FLORIBUS LUMINATA, QUIPPE VERIS DEUM CONSPEXERAT SUBVOLARE MERCURIUM

cum FLORIBUS i. habens <flores>. floribus: per flores *BeLa*, cum] per *Be* floribus *om. Be* flores] fores *Be*

†LUMINA illius temporis. s. sunt flores *La*

Tripliciter i. nam tellus conveniebat cum floribus vel cernebat in floribus deum Mercurium subvolare vel per lumina veris conspexerat deum Mercurium subvolare, quia flores non sine ratione nascuntur, quoniam splendente sole omnia elementa letantur *BeLa*, vel cernebat in floribus] *add. La*[2] deum *om. La*

FLORIBUS i. habens flores s. in vel cum LUMINA *LbVa*

i. luminosa per splendorem illius *LbVaLea*, i. *om. Lea*

Volante Mercurio terra gaudet LUMINA i. luminata *Oxf*

VERIS i. vernalis temporis *LbVa*

SUBVOLARE conscendere *LbVa*

---

* Tironian notes are added to this passage?

Flores enim cum ratione nascuntur, quia fructum indicant se emittere *LbVa*

12.20 ET APOLLINE CONSPICATO AERIA TEMPERIES SUDIS TRACTIBUS RENIDEBAT

APOLLINE sole *LbVa*

[1]Sol diversis utitur nominibus et pro variis significationibus accipitur. [2]Vocatur sol, eo quod prae ceteris elementis plurimum solus luceat vel quod cuncta illuminet. [3]Dicitur Delius, declaratio. [4]Dicitur Apollo id est exterminans, eo quod suo calore omnes humores arborum pellere videatur. [5]Dicitur Phithius id est fidem auferens. [6]Pro diversis etiam ponitur significationibus: nunc pro ipso sole, nunc pro deo, nunc pro divino, nunc pro mundo, nunc pro significatione humanae vocis sicut hic. [7]Cui bene novem musae applicantur, quia humanae vocis officia novem sunt. [8]Omnis enim sermo his novem rebus formatur: primo appulsu quattuor dentium, repercussione duorum labrorum, plectro linguae, cavo gutturis, adiutorio pulmonum. [9]Si enim ex his quid defuerit, vox perfecta non erit.

[10]Et phisica naturarum ratione Apollini musae applicantur. [11]Sicut enim terrestris musicae MHCH id est media principatum in omni tropo tenet, ita sol armoniae caelestis, quae viiii ordines habere dinoscitur scilicet propter vii planetas octavamque spheram caelestem atque ipsam terram. [12]Quae totius consonantiae informitatis vocumque materiei proportionem possidet.

[13]Iuxta vero leges allegoriae apertissimo musarum usus est ordine. [14]Urania igitur primo ponitur id est humanae intellegentiae sublimitas. [15]Post hanc Polimnia id est capacitas memoriae. [16]Tertia hinc ponitur Euterpe, delectatio voluntatis, quia intellegentia memoriae iungitur et voluntati. [17]Inde Erato. [18]Nulla enim re plus voluntas delectatur quam similium collatione. [19]Hoc autem sine profunda cogitatione fieri non potest. [20]Ideo subsequitur Melpomene. [21]Huic Terpsicore quasi artium delectatio sociatur. [22]Cogitationum enim perfectio sine disciplinarum exercitatione fieri non potest. [23]His omnibus omnis humanae locutionis honestas gignitur, quae Calliope signatur. [24]Inde Clio hoc est bona fama nascitur. [25]Sub qua Talia quasi in ultimo loco ponitur, hoc est positio vel germinatio virtutum. [26]Omnis enim virtus propter pulchram appetitur famam. [27]Ideoque albo alite id

est cigno sustollitur stagnisque deponitur, quia bonae famae semina per terram spargi monstrantur.
[28]Totius mundi armoniam viii ptongis constare manifestum est: vii quippe planetarum motibus atque ipsa sphera ultima, quae semper celerrimo volvitur motu. [29]Terra vero in proportione i. in ratione. [30]Sicut sol in armonia et mese in lyra tetracordum faciunt, ita lingua in voce* *BeLaLbVaLea, lin. 10-27 Oxf et lin. 20-29 BeLa: ad* 12.23 QUIPPE MUSAS**.

1 et pro variis significationibus accipitur *om. BeLa*; 2 elementis] elimentis *LbVaLea* solus] sol *La* plurimum solus] solus plurimum *Lea* illuminet] inluminet *BeLa* illaminet *Va*; 3 *post* Delius] id est *add. Lea* declaratio] i. claritas et declaratio *BeLa*; 4 eo quod - pellere videatur *om. BeLa*; 5 Dicitur] et *BeLa* dicit *Va* Phitius] Fitius *La* Pithius *Va*; 7-8 humanae vocis - adiutorio pulmonem] humana vox ut perfecta sit viiii adiutoriis fulcitur i. iiii dentes anteriores, duo labia, linguam, concavitatem palati et gutturis pulmonemque ploalem aerarium habet ad perfectionem *BeLa*; 8 omnes] omnis *Lea* quattuor] quatuor *Va* pulmonum] pulmonis *Lea*; 9 his] is *La*; 1-9 Sol diversis - non erit *om. Oxf*; 10 et *om. Oxf* Apollini] appollini *La* Apollini musae] musae apollini *Lea* applicantur] asscribuntur *Oxf*; 11 enim *om. Oxf* MHCH] mese *LbVa* viiii] iiii *Va* spheram] speram *Oxf*; 12 informitatis] infirmitatis *Oxf* materiei] materie ei *Be* materie *La*; 13 apertissimo] aptissimo *LbVaLeaOxf*; 14 id est] haec est *Oxf* humanae] umanae *La*; 16 hinc *om. BeLa* voluntatis] voluptatis *La* quia intellegentia] qua intellegentia *Lea* qua intelligentia *Oxf* memoriae] meriae *Oxf* voluntati] voluntate *Be* et voluntati *om. Oxf*; 17 Inde] hinc *BeLaOxf*; 18 voluntas] voluptas *La* collatione] conlatione *BeLa*; 19 profunda] perfecta *BeLa*; 19-22 hoc autem - non potest] Haec autem sine perfectione sine disciplinarum exercitatione stare non valet *Oxf*; 20-29 *cf. BeLa* 12.23 QUIPPE MUSAS; 22 enim] etenim *Lea*; 23 omnis] praedictis *Oxf* quae] quae per *Oxf* signatur] significatur *Oxf*; 24 hoc est bona fama] i. fama bona *Oxf*; 25 hoc est positio vel germinatio] i. positio atque germana *Oxf*; 26 pulchram] bonam *Oxf*; 27 cigno] gigno *LbVa*; 28 armoniam] armonia *LbVa*; 28-30 Totius mundi - in voce *om. Oxf*; 29-30 Terra - in voce *om. Lea*

CONSPICATO viso *LbVa*

AERIA Ether ipsum firmamentum quod est super aera *BeLa*
firmamentum superius aere *LbVaLea*

SUDIS TRACTIBUS puris vel lucidis spatiis *BeLa*, vel lucidis *om. Be*
lucidis spatiis *LbVa*
serenis, sub-udis *LbVaLea*

---

* Cf. Isid. Etym 8.11.53-55, 3.20.14; Fulg. Myt. I.15; Myth.Vat.I 112-113; Myth.Vat.II 29, 34; Sed.Scot. Don.art.mai. I.4.65, II.112.94

** In *La* this gloss is placed at the bottom of the preceding page. There is no sign to link the gloss to the proper lemma.

serenis spaciis *Oxf*
RENIDEBAT resplendebat *LbVaLea*
splendebat *Oxf*

12.21-3 SUPERI AUTEM GLOBI ORBESQUE SEPTEMPLICES SUAVIS CUIUSDAM MELODIAE HARMONICIS TINNITIBUS CONCINEBANT AC SONO ULTRA SOLITUM DULCIORE
SUPERI AUTEM GLOBI circuli *LbVa*
Ipsum firmamentum et spheram superiorem vel articum et antarticum *BeLa*, et *om. La*
Firmamentum superius vel spheram caelestem dicit id est anticum (*sic, lege* articum) et antarticum id est arcturum superius et inferius, qui a nobis non videntur *LbVaLea*, superius et inferius] superiorem et inferiorem *Lea* qui] quod *LbVa*
ORBESQUE SEPTEMPLICES septem circuli planetarum *LbVaLea*
i. circuli septem planetarum *Oxf*
ARMONICIS adunatis vocibus *BeLa*
s. adunatis *LbVa*
musicis *LbVaLeaOxf*, tonis *add. Oxf*
TINNITIBUS vocibus *LbVaLea*
SONO ULTRA SOLITUM s. concinebat *LaLea*
Concinebant praeter moram *LbVa*

12.23 QUIPPE MUSAS ADVENTARE PRAESENSERANT
QUIPPE MUSAS Musae namque ut ferunt Iunonis et Iovis filiae sunt, quoniam omnis vox ex aere et aethere formatur secundum phisicam rationem* *LaLbVaLea*, namque *om. Lea* ex] ab *La* *post* et aethere] i. aura et igne *add. La* rationem] arte onem *Va* phisicam rationem] fisicos *La*

[1]Ideo subsequitur Melpomene. [2]Huic Tersicore quasi artium delectatio sociatur. [3]Cogitationum enim perfectio sine disciplinarum exercitatione fieri non potest. [4]His omnibus predictis omnis humanae locutionis honestas gignitur, quae per Calliope<n> signatur. [5]Inde Clio i. bona fama nascitur. [6]Sub quo Talia quasi in ultimo loco ponitur hoc est positio vel germinatio virtutum. [7]Omnis enim virtus propter pulchram appetitur famam. [8]Ideo albo alite i. cigno sustollitur stagnisque

* Sed.Scot. Don.art.mai. II.112.91

deponitur, quia bonae famae semina per terram spargi monstrantur.
[9]Totius mundi armenia (*sic, lege* armoniam) viii ptongis constare manifestum est: vii quippe planetarum motibus atque ipsa sphera ultima, que semper celerrimo volvitur motu.
[10]Terra vero in proportione i. in ratione. [11]Sicut enim in armonia sol et mese in lira tetracordum faciunt, ita lingua in voce *BeLa, cf. ad 12.20. cf. LbVa* 12.20 APOLLINE (lin. 20-30). 3 His] is *La* humanae] umanae *La*; 7 appetitur] apetitur *Be*; 8 sustollitur] sultollitur *Be* stagnisque] per stagnisque *Be*; 9 tetracordum] tethracordum *La*[*]

ADVENTARE advenire *BeLaLbVa*

PRAESENSERANT antesenserant *BeLa*

Praecognoverant ipsi circuli *LbVa*

illi orbes *Oxf*

4-1 QUAE QUIDEM SINGILLATIM CIRCULIS QUIBUSQUE METATIS, UBI SUAE PULSUM MODULATIONIS AGNOVERANT, CONSTITERUNT

QUAE s. musae *LbVaLeaOxf*

SINGILLATIM unaquaeque seorsum *BeLa*

seorsum *LbVa*

METATIS mensuratis *BeLa*

electis mensuratis *LbVa*

i. electis vel mensuratis *Oxf*

UBI SUAE PULSUM postquam. sonum, motum *BeLa*, motum *om. Be*

postquam suam modulationem vel sonum *LbVa*

propriae i. sonum *Oxf*

CONSTITERUNT Unaquaeque musarum sono sibi convenienti hesit *LbVaLea*

---

[*] Ad 9 *in ratione*] *La* adds some lines which are also to be found ad 9.24 FIDIBUS PERSONANTI in *BeLbVaLea*, and also in *La*:

Informitates omnium vocum ponitur, quia dum sit semper stabilis et nullam vocem proferre videatur et nulla vox fieri potest, nisi ex informitate illius formetur. Verbi causa: cum prima corda figitur in cythara, laxissimum et gravissimum sine pulchritudine ulla sonum emittit. Si tamen extenta fuerit, aliquam formam assumit, quae in numero troporum ipodorius vel subdorius positus vocatur.

The underlined words are represented by signs, perhaps those belonging to the Tironian set of abbreviations.

13.1-2 NAM URANIE STELLANTIS MUNDI SPHAERAM EXTIMAM CONTINATUR, QUAE ACUTO RAPTABATUR SONORA TINNITU

URANIA Celestis dicitur *BeLa*

Quae dicitur caelestis *LbVa*

STELLANTIS stellas habentis *BeLa*

stellis obducti *LbVa*

MUNDI SPERAM circulum supernum *LbVa*

i. rotunditatem celi *Oxf*

EXTIMAM in supremum *La*

supremam *LbVa*

†CONCINATUR disponit *BeLa*

Concinando movet *La*

†CONCINAT disponit, modulatur *LbVa*

i. resonare facit *Oxf*

QUAE s. sphera *Be*

quae spera extima *LaLbVa*, quae] s. *La*

RAPTABATUR volvebatur *BeLaLbVa*

SONORA garrula *BeLa*

vel stridula *La*

canora *LbVa*

Eo quod ibi nullus aer est: quo altius fuerit pulsus, acutior erit et sonus *Oxf*

28/
13.3-4 POLYMNIA SATURNIUM CIRCULUM TENUIT, EUTERPE IOVIALEM, ERATO INGRESSA MARTIUM MODULATUR

POLIMNIA Multa memoria quasi polin menen *BeLaLbVaLea*, polin] polyn *Be* menen] nemen *BeLa*

Polimnia quasi polin nemene i. multa memoria *Oxf*

EUTERPE bene delectans *BeLaLbVaLeaOxf*, i.] *add. Oxf*

IOVIALEM s. circulum *LbVa*

ERATO Inveniens similem quasi erin omoeon *BeLaOxf*, inveniens similem *om. Oxf*

Inveniens similem *LbVaLea*

Erato quasi erin omeon, i. inveniens similem *LbVa*

MARTIUM sidus vel circulum *BeLa*, vel circulum *om. Be*

s. circulum *LbVaLea*, s. *om. Lea*

MODULATUR temperat *Oxf*

5 MELPOMENE MEDIUM, UBI SOL FLAMMANTI MUNDUM LUMINE CONVENUSTAT
MELPOMENE Meditationem faciens permanere *BeLaLbVaLeaOxf*, permanere] manere *LaOxf*, *om. Lea* subaudis *add. Lb*
i. quasi melon poieon mene *BeLaOxf*, i. *om. LaOxf* poieon] pieon *La* mene *om. Oxf*
MEDIUM s. ingressa est circulum *BeLa*, ingressa est *om. La*
s. circulum *LbVa*, s. *om. Lb*
UBI in quo circulo *LbVa*
SOL FLAMMANTI ardenti *BeLaLbVa*
CONVENUSTAT decorat, ornat *BeLaOxf*, decorat *om. Oxf* ornat *om. Be*
condecorat *LbVa*

6 TERPSICHORE VENERIO SOCIATUR AURO, CALLIOPE ORBEM COMPLEXA CYLLENIUM
TERPSICORE Delectans instructionem *BeLaLbVaLeaOxf*
AURO splendore *BeLa*
pulchritudini *LbVaLea*
Veneri *Oxf*
CALLIOPE sonoritas *BeLa*
sonoritas i. bona vox *LbVaLea*, i. *om. Lea*
pulcra sonoritas *Oxf*
COMPLEXA s. est *BeLaLbVaLeaOxf*, s. *om. Be*

7 CLIO CITIMUM CIRCULUM, HOC EST IN LUNA COLLOCAVIT HOSPITIUM, QUAE QUIDEM GRAVES PULSUS MODIS RAUCIORIBUS PERSONABAT
CLYO Cogitatio vel bona fama, quia cleos Grece fama *BeLa*
cogitatio *LbVaLea*
bona fama *LbVaOxf*
CITIMUM CIRCULUM ultimum i. terrae proximum *BeLa*, ultimum i. *om. Be*
s. terrae contiguum *LbVaLea*, contiguum *om. Lea*
tenuit *Lb*
proximum *Oxf*
HOSPICIUM conversationem *LbVa*
QUAE s. luna *La*
Clio *LbVaOxf*, s. *add. Oxf*
PULSUS sonoritatis. genitivus *BeLaLb*
MODIS RAUCIORIBUS cantilenis *BeLaLea*

MODIS sonis. cum RAUCIORIBUS gravioribus *LbVa*
Eo quod propior esset aeri *Oxf*

13.8-10 SOLA VERO, QUOD VECTOR EIUS CYCNUS IMPATIENS ONERIS ATQUE ETIAM SUBVOLANDI ALUMNA STAGNA PETIERAT, THALIA DERELICTA IN IPSO FLORENTIS CAMPI UBERE RESIDEBAT

VECTOR Circulum terrae describit *BeLaLbVa*, Circulum terrae] terrae circulum *LbVa*

INPATIENS s. non PATIENS *La*

SUBVOLANDI Nam musae ut ferunt Iunonis et Iovi<s> ficiliae (*sic, lege* filiae) sunt, quoniam omnis vox ab aere et ethere i. aura et igne formatur secundum fisicos *Be, cf. LaLbVaLea* 12.23 QUIPPE MUSAS

ALUMNA nutritoria *BeLaLbVaLea*, i. *add. Be*

TALIA capacitas. Et dicitur a Grega (*sic, lege* Greca) ethimologia quasi titonlia i. ponens germina *BeLa*, Et dicitur a Grega ethimologia *om. La* titonlia] tonlia *La*

i. capacitas quasi titonlia hoc est ponens germina fecunditate. A reliquarum musarum societate euntium *LbVaLea*, ponens germina] germina ponens *Lea* A reliquarum - euntium *om. Lea*

positio *Oxf*

UBERE ubertate *La*

fecunditate *LbVa*

RESIDEBAT manebat *BeLaLbVa*

GLOSSES ADDED TO SECTIONS OF BOOK TWO

105/
30.3-4 NAM ET ILLE, QUOD RATIO PRINCIPIUM, MEDIUM FINEMQUE DISPENSAT, PRO CERTO PERFECTUS EST

ILLE i. ternarius numerus *LbVaLea*, i. *om. Lea*

eo QUOD RATIO causa *BeLaLbVaOxf*, i. *add. Oxf*

30.4-6 QUIPPE LINEAM FACIT PRIMUS ET SOLIDORUM FRONTES INCUNCTANTER ABSOLVIT, NAM LONGITUDINE <LATITUDINE> PROFUNDITATEQUE CENSENTUR

QUIPPE LINEAM [1]Quod dicit *lineam facit*, geometricam tangit. [2]Omnis soliditas tribus speciebus constat: longitudine, latitudine, altitudine. [3]Atque ideo ternarius frontes solidorum reddere monstratur, quae sunt vi. [4]Nam longitudo duas frontes colligit: ante et retro; latitudo alias duas: dextrorsum et sinistrorsum; profunditas i. altitudo duas: sursum et deorsum. [5]Fiuntque sex *BeLaLeaOxf*, 1 geometricam] giometricam *La*; 3 ideo *om. Oxf* ternarius] ternarium *La* quae sunt vi] *La*[2], *om. Be*; 4 duas] duos *Lea* alias duas] alios duos *Lea* i.] unum *La* altitudo duas] altitudo duos *Lea*; 5 Fiuntque] fiunt *La*

SOLIDORUM s. numerorum *BeLa*

corporum *LbVa*

INCUNCTANTER i. indubitanter *BeLaLbVa*, i. *om. BeLa*

ABSOLVIT reddit *BeLa*

i. reddit perficit *LbVa*

perficit *Lea*

LONGITUDINE Dicendo longitudinem et profunditatem, latitudinem subintellegendo dimisit. Ubi enim longitudo et profunditas est, necessario est et latitudo *BeLaLbVaOxf*, longitudinem et profunditatem] longitudine et profunditate *LbVa* subintellegendo] subintellegendam *LbVa*

profunditas] profundita *Oxf* necessario est et latitudo] necessario et latitudo est *LbVa*

CENSENTUR s. illae frontes *BeLaLbVa*, s. *om. BeLa*

numerorum *La*

-7 DEHINC QUOD NUMERI TRIPLICATIO PRIMA EX IMPARIBUS CYBON GIGNIT

DEHINC s. perfectus est *La*

s. perfectus dicitur *LbVaLea*

NUMERI i. ternarii *BeLa*

PRIMA EX IMPARIBUS Hic tangit arithmethicam *LbLea*

CYBON Hoc ideo dicit, quia omnium cuborum nascentium ex imparibus numeris primus ille est, qui ternarii triplicatione consistit i. ter tria ter *BeLaOxf*, *LbVa ad* 30.6-7 TRIPLICATIO. cuborum] cyborum *Oxf* imparibus] impartibus *Va* qui ternarii triplicatione consistit] qui ternarii triplicatio triplicatione consurgit *LbVaLea* qui ternarii triplicatione consurgit *Oxf* i. ter tria ter] i. tria ter *Oxf*

-8 TRES AUTEM SYMPHONIAS QUIS IGNORAT IN MUSICIS?

TRES AUTEM SYMPHONIAS consonantias *La*

diatessaron diapente diapason *LaOxf*, i. *add. Oxf*

i. diapason diapente diatesseron *LbVaLea*, i. *om. Va*

IN MUSICIS Musicam tangit *LbVaLea*

30.8 NUMERUSQUE IMPAR MARIBUS ATTRIBUTUS
IMPAR Impar numerus fortioris naturae est, quia indivisibilis est fortiorique sexui deputatur quam numerus par i. virili, qui propter mutabilitatem quasi infirmiori deputatur sexui i. femineo *BeLaLbVaOxf*, virili] virilis sexui *LbVa* qui] s. par numerus *add. LbVa* quam numerus - femineo] par autem numerus quia dividi potest et mutalilitatem (*sic, lege* mutabilitatem) femineam significat, infirmiori deputatur sexui i. femineo *Oxf*
ATTRIBUTUS s. est *BeLaLbVaLeaOxf*

30.8 OMNE VERO TEMPUS TRIBUS VICIBUS VARIATUR
osten<den>dum perfectionem ternarii *La*
i. praesenti praeterito et futuro *BeLaLbVaLeaOxf*, i. *om. BeLaLeaOxf* *post* praesenti] et *add. Lea* et *om. Oxf*

30.9-10 ATQUE IDEM NUMERUS SEMINARIUM PERFECTORUM, SEXTI VIDELICET ATQUE NONI, ALTERNA DIVERSITATE IUNCTURAE
NUMERUS ternarius *LbVaLea*
SEMINARIUM s. est *BeLaLbVaLeaOxf*
origo *LbVa*
Quia ex ipso seminantur *LbVaLea*
VI sexti *BeLaOxf*
Senarius numerus ideo perfectus est, quia suis partibus impletur. Habet enim tres partes: dimidiam iii, tertiam ii, sextam i. Unum autem et ii et iii fiunt vi *BeLaLbVaLea*, tres partes] partes tres *Lea* tres partes: dimidiam iii, tertiam ii, sextam i] partes: tres dimidia<m>, ii tertiam et sextam i *LbVa*
ATQUE VIIII noni *BeLaOxf*
Novenarius vero numerus ideo perfectus est, quia primi versus numerorum plenitudo est et finis. Et totius musicae efficitur epogdoum i. tonum *BeLaLbVaLeaOxf*, numerus *om. BeLa* dicitur n<umerus> *Oxf* ideo] ide *Be* musicae] musae *BeLa* efficitur] efficit *LbVaLea* epogdoum] epocdoum *LbVaLea* epiogdoum *Oxf* tonum] totum *Va*
ALTERNA DIVERSITATE cum *La*
Ideo cum alterna diversitate ternarius seminarium perfectorum numerorum sexti videlicet et noni efficitur, quia et per suam et per alterius quantitatem ad hoc agendum multiplicatur. Si vero per suam id est ter tres, novenarium, sin autem per alterius i. ter bis, senarium reddit *LbVa*

30.10 RITE IGITUR DEO ATTRIBUITUR RATIONIS

RITE Quoniam sic perfectus est *La*
DEO Mercurio *LaLbVaLeaOxf*, i. *add. Lea*
RATIONIS sapientiae vel sermonis *LbVa*

1-2 PHILOLOGIA AUTEM, QUOD ETIAM IPSA DOCTISSIMA EST, LICET FEMINEIS NUMERIS AESTIMETUR, ABSOLUTA TAMEN RATIONE PERFICITUR
PHILOLOGIA i. numerus Filologiae *Oxf*
QUOD s. et ternarius *La*
FEMINEIS NUMERIS i. quaternario *La*
  divisibilibus *LbVaLea*
  divisionibus *Oxf*
AESTIMETUR iudicetur *LbVa*
ABSOLUTA perfecta *LaOxf*
  disiuncta vel libera *LbVaLea*, vel libera *om. Lea*
  manifesta *Oxf*
RATIONE quaternario numero *LbVaLea*

2-3 NAM QUATERNARIUS SUIS PARTIBUS COMPLET DECADIS IPSIUS POTESTATEM
SUIS PARTIBUS Nam i. ii. iii. iiii. x. fiunt *La*
  in *LbVa*
  Id est i. ii. iii. iiii *LbVaLeaOxf*, Id est *om. Oxf*
DECADIS numeri *LbVa*
  i. denarii *Oxf*

3-5 IDEOQUE PERFECTUS EST ET HABETUR QUADRATUS, UT IPSE CYLLENIUS, CUI ANNI TEMPORA, CAELI CLIMATA MUNDIQUE ELEMENTA CONVENIUNT
IDEOQUE PERFECTUS EST Nam i. ii. iii. iiii. x. fiunt *BeLa*
QUADRATUS ipse numerus *BeLa*
  i. solidus *LbVa*
UT IPSE CYLLENIUS i. ratio *BeLaLbVa*, i. *om. LbVa*
  Ideo Cyllenius quadratus est, quia ei quattuor tempora deputantur *BeLaLbVa*, Cyllenius] cillenius *BeLa* quattuor] quatuor *Be*
TEMPORA Ver, aestas, autumnus et hiemps *LaLbVa*, i. *add. La* hiemps] yemps *La*
CAELI iiii^{or} partes mundi *LaLbVaLea*, i. *add. La* iiii^{or}] quatuor *La*

MUNDIQUE ELEMENTA ignis, aer, aqua et terra *LaLbVa*, aqua et terra] terra et aqua *La*
CONVENIUNT Quia de his omnibus sermo disputat *LaLbVa*

107/
30.15-6 AN ALIUD ILLA SENIS DEIERATIO, QUI MA THN TETPAΔA NON TACUIT, CONFITETUR NISI PERFECTAE RATIONIS NUMERUM?
AN ALIUD numquid aliud *La*
ILLA SENIS i. Pitagorae *LbVaLeaOxf*
DEIERATIO assertio *BeLaLbVaOxf*
QUI MATHEN TETRADAN* doctrinam quaternariam *BeLaLbVaLeaOxf*, i. *add. Oxf*
†ΗΘΙΛΩΜΑΛΟΝ *BeLa*
†ΓΗΤΙΛΟΓΜΑΛΟΝ *LbVa***
Illa enim doctrina in iiii[or] constat: arithmetica, geometrica, musica, astronomia. Ipsum est quadruvium sapientiae *LbVaLea*, doctrina] doctrinam *Lb* quadruvium] quadri avium *Lb*
NON TACUIT sed docuit *Oxf*
CONFITETUR asseverat *LbVa*
RATIONIS sapientiae *LbVa*
NUMERUM Quaternarium, quia illa ars mathematica ex iiii constat disciplinis *LbVa*, iiii] iiii[or] *Va*

30.16-8 QUIPPE INTRA SE UNUM, SECUNDUM TRIADEMQUE IPSUM BIS BINUM TENET, QUIS COLLATIONIBUS SYMPHONIAE PERAGUNTUR
SECUNDUM diapason *BeLa*
TRIADENQUE diapente *BeLa*
BIS BINUM diatesseron *BeLa*
quaternarium *LbVaLea*
QUIS quibus *BeLaLbVaLea*
COLLATIONIBUS comparationibus *BeLa*
partibus *LbVa*
SYMPHONIAE aequalitates vocum *BeLaLbVa*
i. qualitates vocum *Oxf*

---

* MATENTE TRADAN *BeLa* MANTENTE TRADAN *Va*[1] MATHENTE TRADAN *Va*[2]

** ΗΘΙΛΩΜΑΛΟΝ / ΓΗΤΙΛΟΓΜΑΛΟΝ: the meaning of these glosses is unclear. They probably refer to some kind of ethymology with the word ὁμαλόν, meaning even or equal. Cf. 31.1-2 ΕΒΔΟΜΑΔΩΝ

8-9 NAM TRES AD QUATTUOR EPITRITUS VOCITATUR ARITHMETICA RATIONE AC DIATESSARON PERHIBETUR IN MUSICIS
EPITRITUS supertrinus *LaLeaOxf*
supertertius *LbVa*
AC DIATESSARON supertrinus *LbVa*
ipsa est sesquitertia *LbVaLea*
Dia si per *i* nostrum scribitur, nihil aliud nisi *ex* significat, sed per *y* Grecum, duo ostendit* *BeLa*, nihil] nil *La*
Diatesseron: *di* quando per *i* scribitur, pro *de* vel pro *ex* accipi debet, quando per *y*, duo significat* *LbVaLeaOxf*, *di*] *om. LbVaOxf*
post quando] vero *add. Oxf* per *om. Va*
PERHIBETUR dicitur *LbVa*

9-1 ITEM INTRA EUM IACENT TRES AD DUO, QUAE HEMIOLIOS FORMA EST, SYMPHONIAMQUE SECUNDAM, QUAE DIAPENTE DICITUR, REDDUNT
INTRA EUM quaternarium *LaLbVaLeaOxf*, i.] *add. Oxf*
HEMIOLIOS sesqualtera in arithmetica *La*
totum et semis *LbVaLea*
emis semis. olon totum *Oxf*
DIAPENTE i. sesqualtera *LbVaLea*
REDDUNT i. tres ad duo *LaLbVaLea*, i. *om. LaLea*

1-2 TERTIA SYMPHONIA DIAPASON IN MELICIS PERHIBETUR DIPLASIOQUE CONFICITUR, HOC EST UNO DUOBUS COLLATO
DIAPASON ex omni *La*
IN MELICIS in musicis *BeLaLbVaOxf*, i. *add. Oxf* in *om. LbVa*
DIPLASIOQUE duplicatione *BeLaLbVaLeaOxf*, i. *add. Oxf*

2-4 IGITUR QUATERNARIUS NUMERUS OMNES SYMPHONIAS SUIS PARTIBUS PERFECTUS ABSOLVIT OMNIAQUE MELA HARMONICORUM DISTRIBUTIONE CONQUIRIT
OMNES SIMPHONIAS diapason et diapente et diatesseron *BeLaLbVaLea*, et *om. BeLaLea*
diatessaron, diapente, diapason *Oxf*
SUIS PARTIBUS [1]Plenitudo totius musicae tribus simphoniis continetur, quae sunt diatesseron, diapente, diapason, quarum

---

* Cf. Sed.Scot. Don.art.mai. I.10.72

rationem inter quaternarium contineri manifestum est. [2]Nam sicut quaternarius ad ternarium, ita est diatessaron i. maior numerus ad minorem et tertiam partem minoris colligit in se: sic extremi soni simphoniae diatessaron. [3]Similiter ut sunt tria ad duo in ratione sesqualtera, sic diapente in extremis suis sonis. [4]Habet enim maior minorem et dimidiam partem minoris intra se. [5]Quemadmodum autem sunt duo ad unum in dupla proportione, sic extremi soni diapason simphoniae. [6]Habet enim maior minorem bis. [7]Hinc igitur inter quaternarium numerum et ternarium plenitudo artis musicae continetur. [8]Nulla enim musicae ratio extra has tres simphonias protenditur; quicquid enim extra has est, ex his compositum est* *BeLaLbVaOxf*, 1 simphoniis] simphonii *Lb* diatesseron] diatessaron *BeLaOxf*; 2 diatessaron] diatesseron *LbVa* colligit in se] in se colligit *LaOxf* diatessaron] diatesseron *LbVa*; 3 ut sunt] sicut *BeLa*; 5 autem *om. La* autem sunt] somno *Be*; 7 inter] intra *BeLaOxf* numerum *om. Oxf* numerus *Lea*; 8 has *om. BeLa* quicquid] quicquit *La* extra has] extra as *La* ex has *LbVa* ex his] ab his *BeLa* compositum] conpositum *La*

PERFECTUS actus ductus *LbVa*
auctus *Oxf*
ABSOLVIT pandit *BeLa*
reddit *BeLaLbVaOxf*, i. *add. Oxf*
OMNIAQUE MELA omnem sonoritatem *BeLaOxf*, i. *add. Oxf*
dulcedines *La*
sonoritatem vel omnes modulationes. symphoniam *LbVa*
to melos, ta mele, haec mela *La*
ARMONICORUM sonorum *BeLa*
DISTRIBUTIONE divisione *BeLa*
CONQUIRIT ipse quaternarius *LbVa*

30.24-5 HANC IGITUR DISCUTIENS NUMERI CONGRUENTIAM PERITA VIRGO GRATULATUR
DISCUTIENS discernens dinumerans *BeLa*
subtiliter inquirens *LbVaOxf*
CONGRUENTIAM convenientiam *BeLaLbVaOxf*
PERITA VIRGO GRATULATUR docta gaudet *Oxf*

* Cf. Joh.Scot.Er. Exp. 6.187

5-1 DEINDE UTRUMQUE CONSOCIAT, ET TRIAS QUATERNARIO SOCIATA HEPTADEM FACIT

UTRUMQUE CONSOCIAT s. nomen iungit *BeLa*, iungit *om. Be*
nomen vel numerum *LbVaLeaOxf*, s. *add. Oxf*

ET TRIAS i. ternarius numerus *LbVaLeaOxf*, i. *om. Lea*

SOCIATA coniuncta *LaLbVa*
iuncta *Oxf*

EPTADEN septenarium s. numerum *BeLaOxf*, i. *add. Oxf* s. *om. Oxf*
septenarium *LbVa*
i. septem *Lea*

-2 QUI NUMERUS RATIONIS SAPIENTIAEQUE PERFECTIO EST, SICUT EBΔOMAΔΩN ILLA EDOCET PLENITUDO

QUI NUMERUS septenarius *LaLbVa*

†SUPERAE caelestis *LaLbVaOxf*, i. *add. Oxf*

†EΘIΛOMAΛΩN aritmeticus vel astronomicus *LaOxf*, arithmeticus *Oxf* astronomius *Oxf*

†OMAΛΛON i. planarum s. figurarum *LbVaLeaOxf*^{2}, i. *et* s. *om. LeaOxf*^{2}

†DOCET PLENITUDO s. et eiusdem superae perfectionis *LbVaOxf*, s. et eiusdem *om. Oxf*

-4 AN ALIUD FATALIS TEMPERAMENTI CURSUS SIDERUMQUE CIRCULI MOTUSQUE TESTANTUR INTRAQUE LATEBRAS UTERI SEPTIMO MENSE ABSOLUTA MORTALITAS?

AN ALIUD numquid *LbVaOxf*, i. *add. Oxf*

FATALIS TEMPERAMENTI CURSUS constillationem siderum dicit *LaLbVaLeaOxf*, constillationem] constellatione *LbVa* dicit *om. Oxf*

[1]Et per hoc intellegitur et fatale temperamentum i. constellatio siderum et numerus circulorum planetarum motusque earum perfecti esse, quia numero septenario continentur. [2]Non enim recurrit, ut dicam septenarium numerum perfectum esse, quia fatale temperamentum et circuli siderum motusque septenarii sunt *BeLaLbVaOxf*, 1 Et *om. BeLaOxf* intellegitur] intelligitur *Oxf* planetarum] planearum *Va* constellatio] constillatio *La*; 2 enim] autem *BeLa* dicam] dicamus *BeLaOxf* numerum perfectum esse] numerum esse perfectum *BeLaOxf* circuli] circulis *La*

SIDERUMQUE planetarum *LbVa*

MOTUSQUE TESTANTUR s. quam illam perfectionem *Oxf*
septem numerum *Oxf*

LATEBRAS UTERI cunas ventris matris *BeLaOxf*, cunas *om. Oxf*
ventrem matris *LbVa*
ABSOLUTA nata vel vivificata *BeLaOxf*, s. est *add. Oxf*
vivificata vel nata *LbVaLea*
MORTALITAS s. aliud testatur? *LbVa*

31.4-5 DEHINC QUOD TRIAS TRIPLICATA NOVEM NUMEROS FACIT
DEHINC docet ille *La*
TRIAS TRIPLICATA ethilomalon *La*
consotiatur (*sic*) *LbVa*
[1]In hoc etiam perfectio ternarii et quaternarii cumulatur, quod ex his primi numeri inter quos epogdous componitur, procreantur. [2]Novenarius enim ex ternario i. se ipso multiplicato. [3]Octonarius vero ex quaternario geminato fit. [4]Inter novenarium vero et octonarium epogdoum esse manifestum est. [5]Epogdous autem in numeris, tonus in musica vocatur *LbVaLeaOxf*, *BeLa ad* 31.5-6 ΕΠΟΓΔΟΙ EFFICIUNT IUNCTIONEM. 1 his] is *La* epogdous] epiogdous *Oxf*; 2 ipso] ipsum *La*; 3 ex quaternario] et quaternarius *Oxf* fit] gignitur *BeLaOxf* novenarium] novennarium *La*; 4 epogdoum] epocdoum *LbVa* epiogdoum *Oxf* manifestum est] non dubium est *BeLa*; 5 epogdous] epocdous *LbVa* epiogdous *Oxf* tonus in musica vocatur] totus in musica *Be* totus in musica vocatur *Oxf* in musica tonus vocatur *La* tonus vocatur in musica *Lea*

31.5-6 QUATERNARIUS AUTEM PER ΔΙΠΛΑΣΙΟΝ GEMINATUS OCTO REDDIT, NOVEM VERO AD OCTO EPOGDOI NUMERI EFFICIUNT IUNCTIONEM
ΔΥΠΛΑCΙΟΝ* duplam proportionem *La*
duplicationem *LbVa*
i. geminationem *Oxf*
NOVEM VERO iuncti AC OCTO *BeLa*
ΕΠΟΓΔΟΙ** EFFICIUNT IUNCTIONEM sesquioctavi numeri. In hoc etiam perfectio ... *BeLa*, *cf. LbVaLeaOxf* 31.4-5 TRIAS TRIPLICATA
rationis *LbVaOxf*

31.7-8 TANTUMQUE PENSAT IN NUMERIS, QUANTUM SYMPHONIA [DIAPASON] IN MELICIS QUAE TONON FACIT, QUI EST CONSONAE UNITATIS CONTINUA MODULATIO

---

* ΔΙΑΠΛΑ SION *La*
** POGDOI *La* EPODOI *Be*

PENSAT epogdoi ratio *BeLaOxf*, valet s. *add. Oxf* epogdoi] epiogdoi *Oxf*
IN NUMERIS in arithmetica *BeLaOxf*, arithmetica] aritmetica *La*
SYMPHONIA ex integro *La*
TANTUMQUE i. tantum valet epocdoi ratio *LbVaLea* QUANTUM i. quantum in arithmetica *LbVa*
IN MELICIS in musicis *BeLaLbVaLeaOxf*, in *om. LbVaOxf*

[1]Sunt enim in melicis consonae et dissonae. [2]Dissonae vero sunt cum a se separantur aut emitonio aut tritemoria i. tertia parte toni aut tetratemoria i. diesi, quarta scilicet parte toni, quae spatia sunt in diatesseron et diapente. [3]Consonae vero voces sunt, cum uno eodemque spatio toni separantur, ut est in simphonia diapason *BeLaLbVaOxf*, 1 enim *om. BeLa*; 2 vero] ei *Be*, *om. La* emitonio] emitenio *Va* i. *om. Va* i. tertia parte toni *om. Be*, *add. La*[2] tetratemoria] tetratemoriai *Be* tetractemoria *LbVaOxf* quarta] quara *BeVa* scilicet *om. BeLaOxf* diatesseron] diatessaron *BeLa*; 3 eodemque] eodem *LbVa* *post* diapason] conveniebat *add. Be*

QUAE TONON FACIT simphonia *Be*
simphonie *La*
simphonia diapason *LbVaLeaOxf*, s. *add. Oxf*
QUI EST CONSONAE s. tonus *BeLaLbVa*, s. *om. LbVa*
Difinitio est toni *LbVa*, diffinitio *Va*
CONTINUA i. equalis *BeLa*
propinqua *LbVa*
Ubi non est distantia *LbVaLea*

9-10 EX QUO NIHIL EST, QUOD DISCREPET AUT RESULTET IN MEDIO, CONSENTANEAQUE CONGRUIT IUGITATE
EX QUO s. tono *LbVaLea*
s. toto *Oxf*
DISCREPET dissonet *LbVaOxf*
RESULTET contradicat *LbVaLeaOxf*
s. cum CONSENTANEAQUE consimili *BeLaLbVaLeaOxf*, s. cum *om. BeLaOxf* s. *om. Lea*
CONGRUIT convenit *Oxf*
10 IUGITATE societate *BeLa*
iunctione *Lea*
iunctione vel societate *Oxf*
iunctione consequentia *LbVa*, iunctione] coniunctione *Va*

33.16 ECCE ANTE FORES QUIDAM DULCIS SONUS MULTIFIDIS SUAVITATIBUS SUSCITATUR

FORES s. virginis *BeLaLbVaLeaOxf*, s. *om. BeLaOxf* virginis] virgines *Be*

MULTIFIDIS SUAVITATIBUS s. cum multiplicibus modulationibus *BeLaLbVaOxf*, s. cum *om. BeLa* s. *om. Oxf*

†CIETUR concitatur *BeLaOxf*

33.17-8 QUEM MUSARUM CONVENIENTIUM CHORUS IMPENDENS NUPTIALIBUS SACRAMENTIS MODULATIONIS DOCTAE TINNITIBUS CONCINEBAT

QUEM sonum *La*

s. sonum dulcem *LbVaLeaOxf*

MUSARUM muse novem *La*

IMPENDENS reddens illum *BeLaLbVa*, illum *om. BeLa*

reddens vel tribuens *Oxf*

DOCTE i. alterne *LaOxf*, i. *om. Oxf*

Docte quia primum omnes post secretim cantaverunt *BeLa*

TINNITIBUS cum sonitibus *LbVa*

33.18-9 NAM NEC TIBIARUM MELA NEC EX FIDIBUS SONITUS NEC HYDRAULARUM HARMONICA DEERAT PLENITUDO

MELA dulcedines *BeLa*

suavitas *Oxf*

EX FIDIBUS ex cordibus (*sic, lege* cordis) *BeLa*

cordis *LbVaLeaOxf*

HYDRAULARUM ARMONICA coadunatio vocis *BeLaLbVaLea*, vocis *om. Lea*

organum de aqua factum *BeLa*

hydrocula organarum de aqua factum *LbVa*

hydraula organum aqua compositum *Oxf*

33.19-1 SED IN BLANDUM COLLATA CANTUM AC MODIFICATO FINE COMPACTUM VOCI VIRGINUM COMPLEMENTI SPATIO RATUM FECERE SILENTIUM

BLANDUM suavissimum *LbVaOxf*

COLLATA coniuncta *BeLaLbVaLeaOxf*

CANTUM s. illa genera modulationis *LbVaOxf*

MODIFICATO temperato *BeLaLbVaOxf*

COMPACTUM compositum *LaLbVaLeaOxf*, conpositum *La*

VIRGINUM musarum *LbVa*
COMPLEMENTI SPATIO cum. convenienti *BeLa*, cum *om. La*
ut consimil<...> (*consequens illegibile*) *La*
s. cum. Donec illae musae sua carmina complerent *LbVa*
congruenti *Oxf*
RATUM aptum *BeLaLbVaOxf*

1-2 AC TUNC ILLE OMNIS CHORUS CANORIS VOCIBUS DULCIQUE MODULATU PRAEVERTIT OMNES ORGANICAS SUAVITATES
CANORIS VOCIBUS s. cum *LbVa*
MODULATU modulatione *Oxf*
PRAEVERTIT superat *BeLaLbVaLeaOxf*

3 ET CUM SACRAE NUMERIS CANTILENAE HAEC DICTA FUNDUNTUR:
NUMERIS rithmis *LbVa*
CANTILENAE s. versuum. Vel quia novem erant *LaLbVaLeaOxf*, s. versuum] s. metris vel versuum *Oxf*
FUNDUNTUR dicuntur *BeLaLbVaOxf*

-2 SCANDE CAELI TEMPLA, VIRGO, DIGNA TANTO FOEDERE;
TE SOCER SUBIRE CELSA POSCIT ASTRA IUPPITER
METRUM TROCHAICUM TETRAMETRUM. RECIPIT ETIAM SPONDEUM ALTERNIS CURRENS VERSIBUS. UNUS RECIPIT QUATTUOR, ALTER TRES SEMIS *BeLaLeaOxf*, trochaicum] dactilicum *Oxf*[1] trocaicum *Oxf*[2] recipit] quod recipit *Lea* unus] et unus *Lea* quattuor] quatuor *Be* semis] et semipedem *Oxf*
METRUM TROCHAICUM TETRAMETRUM ALTERNIS CURRENS VERSIBUS. RECIPIT ETIAM SPONDEUM ET IAMBUM. UNUS VERSUS RECIPIT QUATTUOR PEDES, ALTER TRES ET CATALECTON *LbVa*, quattuor] quatuor *Va*
copula *BeLa*
copula amicitia *Oxf*[*]
Generaliter omnes muse hoc cantabunt i. scande caeli *La*
trocheum ipponactium tetrametrum *Va*[1]
TANTO FOEDERE tanta copula *LbVaLea*
vel coniunctione *LbVa*

---

[*] The copula is only in this manuscript provided with music notation.

SUBIRE conscendere *BeLaLbVaOxf*
POSCIT vult *BeLaLbVaOxf*

119/
34.16 TUNC CALLIOPE
s. canere coepit *LbVa*
LAUS PHILOLOGIAE DE MUSICA *Lb*
METRUM FALLEUCIUM CONSTANS EX SPONDEO, DACTILO ET TRIBUS TROCHEIS *BeLaLbVaLea*
constans] constat *Va* ex *om. LbVaLea* tribus] duobus *BeLa*

34.17-8 SEMPER COMPLACITIS AMICA MUSIS, CUI PERMESIA POCULUM FLUENTA
COMPLACITIS Tibi placatis i. in tuo honore canentibus *BeLa*, i. *om. La*
Tibi placitis o Filologia, quae es *Oxf*
CUI i. in cuius honorem poetae cecinerunt qui Libertim (*sic, lege* Libetrim) fontem hauriunt *LbVa*
i. tibi tulerunt *Oxf*
†MAGNESIA* Magnesia urbs est vicina Tessaliae, cui contiguus est Libetris fons poetis dicatus *BeLa*, Magnesia *om. La*
[1]Magnesia est civitas vicina Thessaliae habens fluenta id est copiosa est de fontibus. [2]Cuius fons Libetris poetis est ad bibendum consecratus. [3]Et est ordo: Cui Magnesia civitas est habens fluenta poetis sacrata, tulit poculum i. in tuum honorem poetas canere fecit** *LbVaLeaOxf*, 1 Thessaliae] tessalie *Oxf*; 3 civitas est] civitas *Lea* poetas] poetis *LbVa*; 3 Et est - canere fecit *om. Oxf*
POCULUM instructionem *LbVa*
i. carmen *Oxf*
FLUENTA s. habens *BeLaLbVaLea*, s. *om. LbVaLea*

34.19-0 ET FONS GORGONEI TULIT CABALLI, VERTEX AONIDUM VIRENS COROLLIS
FONS GORGONEI s. cui attulit poculum *LaLbVaOxf*, s. *om. La* s. cui] cui s. *Oxf* poculum] potum vel carmen *Oxf*

---

* PERINNESIA *Va*, vel magnesia *Va*[2]
** Cf. Serv. Com.Verg.E. 7,21; Plin. Nat.Hist. 4,9

[1]Hic fabulam tangit. [2]Forcus rex tres filias Gorgonas fertur habuisse: Stenno, Euriale, Medusa. [3]Locupletes nimis, unde Gorgones dicuntur quasi george vel georgi i. cultrices terrae. [4]Sed mortuo patre successit ei Medusa in regno, quam auxilio Minervae interfecit Perseus rex Asiae. [5]Et de sanguine eius egressus est Pegasus equus, qui pede suo terram percutiens produxit fontem, sicut fabula finxit Grecia, qui fons dicatus est poetis et musis, hinc Martianus *et fons Gorgonei tulit caballi.* [6]Hoc fabulosum, veritas tamen est inquirenda. [7]Gorgo terror, Stenno debilitas, Euriale lata profunditas, Medusa oblivio. [8]Haec omnia terror in hominibus facit, quae omnia Perseus occidit. [9]Perseus Grece, Latine virtus. [10]Cum auxilio Minervae, quia virtus auxiliatrice sapientia omnes terrores vincit. [11]De cuius sanguine equus Pegasus egressus: Pegasus enim fama dicitur, quia virtus omnia superans famam sibi conquirit. [12]De quo potant poetae, quia in laudem virtutis videntes eam victricem prosiliunt* *BeLaLbVaLeaOxf, Va ad* 34.21-2 CUI FRONDIT VIOLAS. 1 Hic fabulam tangit *om. Be*; 2 Forcus rex] Forcus quidam rex *BeLa* tres filias Gorgonas fertur habuisse] habuit tres filias *BeLa* Medusa] et Medusa *Oxf*; 3 *ante* Locupletes nimis] Fuerunt enim *add. LaOxf* dicuntur] dictur *Lb* dicitur *Va* vel georgi *om. LaOxf* i. cultrices terrae *om. LbVaLea* i. terre cultrices *La* i. terae cultrices *Oxf*; 5 mortuo patre] patre mortuo *Lea* successit] succes (sic) *Oxf* egressus] egressus est *Oxf* Pegasus equus] equus Pegasus *Be* *post* fontem] nomine Pegasum *add. LaOxf* sicut fabula finxit Grecia *om. Be* fabula] fabulosa *Oxf* Grecia] gretia *La* qui fons dicatus est] qui est dicatus *Be* hinc Martianus] hoc est quod dicit *Be* hinc Marcianus *Va*; 7 lata] leta *Be* oblivio] profunda vel propria obliviosa *Be*; 8 terror] terrorem *LaOxf* faciunt] facit *Oxf* quae omnia] quam *Be*; 10 Minervae] Minerve interfecit *LaOxf*; 11 equus] equs *La* equus Pegasus egressus] egressus pegasus equus *Va* egressus] egressus est *LaLeaOxf* fama] fama bona *Be* conquirit] adquirit *Oxf*; 12 laudem] laude *La*

VERTEX AONIDUM montes Boetiae *BeLa*

cacumen Boaetiae montis. Aon mons *LbVa*

Ahon mons est Boetiae *Oxf*

VIRENS florens *LaLbVaOxf*

†CORAULIS poetis *LaLbVaLeaOxf*, de *add. Oxf*

Coraula est cantor, vel a choro vel ab allechore i. laude chori dictus *BeLaLbVaOxf*, a choro] alchoro *LbVa* a coro *LaOxf* allechore]

* Cf. Serv. Com.Verg.A. 2.616, 6.289; Fulg. Myt. I.59-62; Myth.Vat.I <127>; Myth.Vat.II <135>, <136>

allecoro *La* allechora *Lb* allechoro *Va* allecora *Oxf* laude] a laude *LaOxf* dictus] poetis *Be*

34.21-2 CUI FUNDIT VIOLAS PARANTE CIRRA; TU VATUM MELA DULCIBUS CAMENIS

CUI FRONDET* s. in tuo honore *BeLa*

s. vertex *LbVaLea*

floret *Oxf*

CYRRA Cirra mons Indie dicatus Phoebo *BeLa*, Cirra *om. La*

i. pars Parnasi montis *Lea*

Cyrra et Nisa duo Parnasi montis cornua. Sed in Cyrra Apollo, in Nisa Liber pater colitur** *LbVaLeaOxf. Cf. Be ad* 6.16-7 LOQUACIA, *La ad* 6.16-7 CHIRREOS. Cyrra] cirra *LaVaLeaOxf* colitur] colebatur *Oxf*

TU scis *Oxf*

VATUM poetarum *LaLbVaOxf*

MELA cum DULCIBUS *Be*

MELA dulcedinem cantilenae *LbVa*

DULCIBUS sonoritatibus *LaLbVaOxf*

CAMENIS vel cantibus *Oxf*

34.23-1 ET SCIS PINDAREAM CHELYN REFERRE; TE DICTANTE FIDES SACRUMQUE PLECTRUM

CHELIM citaram *Be*

CHELIN lyram *LbVa*, lyram] liram *Va*

Cantilena Pindari non te latet *LaLbVaLeaOxf*, i. *add. LaOxf*

Pindareus optimus citharista *BeLa*, optimus] obtimus *La*

Pindarus musicus fuit *LbVaOxf*, Pindarus] pindarius *Oxf*

Et est species pro genere *LaLbVaOxf*, genere] gere *Oxf*

REFERRE rememorare *La*

rememorari *Oxf*

rememoramini *LbVa*

DICTANTE i. monstrante *BeLa*

agente *LbVaOxf*

vel docente[m] *Oxf*

FIDES cordas *BeLaLbVaLeaOxf*, corda *Be*

---

* *BeLaLbLeaOxf* FRONDET, *Va* FRONDIT

** Cf. Isid. Etym. 14.8.11

SACRUMQUE PLECTRUM percussionem *LaOxf*, i. sacram *add. Oxf*
sancta percussio. Instrumentum musicum quo percutiuntur *LbVa*

NOVIT THREICIUM SONARE CARMEN
†MOVIT THREICIUM adgressa est Orfaicum *BeLaLbVaLeaOxf*, adgressa] agressa *LbLea* agress *Va* agressum *Oxf* Orfaicum] orphaicum *LbVaOxf*, *om. Lea*
Qui Traicus fuit *BeLa*
Pro auctore posuit *La*
Carmen pro auctore carminis Orpheo posuit *LbVaLeaOxf*, carminis *om. Oxf* Orpheo posuit] posuit i. Orpheo *Lea* posuit Orpheo *Oxf*
SONARE Ut sonarent *Oxf*

-4 O LUX NOSTRA, SACROS PROBARE CANTUS SUESCE ATQUE ORGANICIS BEARE CIRCIS
NOSTRA i. nostra princeps s. scis *Oxf*
PROBARE laudare *LaVaOxf*
SUESCE illos cantus *BeLa*
disce *LaLbVaOxf*
imperativus *LbVa*
ORGANICIS BEARE CIRCIS cum dulcibus *BeLa*, cum *om. La*
s. cum *LbVaOxf*, s. *om. Oxf*
melodiis *Be*
beatificare modulationibus *LaLbVaOxf*, modulationibus] i. modulationibus *Oxf*
Circus est proprie in melodia cantus, qui in eundem tonum finit, a quo incipit *BeLa*
Circus est ambitus musicae modulationis i. ad eundem sonum a quo inchoat reversio *LbVaOxf*, ambitus] habitus *Oxf* i. ad] atque *LbVa*

AC SIC POLIMNIA[*]
EST FIGURA EFEXEGIS (*sic*, *lege* ἐπεξήγησις). I. PRAEDICTAE REI EXPOSITIO *BeLa*, est figura efexegis] efexegis est *LbVa* efexegis] efexeis *La* predicte rei] predicta herei *Be*
IAMBICUM SENARIUM HYPONACTIUM CONSTANS

[*] To this stanza music notation is added in *Oxf*.

TRIMETRO ACATALECTO *BeLaLbVaLea*[2], hyponactium] iponactium *La* yponacticum *LbVa* acatalecto] catalecto *BeLaLea*[2]

35.8-9 TANDEM LABORIS FRUCTUS, AETHRAM FULGIDAM DIVUMQUE SEDES AC IOVIS CONSORTIA

TANDEM nunc vel ad ultimum *BeLaLbVa*

FRUCTUS tui *BeLa*

disciplina *BeLaLbVa*

disciplinae *LbVa*

DIVUMQUE deorum *LaLbVaOxf*

35.10-1 PROVECTA CARPIS INDITOQUE NUMINE, CRUENTA DUDUM, QUAE IUGARE RHYTHMICA

PROVECTA exaltata in altum CARPIS accipis *BeLa*

ó PROVECTA ducta *LbVa*

[1]O provecta virgo, carpis tu tandem fructus laboris tui, divum sedes et Iovis consortia aethramque fulgidam, quae sueta es iugare cum indito numine cruenta rithmica ac mixta cum dispari regula. [2]Et quae sueta es probare, quid torqueat circulus in iacente linea et quid recurvet trigonus in iugata linea. [3]Ac solita es probare melos et tonos et crusmata artesque cunctas. [4]Et solita es probare mente adacta, quae possunt parare culmina caelitum *BeLaLbVaLeaOxf*, *LbVa ad* 35.16-7 CULMINA. 1 Sensus est] *add. Lea* laboris tui] tui laboris *Lea* tui] tu *Oxf* sedes et] sedet *Oxf* aethramque fulgidam *om. Oxf* fulgidam] fulgentem *LbVaLea* sueta es] suetes *La* cum *om. Oxf* rithmica] ritmica *La* ac mixta cum dispari regula] cum mixta regula ac dispari *LbVaLeaOxf*; 2 sueta es] suetes *La* quid] quit *La*; 3 solita] sueta *LbVa* es] ses *Va* et tonos] ac tonos *Lea*; 4 Et] ac *LbVaLea* parare] rare *Be*

INDITOQUE insita *BeLaLbVaOxf*

vacat *LbVaLea*

NUMINE maiestate *BeLaLbVa*

divinitate *Oxf*

CRUENTA nova *BeLaOxf*

pulchra et rubicunda. imperfecta *LbVa*, imperfecta] imperta *Va*

IUGARE RYTHMICA iungere modulatione *BeLaLbVaLeaOxf*, modulatione] modulatio *La* modulationem *Lea*

Nunc laudat Philologiam [a] musica *LaOxf*, Filologiam *Oxf*

Rithmus numerus, rithmicum opus consistens ex numero pedum syllabarumque *LbVaLeaOxf*, rithmus] rithmus est *Lea* *post* rithmicum opus] i. coronas quae poetis debentur *add. Oxf* consistens] consistit *Be* constat Va^{1} constans Va^{2}*Oxf* syllabarumque] syllabarum *LbVa* et syllabarum *Lea*

Per cruenta ritmica musicam humanae vocis astruit *BeLaLbVaOxf*, ritmica] rithmica *LbVaLeaOxf* musicam] musica *La* musicam humanae vocis astruit] humanae vocis musicam ostendit *LbVaLea* humanam vocem instruit *Oxf*

2-3 AC DISSIPARE MIXTA SUETA REGULA, MOX QUID IACENTE, QUID IUGATA LINEA

†DISPARI cum *Be*

variata *BeLa*

varia *LbVa*

propter dissimilitudinem sonorum in musica *LbVaLea*

MIXTA cum *LbVa*

SUETA consueta *BeLa*

s. es *LbVaLeaOxf*

MOX abhinc *La*

QUID IACENTE Laudat illam a geometrica *BeLa*, geometrica] giometrica *La*

sueta in IACENTE *LbVa*

QUID IUGATA LINEA in angulum ducta *BeLaLbVaLeaOxf*, in] in in *Va*

i. totam scis geometricam *LaLbVaLeaOxf*, totam] tu *Oxf* geometricam] geometriam *Oxf*

4-5 TRIGONUS RECURVET CIRCULUSQUE TORQUEAT, MELOS PROBARE AC TONOS ET CRUSMATA

TRIGONUS Δ *BeLaOxf*

▽ *LbVa*

CIRCULUSQUE ○ *BeLaOxf*

MELOS tropos *BeLaLbVaOxf*

Item laudatur a musica *BeLa*

PROBARE Sueta es tu *BeLbVaLea*, s. *add. Lea* tu *om. Be*

i. suetes *La*

AC TONOS i. spatium corda *BeLa*, i. *om. La*

spatia cordarum *LbVaLeaOxf*

CRUSMATA diversos colores *BeLa*

Croma varietas, unde crusmata quasi cromata i. crusmata i. colores varii scilicet motus *BeLaLbLea*, croma] cromata *La* i. crusmata *om. La* i. crusmata ... scilicet motus *om. BeLea* varii] variis *La*

Per cromaticum tria genera musice vult intellegi, quae sunt

diatonicum, cromaticum et enarmonium *BeLa*, genera] genere *La* musice] misice *Be* quae sunt] quasi *Be*
tractus cordarum *Oxf*

35.16-7 ARTESQUE CUNCTAS SOLITA, QUAEQUE CAELITUM POSSUNT PARARE MENTE ADACTA CULMINA
ARTESQUE CUNCTAS accusativus *La*
omnes *LbVa*
SOLITA s. es *BeLaLbVaLeaOxf*
probare artes *LbVaLea*, artes *om. Lea*
QUAEQUE munera vel itinera *BeLa*
CAELITUM deorum *BeLa*
PARARE perficere *BeLaLbVaLeaOxf*
MENTE ADACTA vel mente adacta i. coacta s. tua *BeLaLbVaLea*, adacta] adtacta *La*
tua *Oxf*
CULMINA summi dii *BeLaLbVaLea*
Per culmina celitum astrologiam vult intellegi, tamquam diceret solita es probare astrologiam *BeLaLbVaOxf*, astrologiam] arithmeticam *Be* aritmeticam *La*

121/
35.20 TUNC MELPOMENE
LAUS PHILOLOGIAE DE ARTE POETRICA *BeLaLbVaLea*,laus] palaus *Be* Philologiae] philogiae *Be*
laus Filologiae a poesin *Oxf*
DIMETRUM IAMBICUM HEROICO IUNCTUM *LbVa*

35.21-2 SUETA COTHURNATOS SCENIS DEPROMERE CANTUS SOCCUMQUE FERRE COMICUM
SUETA COTURNATOS o virgo s. es *Oxf*
Coturnus calciamentum poeticum utroque (*sic, lege* utrique) habile pedi *BeLa*, habile] abile *La*
Coturnus est ocrea poetria *LbVaLeaOxf*, poetria] poetica *Oxf*
SCENIS in amoenis locis *LaLbVaLeaOxf*
DEPROMERE CANTUS manifestare *LaLbVaOxf*
i. carmina *Oxf*
SOCCUMQUE habitum *BeLaLbVaOxf*, i. *add. Oxf*
i. calciamentum comicum *Lea*
COMICUM Comicus est qui comediam scripsit *BeLaLbVaLeaOxf*,

Comicus est *om. La* sicut est *Oxf* est *om. Be* comediam] comedia *LbVaLea*

Comedia sunt carmina quae in conviviis cantantur, quasi commessationum odae: comedere enim apud Grecos comos dicitur. Vel comedia dicuntur quasi comitum ode: sunt enim carmina, que in ordinatione comitum cantantur. Est ergo comicus qui comedias componit et sicut poete coturno, sic comici socco utuntur* *La*

3-4 ET REBOARE TUA TULIMUS QUAE CARMINA CURA MELO FAVENTE RHYTHMICO

REBOARE s. sueta *La*
s. solita es *Oxf*
resonare *LaLbVaLeaOxf*

TUA TULIMUS in s. nos *BeLa*
protulimus *Oxf*

CARMINA CUNCTA Quae cantavimus in honore tui *La*

CURA amore *VaLea*
i. per tuam curam *Oxf*

MELO modulatione *LbVaOxf*

FAVENTE modulante *BeLa*
annuente *LaLbVaOxf*

RITHMICO rithmica *LbVa*
metrico *Oxf*

1-2 NUNC TIBI, VIRGO, CANO, SPES ATQUE ASSERTIO NOSTRI, TENORE VERSA CARMINIS

NUNC TIBI, VIRGO i. in tuum amorem *BeLaLbVaOxf*, amorem] honorem *Oxf*

CANO ego *LbVa*

SPES s. quae es *LbVaLeaOxf*, s. *om. Oxf*

ATQUE ASSERTIO prolatio vel affirmatio *BeLa*
prolatio robur *LbVa*
probatio *Oxf*

TENORE fomcsquc *BeLa*
i. laude *Oxf*

VERSA adplausa *BeLaLbVaLeaOxf*

---

* Cf. Isid. Etym. 8.7.6; Beda Orth. 18; Smar. Lib.part.Don. *De genere, textus, lin. 57*

mota *LbVaLea*
frequentata *Oxf*

36.3-4 NAM THALAMUM REDIMIRE IUVAT, TU SERTA PROBATO TUIS PLACERE RITIBUS
THALAMUM tuum *LaLbVaOxf*
REDIMIRE ornare *BeLaLbVaOxf*
IUVAT delectat *BeLaLbVaOxf*
SERTA PROBATO i. coronam quae poetis debetur *BeLa*
SERTA corona<s> PROBATO laudato *LbVa*
coronas eligite vel laudato *Oxf*
Epithalamium componendo i. serta i. coronas quae poetis debentur *LbVaLea*
PLACERE RITIBUS i. ut placeant *Oxf*
usibus *BeLaLbVaOxf*

36.5-6 DIGNA MARITALI SEMPER VIDEARIS OLYMPO DECENTIORQUE CAELITUM
MARITALI coniugio *BeLaLbVaOxf*
deo Cillenio *BeLa*, decillenio *La*
SEMPER VIDEARIS ut *BeOxf*, s. *add. Oxf*
OLYMPO caelesti *BeLaLbVaOxf*
Id est deo Cyllenio *LbVa*
a more Grecorum dativus pro ablativo *BeLa*, a *om. La*
DECENTIORQUE CAELITUM honestior omnium dearum *BeLaLbVaOxf*, onestior *Oxf* omnium *om. BeLa* omnium dearum *om. Oxf*
more Grecorum genitivo pro ablativo caelitibus *LbVa*

127/
40.20-1 DUM HAEC IGITUR MUSAE NUNC SOLICANAE, NUNC CONCINENTES INTERSERUNT VICISSIMQUE MELA DULCIA GEMINANTUR
HAEC carmina supradicta *Be*
s. supradicta carmina *LbVaLea*, s. *om. Lea*
MUSAE s. praedictae *Oxf*
NUNC aliquando *BeLaLbVaOxf*
SOLICANAE singillatim canentes *BeLa*, canentes *om. Be*
solae canentes *Oxf*
NUNC aliquando *LbVaOxf*

CONCINENTES simul canentes *BeLaLbVaLea*
INTERSERUNT promunt *BeLa*
simul edunt *LbVaLea*
VICISSIMQUE alternatim per vices *LbVaLeaOxf*, per vices *om. LeaOxf*
MELA DULCIA GEMINANTUR Dulces modulationes duplicantur *BeLaLbVaLeaOxf*, Dulces modulationes] modulationes dulces *LbVaLea* duplicantur *om. Lea*
iiii *Va*

1-4 ECCE QUAEDAM MATRONAE SOBRIO DECORE LAUDABILES NEC CONQUISITIS FIGMENTIS CIRCA FACIEM VULTUOSAE, VERUM SIMPLICI QUADAM COMITATE PRAENITENTES, IN PENATES VIRGINIS THALAMUMQUE CONVENIUNT
MATRONAE Descriptio virtutum *BeLa*
Descriptio iiii^or virtutum *LaLbVa*
SOBRIO DECORE Duobus modis pulchritudo adquiritur: naturaliter et conquisitive *BeLbVaLeaOxf*, modis] modus *Oxf* pulchritudo adquiritur] sit pulcritudo *LbVaLeaOxf* conquisitive] conquisitae *LbVaLeaOxf*
SOBRIO honesto DECORE pulchritudine *BeLa*
honesta pulchritudine *LbVaOxf*, pulcritudine *Oxf*
CONQUISITIS de *BeLaLbVaOxf*, s. *add. Oxf*
adeptis *LaOxf*
inventis *LbVa*
FIGMENTIS falsitatibus *BeLaLbVaLea*
facultatibus *Oxf*
VULTUOSAE ornate *La*
ornatae, honestae *LbVa*
SIMPLICI naturali *LbVa*
QUADAM COMITATE pulchritudine *BeLaLbVaLeaOxf*, pulcritudine *BeVaOxf*
PRAENITENTES valde nitentes *LbVa*
valde splendentes *Oxf*
IN PENATES in secretis *BeLa*
secreta *LbVaLeaOxf*, in *add. Oxf*
VIRGINIS Philologiae *LbVaLeaOxf*
THALAMUMQUE CONVENIUNT s. in *LbVa*
de iiii^or virtutibus dicit *LbVaLea*

4-5 QUARUM UNA INTENTA CIRCUMSPECTIONE CAUTISSIMA ET

OMNIA RERUM VIGILI DISTINCTIONE DISCRIMINANS DICEBATUR PRUDENTIA VOCITARI
QUARUM s. matronarum *LaLbVaOxf*, s. *om. Oxf*
UNA vigil *LbVa*
INTENTA vigili *Oxf*
CIRCUMSPECTIONE consideratione *LaLbVaOxf*
CAUTISSIMA s. illa una *LbVaLea*
RERUM argumenta cum *BeLa*
VIGILI s. cum sagaci *LbVaOxf*, s. cum *om. Oxf*
DISCRIMINANS discutiens *LaLbVaOxf*
  discernens *LbVaLeaOxf*, vel *add. Oxf*
PRUDENTIA Prudentia est omnium rerum cum cautissima circumspectione discriminans distinctio *BeLaLbVaLeaOxf*, cum *om. La*

131/
41.19-0 SED ECCE MAGNO TYMPANI CREPITU CROTALORUMQUE TINNITU UNIVERSA DISSULTANT
TYMPANI timpanum pellis extenta* *LbVaLea*
CREPITU sonitu *BeLaLbVaOxf*
CROTALORUMQUE species cithare *BeLaOxf*, cymbalorum vel *add. Oxf*
  species cimbalorum vel genus cythare *LbVaLea*, cimbalorum *om. Lea*
TINNITU sonitu *LbVa*
UNIVERSA s. loca *BeLaLbVa*
DISSULTANT resonant *BeLa[1]Oxf*
  resonabant *La[2]LbVa*

41.20-1 EO USQUE UT MUSARUM CANTUS ALIQUANTO BOMBIS TYMPANI OBTUSIOR REDDERETUR
EO USQUE in tantum *BeLaLbVaOxf*
ALIQUANTO paululum *BeLa*
  aliquantulum *Oxf*
BOMBIS TYMPANI sonitibus clamoribus *Be*
  clamoribus vel sonitibus *LaOxf*
  vel sonitibus *LbVaLea*, vel *om. VaLea*
OBTUSIOR obscurior *BeLa*
  retunsior attritior *LbVaLea*, retunsior *om. Lea* atritior *Lea*

---

* Cf. Isid. Etym. 3.22.9; Beda Luc.evang.exp. 5.19.1665

retonsior *Oxf*

1-2 ET CUM SONITU INTROFERTUR LECTICA INTERSTINCTA SIDERIBUS, CUI RITU MYSTICO CREPITUS PRAECINEBANT

LECTICA capulum gestatorium *LaOxf*, gestatorum *La*

Lectica est capulum gestatorium *LbVaLea*

lectus in curru *Oxf*

INTERSTINCTA celata *BeLaLbVaLea*

sculpta *LbVaLea*

i. variata *Oxf*

CUI RITU MYSTICO antequam more spiritali *BeLaLbVaLea*, antequam] proquam *Be* spiritali] spirituali *Lea*

lecticae more spiritali *Oxf*

CREPITUS voces *LaOxf*

sonitus *LbVa*

PRAECINEBANT antecanebant *LbVaOxf*

2-3 QUA MOS FUERAT NUBENTES DEAS IN CAELESTIS THALAMI PERVENIRE CONSORTIA

QUA s. lectica *LbVaLea*

per *Be*

i. per quam *Oxf*

†QUAM per quam *La*

QUA MOS FUERAT* oportuerat *BeLa*

consuetudo *Oxf*

9-1 SED DUM TALIA VIRGO UNDANTER EVOMERET, PUELLAE QUAMPLURES, QUARUM ARTES ALIAE, ALTERAE DICTAE SUNT DISCIPLINAE, SUBINDE, QUAE EX ORE VIRGO EFFUDERAT, COLLIGEBANT

TALIA s. volumina *LaLbVaLeaOxf*, s. *om.* *LaLeaOxf*

VIRGO Philologia *LbVaLea*, phillogia *Va*

UNDANTER copiose *BeLaLbVaOxf*

QUARUM puellarum *LbVaLea*, s. *add.* *Lea*

ARTES Pos<t>quam ad perfectionem venitur *La*

Quia latior est [p]ars quam disciplina *LbVaLea*

---

* QUAM OFFUERAT *La*

ALTERAE puellae *LbVa*

Quia disciplina continetur in artibus *LbVaLea*

DISCIPLINAE Inter artem et disciplinam hoc interest, quod disciplina dicitur dum docetur, ars vero cum ad perfectam animi intelligentiam perducitur* *BeLaLbVaLea*, dicitur dum] est cum *LbVaLea* perducitur] pervenitur *LbVaLea*

SUBINDE actutum statim *BeLa*

statim saepe *LbVa*

statim *LeaOxf*, i. *add. Oxf*

ORE suo *LaLbVa*, s. *add. La*

EFFUDERAT emiserat *LbVa*

42.21-2 IN SUUM UNAQUAEQUE ILLARUM NECESSARIUM USUM FACULTATEMQUE CORRIPIENS

IN SUUM UNAQUAEQUE ILLARUM NECESSARIUM "Quia non omnia possumus omnes"** *BeLa*, possumus] posumus *Be*

vel artibus vel disciplinis *LbVaLea*, artibus] artium *Lea*

Bene dicit in suum necessarium usum et facultatem, quia dum advivimus non omnia capere possumus *La*

FACULTATEMQUE possibilitatem *LaLbVaOxf*

CORRIPIENS accipiens *LaLbVa*

violenter capiens *Oxf*

42.22-3 IPSAE ETIAM MUSAE, PRAESERTIM URANIE CALLIOPEQUE, INNUMERA GREMIO CONGESSERE VOLUMINA

PRESERTIM maxime *LbVa*

URANIE nominativus Grecus *LbVaLea*, uranie *add. Lea*

CALLIOPEQUE Per Uraniam ratio caelestium elementorum, per Calliopen, quae bona sonoritas dicitur, musica totius naturae, et celestis et terrestris, significatur*** *BeLaLbVaLea*, significatur *om. Be*

INNUMERA copiosa in GREMIO in sinu suo *BeLaLbVa*, in sinu *om. BeLa*

plura *Oxf*

†CONCESSERE collegere *BeLaOxf*, conllegere *Oxf*

collegerunt *LbVa*

---

* Cf. Isid. Etym. 1.1.3

** Verg. Ecl. 8,63

*** Cf. Fulg. Myt. I.15 etc. (cf. ad 12.20 APPOLINE)

3-1 IN ALIIS QUIPPE DISTINCTAE AD TONUM AC DEDUCTAE PAGINAE

IN ALIIS QUIPPE s. voluminibus *BeLaOxf*, s. *om. Oxf*

DISTINCTAE s. erant *BeLaLbVaLeaOxf*

AD TONUM usque ad tropum *BeLaOxf*

i. ad proprietatem uniuscuiusque toni *BeLaLbVaLeaOxf*, i. *om. LbVaLea*, usque *add. LaOxf* uniuscuiuscumque *Lea*

DEDUCTAE extentae *BeLaLbVaOxf*

PAGINAE Musicam tangit *Be*

3 IN ALIIS CIRCULI LINEAEQUE HEMISPHAERIAQUE CUM TRIGONIS ET QUADRATIS MULTIANGULAEQUE FORMAE PRO THEOREMATUM VEL ELEMENTORUM DIVERSITATE FORMATAE

IN ALIIS s. voluminibus *LaLbVaLeaOxf*, s. *om. Oxf*

Geometricam tangit *Be*

Nunc tangit geometricam et astrologiam *LaLbVaOxf*, tangit geometricam] tangeometricam *Va* giometricam *La* astrologiam] astralogiam *Va*

CIRCULI s. planetarum *BeLaOxf*, s. *om. Oxf*

planetarum cursus *LbVaLea*

LINEAEQUE celestes semitae *LaOxf*, celeste *La*

Lineae sunt in caelo clxxxiii: duo hemispheria, unum superius a signifero usque ad polum articum, alterum vero inferius sub signifero usque ad antarticum i. usque ad austrinum polum *BeLaLbVaOxf*, duo hemispheria] hemisperia ii *LbVa* duo emispheriam *Oxf* articum] artuum *Oxf* antarticum] antartiva *Oxf*

CUM TRIGONIS Trigonus quasi tergona i. genua flectens, quia gonum dicunt Greci genua *BeLaLbVaLeaOxf*, i. *om. Oxf* gonum] gunum *Lb* genua] genu *LbVaLeaOxf*

MULTIANGULAEQUE Multiangulae sunt omnes figurae, a quadrato usque ad infinitum, ut sunt pentagoni, exagoni, eptagoni ceterique numeri in plures angulos ducti *BeLaLbVaLeaOxf*, sunt *om. Oxf* pentagoni, exagoni, eptagoni] pentagonae, exagonae, epdagonae *Oxf*

THEOREMATUM Theorema est contemplatio rationis, cum lineae lineis iunguntur, anguli lateribus, latera angulis et anguli angulis et latera lateribus etcetera *BeLaLbVaLeaOxf*, Theorema] Pro theorema *Oxf* anguli] angulli *Va* latera angulis] littera angulis *Be* latera anguli *La* et *om. Lea* et anguli angulis *om. BeLa* et latera lateribus] et latera litteris *BeLa*

contemplationum *Oxf*

ELEMENTORUM Initiorum, per quae incipiunt ipsae disciplinae *LbVaLea*

Elementa vero geometriae sunt punctus, linea, angulus

BeLaLbVaOxf, geometriae] geometrice *La* geometriae sunt] sunt geometriae *LbVaOxf* angulus] et angulus *LbVaOxf* *post* angulus] sicut literae elementa artis *add. Oxf*

43.3-4 DEHINC PICTURA ANIMALIUM MEMBRA MULTIGENUM IN UNAM SPECIEM COMPLICABAT
DEHINC postea *Oxf*
PICTURA ANIMALIUM cum. Tangit cosmographiam *Be*
In similitudinem animalium erant factae *BeLa*
Cosmographiam tangit hic id est descriptionem mundi *LaLbVaOxf*, cosmographiam] cosmagraphiam *La* cosmografiam *Oxf* hic *om. LaOxf* descriptionem] descriptione *La*
IN UNAM SPECIEM in unam paginam *BeLa*
i. paginam *LbVaLeaOxf*, i. *om. LeaOxf*
COMPLICABAT depingebat *BeLaOxf*
conpingebat *LbVaLea*, compingebat *Lea*

43.4-6 ERANT ETIAM LIBRI, QUI SONORUM MELA SIGNAQUE NUMERORUM ET CANTANDI QUAEDAM OPERA PRAEFEREBANT
ERANT ETIAM LIBRI Musicam tangit *LaLbVaLeaOxf*
SONORUM MELA dulcia *Be*
dulcedinem soni *La*
i. dulcedines sonorum *Oxf*
dulcedinem *LbVaLea*
SIGNAQUE NUMERORUM Proprie numerus dicitur concordia humanae vocis *BeLaLbVaLeaOxf*, numerus dicitur] dicitur numerus *Lea* concordia humanae vocis] humanae vocis concordia *Be*
OPERA studia *LbVa*
PRAEFEREBANT demonstrabant *BeLaLbVaOxf*

139/
43.6-7 POSTQUAM IGITUR ILLAM BIBLIOTHECALEM COPIAM NIXA IMITUS VIRGO DIFFUDIT
[1]Omnia quae vomuit Philologia, ad significationem mundanorum studiorum posita sunt. [2]Humana enim studia aliquando fallunt, quia in hac vita artium veritas apparere non potest. [3]Ideoque inflantur quasi quodam superbiae humore illi, qui illis utuntur. [4]Per illud vero sorbillandum i. poculum, quod Philologiae dederat, veritas artium, quae solis immortalibus et veritatem contemplantibus manifestabitur, apertissime significatur

BeLaLbVaLeaOxf, 2 humana] humane *La* studia *om. Lea* aliquando] aliquid *Lea* apparere] aparere *La* operari *LbLea* operare *Va* aperire *Oxf*; 3 humore] tumore *LbVaLea*; 4 Per illud] illud *BeLa* dederat] dedit Atanasia *Lea* manifestabitur] manifestabetur *Lea* significatur] significat *BeLaLbVa*

4 TUNC PORTITORES DIVAE CORREPTA LECTICA MAGNO EAM MOLIMINE SUBVEXERE

TUNC PORTITORES Amor et Labor *BeLa*

DIVAE Pimelia et Agripinia *BeLa*
    Philologiae *LbVaLea*

CORREPTA arrepta *BeLaLbVaOxf*

LECTICA [1]Per lecticam Philologiae corpus omnium ratiocinantium significatur, quod quattuor elementis naturali iunctione componitur. [2]Ideoque quasi sexibus duobus illa lectica portatur, quia quattuor elementorum duo i. ignis et aer quasi masculino sexui comparantur; agunt enim et numquam patiuntur. [3]Alia vero duo i. terra et aqua feminino; patiuntur enim et non agunt *BeLaLbVaLeaOxf*, 1 significatur] signatur *BeLa* quod] quot *La* quattuor] quatuor *Oxf* iunctione] coniunctione *BeLa*; 2 sexibus] sexubus *Oxf* lectica] lectico *La* quattuor] quatuor *Oxf*

MOLIMINE conamine fortitudine *BeLaOxf*, conamine] conamine vel *Oxf*
    fortitudine conatu *LbVaLea*, fortitudine *om. Lea*

5-6 SED POSTQUAM CENTUM VIGINTI SEX MILIA STADIORUM AERIA SUBVECTI LEVITATE CONSCENDERANT AC TONUM PRIMUM EX PTHONGIS COMPLEVERE CAELESTIBUS

SUBVECTI portati *BeLaLbVa*
    sursum vecti *Oxf*

LEVITATE aura *BeLaLbVaOxf*

AC TONUM PRIMUM spatium *LbVaOxf*

In hoc loco colligit et spatium primum a terra usque ad lunam et naturam lunaris corporis i. quod ex superna aquarum natura sit factum et quod eadem luna nihil proprii luminis habeat, sed iaculis radiorum solarium inluminata revibrat (*sic, lege* revibret) *BeLaLbVaLeaOxf*, spatium primum] primum spatium *LbVaLea* usque *om. LbVaLeaOxf* lunam] luna *BeLa* i. *om. LbVaOxf* sit factum] factum est *LbVaLea* est factum *Oxf* nihil] nil *BeLa* inluminata] illuminata *LbLeaOxf* illumina *Va*

EX PTONGIS sonis *LbVaLea*

CAELESTIBUS Id est ego Iuno, quae aeris huius potestatem habeo, diva cum (*sic, lege* divina) in aethere sum *LbVa*

49.17-0 LUNAREM INGRESSA CIRCULUM VIRGO DIVAE CONGRUIS NIDORIBUS SUPPLICABUNDA DE PROXIMO CONSPICATUR GLOBOSUM QUODDAM TENERUMQUE CORPUS E SUPERNI RORIS LEVITATE COMPACTUM INSTAR SPECULI PRAENITENTIS ADIACULATI FULGORIS RADIOS REVIBRARE

LUNAREM [1]Describit, quae continentur in corpore lunari: per sistra Niliaca fontes omnium fluminum, per Eleusinam lampadem aeternum splendorem. [2]Salmoneus fuit rex Eleusinae civitatis, qui fecit pontem aereum volens imitari tonitruum Iovis et fulgora. [3]Nam faculam emittebat ad quemlibet sibi concursum. [4]Quo percusso a Iove facula illius ad caelum est translata, quae modo est luna* *LbVaLeaOxf, cf. BeLa ad* 49.20 SISTRA NILIACA *et* 49.20 HELEUSINAQUE LAMPAS. 1 fluminum] flumina *LbVa* lampadem *om. LbVaLea post* splendorem] quem a sole accipit *add. Oxf*; 2 *post* volens] currendo super eum *add. Oxf* et fulgora] ac fulgura *Oxf*; 3 emittebat] emitebat *Oxf* quemlibet] quamlibet *LbVa* concursum] offensum *Oxf*; 4 quo percusso a Iove] quem perimebat quasi percussus est et fulmine Iovis *Oxf* ad caelum] in caelum *Oxf* est *om. Oxf*

DIVAE lunae *BeLaOxf*
Iunoni *LbVaLea*

CONGRUIS cum *BeLa*

NIDORIBUS odoribus *BeLaLbVaLeaOxf*
sacrificiis *LbVa*

GLOBOSUM rotundum *BeLaLbVaLeaOxf*

TENERUMQUE Ex natura aquae et aeris describit lunare corpus *BeLaLbVaLeaOxf*, aquae] aqua *La*

LEVITATE subtilitate *BeLaLbVaLeaOxf*

COMPACTUM compositum *BeLaLbVaLeaOxf*, conpositum *La*
solidatum *LbVa*

INSTAR SPECULI "Instar speculi" dicit, quia quicquid naturarum speculo applicatur, quasi reddendo imaginem revibrata ostendit *Oxf*

ADIACULATI inmissi splendoris solis ad lunam *BeLa*
repercussi inmissi *LbVa*
a sole ad lunam *Oxf*

REVIBRARE splendere s. illud corpus *BeLa*

---

* Cf. Serv. Com.Verg.A. 6.585; Myth.Vat.I <81>; Myth.Vat.II <74>

illud corpus splendere *LbVaLea*, splendere *om. Lea*
resplendere *Oxf*

IN EO SISTRA NILIACA ELEUSINAQUE LAMPAS ARCUSQUE DICTYNNAE TYMPANAQUE CYBELEIA VIDEBANTUR

IN EO s. corpore *BeLaLbVaLeaOxf*, s. *om. Lea* lunari *add. Oxf*

SISTRA NILIACA Descripsit ea, quae continentur in corpore lunari: per sistra Niliaca fontes omnium fluminum, per lampadem Eleusinam aeternum splendorem. *BeLa, cf. LbVa ad* 49.17-0 LUNAREM

Sistrum species indumenti (*lege* instrumenti?) ab Iside regina inventum, a qua et nomen accepit, et reperitur maxime circa Nilum[*] *BeLa*, species] speties *La* regina] reginae *La* accepit] acepit *Be*

Sistra species est indumenti (*lege* instrumenti?), quae maxime circa Nilum sive Aegyptum invenitur[*] *LbVaOxf*

ELEUSINAQUE LAMPAS Musicam describit in luna *BeLaLbVa*

[1]Salmoneus fuit rex Eleusinae civitatis, qui fecit pontem aereum volens imitari tonitruum Iovis et fulgura. [2]Nam faculam emittebat ad quemlibet sibi offensum. [3]Quo percusso a Iove facula illius ad caelum translata est, quae modo luna dicitur. [4]Et hoc est quod dicit Eleusina lampas, quasi regis Eleusinae lampas[**] *BeLa, cf. LbVa ad* 49.17-0 LUNAREM. 1 aereum] aeream *La*; 4 Eleusina] heleusina *Be* Eleusinae] heleusinae *Be*

ARCUSQUE DICTINNAE i. Dianae *BeLa*

Quia ve[r]natrix fuit *LbVa*

dominae *Va2*

A Dictinno oppido, in quo colebatur *BeLa*

A Dictinno oppido: dictae i. rete, dictinnae i. retialis ipsa est Diana[***] *LbVaLea*, dictinnae] dictinae *Va*

Dianae i. lunae, quae praeest venatoribus, qui aliquando ad lunam venantur *Oxf*

CYBELEI mater Veneris *BeLa*

vel ipsa *Be*

Veneris *LbVa*

Cybeles tympana lunae deputantur, quod omnia corpora

---

[*] Cf. Isid. Etym. 3.22.12

[**] Cf. Serv. Com.Verg.A. 6.585

[***] Cf. Myth.Vat.II <36>; Lact. Theb. 9.632

maxima dum in motu sunt, musicam de se emittunt, sola vero terra, quae in statu est, nullum sonum de se reddit vel promit, proximo tamen sono i. lunari utitur *BeLaLbVaLea*, Cybeles tympana] cibeles timpana *Va* quod] quia cum *Lea* omnia corpora] cum omnia dicit corpora *LbVa* in motu] immota *LbVa* emittunt] emittant *Lea* vero *om. Lea* quae] qui *Be* reddit vel *om. BeLa*

Cybeleia timpana lunae deputantur, quia dum omnia corporea (*id est, planetae*), cum in motu sunt, musicam de se emittunt, solum vero corpus (*id est, terra*), cum non movetur, minime est sonorum *Oxf*

49.21-1 TRIFORMIS ETIAM DISCOLORQUE VIRAGO TERRIBILI QUADAM MAIESTATE RUTILABAT

TRIFORMIS i. cornicularis, medialis, plena. *LbVaLea*, cornicularis] tricornicularis *Va*

[1]Triformis est luna propter varietatem motus. [2]Prima forma cornicularis usque ad octavam, cornua habens ad orientem. [3]Secunda forma ab octava usque ad vicesimam secundam, in quo spatio aut plena est aut prope plena. [4]Tertia autem a praedicto termino usque ad nonam, et est tunc cornicularis, sed versis cornibus ad occasum. [5]Item triformis dicitur luna, cum aut cum sole est aut ex latere solis est aut ex adverso. [6]Cum sole est, cum in coitu est, et est nova. [7]Dum de latere est octava est, vel vicesima secunda et semiplena. [8]Semiplena est octava et xxii. [9]Cum ex adverso est, quinta decima est, et plena est *BeLaLbVaLeaOxf*, 1 propter] ob *LbVaLea*; 2 usque ad octavam *om. LbVaLea*; 3 vicesimam secundam] xxx^{mam} ii^{dam} *LbVa* aut plena est] est iam plena *LbVaLeaOxf*; 4 nonam] novam *LbVaOxf* tunc] tunc etiam *BeLa* et est] est etiam *Oxf*; 5 cum aut cum sole est] cum aut cum est cum sole *Oxf* solis est] solis *BeLa* ex adverso] ex adverso solis *Oxf*; 6 cum in coitu] dum in coitu *Oxf* est nova] nova *BeLaOxf*; 7 semiplena] semis plena *LbVaLea*; 8 Semiplena - xxii *om. BeLaOxf*; 9 adverso] adversa *BeLa* quinta decima est] quinta decima *LeaOxf* plena est] plena *BeLa* est plena *Oxf*

DISCOLORQUE i. varia *BeLbVaLeaOxf*, i. *om. BeVaLeaOxf*

Quia ex illa parte, qua accenditur, lucet et ex altera pallescit *LbVa*, altera] altara *Va*

VERTIGO motio *Be*

motus conversio *LbVaLea*

rotatio lunae *Oxf*

50.1-3 QUAE LICET CORNIGERA ET ASPERA CREDERETUR, EGESTIONIBUS

OPPORTUNA TAMEN ET FELEM ET CERVAM ET CONVERSIONES BIS BINAS VULTIBUS PRAEFEREBAT

ASPERA prop<ter> sonum *Be*
　propter cornua *Oxf*

CREDERETUR quia cornuta *LbVa*

†AGGESTIONIBUS Ab his quae gerit *BeLaLbVaLea*, que] quasi *Be*
　crementis *LbVa*
　mutationibus vel operationibus *LbVa*
　aucmentationibus i. de incremento sui luminis *Oxf*

OPORTUNA moderate *Be*
　aptata et moderata *La*
　ipsa *LaLbVa*
　secundum oportunitatem *LbVa*, *cancell. Lb*[2]*Va*[2]
　s. moderata *Oxf*

FELEM bestiam cornutam *BeLa*
　propter cornua *Be*
　capram *Oxf*

CERVAM Quae est bestia cornuta i. damma vel simia (*sic, lege* similia?) *LbVaLea*
　propter velocitatem sui cursus *Oxf*

CONVERSIONES mutationes *BeLaLbVaLea*, mutaciones *LbVa*

BIS BINAS [1]Conversio lunaris bis bina dicitur propter quadrifariam mutationem. [2]Primus enim motus est a nova usque ad octavam, qui motus cornicularis vocatur. [3]Secundus ab octava usque ad plenilunium, in qua mutatione amphicirca vocatur. [4]Tertius a plenilunio usque ad vicesimam secundam, in qua etiam simili modo amphicirca vocatur. [5]Quartus usque ad xxx, in qua etiam cornicularis paret *BeLaLbVaOxf*, 1 quadrifariam] qqadrifariam *Be*; 2 Primus enim motus est] prima enim conversio est *LbLea* prima enim est conversio *Va* a nova] a nona *Lea* motus *om. Oxf* modus *Va* octavam] octava *La*; 3 Secundus] secunda *LbVaOxf* amphicirca] amficirca *Oxf*; 4 Tertius] deinde *LbVaLeaOxf* ad *om. Va* etiam *om. BeLa* et *Oxf* similimodo amphicirca] similiter; amficirca *Oxf*; 5 Quartus ... paret] Hinc usque ad xxx^{mam} cornicularis *LbVaLeaOxf*

VULTIBUS in *BeLbVa*
　s. suis *BeLbVaLea*, s. *om. Lea*

PRAEFEREBAT ostendebat *LbVa*

-4 EXHINC MEDIO QUAM AD LUNAM CONSCENDERAT, AD CYLLENII

CIRCULUM VENIT
EXHINC MEDIO Que est in emitonio *La*
de medietate *BeLbVa*, de] *cancell. Lb*[2]*Va*[2]
De medietate spatii, quo venerat ad lunam, venit ipsa ad siderem (*sic, lege* sidus) Mercurii *Oxf*

50.4-5 QUO HEMITONIO PERMEATO MULTIPLEX EI, UTPOTE NUBENTI DOMINAE, MINISTRORUM POPULUS LAETABUNDUS OCCURRIT
QUO EMITONIO medietas toni *Oxf*
PERMEATO transcurso *Be*
transacto *LbVa*
UTPOTE veluti *La*
sicuti *LbVa*
LETABUNDUS similis laetanti *LaLbVa*
OCCURRIT Filologiae *Oxf*

181/
51.7-8 HINC FESTINATUR ASCENSUS, ET USQUE IN VENERIS CIRCULUM HEMITONIO TRANSVOLATUR
HINC deinde vel ab illo loco *Oxf*
FESTINATUR ASCENSUS Filologia *Oxf*
HEMITONIO cum *BeOxf*
spatio emitonii *LbVa*
vel per emitonium *Oxf*
TRANSVOLATUR transcurritur s. ab illis *Be*
curritur *LaLbVaOxf*

51.8-10 IPSAQUE VENERE, QUAE NUPTIIS ALLUBESCEBAT, QUANTUM DECEBAT ORATA, HOC IN EA PERHIBETUR INTUITA
ALLUBESCEBAT favebat *BeLaLbVaLeaOxf*
ORATA HOC adorata *La*
i. deprecata *LbVa*
precata illam rem *Oxf*
PERHIBETUR fertur *BeLbVa*
INTUITA cognovisse *Oxf*

51.10-1 QUOD ADMODUM PULCHRA, TAMEN ANTIAS DRACONIBUS CIRCUMFLEXA CREBROQUE CAPILLITIO VULSA, AMBIFARIUMQUE †NITAL SECUM CONGRESSA MITIFICAT

ADMODUM valde *BeLaLbVaOxf*, i. *add. Be*

PULCHRA s. sit *BeLaLbVaOxf*, s. *om. Oxf*

ANTHIAS contraria *LbVaOxf*, i. *add. Oxf*

dea auxiliatrix virginum *BeLaLbVaLeaOxf*, vel *add. Oxf*

Anthias s. <ut> pateret virgo *La*

Quale ille voluerat ipsi Philologie stuere (*sic, lege* (ob)struere?) *La*

CIRCUMFLEXA cincta *Be*

CREBROQUE spisso *Oxf*

CAPILLITIO VULSA i. habens capillum evulsum VULSA est s. <ut> virgo pateret Philologia *Be*

soluta *La*

habens vulsum capillum *LaLbVaLea*

Soluta ut virgo pateret *LbVaLea*, soluta] et solutum *Lea*

Vel pallium (*sic, lege* capillum?) habens vulsum. capillatura soluta *Oxf*

AMBIFARIUMQUE Quia binis partibus favet: et superis et inferis *BeLaLbVaLeaOxf*, quia binis partibus] ex bis postibus *Oxf* et inferis] et sedat inferis *La*

†NI ut non *Be*

†NITALE i. rorem, quia delectatio quasi ros suavis videtur *Oxf*

SECUM cum illo *La*

CONGRESSA confligens *BeLaOxf*

MITIFICAT sedat *Be*

decertabat *La*

2-3 MOX STUDIUM IN SOLAREM LABORARE CIRCUM; QUIPPE SESCUPLO FATIGABAT ASCENSUM, QUI TONOS AC DIMIDIUS HABEBATUR

STUDIUM s. est illi *BeOxf*

s. fuit *LbVaLea*

LABORARE conscendere *BeLaOxf*, i. laborando *add. Oxf*

contendere *LbVaLea*

SESCUPLO i. tono et dimidio *LbVaLea*

FATIGABAT s. itinere *Be*

ipse circus *La*

circus *LbVaLeaOxf*

ASCENSUM ascendentes *BeLaLbVaLea*, s. circus *add. Be* i. *add. La*

Filologia *Oxf*

TONUS AC DIMIDIUS Duobus emitoniis, quae sunt a luna ad Mercurium et a Mercurio ad Venerem. Est comparatio ad tonum infra solem et emitonium supra *LbVaLea*, et emitonium] et ad emitonium *Lea*

s. ascensus *Oxf*

187/
52.13 NONNE A PRINCIPIO GEMINUM TU DAS TETRACHORDON?
TETRACORDUM* [1]Inter caelum et terram gemina sunt tetracorda i. duo. [2]Septenarius enim planetarum numerus duo tetracorda efficit. [3]Prima corda lunaris circulus, secunda Mercurialis, tertia Veneria, quarta solaris, quae etiam finem prioris tetracordi et initium sequentis obtinet, sicut mese in musica. [4]Est igitur secundum tetracordum, cuius quasi prima corda solaris circulus, secunda Martis, tertia Iovis, quarta Saturni *BeLaLbVaLeaOxf*, 2 Septenarius] septinarius *Va*; 3 Veneria] venerialis *Oxf* prioris] priuris *BeLb* primi *Oxf* obtinet] facit *Oxf*; 4 corda] pars *Oxf*[1] corda *Oxf*[2] tertia] tercia *LaVa*

194/
53.12 HIS AUDITA DEORUM SEDES IUSSA EST PERMEARE
HIS AUDITA precibus exaudita *BeLaLbVaLea*

Philologia ab Apolline *BeLaLbVaLeaOxf*, Philologia *om. BeLa*

s. precibus s. es Philologia *Oxf*

DEORUM SEDES ad *BeLaLbVaLea*, s. *add. LbVaLea*

IUSSA** vel ab Apolline vel ab Atanasia *BeLaLbVaLea*, vel *om. LbVaLea*

s. est illa *Oxf*

PERMEARE transire *LbVaOxf*

pertransire *Lea*

53.12-3 VERUM HEMITONIO SUBLEVATAM PYROIS CIRCULUS IMMORATUR, IN QUO IOVIS FUERAT MAXIMUS FILIORUM
VERUM sed *Be*

s. tam *La*

HEMITONIO in *Be*

---

* TETRO CORDO *Va*[1], TETRACORDUM *Va*[2].

** VISA *Va*[1]

a sole usque ad Martem *LbVaLea*
cum *Oxf*
SUBLEVATAM s. illam Philologiam *Be*
illam *Oxf*
PYROIS CIRCULIS i. Martis *BeLaLbVaLea*, i. *om. Be*
igneus circulus Martis *Oxf*
INMORATUR tardat *BeOxf*
tradatur (*sic, lege* tardatur) *La*
illam tardat *LbVaLea*
detinet *Oxf*
MAXIMUS FILIORUM Mars *BeLaLbVaLeaOxf*, i. *add. LaLea*

4 EX QUO CIRCULO VISUS PYRPHLEGETON AMNIS AD INFERA DEMEARE
EX QUO CIRCULO visus s. est *BeLa*
s. Martis *LbVaLeaOxf*, s. *om. Oxf*
usus est *Oxf*
†PYRYFLEGETHON igneus vel sulphureus *BeLaLbVaLea*
AMNIS fluvius *BeLbVaLea*
DEMEARE descendere *BeLbVa*
de alto meare *Oxf*

4-6 QUO TRANSGRESSO (NEQUE ENIM LABOR FUERAT HEMITONII INTERIECTA TRANSCURRERE), IN IOVIALIS SIDERIS PERVENERE FULGORES, CUIUS CIRCULUS PTHONGO <PHRYGIO> PERSONABAT
QUO TRANSGRESSO emitonio Martis *LaLbVaLea*
circulo *BeLaLbVa*, vel *add. La*
semitonio vel circulo *Oxf*
LABOR illi Philologiae *LbVaLea*
HEMITONII INTERIECTA interposita *Be*
s. spatia *BeLaLbVaOxf*, s. *om. Oxf*
SIDERIS splendoris *LaLbVaOxf*
FULGORES splendores *BeLbVaLea*
CUIUS CIRCULUS In hoc differt musica caelestis a terrestri: quia in quantum sunt breviores fistulae sive cordae in terrestri, tantum

acutiorem sonitum reddunt, et quantum sunt longiores, tantum graviorem; in caelesti vero musica quantum sunt angustiores circuli planetarum, tantum sunt e contrario graviores, ut est in circulo lunae, quantum autem longiores vel vastiores, tantum acutiores, ut in circulo Saturni *LbVaLea*, et quantum sunt longiores] et quantum longiores *Lea*

Iovialis *LaLbVaLeaOxf*, i. *add. La*

†PHTONGIO integro sono *BeLaLbVaLeaOxf*, i. *add. Oxf* sono integro *Be*

54.1-2 ILLIC SIDUS ERAT TEMPERAMENTI VIVIFICI AC SALUBRIS EFFULGENTIA BLANDISQUE VIBRATA CANDORIBUS

ILLIC in Iovis circulo *LaLbVaLeaOxf*

TEMPERAMENTI Omne temperamentum philosophi dicunt a Iove descendere *BeLaLbVaLea, BeLa ad* 54.4-5 NAM IPSE. Omne temperamentum philosophi dicunt] philosophi dicunt omne temperamentum *BeLa*

Quia iuxta paganos bona co<n>stellatio ibi capiebatur *Oxf*

VIVIFICI vivificantis *BeLaLbVaLeaOxf*

EFFULGENTIA lux. nominativus *La*

s. et erat ibi lux *BeLbVaLea*, et *om. Lea* ibi *om. Be*

i. lux s. erat *Oxf*

BLANDISQUE temperatis *BeLaLbVaLea*, temperata *La*

cum *LbVa*

VIBRATA radiata *Be*

radiata ipsa lux *LaLbVaLea*, radiata *om. Lea* ipsa *om. La*

irradiata *Oxf*

CANDORIBUS splendoribus *LbVa*

fulgoribus *Oxf*

54.2-3 CUIUS QUIDEM LUCIS NATURA EX CALIDIS UMIDISQUE COMMIXTIONIBUS CANDENS QUADAM PROSPERITATIS TRANQUILLITATE RUTILABAT

CUIUS i. Iovis *LaOxf*

circuli Iovialis *LbVa*

EX CALIDIS UMIDISQUE Ex his duabus rebus optime fit temperatio *La*

CALIDIS ex parte Martis UMIDISQUE ex parte Saturni, qui est frigidus *LbVaLea*

CANDENS lucens *BeLaLbVaOxf*

TRANQUILLITATE mitificatione *BeLbVaOxf*

mitigatione *La*

RUTILABAT fulgebat *Be*
resplendebat *LaLbVa*, splendebat *Va*

-5 VERUM IBI SIDUS IOVIS, NAM IPSE TOTIUS MUNDI MEMBRA COLLUSTRANS AD DEORUM DICEBATUR IMPERIUM ET SENATUM CAELITEM COMMEASSE

VERUM IBI SIDUS s. erat *BeLaLbVaLeaOxf*, s. *om. LaOxf*
IOVIS Noluerunt dicere philosophi transire deos in sidera *LaLbVa*, dicere philosophi] philosofi dicere *La* deos] dee *La*
NAM IPSE Iovis *LbVa*
Iupiter *Oxf*
TOTIUS MUNDI caeli *LbVa*
MEMBRA CONLUSTRANS partes circuiens *BeLaLbVaLeaOxf*, partes *om. LaOxf* circumiens *LeaOxf*
IMPERIUM ut imperaret *LbVaLea*
SENATUM CAELITEM curiam caelestium *LaLbVa*
COMMEASSE transisse *LbVa*

-7 HINC ETIAM PRAETERGRESSA CIRCUM AC PARILI INTERIECTIONE SUBLIMIS DEORUM RIGIDISSIMUM CREATOREM IN ALGIDO HAERENTEM PRUINISQUE NIVALIBUS CONSPICATA

HINC ex hoc loco *BeLaLbVaOxf*, i. *add. LaOxf*
PRAETERGRESSA Philologia *Be*
CIRCUM Iovialem *BeLaLbVaLeaOxf*, s. *add. BeOxf*
PARILI INTERIECTIONE simili *La*
Hoc est de consimili itinere *BeLa*, hoc] haec *La*
i. tono *Be*
spatio vel tono *LaLbVaLeaOxf*
SUBLIMIS elata *BeLaLbVaLeaOxf*
RIGIDISSIMUM frigidissimum *BeLa*
asperrimum *LaLea*
Saturnum asperrimum *LbVaOxf*, asperrimum Saturnum *Oxf*
CREATOREM Saturnum *BeLa*
i. patrem Iovis vel Vulcani *LbVaLea*
IN ALGIDO s. loco *BeOxf*, s. *om. Oxf*
s. in loco frigido *La*
s. circulo *Lea*
s. circulo i. loco frigido *LbVa*

HAERENTEM fixum *BeLaLbVaLea*
PRUINISQUE NIVALIBUS et in *Be*
cum *La*
Quia longe est a sole *LbVa*
CONSPICATA s. est *BeLaLbVaOxf*, s. *om. Oxf*
videt *LbVa*

54.8 VERUM IDEM, QUEM CIRCUMIRE NITEBATUR, ORBIS MELO DORIO TINNIEBAT
VERUM sed *Be*
IDEMQUE circus *Be*
s. circulus Saturni *LaOxf*, Saturni *om. Oxf*
NITEBATUR Saturnus *Be*
conebatur *LbVaLea*
ORBIS mundi *Be*
circulis *LbVaLea*, circulus *Lea*
MELO DORIO sono gravissimo *BeLaLbVaLeaOxf*, de *add. Be* gravissimo sono *LaOxf*
Dorus enim trophos in musica graviter sonat *Oxf*
TINNIEBAT resonabat *Be*

54.8-10 SED IPSI PRAESULI NUNC DRACONIS FACIES, NUNC RICTUS LEONIS, NUNC CRISTAE CUM APRUGNIS DENTIBUS VIDEBANTUR, TOTOQUE EXITIALIS SAEVIEBAT HORRORE
SED IPSI PRESULI Saturno *BeLaOxf*
principi *LbVaOxf*
NUNC DRACONIS FACIES Quia docent omnia *La*
pro ipsa ferocitate frigoris *LbVaLea*
propter frigus. Quia cum tangit alios circulos, igneus efficitur *Oxf*
CRISTAE comae *La*
CUM APRINIS propter intemperantiam *Oxf*
TOTOQUE s. cum magna *LbVa*
EXITIALIS periculosus *BeLaLbVaLeaOxf*
mortifer *Oxf*
SAEVIEBAT insanibat *Be*
crudescebat *LbVa*
HORRORE asperitate *BeLaLbVaLeaOxf*, tota *add. Oxf*
Haec omnia pro asperitate frigoris *BeLa*

0-1 CUI TAMEN POTESTAS PRO CIRCI GRANDITATE MAIOR AC PRAELATA CETERIS HABEBATUR
CUI i. Saturno *LbVaLea*
  i. a quo Saturno *Oxf*
PRO CIRCI GRANDITATE Quia in xxx annis ipsum percurrere fatetur *LbVaLea*
CETERIS s. planetis *Be*
  stellis *LbVaLea*

1-2 DENIQUE HARPIS BOMBISQUE PERTERRITA TAM INTOLERANDAM CONGRESSIONEM VIRGO DEFUGIT
ARPIS ab *Be*
  sonitibus *LaLea*
  sonitibus gravibus *LbVa*
  i. armoniis melodiis *BeLa*, i. *om. Be*
  Pro hominibus melodias accipe *LaLbVaLea*, hominibus] omnibus *LaLea* melodias] melodiis *La*
BOMBISQUE et a *Be*
  clamoribus *BeLaLbVaLea*
TAM †INTOLERANDA non ferenda *BeLaLbVaLea*
  pro magna *La*
†CONGRESSIONE i. difficili itinere *Be*
  conscensione vel itinere *LaLbVaLea*
  de *LbVa*
  Quasi congressio esset in illa musica *LbVaLea*
†DIFFUGIT celeriter transivit *Oxf*

3 INDE MAXIMIS CONATIBUS SESCUPLO ITINERIS EVEHUNTUR
INDE circulo Saturni *LbVaLea*, ex *add. Lea*
MAXIMIS CONATIBUS cum *BeLaLbVa*, s. *add. LbVa*
SESCUPLO tono ac dimidio *LaLbVaLeaOxf*, i. *add. Oxf* ac] et *Oxf*
EVEHUNTUR s. cum pedissequis *LbVa*

13-5 NAM TONO AC DIMIDIO AD IPSIUS CAELITIS SPHAERAE GLOBUM AC LAQUEATUM STELLIS AMBITUM PERVENITUR
TONO AC DIMIDIO cum *Be*
SPHERAE i. ad firmamentum aetherium *Be*
  rotunditas *La*

rotundititatis *LbVaLea*

firmamenti *Oxf*

AC LAQUEATUM STELLIS ornatum *BeLaLbVa*

de *Be*

AMBITUM PERVENITUR circumdatum *Be*

circulum *LaLbVa*

i. a Ioviali circulo, quia ibi sunt omnes stellae fixae *LbVaLea*, a Ioviali circulo] iovialem circulum *Lea*

199/
54.15-6 SICQUE SEX TONORUM CONSCENSIONIBUS <ET> STADIORUM DEFECTA LASSITUDINE FATIGATI

SICQUE SEX TONORUM [1]A terra secundum istum usque Saturnum sex toni sunt, i. a terra usque ad lunam tonus, a luna usque Mercurium hemitonium, ab illo usque Venerem similiter, a Venere usque ad solem tonum et semissem, ab isto Marte tenus hemitonium, a Marte usque Iovem tonum et hinc usque ad Saturnum tonum. [2]Fiunt simul sex toni. [3]Alii doctores aliter dicunt *BeLa*

Unus tonus habet cxxvi stadiorum. Perfectio musicae id est plenitudo diapason *La*

spatiorum *Oxf*

DEFECTA affecta *LbVaLea*

Eo quod deficere lassitudo faciat homines *LbVaLea*, deficere lassitudo] lassitudo deficere *Lea*

LASSITUDINE Ypallage: eo quod ascendentes deficere faciat *BeLa*, ascendentes] ascendente *La*

ypallage *LbVaLea*

i. consumpta, quia deffectos facit *Oxf*

54.16-7 CUM DIAPASON SYMPHONIAM QUICQUID EMENSI ERANT ADVERTERENT CONSONARE PERFECTIONE ABSOLUTAE MODULATIONIS

per DIAPASON SYMPHONIAM equalitate<m> sonoritatis *Be*

DIAPASON ex toto, SYMPHONIAM duplicitate<m> sonoritatis, consonantiam *La*

duplicitatem sonoritatis *LbVa*

duplicem sonoritatem *Lea*

DIPASON i. ex omni, SYMPHONIAM i. perfectam modulationem musicae sonoritatis *Oxf*

consonantiam *Oxf*
EMENSI ERANT transierant *Be*
mensuraverant *LaLbVa*
s. tonorum ascenderant *Oxf*
ADVERTERENT conspicerent *Be*
PERFECTIONE ABSOLUTAE plenitudine perfecte *La*
cum PERFECTIONE *BeLaLbVa*, s. *add. LbVa*
ABSOLUTAE liberae, non servilis *LbVaLea*

8 POST LABORES MAXIMOS RECREATI PAULULUM CONQUIERUNT
RECREATI refecti *Be*
refocilati *LbVa*
CONQUIERUNT requierunt *Be*
requieverunt *LaLbVa*

2-3 ARDEBAT HERACLITUS, UDUS THALES, CIRCUMFUSUS ATOMIS DEMOCRITUS VIDEBATUR
ARDEBAT HERACLITUS splendebat *BeLaLbVa*
Quia de igne<a> disputavit vi[*] *Be*
Quia de ignea vi disputaverunt[*] *LaLbVaLea*, quia] qui *La* ignea vi] igne et vi *La*
Summus medicus, qui dixit constare omnem materiam igne[*] *Oxf*
UDUS TALES Qui dixit omnia aqua creari[**] *Oxf*
CIRCUMFUSUS ATOMIS repletus *BeLa*
Venit de athomis *LbVa*
Quomodo iste mundus venit de athomis *LbVaLea*
DEMOCRITUS Quia de invisibilibus disputavit rebus et de elementorum natura *BeLa*
Qui de re invisibili i. de athomis scripsit *LbVaLea*
Qui de re invisibili disputavit *Oxf*

14-5 SAMIUS PYTHAGORAS CAELESTES QUOSDAM NUMEROS

---

[*] Cf. among others Tert. Adv.Marc. 1.306.26; Tert. An. 5.5; Aug. Civ.Dei 8.2.25; Isid. Etym. 8.6.20, 8.6.23

[**] Cf. among others Tert. Adv.Marc. 1.306.26; Tert. An. 5.5; Aug. Civ.Dei 8.2.25, 8.5.36

REPLICABAT, ARISTOTELES PER CAELI QUOQUE CULMINA ENTELECHIAM SCRUPULOSIUS REQUIREBAT
SAMIUS a Samia civitate Beneventi* *BeLa*
Beneventanus *LbVaLeaOxf*
PYTAGORAS arihtmeticus** *La*
Qui de arithmetica disputavit** *LbVaLea*, aritmetica *Va*
REPLICABAT numerabat *LaLbVaLeaOxf*
ARISTOTELES musicus *LaLbVaLea*
Quia de natura locutus est animae*** *BeLa*
†ENDELICHIAM summam partem animae *BeLaLbVaLeaOxf*
SCRUPULOSIUS subtilius *BeLaLea*
subtilius, sollertius *LbVa*

56.16 EPICURUS VERO MIXTAS VIOLIS ROSAS ET TOTAS APPORTABAT ILLECEBRAS VOLUPTATUM
MIXTAS VIOLIS ROSAS voluntate (*sic*, *lege* voluptate?) cum castitate *La*
amores cum castitate *LbVaLea*
ILLECEBRAS VOLUPTATUM voluptates *LbVa*
delitias voluptatum *Oxf*

56.16-8 ZENO DUCEBAT FEMINAM PROVIDENTEM, ARCESILAS COLLUM INTUENS COLUMBINUM, MULTUSQUE PRAETEREA PALLIATORUM POPULUS STUDIIS DISCREPANTIBUS DISSONABAT
ZENO Quia de nuptiis scripsit *LaLbVaLea*, scripsit] disputavit *La* scribit *Lea*
magister Stoicorum**** *LbVaLea*
Stoicus, qui de nuptiis prudentis mulieris disputavit *Oxf*
PROVIDENTEM prudentem *BeLa*
ARCHESILAS COLLUM INTUENS Qui de avibus disputavit *LbVaLea*, qui] quia *Lea*
narrator avium ac disputator *Oxf*
PRAETEREA praeter praedictos *LbVaLea*
PALLIATORUM POPULUS s. philosophorum, sicut episcopi

---

* Cf. Isid. Etym. 14.6.31
** Cf. Isid. Etym. 3.2.1, 8.6.20
*** Cf. Isid. Etym. 8.6.13
**** Cf. Hier. Hier.proph. 1.21; Hier. Trac.Marc. 6.148; Isid. Etym. 8.6.8

habebant pallios suos *LbVaLea*, philosoforum *Lb*
Qui enim in artibus studium perpetuum haberent, paliis utebantur *Oxf*
DISCREPANTIBUS dissimilibus *BeLbVaLea*
DISSONABAT Sicut fuit unicuique suus sensus *LbVaLea*, suus sensus] suum sensum *LbVa*

9-0 QUI QUIDEM OMNES INTER MUSARUM CARMINA CONCINENTUM AUDIRI, LICET PERSTREPERENT, NULLO POTUERE RABULATU
PERSTREPERENT personarent *Oxf*
RABULATU Rabulatus est altercatio cum arma (*sic, lege* armis) et est rethorum propri[a]e *La*
sonitu rabido et nimio *LbVaLea*
sonitu *Oxf*

GLOSSES ADDED TO BOOK NINE, *DE HARMONIA*

.5 EXPLICIT LIBER ASTROLOGIAE, INCIPIT DE MUSICA LIBER NONUS *BeLa*
MARTIANI MINNEI FELICIS CAPELLAE AFRICARTAGINIENSIS DE ASTROLOGIA LIBER VIII EXPLICIT, INCIPIT LIBER VIIII DE MUSICA *Leb*

.7 IAM FACIBUS LASSOS SPECTANS MARCENTIBUS IGNES
FACIBUS flammis .*BeLa*
LASSOS de sole vergente ad occasum *BeLa*
deficientes *Par*
cum MARCENTIBUS deficientibus *BeLaLebPar*, cum *om. LebPar*
SPECTANS Iam nocte inminente aspiciens *BeLa*
IGNES lucernas *Par*

.8 INSTAURARE IUBET TUNC HYMENEA VENUS
INSTAURARE IUBET i. aliam <artem> dotalium inducere s. Iuppiter Apollini *BeLa*
Pallas *Leb*

HYMENEA i. mater Hymenei *BeLa*

337.9 'QUIS MODUS' INQUIT 'ERIT? QUONAM SOLERTIA FINE
MODUS terminus vel mensura *BeLa*[1]*La*[2]
SOLLERTIA i. artes impediunt nuptias *BeLa*
ipsa Pallas *LebPar*, ipsa *om. Leb*
FINE per finem *BeLa*
hyperbaton est *Leb*

337.10 GYMNASIA exercitationes sollicitae *BeLa*
exercitationes liberalium artium *Leb*

337.11 DERIGUIT COMIS BLANDISQUE ASSUETA VOLUPTAS
DERIGUIT exhorruit vel induruit *BeLa*, exorruit *La* vel *om. Be*
obstipuit *LebPar*
COMIS per comas *BeLa*
pulchra *Leb*
comum pulchrum (*sic, lege* per comam pulchram) *Leb*
pulcritudinibus *Par*
ASUETA solita vel consueta *BeLa*, vel consueta *om. Be*

337.12 CONTRAHIT PUER
CONTRAHIT torpet *La*
PUER filius *BeLa*
Haec signa irae sunt et tristitiae *BeLa*
In unum colligit quod factura *Par*
PUER Cupido *Par*

337.13 IPSA ETIAM FULCRIS REDIMICULA NECTERE SUETA
FULCHRIS toris *La*
toris nuptiarum *Par*
lectis *Leb*
REDIMICULA ornamenta *Par*
NECTERE SUETA coniungere solita *BeLa*

337.14 FLORA DECENS TRINA ANXIA CUM CHARITE EST
FLORA DECENS TRINA dea florum honesta *BeLaLebPar*, honesta *om. LebPar*
pulcra dea *Par*

CARITE Nomen nimphae de famulitio Veneris i. cum amore nuptiali *BeLa*, famulitio Veneris] famuliti *BeLa*[1], -o Veneris *suprascr.* *La*[2]
caris gratia *LebPar*, caris *om.* *Par*
ablativus *Leb*

15 NEC MELICUM RETICENS
canere voce, succinere corde *BeLa*
MELICUM dulce *La*
†RECINENS canens *BeLa*

16 NEC DULCIS TEMPTAT PSALLERE MELPOMENE
TEMPTAT indicat, orditur *La*
MELPOMENE nona musarum *BeLa*
cantus perseverans *LebPar*, perseverans cantus *Par*

17 OMNIA, QUAE TENERO MORIS FERVESCERE LUDO
OMNIA s. sunt vel fuerant *BeLa*
in †TENERE novo, dulci *BeLa*
MORIS FERVESCERE consuetudinis ardere *BeLa*

18 IN CUMULUM DOCTAE VOCIS HONORE TACENT
IN CUMULUM DOCTAE aucmentum artium *BeLa*
simul *Par*
VOCIS HONORE TACENT in *BeLa*
propter liberales artes *Par*

19 NEC SUADA ILLECEBRIS SPONSALIA PECTORA MULCET
SUADA Iuno *BeLaLebPar*
Suada dicitur Iuno, quia suadet nuptiis *BeLa*
INLECEBRIS voluptatibus *BeLa*
MULCET lenit *BeLa*, linit *Be*

20 STIMULA INCENSO ALLICIT ACULEO
i. amores spiculo provocat acuto *BeLa*
Alia dea irata provocat *LebPar*, alia *om.* *Leb*
STIMULA s. amore *Par*

1 QUIN ETIAM INTERULOS GAUDENS DISSOLVERE NEXUS
QUIN adfirmative *BeLa*

INTERULOS interiores *BeLaPar*
DISSOLVERE NEXUS i. dona virginitatis *BeLa*
nodos amoris *Par*

338.2 BLANDIFICAQUE LIBENS STRINGERE CORDA FACE
cum BLANDIFICAQUE voluptuoso vel pio *BeLa*
STRINGERE iungere *La*
CORDA FACE i. sponsali amore *BeLa*
dea *LebPar*
Et est nominativus singularis *Leb*

338.3 NICTANTES OCULOS REPRIMIT STUPEFACTA PAVORE
†NECTANTES vigilantes *BeLa*, vigilantis *La*
REPRIMIT retinet *BeLa*
†VAPORE lumine solis *BeLa*

338.4 NEC PERFERRE VALET GORGONOS ORA TRUCIS
PERFERRE i. sustinere s. Iuno *BeLa*
GORGONOS ORA †TRUCES per i. vultus terribiles *BeLa*
Quia impedit nuptias propter Palladem *BeLa*, inpedit *La* Paladem *La*
Quia mutat homines [hom] in lapides *Par*
GORGONOS genitivus *Leb*

338.5 SI ERUDITA PLACENT CERTE SPONSALIA DIVI
ERUDITA docta *BeLa*
DIVI ó *La*
Mercurii *LebPar*

338.6 SALTEM DOCTA FERAT CARMINA CALLIOPE
i. si non est locus nuptiarum, saltem concedite carmina *BeLa*
CARMINA s. et non artes CALLIOPE bona vox *Par*

338.7 OBLECTANS
deliniens *BeLa*, daeli<ni>ens *La*

338.8 TEDIA DULCISONIS AUFERET ILLA TONIS
TEDIA i. tedia auferre potest et oblectare mentes *BeLa*
DULCISONIS cum *Be*

ILLA TONIS Calliope carminibus *BeLaPar*, carminibus *om. Par*

9 QUIPPE SCRUPOSIS (FATEOR) LASSATA PUELLIS
SCRUPOSIS a mirabilibus *BeLa*, a *om. La*
fastidiosis *Leb*
FATEOR dico PUELLIS artibus *BeLa*
LASSATA s. sum *Leb*
Venus *Par*

10 INSUETIS LAEDOR MAESTIFICATA MORIS
IN- ab *BeLa*
LAEDOR quod offendor *BeLa*
MESTIFICATA tristis facta *BeLa*
MORIS tarditatibus *BeLa*

11 PRONUBA, SI VOLUPE EST HAEC SERIA CARPERE, IUNO
PRONUBA Taliter instat dea voluptatis propria iura copulae audire *BeLa*
ó *BeLeb*
Iuno *Leb*
VOLUPE EST i. cum voluptate *BeLa*
nominativus voluptas *LebPar*, nominativus *om. Par*
SERIA longa gaudia IUNO Vocativus est. YRONIA *Par*

12 NEC CURA ASTRIFERI TE STIMULAT THALAMI
CURA amor *BeLa*
STIMULAT compungit, incitat *BeLa*

13 AST EGO SUBCUBUI, LEPIDISQUE ASSUETA CHOREIS
SUBCUBUI i. ferre non valeo *BeLa*
LEPIDISQUE lascivis *BeLaLebPar*
CHOREIS conventibus virginum ludentium *BeLa*
choraulis *Leb*
modulationibus *Par*

14 NON VALEO TRISTES CERNERE CECROPIDAS
TRIS CERNERE* videre *BeLa*

---

* TRIStes *Be*

s. quia nuptias impediunt *BeLa*
Quia tristitiam gignunt nescientibus *BeLa*
Quia tristes faciunt *Leb*
CECROPIDAS artes *BeLaPar*

889/
338.15-6 ET CUM DICTO RESUPINA PAULULUM RECLINISQUE PONE CONSISTENTIS SESE PERMISIT AMPLEXIBUS VOLUPTATIS
ET CUM DICTO hoc *Be*
i. cum hoc dixisset *BeLa*
RESUPINA s. Venus *La*
PONE iuxta *BeLa*
retro *Par*
PERMISIT s. Venus *BeLa*
AMPLEXIBUS manibus *BeLa*

338.17 QUAMPLURES ETIAM ASTRILUCI ASSENSERE DIVI
ASTRILUCI Eo quod in astris sunt *BeLa*, in astris] inter astra *La*

338.18 PRAESERTIMQUE LEMNIUS MULCIBER, FABRILIUM TANTUM OPERUM SOLLERS MARITUS
LEMNIUS MULCIBER Vulcanus mulcens ignem* *BeLa*
SOLLERS sagax *BeLa*
MARITUS s. illius Veneris *BeLa*

338.19 PROMPTIORE ATTESTATIONE COLLAUDAT
PROMPTIORE cum *Be*
faciliore ante omnes *BeLa*
ADTESTATIONE COLLAUDAT s. verba Veneris *BeLa*

338.19 TUNC DENIQUE, QUOD DECENTER INNIXA
QUO pro quia *BeLa*, pro *om. Be*
DECENTER honesta *BeLa*
INNIXA suffulta a Voluptate vel famulitio *BeLa*

338.20 ATQUE IPSA RELABENTEM LASSITUDO DECUERAT

---

* Cf. Serv. Com.Verg.A. 8.724

RELABENTEM reclinantem *BeLa*
DECUERAT ornaverat *BeLa*

20-1 MARS EMINUS CONSPICATUS TENERE CUM ADMIRATIONIS OBTUTU LANGUIDIORE FRACTIOR VOCE LAUDAVIT
MARS EMINUS corruptor Veneris *BeLa*
TENERE blande voluptarie *BeLa*
cupide *Par*
OBTUTU aspectu *BeLa*, aspectucu *Be*[1] aspectacu *Be*[2]
LANGUIDIORE infirmiore *BeLa*, infirmore *Be*[1]
LANGUIDIORE FRACTIOR VOCE propter tristitiam infirmior *Par*

22 NEC BROMIUS IN FAVORIS GRATIAM DISPAR FUIT
IN FAVORIS in laudis *BeLa*
†GRATIA dono, laude *BeLa*
NEC ... DISPAR litotes *BeLa*
dissimilis *BeLa*, dissimiles *Be*

23 QUIN ETIAM IPSUM ATLANTIADEM TAM FLAMMATAE CUPIDITATIS CURA CONCUSSIT
†QUI NON insuper *BeLa*
ADLANTIADEM Mercurium *BeLaPar*
FLAMMATAE voluptuosae *BeLa*
CONCUSSIT movit *BeLa*

24 UT OMMITTERE VELLET, QUAE CIRCA SPONSALEM CONVENTUM ORNATIORA DISPOSUIT
OMITTERE dimittere *BeLa*
SPONSALEM s. Philologiam *BeLa*
†CENTUM* vel numerum. Multa, sive finitum pro infinito posuit *BeLa*
ORNATIORA ornamenta *BeLa*
Id est musicam in qua sunt plura ornamenta *Par*

25 TANTI QUIPPE VISUM NUMQUAM VENERI DISPLICERE
TANTI s. viri i. Mercurii *BeLa*
VISUM s. est *BeLa*, est *La*[t]

---

* .C. *La*

890/
338.25-6 VERUM SUPERUM PATER, LICET INSINUATIONE GERMANAE UT PROPERARET ADMONITUS
SUPERUM PATER i. Iuppiter *BeLa*, i. *om. La*
GERMANAE Iunonis *BeLa*
AMMONITUS s. esset *Be*

338.26-1 NE QUID NUPTIALIBUS DEROGARET ORNATIBUS
QUID aliquid *BeLa*, aliquida *Be*
DEROGARET subtraheret *BeLa*

339.1-2 AUT TANTAE ERUDITIONIS EXAMEN DELICIOSA FESTINATIONE CONVELLERET
HERUDITIONIS EXAMEN doctrinae inquisitionem *BeLa*
DELITIOSA cum voluptuosa *BeLa*, cum *om. La*
CONVELLERET solveret *BeLa*

339.2-3 QUI PROBANDARUM NUMERUS SUPERESSET, NIHIL AFFERENS FESTINATIONIS EXQUIRIT
PROBANDARUM s. feminarum *BeLa*
AFFERENS addens FESTINATIONIS accelerationis *BeLa*

891/
339.3-4 CUI DELIUS MEDICINAM SUGGERIT ARCHITECTONICAMQUE IN PRAEPARATIS ASSISTERE
CUI s. Iovi MEDICINAM puellam SUGGERIT insinuat *BeLa*
IN PRAEPARATIS s. ornatibus *BeLa*

339.4-5 SED QUONIAM HIS MORTALIUM RERUM CURA TERRENORUMQUE SOLLERTIA EST
SED quoniam *La*
HIS s. virginibus *BeLa*
SOLLERTIA cura sollicitudo *BeLa*

339.5-6 NEC CUM AETHERE QUICQUAM HABENT SUPERISQUE CONFINE
AETHERAE caelo *BeLa*
SUPERISQUE diis *Be*
CONFINE propinquum *BeLa*, propinqum *Be*
consimile *Par*

6-7 NON INCONGRUAE, SI FASTIDIO RESPUUNTUR, IN SENATU CAELITE RETICEBUNT
NON INCONGRUAE s. rationabiliter *BeLa*
SI FASTIDIO i. si propter fastidium *BeLa*
CAELICO celesti *BeLa*, celaesti *Be*

7 AB IPSA DEINCEPS VIRGINE EXPLORANDAE DISCUSSIUS
DEINCEPS postea *BeLa*
VIRGINE i. Philologia *BeLa*, i. *om. La* philologia] philologiae *La*
Pallade vel Filologia *Par*
EXPLORANDAE DISCUSSIUS probandae vel inquirendae diligentius *BeLa*, diligentius] dilengitius *La*

7-8 UNA VERO, QUAE POTISSIMA CAELI SIDERUMQUE DILECTIO EST
Quia omnis armonia resonat in vertigine caeli siderumque *BeLa*

8-9 EXAMINIS HUIUS TAM FAVORE QUAM VOLUPTATE DISQUIRITUR
EXAMINIS exquisitionis *BeLa*
iudicii *Par*
HUIUS Palladis *Par*
FAVORE laude *BeLa*
VOLUPTATE DISQUIRITUR Amore desiderio queritur *BeLa*

9-10 NON POTERIT SINE SCELERE VIDUARI
NON POTERIT ó Iuppiter *BeLa*
o Iovis *Par*
SCELERE crimine *BeLa*
VIDUARI separari *BeLa*

11 PRAE CUNCTIS INTIMATUM VELIM
CUNCTIS omnibus INTIMATUM VELIM persuasum s. fieri *BeLa*

11-2 QUOD DOTIS OFFERENDAE CUMULANDAEQUE RECIPROCO ALIAS ADVEXIT VIRGINIS MATER PUELLAS
OFFERENDAE s. illae virgines *BeLa*
RECIPROCO ADVEXIT In mutua intentione adduxit *BeLa*
mutuo sicut Mercurius *Par*
ALIAS istas *Par*

VIRGINIS Philologiae *BeLa*
MATER i. Fronesis *BeLaPar*, i. *om. Par*
Phronesis i. prudentia *Leb*
PUELLAS Trium illarum puellarum tantum in hoc libro nomina ponuntur. Caeterarum nomina beatus Augustinus in libris de civitate dei commemorat *BeLa**

339.12-3 QUAS OPULENTIS COLLOCATURA MUNERIBUS EXAMINE ISTO EXPLORARE CONSTITUIT
OPULENTIS cum maximis *BeLa*, cum *La*[1]
COLLOCATURA in dotem datura Mercurio *BeLa*
MUNERIBUS donis *BeLa*, donis] dotis *La*[1]
EXAMINE in *Be*
EXPLORARE probare vel iudicare *BeLaPar*, vel iudicare *om. Par*

339.13-4 HAE IGITUR NON DISPARES NUMERO NEC DISGREGAE VENUSTATE
s. sed tot sed quot et Mercurii i. vii *BeLa*
DISGREGAE dissimiles vel diversae *BeLa*
VENUSTATE honestate *BeLa*, honeste *La*

339.14-5 ERUDITIONIS ETIAM DIGNITATE PAUCIS ADSTANTIUM CONFERENDAE
PAUCIS Quod enim optimum est, paucis comparatur *BeLa*
s. non *Leb*
CONFERENDAE comparandae s. sunt *BeLa*

339.15-6 IN PENETRALIBUS QUOQUE VIRGINIS [SECRETIS SANTIORIBUS] ALUMNATAE
IN PENETRABILIBUS i. in profundioribus disciplinis, s. doctrinis *BeLa*

---

* I did not find the names of the other "prophetic arts" in Augustin's *De civitate Dei*. Willis and Lutz (edition Remigius II, 472.10) both refer to *De civitate dei* 8, 18, but although this section concerns the artes magicae, it does not give the names of these arts. Martianus mentions Genethliace (the art having to do with birth dates, horoscopes), Symbolice (the symbolic, riddling art) and Oeonistice (the art having to do with bird omens) by name, and introduces the other four as the "trigarium", the triple sisterhood, and the "epistularis" of Jupiter.

VIRGINIS Filologiae *Par*
ALUMPNATE doctae sunt *BeLa*

16 QUONE interrogative *BeLa*

16-7 AC PRORSUS ENERVI MOLLITUDINE SIDERALIS CURIAE TRANSEANTUR
ENERVI MOLLITUDINE infirma *BeLa*
SIDERALI CURIA TRANSEUNTUR a divina *BeLa*
Praetermittuntur sine examine *BeLa*, praetermituntur *La*

17 IOVIALIS VIGORIS MAIESTATE PERCENSAE
VIGORIS MAIESTATE potestatis potentia *BeLa*
PERCENSAE numerate sunt *BeLaPar*, numerantae *Be* sunt *om. Par*

18 HIC TRITONIDA 'ATQUIN' AIT 'VIRGINES, QUAS PHRONESIS EDUCAVIT
HIC in hoc loco *BeLa*
TRITONIDA Pallas *BeLa*
ATQUIN certe *BeLa*
FRONESIS Prudentia *Par*
EDUCAVIT nutrivit *BeLa*

18-9 SUPERUM INCESSABIBLITER PECTORUM ARCANA TENUERUNT
SUPERUM divinorum *BeLa*
INCESSABILITER insolubiliter *BeLa*
PECTORUM ARCHANA secretorum *BeLa*

19-0 NEC FUIT VESTRUM QUISQUAM, QUI NON ILLIS VOLUNTATIS ABDITAE INTERPRETAMENTA COMMISERIT
VESTRUM ó divi *BeLa*, divi] dii *La*
ABDITAE secretae *BeLa*
INTERPRETAMENTA publicationem *BeLa*

20-1 DENIQUE EX HIS QUAMPLURES ANTISTITIA PROFESSAE DIVUM
EX HIS s. puellis *BeLa*
ANTISTICIAM principalitatem *BeLa*
principium *Leb*

principatum *Par*
PROFESSAE confessae s. sunt *BeLa*
DIVUM s. habuisse *BeLa*

340.1 IN TERRIS SANCTIS LIBARE NUMINIBUS, CERTE LITARE PENITUS ABNEGATUM
†LIBRARE s. suspendere vel auferre *BeLa*, s. *om. Be*
†LIBERARES ó Iupiter *Leb*
LITARE sacrificare *BeLa*
ABNEGATUM s. esset *BeLa*, esset *La*[t]

340.1-3 VERUM TRANSACTAE VIRGINES ERUDITIONIS CASCAE PRAESTITERE MIRACULA, HAE FAMILIARITATIS VESTRAE PRAEFERUNT DOCUMENTA
VIRGINES ille dotales *BeLa*
ERUDITIONIS †CAUSA ablativus *Leb*
HAE s. quas Fronesis educavit *BeLa*

340.3 NAM INTER DIVINA HUMANAQUE DISCIDIA Quando fuit bellum inter deos et homines *BeLa*

340.4-5 HAS IGITUR INGRESSAS AGNOSCETIS PROBABITISQUE, DIVI
†PROBETISQUE laudetis vel laudabitis *BeLa*

894/
340.5 EARUM DISQUISITA VOCABULA
EARUM s. puellarum DISQUISITA exquisita *BeLa*
VOCABULA nomina s. sunt *BeLa*

340.6 TUNC PHOEBUS 'GENETHLIACE PRIOR' INQUIT 'ASTABIT
PHOEBUS Apollo *BeLa*
GENETLIACE i. mathematica, quae de genituris hominum deorumque disputat* *BeLa*, genituris] genitoris *Be*[1]
librамina *BeLa*
i. genitura vel constellatrix quae curat de genituris hominum* *LebPar*, i. *om. Par* quae curat de genituris hominum] de genituris hominum curam habens *Par*

---

* Cf. Aug. Quaest.ad Simpl. 1.2.3

I *Leb*
ASTABIT s. in ordine puellarum *BeLa*

6-7 QUAE RATIONIS AETHERIAE CONSCIA PENSA LACHESIS ADAPARIT
AETHERIAE celestis CONSCIA gnara *BeLa*
PENSA determinationes fatalis *BeLa*
fila *LebPar*
LACHESIS dea *BeLa*
sortis *LebPar*
†ADOPERTAE occulte *BeLa*
obscure. Filant vitae longitudinem *Par*

7-8 ATQUE INSTANTIUM SAECLORUM GERENDA DENUNTIAT
INSTANTIUM futurorum *BeLa*
GERENDA agenda *BeLa*

8-9 DEINDE SYMBOLICE, QUAE VARIETATIBUS OMINATIS VENTURA COMPONENS
SYMBOLICE Consiliatrix, quae consilia deorum habet *BeLa*
consiliatrix deorum *LebPar*
II *Leb*
OMINATIS cum significatis *BeLa*
VENTURA COMPONENS ordinans sive praesciens *BeLa*

9 AUSPICIORUM PROVENTUS PENSAT
AUSPICIORUM miraculorum *BeLa*
PENSAT perpendit, iudicat *BeLa*
numerat *Leb*

10 OEONISTICE TERTIA EST, PER QUAM TRIPUS ILLA VENTURI DENUNTIA
OEONISTICAE Quae de statu omnium seculorum disputat *BeLa*, statu] statutu *La*
auguratrix *LebPar*
III *Leb*
†TRIPOS s. Apollinis *BeLa*
laurus; propter tres radices *Leb*
altare Apollinis de lauro *Par*
†VENTURA futura *BeLa*

340.10-1 ATQUE OMNIS EMINUIT NOSTRA CORTINA
ATQUE s. per quam *BeLa*
EMINUIT i. ap<p>aruit vel revelata est *BeLa*
NOSTRA CORTINA Quasi cortina, que vera et occulta tenuit, vel a corio Pitonis[*] *La*

340.11-2 MIHI CORVUS ALLUDIT
MIHI i. Apollini *BeLa*
CORVUS Corvus Apollini datus noctem significat, cignus diem, quia ille et obscuram et apertam prophetiam dicit *BeLa*, prohetiam *Be*

340.12 CYCNUS ETIAM SOCIATUR s. michi *BeLa*

340.12-3 UT DIEI NOCTISQUE PROVENTUS NOS AVIBUS PRAESENTIRE CONCOLOR TEMPORUM PLUMA TESTETUR
PROVENTUS eventus NOS s. deos *BeLa*
AVIBUS cum *BeLa*
PRAESENTIRE praescire CONCOLOR alba et nigra *BeLa*

340.14 PRAESAGIA POLLICETUR, HOC EST EXSTANTIS, INSTANTIS ET RAPTI
PRAESAGIA profetiam POLLICETUR promittit *BeLa*
HOC EST EXSTANTIS INSTANTIS ET RAPTI i. presentis futuri praeteriti *BeLaLebPar*, i. *om. LebPar*

895/
340.14-5 DEMUM TRIGARIUM SUPPLICANTIS SEMPER GERMANITATIS ADVENIET
DEMUM deinde *BeLa*
TRIGARIUM quasi una ars tripertita *BeLa*

> Trigarium tres puellae unam potestatem habentes, vel trigarium quasi cultum deorum, eo quod ibi tria considerantur: colitur quia deus est, vel etiam laudatur vel pro peccatis oratur *BeLa*
> tres grates *LebPar*
> IV<, V, VI> *Leb*

SUPPLICANTIS i. orantis *BeLa*

---

[*] Cf. Serv. Com.Verg.A. 3.92, 6.347; Myth.Vat.III 8.5; Lact. Theb. 1.509

ut simul semper sint *Par*

15-6 QUOD VESTRA VULGO SUFFRAGIA POLLICENDO
QUOD s. trigarium *BeLaPar*, s. *om. Par*
VULGO ubique *La*
SUFFRAGIA auxilia *BeLa*

16-7 IN SPEM OPIS SUPERAE PRORITAVIT
OPIS auxilii *BeLa*
SUPERAE PRORITAVIT caelestis provocavit *BeLaLebPar*, caelestis *om. LebPar*
Illae tres puellae suadent hominibus colere deum, quia indigent auxiliis eorum *BeLa*

17 PRAESIDIIQUE INDIGAM IN CULTUM FECIT PROCEDERE NATIONEM
PRAESIDIIQUE adiutorii *BeLa*, aditorii *La*
INDIGAM indigentem *BeLa*
IN CULTUM FECIT s. deorum s. illud trigarium *BeLa*
NATIONEM s. hominum *Be*

18-1 SIVE TRIGEMINAM FEMINAM SIVE TRES IN UNIUS NOMINIS VOCABULUM CONSPIRANTES
Trigarium quasi triformis femina, vel tres puellae sub una forma consistentes *BeLa*
<IV,> V, VI *Leb*
CONSPIRANTES consentientes *BeLa*

2 EXPLODERE i. expellere vel evacuare *BeLa*
expellere *LebPar*
separare *Par*

2 CUM PER EAS IN TERRAS NOS CERTUM SIT DEMEARE?
NOS CERTUM SIT Quia cultura deorum per has invenit<ur> *BeLa*, invenis *Be*
NOS deos *Par*
DEMEARE ad terras transire *BeLa*

896/
341.3 DECENS ILLA SIDEREIS FULGORANSQUE LUMINIBUS
DECENS ILLA SIDEREIS pulchra cum splendidis *BeLa*, pulcra *La*
LUMINIBUS oculis *BeLa*

341.3-4 QUAE EPISTULARIS TUA ET DICTA EST ET PROBATA
EPISTULARIS* Scriptrix, quae futura describit *BeLa*, scribit *La*
scriba tua Cloto et aliae *Leb*
VII *Leb*
Cloto cancellaria *Par*
DICTA annuntiata *BeLa*

341.4-5 HUIC SEMPER, PATER, TUAS ASSERERIS CREDIDISSE MANUBIAS
HUIC s. deae *BeLa*
PATER ó Iuppiter *La*
ASSERERIS pronuntiaris *BeLa*
MANUBIAS i. vestimenta, que secreta occolta significant *La*
vestes i. secreta occulta *Be*
anulum et sigilla *LebPar*, et sigilla] sigillum *Par*

341.5 AC TRISULCAE LUCIS COMMISISSE FULGOREM
TRISULCAE LUCIS fulguris *Leb*
Sic videtur fulgur propter celeritatem *Par*
COMMISISSE dedisse *BeLa*
i. dedisti ei potestatem fulgurum *BeLa*

341.5-6 ADVENTUM EXITUSQUE IGNIVAGAE DENUNTIATIONIS AGNOSCIT
EXITUSQUE terminos *BeLa*
AGNOSCIT Scit unde veniant fulgura, quoque pergant *BeLa*, fulgur *La*

341.6-8 HUIC IGITUR FEMINAE COLLOQUIUM DENEGAMUS, PER QUAM EDICTORUM TUORUM ADMONITUS MORTALIBUS INNOTESCUNT?
HUIC IGITUR numquid *BeLa*
AEDICTORUM praeceptorum *BeLa*
INNOTESCUNT manifestantur *BeLa*, manifestetur *Be*[1]

---

* PISTOLARIS *BeLa*

9 ALACER laetus *BeLa*

10 LATOIUS Apollo *BeLaPar*

11-2 TUNCQUE ALIAS IN ORDINEM CONTINARI
TUNCQUE ALIAS s. quae laudabantur *BeLa*
vii mechanicas *Leb*
†CONTINUARI i. adducere *BeLa*
iungi *Par*

12 HIC LUNA IAM GEMINA EMENSA DIEI PORTIONE COMMONITA
HIC LUNA IAM GEMINA i. post haec adverbium ordinis *BeLa*
inter duas. pluralis *Par*
†MENSI transacti *BeLa*

13 EXPLORARE probare, inquirere *BeLa*

13-4 CETERUM PROPINQUANTE CONFINIQUE NOCTE CONSEQUENTIBUS
CAETERUM sed *BeLa*
†CONFINIO termino *BeLa*
CONSEQUENTIBUS s. feminis *BeLa*
feminis vii mechannicis *Par*

14-5 TEMPORA INPERTIRE NON POSSUM spatia dare *BeLa*

15-6 QUIPPE PLAUSTRI MIHI TAURIQUE AC MUNDANAE DISCURSIONIS ITINER SUBEUNDUM
TAURIQUE Ubi est absis lunae *Leb*
TAURUSQUE In quo absis est *Par*
DECURSIONIS revolutionis *Par*
SUBEUNDUM s. est *Be*

16-7 COMMENTIS STELLIGERI RAPTUS META PERMITTIT
COMMENTIS doctrinis *BeLa*
META terminus *BeLa*

17 ET FATEOR VELLEM s. eam audire *BeLa*

341.17-8 SI QUA EXAMINATIONEM VIRGINUM PROROGARET AUCTORITAS
PROROGARET i. vocaret *BeLa*
AUCTORITAS praeceptum *BeLa*

341.18 IPSA QUOQUE TAM PRAECLUIS ERUDITIONIS ASSERTA COGNOSCERE
IPSA ego luna *BeLa*
TAM PRAECLUIS nobilis *BeLaLeb*, nobiles *Be*
HERUDITIONIS ASSERTA COGNOSCERE aliter praecluis herudi-tionibus *BeLa*
Vel ita: ego, quae sum aliarum doctrinis praecluis, vellem cognoscere asserta tantae heruditionis *BeLa*

341.19-1 PRAESERTIM QUIA COMPERENDINATIO RATIONABILITER EXPECTATUR, NE LASSATA COGNOSCENTIS CURAE FATIGATIONE FASTIDIA OMNEM DOCTAE INTIMATIONIS EXCURSUS GRAVATAE LABORIBUS INTENTIONIS EXCLUDANT
CUMPERENDINATIO crastinatio *BeLa*
aliter breviter vel brevis continuatio *BeLa*
brevitas illarum *LebPar*, illarum *om. Leb*
FATIGATIONE propter fatigationem *BeLa*
LABORIBUS cum *BeLa*

342.1-2 SUBTILITAS IN ODIUM NOSCENDORUM OBTUSA MULTIPLICI PROLIXITATE VERTATUR
OBTUSA illa *BeLa*
PROLIXITATE extensione *BeLa*

342.2-3 AEQUIUS IGITUR DUCO AMPLIANDAM DISCUSSIONEM TANTAM IN ARDENTEM SITIM AURIUM AVIDARUM
AEQUIUS iustius *BeLa*
†VIVIDARUM i. voluptuosarum *BeLa*
desiderantium *Par*

898/
342.3-4 QUAE CUM LUNA DISSERERET, OMNES CERTATIM ASSENSERE DIVI
QUAE superius dicta *BeLa*
OMNES s. dii CERTATIM studiose *BeLa*

4-5 DEHINCQUE UTRUM REPENSATRIX DATA DIESQUE CONFERENDAE DOTIS PROROGARI IURE PUBLICO POSSET INQUIRITUR
†DEINDE Quod dederat Mercurius *BeLa*
UTRUM REPENSATRIX i. Fronesis mater Philologiae *BeLa*, filologiae *Be*
  Utrum Fronesis repensaret diem *Par*
PROROGARI inveniri *BeLa*
  provocari *Leb*
IURE PUBLICO lege generali *BeLa*
  consilio communi *Par*
INQUIRITUR s. a Iove *BeLa*

5-6 QUO DICTO ARCANUS ILLE PRISCI IURIS ASSERTOR MAGNA NEPOTUM OBSECRATIONE CONSULITUR
QUO DICTO s. iure allato *BeLa*
ARCANUS i. Saturnus *BeLaPar*, i. *om. Par*
PRISCI antiqui IURIS legis ASSERTOR affirmator *BeLa*
NEPOTUM filiorum *BeLa*
OBSECRATIONE praece CONSULITUR interrogatur *BeLa*

7-8 RESPONDITQUE REGULARITER ETIAM MATRIMONIO COPULATO DOTEM DICERE FEMINAM VIRO NULLIS LEGIBUS PROHIBERI
REGULARITER iuste *BeLa*
†DYCERE dare, readperere (*sic, lege* reddere?) *Be*
  dare, reddere *La*
VIRO Mercurio *BeLa*
  dativus *Par*

8-9 AC TUNC IUPPITER PERITI PATRIS ERUDITIONE COMMONITUS
PERITI PATRIS HERUDITIONE docti i. Saturni doctrina *BeLa*
COMONITUS instructus doctus *BeLa*

9-10 'HABEO' INQUIT, 'MEA PIGNERA, GRATULANDUM, QUOD FAS ASSERTUR QUICQUID VOS VELLE COGNOVI
HABEO ... GRATULANDUM i. habeo gratias agere *BeLa*
MEA †PIGNORA ó *La*
  vocativus *Par*
FAS lex †ASSERIT docet vel affirmat *BeLa*, vel *om. Be*

342.10-1 NEQUE ENIM FASTUOSA RIGIDUS ELATIONE DEFUGIAM AUT VESTRUM CREBRIUS ACCELERARE CONSILIUM
FASTUOSA cum tumida *BeLa*
RIGIDUS superbus *BeLa*
durus *Leb*
ELATIONE cum inflatione *BeLa*, cum *om. Be*
AUT nec *Be*
†ADCELEBRARE implere *BeLa*

342.12-3 AUT DOCTARUM VIRGINUM EXAMINARE SOLLERTIAM DESIDIS IGNAVIAE DISSIMULATIONE PIGUERIT
EXAMINARE SOLLERTIAM probare disciplinam *BeLa*
DESIDIS IGNAVIAE ociose pigritiae *BeLa*, otiosi *La*
DISSIMULATIONE cum negatione *BeLa*, cum *om. La*
PIGUERIT pro pigebit *BeLa*

342.13 CUM TERRIS INDECENTER EXPULSAS SOLIS OPORTEAT ADHAERERE SIDERIBUS
INDECENTER inhoneste *BeLa*
EXPULSAS s. artes *BeLa*
virgines *Par*
OPORTEAT conveniat *BeLa*
SIDERIBUS i. diis *BeLa*

899/
342.14 NUNC IGITUR PRAECELLENTISSIMAM FEMINARUM HARMONIAM
PRAECELLENTISSIMAM valde celsam *BeLa*

342.15-6 HAEC QUIPPE ET SUPERUM CURAS PRAE CUNCTIS POTERIT PERMULCERE
CURAS i. dolores *BeLa*
PRAE CUNCTIS ante omnes feminas *BeLa*
PERMULCERE i. consolari sedare *BeLa*

342.16 AETHERA CANTIBUS NUMERISQUE LAETIFICANS
AETHERA CANTIBUS i. caelum cum s. suis *BeLa*
†NUMERISVE i. modis *BeLa*

342.17 EXOSA TERRIGENAE STOLIDITATIS IGNAVIAM

EXOSA spernens *BeLa*
STOLIDITATIS IGNAVIAM ineptiae pigritiam stultitiam *BeLa*, pygritiam *La*

18 QUAM MELIORUM INDOCILIS AUGET SINE FINE MORTALITAS
†MELICORUM ignavia sonorum *BeLa*, ignavia *om. Be*
INDOCILIS ignara *BeLa*

18-9 DENIQUE IAM PRIDEM HOMINES DIRUTAQUE GYMNASIA ABSCEDENS ORBE TERRISQUE DAMNAVIT
IAM PRIDEM longe ante *La*
DIRUTAQUE GYMNASIA destructa loca exercitationis *La*
DAMNAVIT sprevit vel deseruit *BeLaLeb*, vel deseruit *om. Leb*

20-1 AC VIX CYLLENIDAE INDAGANTIS EXCURSIBUS NUNC COMPREHENSA POST LONGAE OCCULTATIONIS OBLIVIA DE FUGAE REVOCATUR REDUCITURQUE LATIBULIS
INDAGANTIS EXCURSIBUS requirentis lustrationibus gressibus *BeLa*
†COMPENSA i. numerata inter artes *BeLa*
Numerata i. musica, et est nominativus singularis *Leb*
musica *Par*
LATIBULIS A tenebris, ubi diu latuit *BeLa*

21-3 HANC IGITUR REPERTAM POST SAECULA NUMEROSA ET TANDEM IN USUM MELICUM CARMENQUE RENOVATAM
NUMEROSA multa *BeLa*
ET TANDEM postremum *BeLa*
MELICUM sonorum *BeLa*

23 TAM VOLUPE EST QUAM CONDUCIT AUDIRE
TAM VOLUPAE EST i. tam est voluptas *BeLa*, voluptates *La*
admirative *Par*
CONDUCIT advocat *BeLa*
Mercurius *LebPar*
AUDIRE s. eum *BeLa*

1 CUM RENOVATA LUX FUERIT, INTROMISSAE INTENTIORIS CURAE EXAMINE PROBABUNTUR

RENOVATA LUX i. crastina dies *BeLa*, i.] in *La*
FUERIT INTROMISSAE s. fuerint *BeLa*
EXAMINAE cum inquisitione *BeLa*

901/
343.2-3 HIS ORSIS IOVIALIBUS PHOEBUS ADMONITUS VIRGINEM SIBI PRORSUS ACCOMMODAM ADMISSURUS EGREDITUR
ORSIS coeptis *BeLa*
PRORSUS omnino *BeLa*
ACCOMMODAM aptam *BeLa*
Bene dicit accomodam Phoebo, quia omnis armonia concors est soli *BeLa*
ADMISSURUS suscepturus *BeLa*

343.4 VERUM PAPHIE REMEANTIS FILIAE GRATULATIONE CONCUSSA ANNUIT PUERO
VERUM PAPHIAE sed Venus *BeLa*, sed] *La*[t]
Armoniae *Par*
REMEANTIS intrantis FILIAE Iovis non suae <i. Veneris> *BeLa*
Iovis *Par*
GRATULATIONE CONCUSSA gaudio mota *BeLa*
CONCUSSA irata *LebPar*
PUERO Hymeneo *BeLa*, himeneo *La*

343.5 PRAECINERET NUPTIALE CARMEN
†PROCINERET s. ut *BeLa*
NUPTIALE CARMEN i. epitalamium *BeLa*

343.5-6 ITAQUE HYMENAEUS ALACRI TANDEM VIGORE LUMINATUS
HYMENEUS frater Cupidinis *Leb*
TANDEM ad ultimum *BeLa*

343.6 NEC IPSA IAM RENUENTE TRITONIDE SIC COEPIT
RENUENTE contradicente TRITONIDE Pallade *BeLa*
SIC COEPIT s. canere *BeLa*

902/
343.7-8 AUREA FLAMMIGERUM CUM LUNA SUBEGERIT ORBEM, ROSIS IUGABO LILIA

HEROICUM IUNCTUM CUM DI[A]METRO IAMBICO *BeLa*, herocum *Be*[1]
HEROICUM CONIUNCTUM DIMETRO IAMBICO *LebPar*, heroicum] aeroicum *Leb*
AUREA splendida *BeLa*
†SUBREGERIT* elevaverit vel ostenderit *BeLa*, vel *om. La*
illuminaverit *LebPar*, illuminarit *Par*
IUGABO iungam *BeLa*

9-0 VIRGO DEUSQUE SACRO SOCIABUNT FOEDERA LECTO: FULCRIS PARATE CINNAMA
VIRGO DEUSQUE i. Philologia et Mercurius *BeLa*
SOCIABUNT s. quia *BeLa*
FOEDERA LECTO amicitiam copulae *BeLa*
FOE- longe *Leb*
FULCRIS lectis *BeLa*
PARATAE praePARATAE o Napeae *BeLa*
CINNAMA Amomum cinamomum cinnamum unum est, sicut balsamum xirobalsamum opobalsamum carpobalsamum unum est** *BeLa*, cinamomum] cinnamum *Be*
aramata (*sic, lege* aromata) *Par*

11-2 HESPERUS INTACTAM SERVET LICET USQUE PUELLAM, NUPTAM VIDEBIS, PHOSPHORE
HAESPERUS vesper INTACTAM inviolatam *BeLa*
USQUE donec *BeLa*
PHOSPHORE s. tuo Lucifer vel Venus *BeLa*
lucis sedes *Leb*
Aurora *Par*

13-4 NEC MATRIS LACRIMAE, PRESSI NEC CRINIBUS UNGUES, NEXUS VALEBUNT RUMPERE
NEC MATRIS Fronesis *BeLaLeb*
NEC †VIRIBUS cum *BeLa*
NEXUS i. copulam i. Mercurii et Philologiae *Be*
i. Mercurius (*sic, lege* Mercurii) et Philologiae *La*

---

* SUREGERIT *LaBe*[1]
** Cf. Isid. Etym. 17.8.11, 17.8.14

amicitias *Leb*
RUMPERE separare *BeLa*

343.15-6 NE THALAMOS METUAS; ERIS HOC, QUOD IUNO TONANTI EST, QUAE NUNC SORORE DULCIOR
NE THALAMOS METUAS ó Philologia, verearis *BeLa*
ERIS HOC s. Mercurio *BeLa*
QUOD IUNO TONANTI i. o uxor *BeLa*, o *om. Be*
QUAE s. Iuno *BeLa*
DULCIOR s. est *BeLa*[t]

343.17-8 SI PLACUIT DOCTI SOLLERTIA SACRA MARITI, MAGIS PLACEBUNT OSCULA
SI PLACUIT tibi ó virgo *BeLa*
SOLLERTIA SACRA studium sacrum *BeLa*
MARITI Mercurii *BeLa*
PLACEBUNT tibi *BeLa*
OSCULA illius *BeLa*
Hactenus expectat virginem, sed deinceps iam contemplabitur nuptam *BeLa*

343.19-0 AURORA EXORIENS ROSEIS SPECTABIT OCELLIS, FLORIS RESECTI PRAEMIA
AURORA EXORIENS i. sol *BeLa*
ROSEIS OCELLIS cum pulchris radiis *BeLaPar*, pulcris *La* cum pulchris *om. Par*
RESECTI respersi s. erant *BeLa*
PRAEMIA per s. nuptiarum *BeLa*

343.21-2 IPSE EGO PALLENTEM SPECULABOR LUCE PUELLAM SESE OCCULENTEM VISIBUS
IPSE EGO i. Hymeneus, qui praesum thalamis *BeLa*, talamis *Be*
SPECULABOR videbo cum LUCE *BeLa*
PUELLAM Philologiam *BeLa*
SESE OCCULENTEM occultare volentem *BeLa*
VISIBUS s. hominum *BeLa*

903/
343.23-4 CONSCIA IAM VENERIS NOVA SERTA PARATE, NAPEAE,

CROCUMQUE LECTO ASPARGITE
CONSCIAE gnarae *BeLa*
NOVA SERTA novas coronas *BeLa*
NAPEAE ó nym<phae> saltuum *BeLa*
Oros mons, inde Oreades nymphae. Calchos aes, inde auricalcum* *BeLa*
Musae silvarum vocantur *Leb*
CROCUMQUE LECTO i. super lectum *BeLa*

1-2 INQUE TORUM VIOLAS CERTATIM FUNDITE CONCHIS; PULVINAR ALMUM COMITE
TORUM i. lectum illorum *BeLa*
CERTATIM studiose *BeLa*
CONCHIS cum *BeLa*
vasis *BeLa*, vassis *La*
concha quasi concava *BeLa*
calathis *LebPar*
PULVINAR capitale *BeLa*
COMETIS ornetis. pro comite i. ornate *BeLa*, i. ornate *om. La*

3-4 CRINALE SPICUM PHARETRIS DEPROME, CUPIDO, LIBENS CAPILLUM SOLVERE
CRINALE SPICUM muliebre instrumentum ad ordinandos capillos *BeLa*, ordinando *Be*
spinulam *LebPar*
resdutori c<r>inium *Par*
DEPROME adfer *BeLa*
CUPIDO ó *Be*
LIBENS qui es placens CAPILLUM s. mulierum SOLVERE disiungere *BeLa*

.5-6 FLAMMEA VIRGINEUM QUAE OBNUBERE SUETA PUDOREM, REGINA DEME PRONUBA
VIRGINEUMQUE s. es *BeLa*
OBNUBERE defendere *BeLa*
SUETA solita *BeLa*

* Cf. Serv. Com.Verg.E. 10.9; Isid. Etym. 8.11.97; Myth.Vat.II <64>; Joh.Scot.Er. Exp. 15.748

REGINA et ó *BeLa*
DEME PRONUBA i. solve ipsum pudorem i. Iuno *BeLa*

344.7-8 TE BLANDUM, TENERI QUAE CONSCIA SOLA DOLORIS, VENUS DECENS SOLABITUR
TE BLANDUM apostrofam ad Philologiam *BeLa*, apostropham *La* ad] a *La*
ó Philologia *BeLa*, filologia *Be*
pro blande *BeLaPar*
adverbium *Leb*
TENERI novi *BeLa*
CONSCIA s. est *BeLa*
DOLORIS s. nuptiarum *BeLa*
DECENS pulchra Philologia *BeLa*, pulcra *La* Philologia *om. Be*

344.9-0 QUAE NUNC DURA NOVO NESCIS COMMITTERE AMORI, PARABIT ILLA PECTORA
ILLA PECTORA i. Venus *BeLa*
i. Philologiae et Mercurii, quae <i. pectora> dura fuerunt olim in amore castitatis *BeLa*

344.11-2 TU MODO, QUAE MANDO, OCULOS DEIECTA DECOROS RECONDE MENTIS INTIMO
TU MODO ó Philologia *BeLa*
QUAE MANDO praecipio commendo ego Hymeneus *BeLa*, praecipuo *La*
OCULOS per *Be*
DELECTA valde dilecta *BeLa*
RECONDE retinendum accipe MENTIS cordis *BeLa*, retitinendum *La* cordis] in cordis *Be*
INTIMO i. in secreto *BeLa*

344.13-4 DOCTILOCUM PULCHRIS IUVENEM COMPLEXA LACERTIS FACUNDA REDDE PIGNERA
†CHILIS cum labiis *BeLa*
chylae labia, chaelae bracia *BeLa*, chyloe *La* brachia *La*
IUVENEM Mercurium *BeLa*
COMPLEXA LACERTIS stringens cum brachiis *BeLa*, bracheis *La*
nodosis brachiis *LebPar*, brachiis *om. Par*
REDDE genera vel enitere *BeLa*, vel enitere *om. Be*

†FECUNDA ... PIGNORA multos ... filios *BeLa*

15-6 HIS HYMENAEO DIUTIUS VELUT FESCENNINA QUADAM LICENTIA PERSONANTE [GEMINANTE CREBRIUS], QUAE PLACUISSE RATUS
cum HIS s. versibus *BeLa*
FESCENNINA nuptialia festa *BeLa*, festis *BeLa*[2]
nuptiali *Par*
PERSONANTE canente *BeLa*
GEMINANTE CREBRIUS multiplicante sepissime *BeLa*
QUAE PLACUISSE RATUS s. diis arbitratus est *BeLa*

16-7 MAXIMA CIRCUMSTANTIUM MULTITUDO TRITONIDEN DIONENQUE COMITATA DEGREDITUR
†DIONAEQUE mater Veneris *La*
quasi dia noia i. sensus dilectio *LebPar*, sensus dilectio] delectatio sensus *Par*
COMITATA illa multitudo *La*

17-8 ET AD OS REDUCTAE TANDEM VIRGINIS STUDIO PROPERANTE CONCURRITUR
†NOS Iungit se Martianus cum caeteris, qui extra deorum erant consilium *BeLa*, concilium *La*
VIRGINIS musicae *Leb*
STUDIO cum *Be*
PROPERANTE festinante *BeLa*
CONCURRITUR ab illa multitudine *BeLa*

18-9 VERUM TAM PUELLARUM, QUAE DEAS DOMINAS CONSECUTAE PONE FORTE CONSTITERANT
VERUM sed *Be*
PUELLARUM s. illae *BeLa*
DEAS i. Tritonidam Dianamque *BeLa*, que *om. La*
†DOMINAE i. armoniae *BeLaLebPar*, i. *om. LebPar*
PONE retro *BeLa*

19-0 QUAM HEROUM POPULUS, QUI IUSSUS ADVENERAT, PROPERABAT
HEROUM POPULUS virorum fortium *BeLa*
filosophorum *Par*

PROPERABAT festinabat *BeLa*

344.20-1 ILLAE UT PSALLENDI MATERIEM DISCERENT, HI UT SACRI VULTUS MEMORIAM RECENSERENT
ILLAE s. puellae *BeLaPar*, s. *om. Par*
UT PSALLENDI s. ad hoc properabant *BeLa*, properabat *Be*[1]
†HII i. populus heroum *BeLa*
heroi *Par*
UT SACRI VULTUS armoniae *Leb*
MEMORIAM RECENSERENT Recordarentur vel ad memoriam reducerent *BeLa*, memora *Be*

905/
344.21-2 NEC MORA, ET ECCE QUAEDAM SUAVITAS INTEMPTATA INAUDITAEQUE DULCEDINIS CANTUS INSONUIT
NEC MORA i. sine mora *BeLa*
INTEMPTATA i. ignota *BeLa*
inaudita *LebPar*
AULIDEQUE i. regiae ab aula *BeLa*
fistularis *LebPar*
aulis vel aula fistula *LebPar*, vel aula *om. Par*
DULCEDINIS CANTUS s. quidam *BeLa*

344.22-2 AC MELODIAE ULTRA CUNCTA RERUM OBLECTAMINA RECINENTES AUDITUM MIRANTIUM COMPLEVERE DIVUM
MELODIAE suavissimae cantilenae *BeLa*
OBLECTAMINA delectamenta *BeLa*
DIVUM deorum *BeLa*

345.2-3 NON ENIM SIMPLEX QUIDAM ET UNIUS MATERIAE TINNITIBUS MODULATUS
NON ENIM SIMPLEX s. fuit *BeLa*
MATERIAE i. tropicum *BeLa*
MODULATUS cantus *BeLa*

345.3-4 QUANDAM PLENITUDINEM CUNCTICINAE VOLUPTATIS ADMISIT
QUANDAM omnem *BeLa*
CUNCTICINAE cunctae canentis *Be*
cuncti canentes *La*

4-5 QUO CANORE DIUTIUS CIRUCUMSTANTIUM PECTORA DEORUMQUE MULCENTE
CANORE cantu *BeLa*
CIRCUMSTANTIUM s. hominum *BeLa*
MULCENTE blandiente *BeLa*

6-7 AC TANTI COMITATUS PRAEAMBULAE REVISUNTUR
COMITATUS pluralis nominativus *Leb*
PRAEAMBULARE praecedere *BeLa*

7 SED NON CASSAE DULCEDINIS NEC SINE ALIQUO SONORI MODULAMINIS ARGUMENTO
CASSAE vacue *BeLa*
superfluae *Leb*
DULCEDINIS Quia omnes, qui armoniam praecesserunt, cum magna dulcedine reversi sunt *BeLa*, reversi] versi *La*
NEC SINE ALIQUO sed cum aliquo *BeLa*

8-9 NAM ERATINE, CYPRIDIS FILIA, ET HIMEROS, CUPIDINIS ASSECUTOR, ITEMQUE TERPSIS E FAMULITIO DIONAE
†ERATIME Ero amo. Callimera, bona dies *LebPar*, kallimera *Par*
Imeron serenitas. Terpso de<le>cto. Terpsice delectans *Leb*
Amabilis interpretatur *Par*
IMEROS frater Eratime *BeLa*
serena *Leb*
serenus interpretatur *Par*
ASSECUTOR i. imitator *BeLa*, i. *om. La*
†TERPSICE delectans *Par*
DIONAE i. Veneris *BeLa*

9-10 CONCINENTES GRATISSIME INTRAVERE PRIMUM
CONCINENTES simul canentes *BeLa*
PRIMUM ante omnes *BeLa*

10 SED PUER MONAULETER SONABAT
PUER Hymeneus *BeLa*
Cupido *LebPar*
†MONALITER singulariter vel prae cunctis *BeLa*
vel monauliter i. cum una fistula *LebPar*, vel monauliter i. *om. Par*

906/
345.11-2 POST HOS PSALLENTES PITHO, VOLUPTAS ET GRATIAE ADMIXTIS LYRAE VOCIBUS ATQUE IPSAE HARMONICIS DISSULTANTES MOTIBUS ADVENERE
PSALLENTES s. erant *BeLa*
PITO philosophus *Leb*
quidam musicus *Par*
GRATIAE Tres virgines uno utentes oculo* *BeLa*
ADMIXTIS adiunctis *BeLa*, adiuctis *La*
IPSAE s. gratia et voluptas *BeLa*
ARMONICIS cum *Be*

345.12-3 DEXTRA LAEVAQUE INTEREA PRAECEDEBAT NUMERUS HEROUM CRINITORUMQUE SAPIENTUM
DEXTRA LEVAQUE i. ex utraque parte armoniae *BeLa*
PRAECEDEBAT s. armoniam *BeLa*
HEROUM virorum fortium *BeLa*
CRINITORUMQUE SAPIENTUM intonsorum philosophorum *BeLa*, philosoforum *La*

345.14 MURMURANTES vel pro murmurabant s. erant *BeLa*, murabant *La*

345.15-6 SED ALII LAUDES DEORUM HYMNOSQUE QUAMPLURIMI, ALII MUSICOS TONOS, QUOS MODO COMPERERANT, RETEXEBANT
QUAMPLURIMI multi s. consonabant *BeLa*
MUSICOS TONOS Quia spiritus armoniae illos repleverat *BeLa*, illo *Be*[1] repleverant *Be*
RETEXEBANT recinebant *BeLa*

345.17 VERUM PER MEDIUM QUIDAM AGRESTES CANORIQUE SEMIDEI, QUORUM HIRCIPEDEM PANDURA
†ACRI ingeniosi CANORIQUE striduli *BeLa*
SEMIDII s. egrediuntur vel visuntur *BeLa*
QUORUM deorum *BeLa*
HYRCIPEDEM quasi hirci Pan *BeLa*
Unus deorum qui setosos dicitur habere pedes *BeLa*

---

* Cf. myth of the Gorgones: Serv. Com.Verg.A. 6.289; Myth.Vat.I <127>; Myth.Vat.II <135>

Pana *Leb*
i. Pan habens hircinos pedes *Par*
PANDURA s. decuit *BeLa*
instrumentum musicum s. decuit *BeLa*
Genus est organi vel capsa *Leb*
genus organi *Par*

.17-8 SILVANUM HARUNDINIS ENODIS FISTULAS SIBILATRIX, RURESTRIS FAUNUM TIBIA DECUERUNT
SILVANUM apud Aegiptios *BeLa*
ENODIS sine nodis *BeLa*
FISTULA SIBILATRIX vel bilatrix i. bis latrans stridula *BeLa*
Declinatur hic et haec bilatrix et hoc bilatre *BeLa*
FAUNUM et decuit *Be*
DECUERUNT i. ornaverunt *Be*

.18-9 VERUM SEQUENS HEROUM PRAECLUI ENITUIT ADMIRATIONE CONVENTUS
VERUM sed *Be*
PRAECLUE ENITUIT i. nobiliter refulsit *BeLa*
nobili *Par*

.19-1 NAM ORPHEUS, AMPHION, ARIONQUE DOCTISSIMI AURATA OMNES TESTUDINE CONSONANTES FLEXANIMUM PARITER EDIDERE CONCENTUM
ORPHEUS orios phone, pulchra vox *Leb*
Orpheus dicitur quasi orios phau id est pulcra vox *Par*
AMPHION ambitus *Par*
ARION virtus *Par*
TESTUDINE cithara *BeLa*, citara *La*
lyra *LebPar*, lira *Par*
FLEXANIMUM i. dulcem *BeLa*
testudine *Leb*

.22-3 NAM THRAX QUO DURI RUMPERE REGNA EREBI QUOQUE SUAM MERUIT IMMEMOR EURYDICEN
ELEGIACUM CARMEN PER SE POSITUM *BeLa*
ELEGIACUM SIMPLEX NULLO AEROICO INTERPOSITO

Leb

TRAX i. Orpheus de Tracea[m] *BeLa*, Traceam] tractio *La*

Orpheus *LebPar*

s. edidit *Par*

QUO aliquando *Be*

RUMPERE pro rumpebat REGNA HEREBI claustra inferni *BeLa*

QUOQUE s. melo *BeLa*

MERUIT s. vocare *BeLa*, s. *om. Be*

INMEMOR s. ab inferis *BeLa*

promissionis *BeLa*

obliviosus *Par*

EURIDICEN se s. Dianae *BeLa*, se s.] res *La*

coniugem *BeLa*

Euridicen profunda intentio, et intelligitur musikη *Par*

profunda intentio *Leb*

346.1-2 QUO CANTU STUPIDAE TIGRIDIS IRA RUIT,
QUO FERTUR RABIDAS PERDOMUISSE FERAS

†TUPIDAE in admirationem versae *BeLa*

TIGRIDES Quodam tempore voluerunt illum devorare tigrides, sed eas cecidit: suis mitificavit carminibus *BeLa*

QUO FERTUR melo dicitur *BeLa*

†RAPIDAS veloces PERDOMUISSE valde mitigasse FERAS bestias *BeLa*, mitigasse] mitigastigasse *La*

346.3-4 QUO VIDIT RIGIDAS GLANDIBUS IRE COMAS
ISMAROS ET SILVAS CURRERE MONTE SUAS

RIGIDAS duras *BeLa*

GLANDIBUS cum *BeLa*[t]

vel cladibus i. ramis *Par*

†CLADIBUS Cladis Graece ramus dicitur *Leb*

IRE movere s. quercuum *BeLa*

†ISTMARUS mons Traciae *BeLa*

SILVAS stultos *Par*

CURRERE ad Orpheum *BeLa*, orpheus *La*

Voluit currere ut potuisset Orphei cantum audire *BeLa*

346.5-6 CARMINE QUO STRYMON CONTINUIT LATICES
ET TANAIS VERSIS SAEPE RELATUS AQUIS

STRIMON fluvius Traciae* *BeLa*
palus Tracie* *LebPar*, tracie *om. Leb*
LATICES suas aquas *BeLa*, suas] re *Be*
†ITTANAIS fluvius Scittiae** *BeLa*, Scittiae] traciae *Be*[1] scithiae *Be*[2]
†SEPERFLATUS AQUIS reductus *BeLa*
i. retrorsum conversus est *Be*
i. concussus est retrorsum *La*
reflexis *Par*

7-8 QUO IMPUNE ACCUBUIT RICTIBUS AGNA LUPI
ET LEPUS IMMITI CONTULIT ORA CANI
QUO cantu *La*
INPUNE ACCUBUIT Sine sui poena iacuit *BeLa*
RICTIBUS aperto ori *BeLa*
ET et hoc cantu *BeLa*
CONTULIT ORA Osculum ei prebuit *BeLa*
oscula *Leb*

.9-0 HOC NUNC PERMULSIT INSONUITQUE MELO
ACCUMULANSQUE MAGIS CARMINA SACRA IOVI <...>
HOC cum *BeLa*[1]
PERMULSIT placavit *La*
MELO sono *BeLa*
ACCUMULANSQUE augens *BeLa*
HACTENUS ORPHEUS *BeLa*

.11-2 AMPHION RIGIDO IN CORPORE MENTIS ITER
QUOQUE DEDIT MONTES SUMERE POSSE ANIMAS
AMPHION i. circularis *LebPar*
RIGIDO duro *BeLa*
mortuo *Par*
QUOQUE cantu *La*
MONTES stultos *Par*

.13-4 QUO SENSUS DURIS CAUTIBUS INSERUIT

---

* Cf. Serv. Com.Verg.A. 10.265, 11.580; Serv. Com.Verg.G. I.120, 4.507
** Cf. Serv. Com.Verg.A. 11.659; Serv. Com.Verg.G. 4.516

ET DOCUIT RUPES CARMINA VELLE SEQUI
SENSUS carmina DURIS CAUTIBUS in petris *BeLa*, carmine *Be*
RUPES petras *BeLa*

346.15-6 NAM MUROS THEBIS DULCISONIS FIDIBUS
PRAESIDIUMQUE DEDIT CARMINIS ARBITRIO
THEBIS De Thebis, quae est in Aegipto, dicitur; quae ne ab hostibus expugnaretur, sua cantilena obtinuit. Habuit autem centum portas *BeLa*, quae est] quae est quae est *La*
civitas Egipti *Par*
†DULCIS SONIS FIDIBUS cum *BeLa*
PRAESIDIUMQUE munitionem *BeLa*
bis *Par*
CARMINIS ARBITRIO cum *Be*
HUCUSQUE AMPHION *BeLa*

346.17-8 SED NEC ARIONIAM MARMORA SURDA CHELYN
TEMPSERE, EXTREMAM CUM FLAGITARET OPEM
ARIONIAM fortis dicitur *Leb*
fortem interpretatur *Par*
MARMORA i. maria *BeLaPar*, i. *om. Par*
SURDA †CHELIM i. liram *BeLaLebPar*, i. *om. LebPar* lyram *Leb*
vel sonantem *La*
conTEMPSERE EXTREMAM ultimum *BeLa*, ultima *La*
FLAGITARET OPEM exposceret auxilium *BeLa*

346.19-0 ET LICET INDOMITI DIRA PROCELLA NOTI
SPUMEA SCYLLAEI VERRERET IMA MARIS
INDOMITI inmitis *BeLa*
PROCELLA NOTI seviens tempestas venti *BeLa*
†SPUMEAS CINEI* Cineus rex Traciae, a quo mare Cineum *BeLa*, Cineus] scineus *Be* trachiae *Be*
Indici *Par*
VERRERET IMA MARIS Ab imo fundo verteret *BeLa*

346.21-2 DELPHINAS TOTIS SOLLICITAVIT AQUIS,
ET MELICOS CANTUS BELUA GRATA TULIT

---

* SPUMEA SCINEI *Be*

DELFINAS lupi marini *BeLa*
TOTIS SOLLICITAVIT in *BeLa*
  Sollicitas fecit sive ad se vocavit *BeLa*, vocavit] convocavit *La*
GRATA TULIT accepit sive sustinuit *BeLa*
EATENUS ARION *BeLa*

.23-4 O VERE ANTISTANS NUMINA MAGNA DEUM
QUAE ISTORUM LAUDES PROTULIT HARMONIA
VERBA POETAE IN LAUDE ARMONIAE *BeLa*
ANTISTANS superans *BeLa*
  pro attestans *LebPar*, pro *om. Par* adtestans *Par*
per NUMINA *Be*
DEUM deorum *BeLa*
QUAE s. armonia ISTORUM s. hominum *BeLa*

.1-2 QUAE DOMITARE EREBUM, MARMORA, SAXA, FERAS
ET POTUIT RUPES SENSIFICARE TONIS
DOMITARE HEREBUM ut Orpheus *BeLa*
MARMORA ut Arion *BeLa*
ET RUPES ut Amphion *BeLa*
cum TONIS *BeLa*[t]

.3-4 POST HOS HONORATIOR FONTIGENARUM VIRGINUM CHORUS
PEGASEAE VOCIS NECTARE DIFFLUEBAT
HOS cantores vel philosofos *BeLa*
  tres *LebPar*
FONTIGENARUM nympharum *BeLa*, nymfarum *La*
PEGASE Pege fons, Pegasus fontalis, Pegasi equi cornuti alas habentes *Leb*
  Pegasus dicitur equus cornutus alas habens, ad cuius similitudinem fons erumpit, unde hic fontalis intellectus dicitur[*] *Par*
NECTARE cum dulcedine *BeLa*, dulcedine] dulcem *La*
DIFFLUEBAT personabat *BeLa*

.4-5 QUI PHRYGII CUIUSDAM BUPAEDAE CICUTIS GEMINATIS

---

[*] Cf. Serv. Com.Verg.A. 2.616; Fulg. Myt.I.62; Myth.Vat.I < 127 >; Myth.Vat.II < 136 >

INTERSTINCTUS
QUI PHRIGII s. chorus *BeLa*, corus *La*
Pitagorae *LebPar*
†PUBIDAE grandevi vel senis *BeLa*, seni *La*
Pitagoram dicit, qui musicam invenit *BeLa*
iuvenis adulti *LebPar*, iuvenis *om. Leb*
pubida: pube videns* *Par*
CICUTIS GEMINATIS i. fistulis repetitis *LebPar*, i. *om. Par* fistulis *om. Leb*
INTERSTINCTUS distinctus vel variatus *BeLa*

347.5 OMNES PRAECEDENTIUM SUAVITATUM DULCEDINES ANTEIBAT
ANTEIBAT superabat *BeLa*

347.5-6 TANDEM INTER PHOEBUM PALLADEMQUE MEDIA ARMONIA SUBLIMIS INGREDITUR
ARMONIA SUBLIMIS Quod de aliis non legitur *Par*

909/
347.7 CUIUS SONORUM CAPUT AURI CORUSCANTIS BRATTEIS COMEBATUR
SONORUM stridulum *BeLa*
CORUSCANTIS splendentis *BeLa*
†BLATHEIS de, vel bratteis *Be*
coloribus *BeLa*
COMEATUR ornatur vel ambitur *BeLa*

347.7-8 CAESO ETIAM TENUATOQUE METALLO RIGENS VESTIS
CESO de *Be*
CESO inciso vel sculpto, producto *BeLa*, scupto *Be*
TENUATOQUE in fila ducto *LebPar*, ducto in filum *Par*

347.8-9 ET OMNIBUS AD MOTUM GRESSUMQUE RATA CONGRUENTIA TEMPERATUM
OMNIBUS in *BeLa*
AD MOTUM ut moveri *BeLa*

---

* The commentator possibly tries to give an ethymological explanation: pubida < pube video, pube videns.

Ut gradi posse videretur *BeLa*
RATA CONGRUENTIA cum rationabili convenientia *BeLa*
TEMPERATUM illud metallum *BeLa*

.9 BLANDIS LENITER CREPITACULIS TINNIEBAT
BLANDIS CREPITACULIS cum *Be*
sonitibus i. sonabat *BeLa*
sonis vestis *Leb*

.9-10 CUIUS MATER PAPHIE, UT EAM CONTIGUE SEQUEBATUR
CUIUS i. armoniae *BeLa*
CONTIGUE e vicino *BeLa*

.11-2 DEXTRA AUTEM QUODDAM GYRIS MULTIPLICIBUS CIRCULATUM
DEXTRA in *BeLa*
QUODDAM musica celestis *Par*
GYRIS MULTIPLICIBUS cum *BeLa*
Copiam verborum subintellege *BeLa*
propter septem planetas et circulos *LebPar*, et circulos *om. Par*

.12-3 ET MIRIS DUCTIBUS INTERTEXTUM VELUT CLIPEUM GESTITABAT
MIRIS DUCTIBUS de *BeLa*[t]
mirabilibus tropis *BeLa*, trophis *La*
VELUT CLYPEUM GESTITABAT musica caelestis *Leb*

.13-4 QUOD QUIDEM SUIS INVICEM COMPLEXIBUS MODULATUM
QUOD clipeum *BeLa*
†COMPLEXIONIBUS coniunctionibus *BeLa*
propter commissuram absidum *LebPar*, commissuras *Par*

.14 EX ILLIS FIDIBUS CIRCULATIS OMNIUM MODORUM CONCINENTIAM PERSONABAT
OMNIUM MODORUM quindecim troporum *LebPar*
CONCINNENTIAM convenientiam *BeLa*
PERSONABAT clypeum *Leb*

.15-6 LAEVA AUTEM VIRGINIS QUAMPLURES EX AURO ASSIMULATAE PARVAEQUE EFFIGIES THEATRALIUM VOLUPTATUM RELIGATAE AEQUE PENDEBANT

LAEVA in *Be*
terrestris musica *Leb*
de musica terrestri *Par*
ASSIMULATAE PARVAEQUE s. formatae subtiles *BeLa*
illi clipeo *Par*
EFFIGIES figurae *BeLa*
TEATRALIUM De tropis dicit *BeLa*
RELIGATAE iunctae *BeLa*

910/
347.16-7 VERUM ILLE ORBIS NON CHELYS NEC BARBITON NEC TETRACHORDON APPAREBAT, SED IGNOTA ROTUNDITAS OMNIUM MELODIAS TRANSCENDERAT ORGANORUM
ORBIS circuitus *BeLa*
caelestis *Leb*
NON CHELIS non ut lira *BeLa*
nominativus *Leb*
NEC BARBITON instrumentum musicum *BeLa*, musicum] barbiton *La*
cithara vel fistula *LebPar*, vel fistula *om. Leb*
TETRACORDON fidula *LebPar*
IGNOTA incognita *BeLa*

347.18-9 DENIQUE MOX INGRESSA ATQUE EIUSDEM ORBIS SONUERE CONCENTUS
INGRESSA ut s. est *Be*
ORBIS clypei *BeLa*, clipei *La*
CONCENTUS multarum vocum coadunatio *BeLa*

347.19-1 CUNCTA ILLA, QUAE DISSONA SUAVITAS COMMENDARAT, VELUT MUTESCENTIA TACUERUNT, IPSEQUE TUNC IUPPITER CAELESTESQUE DIVI SUPERIORIS MELODIAE AGNITA GRANDITATE
COMMENDARAT Accomòda reddiderat *BeLa*
VELUT MUTESCENTIA TACUERUNT Antiqui dixit (*sic, lege* dixerunt?) *BeLa*
IUPPITER s. surrexit *BeLa*

347.21-2 QUAE IN HONOREM CUIUSDAM IGNIS ARCANI AC FLAMMAE INSOPIBILIS FUNDEBATUR
IGNIS ARCHANI celestis *BeLa*

INSOPIBILIS semper viventis *BeLa*
inex<t>inguibilis *Leb*
aeterne *Par*

.22-2 REVERITI INTIMUM PATRIUMQUE CARMEN PAULULUM IN VENERATIONEM EXTRAMUNDANAE OMNES INTELLIGENTIAE SURREXERUNT
†PATRIMUMQUE i. patris hoc est mundi, quia cum armonia volvitur *BeLa*
Quia caelestis musica a Saturno incipit *Leb*
Quia a Saturno incipit *Par*
EXTRAMUNDANAE Quia ultra mundum est armonia *BeLa*, armoniae *La*
extremae *Leb*

.2-3 TUNC EGERSIMON INEFFABILE VIRGO CONCLUDENS AD IOVEM
EGERSIMON proprium nomen tropi *BeLa*
Tropus subtilissimus, qui est ultra omnes modos, i. modulatio *BeLa*
Carmen ineffabile adultum elevatum. Aegero, exalto *Leb*
altum carmen *Par*
Egero exalto, egersimon musica celestis, quod superemineat omnia et omnem sensum *Par*
†INFABILE i. ineffabile *Be*

.3-4 REVERSA ALIIS MODULIS NUMERISQUE VOCE ETIAM ASSOCIATA SIC COEPIT
cum †ALIS et MODULIS *Be*
SIC CEPIT s. loqui *BeLa*

.5-6 TE NUNC ASTRISONO CARMINE, IUPPITER,
QUO GEMMATA POLI VOLVERE SIDERA
METRUM ASCLEPIADEUM RECIPIT SPONDEUM CORIAMBOS DUOS ET PIRRICHIUM *LebPar*, metrum *om.* *Par*
TE NUNC spondeum ASTRISONUM coriambus CARMINE IUP- s. tali. coriambus -PITER pirrichius *Par*
IUPPITER ó *Be*
QUO carmine *BeLa*, carmina *La*

GEMMATA lucida *Be*
fulgida *La*
VOLVERE movere *BeLa*

348.7-8 SUEVIT LEGE RATA SACRA RECURSIO,
PRAEFANDUM VENEROR, QUIPPE POTISSIMUS
SUEVIT pro consuevit *BeLa*
RATA firma *BeLa*
SACRA RECURSIO ab occasu ad ortum *BeLa*, ab] ad *La*
revolutio mundi *Par*
PRAEFANDUM laudandum ante omnes deos *BeLa*
Qui praefandus est in omnibus *Par*
QUIPPE POTISSIMUS i. maximus genitor s. es *Be*
niteor (*sic*, *lege* genitor?) i. maximus *La*

348.9-0 NECTIS SCEPTRIFERO SUB DIADEMATE
OMNIGENUM GENITOR REGNA MOVENS DEUM
†NEC DI SCEPTRIFERO s. es, reguli *BeLa*
DIADEMATE corona *BeLa*
GENITOR ó *Be*
MOVENS volvens *BeLa*
DEUM deorum *BeLa*
Quia quamvis mundus moveatur, regna tamen non moventur, ut inferiora fiant superiora aut excelsa infima. Tamen mundus <x>xiiii horis volvitur *BeLa*

348.11-2 MUNDUM PERPETUO DUM ROTAT AMBITU
MENS, QUAM SIDEREO SUFFICIS IMPETE
ROTAT vertit AMBITU recursu *BeLa*
ROTAT pro rotatur *LebPar*
MENS i. spiritus *BeLa*
†SUBICIS supponis i. recursu *BeLa*, subponis *La*
iungis *Leb*

912/
348.13-4 TE NAM FLAMMIGERI SEMINA FOMITIS
SPARGENTEM REFERUNT ASTRA MICANTIA
TE NAM i. Iovem vel mundum *BeLa*
FOMITIS solis *BeLa*

stellarum siderum *Par*
SPARGENTEM i. revocantem ab inferioribus ad superiora, quia mundus Iovis dicitur *BeLa*
REFERUNT laudant *LebPar*
ASTRA MICANTIA splendentia *Be*
ab astris potata *Par*

15-6 TE PHOEBEA SACRO MUNERE LUMINA,
TERRIS PURPUREUM DUM RENOVANT IUBAR
TE †FOVEANT i. Iovem i. mundum *BeLa*
vel potata s. referunt *Leb*
SACRO MUNERE cum luce *BeLa*
†FLUMINA i. maria *BeLa*, i. *om. La*
Vel fulmina i. astra, sidera. Hoc tamen ficte propter solem, qui dicitur mari se inmergere *BeLa*
nominativus pluralis *Leb*
PURPUREUM rubicundum *BeLa*
RENOVANT IUBAR solem *BeLa*

17-8 TESTATA AMBROSIUM SPLENDIFICANT DIEM;
CYNTHIA NOCTIS HONOS LAMPADE MENSTRUA
TESTATA i. perhibentia *BeLa*, peribentia *La*
astra *Leb*
AMBROSIUM i. odoriferum *BeLa*, i. *om. La*
Ambrosium dicit odoriferum amenum s. clarum ab ipsa herba, quae cibus deorum est *BeLa*
pulchrum *Leb*
SPLENDIFICANT apparere faciunt *BeLa*
NOCTIS HONOS i. ipsa luna *BeLa*
luna *Leb*
LAMPADE MENSTRUA luce. cum *BeLa*

19-0 AURATIS RUBUIT PRAEVIA CORNIBUS;
SUB TE PLAUSTRILUCIS LUMINAT IGNIBUS
AURATIS cum *Be*
†CREVIT vidit *BeLa*
†PRAEMIA CORNIBUS i. habens cornua aurea *BeLa*, habent *La*
Fabulam videtur tangere, quia Diana ipsa est luna, quae causa virginitatis premium promeruit luminis *BeLa*

SUB TE i. sub tua potestate *BeLa*
PLAUSTRILUCAS s. stellas *BeLa*
  in plaustro lucentes *La*
  Eo quod in plaustro lucent *BeLa*
inLUMINAT cum IGNIBUS *BeLa*[1]

348.21-2 ANGUIS PARRHASIAS DISICIENS FERAS
SIC SOLIDI TENERUM CORPORIS AMBITUM
†PARNASIAS Nam in Parnaso habitaverunt poetae, qui finxerunt ursas in caelo esse *BeLa*
  Ubi habundat ursas *Par*
†DIFFICIENS URSAS separans duas ursas *BeLa*, separas *La*[1]
  dividens ursas *Leb*
SOLIDI crassi TENERUM solubile *BeLa*
SOLIDI terrae TENERUM aeris *Leb*

349.2-3 TELLUS NON PROHIBENS AXIBUS INDITUR
ALTERNISQUE REGIT ET REGITUR POLIS
AXIBUS ab *Be*
INDITUR circumplectitur *BeLa*
  inferitur *Par*
REGIT monstrat *BeLa*
  Regit tellus, quia dividit inter eos; regitur, quia circumscribitur *BeLa*
  tellus *Leb*
POLIS circulis *BeLa*

349.4-5 SIC NEREA FRETI NOSCERE LIMITEM
SICQUE IGNEM SUPERUM LAMBERE PABULA
SIC i. tua potestate *Leb*
NEREA <ani>malibus marinis *Be*
  marina animalia dea maris *La*
  Duobus nominibus Grecis dicuntur animalia *Leb*
  animalia *Par*
FRETI absolute i. tua potestate *Par*
NOSCERE pro noscit, intellegit *BeLa*
LIMITEM terminum *BeLa*
SICQUE taliter *BeLa*
IGNEM per *Be*

LAMBERE pro lambit PABULA terram *BeLa*
PABULA accusativus est. aquarum *LebPar*, est. aquarum *om. Par*

6-7 UT NULLIS SCATEANT DISSONA LITIBUS,
ATQUE ITA PERPES AMENT DISSITA VINCULUM
†NULLI s. <h>omines *La*
SCATEANT DISSONALIBUS emergant a dissimilibus *BeLa*, a *om. La*
i. separentur *Leb*
PERPES AMENT i. transeant de zona ad zonam propter vim fervoris *BeLa*
DISSITA separata s. illa dissonalia *BeLa*, s. *om. La*
disiuncta *LebPar*

8-9 UT SEMPER METUANT FOEDIFRAGUM CHAOS.
TU RECTOR SUPERUM, TU PATER OPTIMUS,
FOEDIFRAGUM CAHOS informem materiem *BeLa*, informe materiae *La*
Quia timent in informem materiem decidere. Sicut Origenis dicit, quod mundus totus destruetur et iterum ex nichilo formabitur, mentitur *BeLa*
Mortem. Caus quattuor significat, informem materiam, mortem, hi<a>tus ter<r>ae, nichil *Leb*
Fedus frangens. Caus significat informem materiem, vel mortem, vel hiatum terrae vel nihil *Par*
Hactenus armonia caelestis *Leb*
TU s. qui es *Be*, qui] *add.* Be^2
s. que *La*

10-1 COMPLEXUQUE PIO SIDERA COLLIGANS
COMPLEXUQUE societate *BeLa*
NATOS PERPETUO CORPORE VIVIDAS
NATOS deos cum *Be*
i. omnes deos qui filii fuerunt Iovis *Be, La (in margine)*; *La (inter lineas):* i. †homines deos qui filii fuerunt Iovi
PERPETUO CORPORE Ut convertantur in deos *Par*
VIVIDAS firmas *BeLa*
verbum est, vel vividos *LebPar*, verbum est *om. Par*

12-3 SALVE, NOSTRA CUI PERFICITUR CHELYS
BIS PLENUM OMNISONA CUI RECINUNT MELA

†SOLVE NOSTRA i. redde s. debita *Be*
i. debita *La*
Solve nostra debita vel vincula, ut possimus ad nuptias festinare *BeLa*
Ne sint semper in corpore, solve in deos *Leb*
CUI in cuius honore armonia *BeLa*
BIS PLENUM OMNISONA i. diapason totum videlicet *BeLa*, totum videlicet] videlicet totum *La*
bis diapason *Leb*

913/
349.14-5 IAM VOS VERENDA QAESO CAELI GERMINA,
QUAE MULTIFORME SCIT CIERE BARBITON
IAMBICUS SENARIUS I. TRIMETRUM *BeLa*
VERBA ARMONIAE CONTINUANTUR PER TOTUM CARMEN *BeLa*
EXAMETRUM SPONDEACUM *LebPar*, exameter *Par*
Nunc convertitur ad filios Iovis et deorum *Par*
CAELI GERMINA filii Iovis *Leb*
QUE MULTIFORMAE germina *BeLa*
CIERE vocare *BeLa*
BARBITON instrumentum musicum *La*

349.16-7 AFFATA NOSTRIS FERTE CORDA CANTIBUS,
MISCILLA SACRAE DUM GERUNTUR CURIAE
AFFATA consentanea *BeLa*
consentientia *LebPar*
NOSTRIS i. nobis *BeLa*
MISCILLA generalia carmina vel confusa *BeLa*, generalia] genalia *La*
dissona consilia *Leb*
Mixta consilia: quidam volunt nuptias, quidam non *Par*
Aliqui deorum diligunt castitatem, alii nuptias *Leb*
†FERUNTUR cantantur *BeLa*

349.18-9 MULCERE VESTRUM QUAE VELIT CONSORTIUM.
IAM VOS VICISSIM PROQUE LEGE NUMINUM
MULCERE blandiri *BeLa*
QUAE miscilla *BeLa*
†VELLENT ad suam voluntatem *Par*

CONSORTIUM iugalitatem *BeLa*
VICISSIM alternatim *BeLa*
PROQUE LEGE i. secundum legem *BeLa*, i. *om. La*
unumquemque *Leb*

20-1 POSTHAC SONABO DISGREGATO PLASMATE,
SUISQUE CUNCTOS ALLUBESCENTES TONIS
POSTHAC s. miscilla *BeLa*
SONABO laudabo *Leb*
DIGREGATO PLASMATE cum disiuncto vel separato carmine *BeLa*
i. factura carminis *Be*
segregato carmine *LebPar*, segregato *om. Par*
plasma *Leb*
CUNCTOS ADLUBESCENTES s. deos placentes *BeLa*
Adlubesco quidam a libeo, quidam †ab illa via† dicunt nasci *BeLa*
TONIS carminibus *BeLa*
quibus tonis differate *Par*

22-3 DEDUCET, †URGET ATQUE CLEBIT LOCIS,†
STIMULOSQUE RURSUM LENE PERMULCET MELOS
DEDUCET movebit *BeLa*
URGUET compellet *BeLa*
CLEBIT Segregabit i. de suis locis *BeLa*, i. *om. La*
Cleo abscondo vel segrego vel discerno vel etiam nutrio, unde cliens *BeLa*
STIMULOSQUE commotos vel pro stimulatos s. deos *BeLa*
amoris *Leb*
LENE PERMULCET MELOS Leniter placat s. nostrum *BeLa*
PERMULCET Mitigabit: praesens pro futuro *Leb*

24-1 NUNC IGITUR ALMA QUAE SENATUM LUMINA
DEUM VERENDO CULMINATIS VERTICE
NUNC IGITUR s. ferte corda vel quaeso *Leb*
s. vos queso *Par*
LUMINA sol et luna *BeLa*
VERENDO cum reverentia *BeLa*
CULMINATIS erigitis extollitis *BeLa*

ordinatis *Leb*
ornatis *Par*

350.2-3 BIS SENA QUAMVIS VOS ETRUSCI NUMINA
RITUS FREQUENTENT ATQUE OPIMENT VICTIMIS
VOS ETRUSCI Quamvis Etrusci numina bis sena colant, nos tamen plurima veneramur *BeLa*
NUMINA s. colant *BeLa*
RITUS FREQUENTENT i. cultus multiplicent *BeLa*, i. *om. La*
ATQUE OPIMENT honorificent sive augeant *BeLa*, onorificent *La*
sive honorifice augeant *BeLa*
VICTIMIS de *Be*
sacrificiis *BeLa*

350.4-5 TAMEN GEMILLIS QUOS REFULGENS CURSIBUS
NATURA HONORO PROTULIT FASTIGIO
GEMELLIS in *Be*
diurni et nocturni *BeLa*
Quia ex diis et hominibus nascuntur dii *Leb*
Tam de hominibus quam de diis *Par*
REFULGENS CURSIBUS s. est *Be*
ortibus *BeLa*
NATURA HONORO cum vel *Be*
i. mater deorum in celesti sede *BeLa*
sublimitate honorabili *BeLa*, honorabili *om. La*

350.6-7 PHOEBEA UTERQUE PRAEFERENS VOCABULA,
PALLAS CORUSCA AC DECENS LATOIE
PHOEBEA diurna *BeLa*
lumina *Par*
†UTERO in secreto *BeLa*
†PRAEFERES VOCABULA preponens potestatem diurnorum *BeLa*
PALLAS s. vocaberis tu *BeLa*, tu *om. Be*
LATOIE Apollo *BeLa*

350.8-9 (LUNARE QUIPPE HAC VENIT CONSORTIUM:
PHAETHONTOS ARDENS SCANDET AXEM DELIUS)
HAC pro huc, adverbium loci *BeLa*
in nostram partem *Leb*

CONSORTIUM terrenorum hominum *BeLa*, hominum] nominum *Be*
†PHOETHONTOS genitivus grecus *BeLa*
pro fetontis *La*
[veniens.] genitivus *LebPar*
DELIUS Apollo *BeLa*
declarans Apollo renovans *Leb*
declarans *Par*
SCANDET ascendit *Be*
ascendet *La*

10-1 NUNC VOS, PUDICI NOSTRA FRATRES CULMINA,
PRECABOR IPSA, NE VERENDOS CONTRAHANT
PUDICI reverendi *BeLa*
FRATRES ó nostra *Be*
filii Saturni *Leb*
CULMINA celsitudo *Be*
numina *La*
PRECABOR IPSA ego Armonia *BeLa*
VERENDOS venerandos *BeLa*
Precabatur, ne in nauseam venirent eius carmina diis et idcirco non venirent ad nuptias *BeLa*
CONTRAHANT retrahant retro *BeLa*, retraant *La*

12-3 VULTUS IUGANDIS QUAE FERENT DIS CARMINA,
HYMENQUE NOSTRUM INVEHAT FASTIDIUM
VULTUS s. deorum
s. deorum Philologiae et Mercurii
Philologiae et Mercurii *Leb*
IUGANDIS coniungendis *Be*
QUAE germaniae *La*
†FERENDOS manifestandos *Be*
†FERENDIS laudandis *LebPar*, laudendis *Leb*[1]
CARMINA i. propter hec carmina *BeLa*, i. *om. La*
nominativus *Leb*
HYMENQUE i. carmen s. precabor *BeLa*, precabat *La*
carmen. nominativus *LebPar*
INVEHAT FASTIDIUM inferat s. auribus deorum *BeLa*
s. ne *Leb*

350.14-5 PSALLENTE PLECTRO CONCINENTUR NUPTIAE,
FIDIBUSQUE IUNCTA FESCENNINA PRODIENT
†CUM CANENTUR i. laudentur (*sic, lege* laudantur) *BeLa*
†LUXA FESCENNINA copiosa *BeLaPar*
carmina alia *La*
copiosa carmina nuptialia *Leb*
PRODIENT nupti apparebunt *BeLa*, nupti *om. Be*

915/
350.16-7 IAM NUNC BLANDA MELOS CARPE DIONE,
DURUS QUIPPE RIGOR CEDIT AMORI
PENTAMETRUM SPONDEACUM CATALECTON IN MEDIO QUAE SYLLABA ACCIPITUR PRO UNO PEDE *LebPar*
pentametrum spondeacum] spondeacum pentametrum *Par*
CARPE DIONE i. tene vel incipe mater Veneris *BeLa*
MELOS accusativus pluralis (*sic, lege* singularis) *Leb*
RIGOR s. castitatis *BeLaLeb*, s. *om. Leb*
CEDIT dat locum *La*
AMORI i. Cupidini *BeLa*

350.18-9 NOVIT NAM TENERUM PROBARE CARMEN
IPSUM SPUMIGENAE SALUM CYTHERES
pentametrum est *Par*
NOVIT s. ipse amor TENERUM amatorium *BeLa*
NOVIT armonia marina *Leb*
PROBARE laudare *BeLa*
CARMEN Hoc carmen matri Veneris cantatur in nuptiis. Aut laudatur Venus aut eius mater aut Hymeneis *BeLa*
IPSUM s. novit probare *BeLa*
SPUMIGENAE spuma libidinis *La*
SALUM mare *BeLa*
Salum liquor maris et hic pro semine Veneris* *Leb*
Semen. Omnis liquor salum dicitur* *Par*
CHITERES Venus *Be*
i. Veneris *La*

---

* Cf. Fulg. Myt. 2.70

20-1 NERINAQUE CHELY MOVET CAMENAM,
DUM CONCHIS GALATEA PERSONANTES

NERINAQUE cum *Be*
dea maris *BeLa*, dea]de *Be*[1]
marina *Leb*
CHELI cithara *BeLa*, citara *La*
CAMENAM i. musam suam *BeLa*
melodiam *Leb*
Quasi canena a canendo *Leb*
CONCHIS in *La*
ostreis *Leb*
GALATHEA filia Oceani *BeLa*
alba dea Aurora *Leb*

22-4 NANTES FLUCTIGENAS CHORUMQUE PHORCI
FLAGRANS CURA TRAHIT, NIVALIBUSQUE
INRESTINCTUS AQUIS TRIUMPHAT ARDOR

FLUCTIGENAS i. nymphas *BeLa*, nymfas *La*
CHORUMQUE FORCI Propter Scillam dicit filiam Forci[*] *Be*
Propter Scilla<m> dicitur, que filia fuit Forci monstruosa[*] *La*
Filias Nerei i. vel omnes bestias et pisces maris *Leb*
FLAGRANS i. accendens ad audiendum *BeLa*, i. *om. La*
CURA s. carminum TRAHIT movet *BeLa*, s. *om. La*
CURA amor musicae *Par*
INRESTINCTUS superabilis *BeLa*
inextinguibilis *Par*
TRIUMPHAT i. letatur *BeLa*
ARDOR i. amor *BeLa*, i. *om. Be*
amoris *Leb*

25 IAM NUNC BLANDA MELOS CARPE DIONE
INTERCALARE VERSUS *BeLa*

1-2 CARMEN MAENALIAE TULERE PINUS,
PERCUSSAQUE MODIS SONAT LYCAEIS

---

* Cf. Serv. Com.Verg.A 3.420; Serv. Com.Verg.E 6.74

CARMEN MENALIAE Menalus mons est Archadiae *BeLa*, acadiae *La*
armonia terrae *Leb*
ut in Atlante *Par*
PINUS abundans pinis *BeLa*
hic in terra *Par*
PERCUSSAQUE s. nemora illa *BeLa*, memora *Be*
MODIS modulationibus *BeLa*
tonis *Leb*
LICHEIS Licheus mons est a lupis dictus *BeLa*
mons Arcadiae *Leb*
Archadicis *Par*

351.3 SOLLERS ARCADIAE NEMUS CICUTA
PERNIX SEMIFERI PUELLA PANOS
SOLLERS sagax *BeLa*
ARCADIAE pro poetis dicit *Be*, pro positis dicit *La*
SEMIPHERI semidei *BeLa*
PUELLA Siringam vel Latmiam seu aliam quamlibet dicit[*] *BeLa*, fevaliam quamlibet dicit *om. Be*
Siringa vel Menala puella proprium est *Leb*
Siringa vel Menalia *Par*

351.5-7 NAM VERSA IN CALAMOS SONAT LOQUACES,
QUAM DUM FORTE DEUS PREMIT LABELLIS,
SUSPIRAT VELUT OSCULIS CANOREM
SUSPIRAT illa *BeLa*
OSCULIS de *BeLa*
CANOREM dulcedinem *BeLa*
cantum *Leb*
VERSA illa *BeLa*[t]
DEUS i. Pan *BeLa*
LABELLIS cum *Be*
suis *BeLa*

---

[*] Cf. Myth.Vat.I <124>: Myth.Vat.II <60>

10-1 PUER IPSE VERSIFORMIS
FIDIBUS STUDET CUPIDO
IAMBICUM DIMETRUM CATALECTICUM EX ANAPESTO DUOBUS IAMBIS ET SYLLABA *BeLa*
IAMBICUM DI[A]METRUM *LebPar*, iambicus diametrus *Leb*
†FACIBUS i. ignibus amorum *BeLa*
amoris *Leb*
amoribus *Par*
STUDET Studium habet de facibus CUPIDO i. amoris *La*

12-3 ARCUSQUE DULCINERVES
ROSEO LIGANS RAMALI
DULCINERVES i. ex nervo habentes cordas *BeLa*
ROSEO pulchro *Leb*
LIGANS illos *BeLa*
RAMALI arbore *BeLa*

14-5 FERIATA LIQUIT ARMA
CALAMOS PARANTE MUSA
†FERIATO pausato vel quieto *BeLa*
†ACTU opere *BeLa*
praePARANTE *BeLa*[t]

16-7 IN CARMINIS LEPOREM
CURAM NEGAT SAGITTIS
†LEPORE in habundantia *BeLa*, abundantia *La*
CURAM amoris *Leb*

18-9 TENERUMQUE HARUNDINETUM
NOSTRUM IN MELOS RELIQUIT
TENERUMQUE s. dulcinerves arcus *BeLa*, s.] i. *La*
HARUNDINETUM amoris *Leb*
amorem *Par*
NOSTRUM IN MELOS i. in nostram musam *La*

20-1 NUNC ERGO CORDA MULCENT
VAPOR ET CANOR IUGATI
CORDA i. amantium *La*

MULCENT placent vel molliunt *BeLa*
VAPOR amor *BeLaLeb*
ET †CANO IUGATI i. simul complexi illi duo *BeLa*, complexi] conplexi *Be*
cantus in voce Mercurio *Leb*

351.22-3 PLACET AC DECET, VICISSIM
CANTEMUS ATQUE AMEMUS
PLACET amor nuptiarum *La*
DECET honestum est *BeLa*
VICISSIM per vices *BeLa*

918/
351.26-7 PRUDENS PUELLA PULCHRAE
MATER FUIT LACAENAE
IAMBICUM DIMETRUM CATALECTICUM *BeLa*
PUELLA i. Leda *BeLa*
MATER Ledae filia Thesei *LebPar*, Thesei] tesei *Par*
LACENAE Helenae *BeLaLebPar*, i. *ad. Be*

351.28-9 ILLECTA SED CANORE
NESCIT DOLUM CAVERE
INLECTA decepta *BeLa*
CANORE cantu *BeLa*
Lede *Leb*
DOLUM s. Iovis *La*
CAVERE declinare *BeLa*

352.1-2 NAM CANDIDUS NIVOSIS
OLOR INVOLUTUS ALIS
NIVOSIS de. candidissimis *BeLa*
OLOR cignus *La*
INVOLUTUS indutus *BeLa*
ALIS s. est *La*

352.3-4 ARGENTEIS MINOREM
PLUMIS VIDENS DECOREM
ARGENTEIS s. esse *Be*
s. est *La*

PLUMIS in suis *Par*
VIDENS sentiens *BeLa*

5-6 NEC POSSE PURPURATIS
NIMIUM PLACERE OCELLIS
NIMIUM valde *La*
OCELLIS s. puellae *LebPar*, s. *om. Leb*

7-8 TENEROS CIERE CANTUS
PHAETHONTIASQUE MUSAS
TENEROS dulces *BeLa*
CIERE vocare i. sonare *BeLa*, vocare] evocaret *Be*
pro ciebat *Leb*
CANTUS s. amoris *La*
PHAETHONTIASQUE MUSAS sorores Phetontis *La*

9-0 COEPIT REPENTE FICTA
FRAUDEM PARANS SENECTA
REPENTE subito in FICTA *BeLa*

11-2 SIC ADMOVETUR ORI
ROSEUMQUE CIRCUMACTAE
ADMOVETUR iungitur *BeLa*
ORI in tali cantu *La*
CIRCUMACTAE i. seductae *BeLa*

13-5 RAPIT OSCULUM PUELLAE,
GREMIOQUE COMPLICATUS
PRETIUM TULIT PUDORIS
GREMIOQUE COMPLICATUS in sinu puellae positus *BeLa*
absTULIT *La*
PUDORIS castitatis *BeLa*

18-0 CARMEN LATMIADEUM
LUCIS DIVA SECUNDAE
SACRIS PRAETULIT ASTRIS
CARMEN †LATMIA Latmia filia Iovis relictos caelos Pana in ter<r>is secuta est *BeLa*, in te<r>ris secuta est] interisse *La*

†DEUM deorum *BeLa*
LUCIS vitae *La*
DIVA Lathmia *BeLa*, latmia *La*
  Luna *Leb*
SECUNDAE prosperae *BeLaLeb*, pro[p]sperae *Leb*

352.21-2 CANTUM QUIPPE SECUTA
LINQUENS CULMINA CAELI
†ANTRUM speluncam Panos *La*
†SECUTUM s. est *BeLa*[1]
reLINQUENS *La*

352.23-4 PASTORALIBUS ARDENS
PALMAM DEDIT CICUTIS
ARDENS i. amans *BeLa*, i. *om. La* armans *Be*[1]
  amore *Leb*
PALMAM laudem CICUTIS fistulis *BeLa*

352.25-6 PULSO ET LUMINIS AURO
PANDIT SEPTA BALLANTUM
LUMINIS i. solis *La*
AURO radio *BeLa*
SEPTA aestiva (*sic*) BALLANTUM i. ovium *BeLa*

352.27-8 SORDENTIQUE ADOPERTA
PRATIS GAUDET AMICTU
in SORDENTIQUE ADOPERTA induta *BeLa*

352.29-0 AC TUNC CYNTHIA NOCTU
PASCIT MONTE CAPELLAS
TUNC deinde *BeLa*
†CONSCIA perita *Leb*
NOCTU per noctem *BeLa*
in MONTE *La*

352.31-2 SPECTANS SIDERA MUNDI
MAGE PERCREPAT FLAGELLO
SPECTANS contemplans *BeLa*
MAGE pro magis PERCREPAT sonat *BeLa*, pro] per *La*

FLAGELLO Abegit, cogit seu in cursu movet capellas *La*

1 NEC CURARE DEORUM
PENSA AUT FERRE SUSURROS
NEC CURARE pro nec curabat *Be*
pro curabat *Leb*
PENSA i. numerum vel honorem *BeLa*, honorem] amorem *La*
fila *Leb*
SUSUROS cantus *BeLa*, canus *La*

3-4 ILLEX CARMINIS AMPLI
AEGRAM CURA COEGIT
ILLEX i. provocatrix ad suam voluptatem *BeLa*
Apollo vel Cupido *Par*
CARMINIS AMPLI carmina Panos *BeLa*

5-6 TEMNIT NOCTIS HONOREM,
PRAEFERT ANTRA SUBULCI
TEMPNIT NOCTIS †ORROREM i. non timet noctis tenebras *BeLa*
PRAEFERT s. orrori noctis *BeLa*
praeponit *Leb*
ANTRA s. caelo *LebPar*, s. *om. Par*

7-9 DURA ET RUPE QUIESCIT
ET POST REGNA TONANTIS
STRAMEN DULCIUS HERBA EST
RUPE petra *BeLa*
HERBA EST s. illi *BeLa*

10 TALIBUS HARMONIAE CARMINIBUS OBLECTATI OMNES
PERMULSIQUE DIVI
ARMONIAE armonia est adunatio vocum[*] *BeLa*
Monos unum, armonia adunatio: haec est adunatio[*] *La*
OBLECTATI delectati *La*
s. sunt *BeLa*

[*] Cf. Boeth. Inst.Ar. 2.32

PERMULSIQUE eblanditi *Be*
blanditi sunt *La*
s. sunt *Leb*

353.11 NEC MINOR QUIPPE EX FIDIBUS SUAVITAS QUAM VOCIS MODULAMINE RESULTABAT
FIDIBUS cordis *La*
MODULAMINE ex *Be*

353.12-3 QUA INDUSTRIA COMPARATUM QUIBUSVE ASSEQUENDUM EDISCENDUMQUE OPIBUS
QUA INDUSTRIA COMPARATUM Disciplina inventum est *La*
compertum *LebPar**
AD SEQUENDUM ad imitandam armoniam *La*

353.13 VIGIL CURA REPROMITTAT
VIGIL CURA s. huius artis *BeLa*
REPROMITTAT fateatur *La*

353.14 LIQUESCAT AFFECTIO, IOVE ADMIRANTE DISQUIRITUR
LIQUESCAT Ut se effundat *La*
AFFECTIO amor *BeLa*
DISQUIRITUR interrogatur *La*

353.14-5 AC TUNC VIRGO, CUM ARTIS PRAECEPTA A SE EXPETI EXAMINANDAE ERUDITIONIS INTENTIONE CONSPICERET
VIRGO armonia *La*
EXAMINANDAE probandae *Be*
Utrum illa bene est et erudita *La*
HERUDITIONIS examinatum est *La*
INTENTIONE cum *BeLa*[t]
voluntate vel sub intuitu *La*

353.16-7 PAULUM MELICIS TEMPERANS EXHORTANTE QUOQUE DELIO PALLADEQUE SIC COEPIT:

---

* In the margin of *Leb* a gloss which is not attached or attachable to a lemma: Patricus patria sunt / Malta emplastrum / Maltare superponere / Petasma velum peri circum.

MELICIS a *BeLa*[t]
modulationibus *BeLa*
consimilem vel concordem motibus *La*
TEMPERANS cessans vel cedans (*sic, lege* sedans) *La*
reticens *La*
parcens *Leb*
EXORTANTE interrogante *La*
s. a DELIO Delius interrogat[ur] Palladem *La*
SIC COEPIT s. canere *Be*

18-9 IAM PRIDEM [QUIDEM] EXOSA TERRIGENAS ET FASTIDIOSA MORTALIUM CAELI ORBES STELLANTIS INCUTIO
EXOSA i. odio habens homines *Be*
IAM PRIDEM QUIDEM EXOSA Dum exosa sum. Ipsa virgo que est EXOSA. despexi. IAM PRIDEM QUIDEM et usque nunc vel olim *La*
TERRIGENAS illos *La*
a terri genis *Leb*
FASTIDIOSA MORTALIUM Fastidio habui mortales *La*
CAELI ORBES circulos planetarum *La*
†STELLANTES lucidos *BeLa*
INCUTIO moveo *Be*
movebo *La*
modulor *LebPar*

.19 IN QUIBUS ARTIS PRAECEPTA EDISSERTARE PROHIBITUM
IN QUIBUS s. orbibus *LaLebPar*, s. *om. LebPar*
EDISSERTARE docere *Par*
PROHIBITUM s. est s. mihi *Be*
s. est *La*
Non licet mihi, ut de celesti armonia loquar *La*

.20-1 CUM MELODIAM OMNISONIS CONVENIENTEM PULSIBUS MODULORUM MACHINAE OBEUNTIS
CUM MELODIAM OMNISONIS Licet habeat suam melodiam, prohibitum tamen est mihi *BeLa*, prohibitum tamen] tamen prohibitum *La* mihi *om. Be*
consimilem vel concordem motibus *Be*, *cf. La* 353.16-7 MELICIS
MACHINAE structurae volubilis *BeLa*, volubilis structurae *La*

OBEUNTIS currentis, volventis *La*

353.21 IPSA RAPIDITAS ET CONCINAT ET AGNOSCAT
RAPIDITAS volutio *BeLa*
ET CONCINAT ET AGNOSCAT conveniat, intellegat *BeLa*
Nullus enim rationem debet dare de armonia caelesti, quia nimiae profunditatis est, quia ibi non est vox humana nisi tantummodo sonus *BeLa*

353.21-2 SED QUONIAM EMERSA TERRIS VIRGO NUPTURA VANESCENTIA INTERCAPEDINATAE PROLIXITATIS OBLIVIA
EMERSA sursum elevata *BeLa*, sursum *om. Be*
ablata s. sum *Leb*
VIRGO De se ipsa dicit *La*
NUPTURA Ut nuptias accipiat in collegio *La*
Philologia *Leb*
VANESCENTIA labentia *BeLa*
Vanescentia oblivia, quia longo tempore oblivioni data est musica in terris *La*
INTERCAPEDINATE spatiosae *BeLeb*
longitudine temporum *La*

353.22-3 IAM SUPERO DEBET VIGORE DISCUTERE, IUSSA PERCURRAM
SUPERO caelesti *BeLa*
divino *Leb*
VIGORE ingenio *BeLa*
divina vi *La*
numine *Leb*
DISCUTERE discernere *Be*
disserere *La*
intendere *Leb*
IUSSA s. ego a diis *BeLa*, ego *om. Be*
PERCURRAM exponam *Be*
breviter narrabo *La*
Ea quae oblivioni data sunt per multa tempora, debet ipsa Filologia exponere *BeLa*, Filologiam *Be*

353.23-4 SI PRIUS INGRATAE MORTALITATIS COMMODA REPETITA RECENSEAM

Si numeravero illa dona musice artis, que tribui mortalibus ingratis *La*

COMMODA utilia, dona musica<e> *BeLa*, utilia *om. La*

†REPUDIATA collata, a me eis, vel repediata id est repetita *Par*

RECENSEAM renumerem *BePar*

Recapitulans iterum numeravero *La*

25-6 DUM ME QUIPPE GERMANAM GEMELLAMQUE CAELO ILLA INCOGITABILIS EFFIGIENTIAE GENUISSET IMMENSITAS

GERMANAM GEMELLAMQUE geminam *La*

Duae germanae fuerunt: una in caelis remansit, altera in terras devenit *BeLa*

Arithmeticam *Leb*

CAELO ablativus *Leb*

INCOGITABILIS in mortalitate *Be*

s. mortalitate *La*

quasi mater *Leb* (*cf. Leb ad* 353.26-1 VOLUMINA)

EFFIGIENTIAE structure formationis vel imaginationis *La*

imaginatione *LebPar*

†GEMUISSET Cum tristis facta est mea germana *La*

IMMENSITAS i. inmensitas celi *La*

divinitas *Par*

26-1 SIDERAE REVOLUTIONIS EXCURSUS ATQUE IPSA TOTIUS MOLIS VOLUMINA COMITATA

EXCURSUS genitivus *Leb*

VOLUMINA illa inmensitas *La*

quasi pater *Leb* (*cf. Leb ad* 353.25-6 INCOGITABILIS)

COMITATA mea germana *La*

coeuntia *Leb*

1-2 SUPEROS INCITOSQUE FULGORES MODIS ASSOCIANS NUMEROS NON RELIQUI

SUPEROS superas musis *BeLa*

INCITOSQUE FULGORES mobilia vel volubilia sidera *BeLa*, mobilia] nobilia *La*

INCITOSQUE veloces *La*

MODIS ASSOCIANS modulationibus rithmos iungens *BeLa*, rihtmos

modulationibus *La*
MODIS tonis *LebPar*
NUMEROS NON RELIQUI non deserui *BeLa*
numeros vel minores i. homines vel deos *BeLa*
arithmeticam *Leb*

354.2-3 SED CUM ILLA MONAS INTELLECTUALISQUE LUCIS PRIMA FORMATIO
MONAS i. unitas *BeLa*
INTELLECTUALISQUE LUCIS i. animae *BeLa*, i. *om. Be*

354.3 ANIMAS FONTIBUS EMANANTES IN TERRARUM HABITACULA RIGARET
FONTIBUS s. suis i. disciplinarum *La*
naturis *Leb*
s. naturae *Par*
EMANANTES venientes *La*
IN TERRARUM HABITACULA i. in corpora *BeLa*
RIGARET i. funderet vel crearet *BeLa*
diffunderet *LebPar*

354.4 MODERATRIX EARUM IUSSA SUM DEMEARE
MODERATRIX temperatrix *Be*
Ego habens potestatem illarum animarum a mea germana vel a diis *BeLa*
IUSSA SUM DEMEARE Quia in terra versatur. Quamquam soror mea i. armonia caelestis doleret, quod in terra manerem *La*
DEMEARE transire *BeLa*
deorsum descendere *LebPar*

354.4-5 DENIQUE NUMEROS COGITABILIUM MOTIONUM
NUMEROS Data est mihi potestas non solum corporum sed etiam animarum: utrum cogitaremus nunc ista nunc illa. Arithmetica in statu, geometrica in figuris, musica in omnium rerum corporalium et incorporalium motu, astrologia in ratione corporum celestium versatur *La*
COGITABILIUM MOTIONUM rationabilium *La*
Quae sola cogitatione cernuntur *BeLa*

5-6 TOTIUSQUE VOLUNTATIS IMPULSUS IPSA RERUM DISPENSANS CONGRUENTIAM TEMPERABAM
VOLUNTATIS IMPULSUS animi *La*
motus animi *BeLa*
IPSA ego Armonia *BeLa*
DISPENSANS ordinans *BeLa*
CONGRUENTIAM concordiam *BeLa*
TEMPERABAM gubernabam *Be*

6-7 QUAM REM DIDERE MORTALIBUS UNIVERSIS THEOPHRASTUS LABORAVIT
QUAM REM i. mee potestatis esse omnes motus *La*
†DEBERE habere *La*
MORTALIBUS s. hominibus *La*
THEOFRASTUS Deum componens vel dei compositor; frasto compono *LebPar*, Deum componens vel *om. Par*
LABORAVIT Laborando invenit *Be*
Per laborem monstravit *La*

7-8 PYTHAGOREI ETIAM DOCUEREUNT, FEROCIAM ANIMI TIBIIS AUT FIDIBUS MOLLIENTES
Hic videtur tangere quod legitur, quod quidam iuvenis dives temulentus in urbe quadam scortum adiit. Cuius domum cum intrare nequiret, domum illius post reveniente (*sic, lege* reveniens) cum urbe flammis cremare [videbatur] conabatur. Unde universi commoti sunt. Quod cum quidam Pytagorae nunciasset, qui tunc forte astris intendebat, ille tibicinam vocavit. Que cum docte tibicinaret, non tamen iuvenis a ferocia sua cessaret, ipsae (*sic, lege* ipsi) Pytagoras fistulam mutari precepit adque aliut modulamen concini iussit. Quod cum actum fuisset, iuvenis insanire cessavit* *La*
TIBIIS cum *Be*
MOLLIENTES mitigantes *BeLa*

.8 CUM CORPORIBUS ADHAERERE NEXUM FOEDUS ANIMARUM
NEXUM colligatum *Be*

---

* Cf. Boeth. Inst.Mus. 1.1

coniunctum *La*
FEDUS concordiam *La*

924/
354.9 MEMBRIS QUOQUE LATENTES INTERSERERE NUMEROS NON CONTEMPSI
MEMBRIS QUOQUE in conexione membrorum *Be*
s. in *La*
Habendi etiam potestatem in conexionem membrorum *La*
NUMEROS NON CONTEMPSI i. motiones pulsus *Be*
Sed accepi in meo ministerio *La*

354.9-10 HOC ETIAM ARISTOXENUS PYTHAGORASQUE TESTANTUR
ARISTOXENUS Virtutis acumen vel xenos hospitium, unde xenodochium domus hospitum *Leb*
Si ab eo quod est aris et oxos virtutis acumen, si ab aris et xenos, virtutis hospicium interpretatur *Par*

354.10-1 DENIQUE BENIGNITATE LARGISSIMA SENSIM IPSA NOTIONEM MEAE OBLECTATIONIS APERUI
DENIQUE Nunc dicit, quod supra dictum est, unde comparatum *La*
SENSIM gradatim *BeLa*
IPSA ego *Leb*
NOTIONEM cognitionem *Be*
intelligentiam *La*
MEAE †OBIECTATIONIBUS APERUI i. meas partes demonstravi *BeLa*
Nunc incipit de inventione *La*
responsionis *BeLa*
interrogationis *LebPar*

354.11-2 NAM FIDES APUD DELPHOS PER DELIACAM CITHARAM DEMONSTRAVI
FIDES cordas *BeLaLeb*
DELFOS civitas est *BeLa*
DELIACAM Apollinis *BeLa*

354.12-3 TIBIAE PER TRITONIDAM NOSTRI COMITEM MARSYAMQUE LYDIUM SONUERUNT

TRITONIDAM i. Palladem *BeLa*, paladem *Be*
COMITEM sociam *Be*
MARSIAM rex Asiae *Leb*
LIDIUM Quia in Lidia repertus est tonus *BeLa*, tonus] sonus *La*
tropus Lidium (*sic*) *La*
Quintum tonum principalem *Leb*

13-4 CALAMOS MARIANDYNI ET AONES IN LAUDES INFLAVERE CAELESTIUM
MARIANDINI nomen gentis *La*
gentes Greciae *Par*
AONES populi Epiri *BeLa*, aepiri *La*
INFLAVERE insonuere *La*
CAELESTIUM deorum *BeLa*

.14-5 PANDURAM AEGYPTIOS ATTEMPTARE PERMISI
PANDURAM genus musicum *Be*
nomen proprium *La*
Dicunt quidam, quod ad similitudinem Panos sit *BeLa*
genus organi *Leb*
pinax id est capsa organi *Par*
ADTEMPTARE probare *La*
PERMISI concessi *BeLa*

.15-6 IPSIQUE ME PASTORALIBUS FISTULIS VEL CANTUS AVIUM VEL ARBORUM CREPITUS VEL SUSURROS FLUMINUM IMITANTIBUS NON NEGAVI
SUSURROS FLUMINUM motiones aquarum *BeLa*
NON NEGAVI s. me *La*

.16-8 PSALTAS, CHORDACISTAS, SAMBUCOS, HYDRAULAS PER TOTUM ORBEM AD COMMODUM HUMANAE UTILITATIS INVENI
†SPELTAS scuta pro peltis dixit *La*
Scutorum sonus, qui in cordis agitur *BeLa*
Paltas (*sic, lege* peltas) dicit aut ipsa scuta, quorum percussione suavis efficitur sonus, aut homines, qui cum ipsis feruntur *La*
Scutum breve <habens> ansulas fer[r]eas cum catenis *Leb*
Scutum breve cum circulis ansulis et catenulis *Par*
CORDACISTAS †CITARISTAS i. *BeLa*[t]

glosa *La*

Qui cordis canunt *LebPar*, canunt] sonant *Par*

SAMBUCOS arbores vel instrumentum musicum *BeLa*, arbores vel *om.* *Be*

cimbalos *LebPar*, cimbala *Par*

YDRAULAS Musica modulatio, que aqua agitur *La*

aquatica organa *Par*

COMMODUM ad utilitatem *BeLa*, ad *om. La* utilytatem *La*

925/
354.18-9 PER ME QUIPPE VESTRUM HOMINES ILLEXERE SUCCURSUM IRASQUE INFERAS PER NAENIAM SEDAVERE

VESTRUM HOMINES ó dii *BeLaLebPar*, s. *add. Leb*

INLEXERE invenere *Be*

<in>hibuere *La*

SUCCURSUM adiutorium *BeLa*

INFERAS inferorum *BeLa*

†NENIAS Nenia carmina super sepulcra i. epitaphia *BeLa*, neniae *Be* epitafia *La*

Nenia sunt carmina, quae in sepulcris mortuorum scribuntur *BeLa*

epitaphia mortuorum *Leb*

carmina sepulcraria (*sic, lege* sepulcralia) *Par*

SEDAVERE placavere s. dicam *BeLa*, placuere *La*

354.20 QUID QUOD BELLA VICTORIAEQUE UNDIQUE MEIS CANTIBUS CONQUISITAE?

UNDIQUE ubique *Par*

MEIS CANTIBUS Non enim minus ex<h>ortatur per cantilenam *La*, minus] meus *La*[1]

CONQUISITAE s. sunt *Be*

Quia apud antiquos nulla hostia fiebat sine musica *La*

354.20-1 NAM CRETES AD CITHARAM DIMICABANT

DIMICABANT pugnabant *La*

354.21-1 NEC ANTE AGGREDIEBANTUR FATA PROELIORUM PRIUSQUAM ILLIS CONTIGERAT LITARE MUSIS

FATA eventus *Be*

LITARE sacrificare *La*

MUSIS diis musarum *La*

.1-2 QUID AMAZONES? NONNE AD CALAMOS ARMA TRACTABANT?
AMAZONES In Scithia erant feminae[*] *Be*
  Femine in Scittie (*sic*, *lege* Scittia) <h>abitantes[*] *La*
  Maza mamma, amazones, sine mammis[*] *La*
  sine mamma[*] *LebPar*
AD iuxta *BeLa*

.3-4 UT MAGNO MUNERE GRATULATA DISCESSIT
pro UT *Be*

.5-6 TUBAS NON SOLUM SONIPEDES ATQUE BELLA
TUBAS propter *BeLa*
  Celica tuba, quae dicitur trumba rustice *Leb*
  cerica *Par*

.6-7 SED AGONAS ACUERE CERTAMENQUE MEMBRORUM NONC QUOQUE COMPERTUM
SED AGONAS i. milites in agone certantes *BeLa*, i. *om. La*
  singulare certamen *Leb*
  Qui agunt palestram *Par*
CERTAMEN palestra *Leb*
  luctam *Par*
COMPERTUM s. est *Be*

/

.7 QUID PACIS MUNIA? NONNE NOSTRIS CANTIBUS CELEBRATA?
MUNIA officia *BeLa*
CELEBRATA s. sunt *Be*

.8-9 AD LYRAM LEGES DECRETAQUE PUBLICA RECITABANT
DECRETAQUE statuta *BeLa*
PUBLICA in curia *Par*
RECITABANT legebant *Be*
  relegebant *La*

---

[*] Cf. Isid. Etym. 9.2.63-64

355.10-1 NAM PHRENETICOS SYMPHONIA RESANAVI, QUOD ASCLEPIADES QUOQUE MEDICUS IMITATUS
PHRENETICOS RESANAVI amantes sanitati reddidi *La*
Fren membranula inter toracem i. ventrem et costas. Quam si aliquis laesam habuerit, in amentiam vertitur *LebPar*, inter toracem i. ventrem et costas] super costas *Par* quam si] quasi *Par* in amentiam vertitur] vertitur in dementiam *Par*
IMITATUS s. est *Be*

355.11-2 CUM CONSULENTIBUS URBIUM PATRIBUS PLEBIS INCONDITAE VULGUS INFREMERET
CUM CONSULENTIBUS Querunt consilium ab eo *La*
INCONDITAE stultae *BeLa*

355.12-3 SEDITIONES ACCENSAS CREBRIOR CANTUS INHIBUIT
INHIBUIT i. prohibuit *BeLa*

355.13 EBRIOS IUVENES PERINDEQUE IMPROBIUS PETULANTES DAMON
PERINDEQUE quia ebrii *La*
IMPROBIUS stultius *BeLa*
PETULANTES lascivientes *Be*
s. iuvenes *La*

355.14 UNUS E SECTATORIBUS MEIS, MODULORUM GRAVITATE PERDOMUIT
MODULORUM tonorum *Par*

355.14-6 QUIPPE TIBICINI SPONDEUM CANERE IUBENS TEMULENTAE DEMENTIAM PERTURBATIONIS INFREGIT
SPONDEUM spondaicum carmen *BeLa*
†TEMULENTIAE i. ebrietatis *Be*
vinolentiae *La*
DEMENTIAM insaniam *BeLa*

355.16 QUID? AFFLICTATIONIBUS CORPOREIS NONNE ASSIDUA MEDICATIONE SUCCURRI?
QUID dicam *BeLa*
†AFFLICTATIONIS laboris CORPOREIS s. rebus NONNE s. multum *La*
Non solum morborum, sed etiam tristitiae *LebPar*

.17-8 ASCLEPIADES ITEM TUBA SURDISSIMIS MEDEBATUR
ASCLEPIADES Asclepiadem interrogaverunt viri civitatis, quid agere deberent, cum vulgus fremeret. Ille †mitis† (*sic?, lege* iussit?) personare symphoniam *La*
†TURBA in *Be*

.18-9 AD EFFECTIONES ANIMI TIBIAS THEOPHRASTUS ADHIBEBAT
EFFECTIONES commotiones *BeLa*
angustias *Par*
ANIMI Quando tristabatur animus, vel ad angustias *Leb*

.19 ISCHIADAS QUIS NESCIAT EXPELLI AULICA SUAVITATE?
†SCIADAS i. morbi umbratilis *Be*, morbium bratilis *Be*[1]
gens morborum *La*
Scia dictum est, quod scia Grece Latine umbra dicitur, eo quod in nocte umbras larvarum videre faciat. Qui autem illud patitur aut sic vocatur [aut sic vocatur] aut sciaticus *La*
Umbratici, qui vertigine capitis laborant. Scia umbra *LebPar*, scia umbra *om. Par*
EXPELLI ab hominibus *LebPar*
AULICA SUAVITATE regali musica *La*
fistulari *LebPar*, fistula *Par*

.19-0 XENOCRATES ORGANICIS MODULIS LYMPHATICOS LIBERABAT
†SENOCRATES synodi caput *Leb*
LYMPHATICOS vel amentes *Be*
hydropicos *BeLaLebPar*, idropicos *LaPar* ydropicos *Leb*
LIBERABAT obliviscebatur potus *Par*

.20-1 THALETEM CRETENSEM CITHARAE SUAVITATE COMPERTUM MORBOS AC PESTILENTIAM FUGAVISSE
COMPERTUM MORBOS s. est *Be*
ex corpore morbos *LebPar*, morbos *om. Par*
AC PESTILENTIAM ex aere *LebPar*
FUGAVISSE expolisse (*sic, lege* expulisse) *BeLa**

.1-2 HEROPHILUS AEGRORUM VENAS RHYTHMORUM COLLATIONE

---

* i.e. expulisse. This Romanistic spelling occurs more often in *Be* and *La*.

PENSABAT
EROFILUS fortitudinis amator *LebPar*
RITHMORUM numerorum *BeLa*
PENSABAT intellegebat vel iudicabat *BeLa*, intellegebat vel *om. La*
Quod sic Bald<..> agit: in numero pulsus manus discernere nititur vitam vel mortem *La*
numerabat *LebPar*

927/
356.2-3 ANIMALIUM VERO SENSUS MEIS CANTIBUS INCUNCTANTER ADDUCI SALTEM THRACIUS CITHARISTA PERDOCUIT
INCUNCTANTER ADDUCI i. non dubitanter moveri *Be*
non cunctanter moveri *La*
SALTEM etiam *Leb*
TRACIUS Orpheus *BeLaLebPar*
In Tracia Orpheo canente dicunt *La*
citharista *Par*

356.3-4 IN QUO NON FABULA, SED VERITAS GLORIAM PROCREAVIT
IN QUO s. actu i. motu *BeLa*
GLORIAM s. animalium, quia optimus fuit citharista *BeLa*, quia] que *La* citarista *La*
s. meam *Leb*

356.4-5 UNDE ENIM CERVI FISTULIS CAPIUNTUR, PISCES IN STAGNO ALEXANDRIAE CREPITU DETINENTUR
UNDE qua de re *BeLa*
PISCES s. nisi mea cantatione *Leb*
CREPITU sonitu, cantu *BeLa*
sono *Leb*
cantu *Par*
DETINENTUR capiuntur *BeLa*

356.5-6 CYCNOS HYPERBOREOS CITHARAE CANTUS ADDUCIT
CIGNOS HYPERBOREOS i. qui sunt in Hyperboreis montibus *BeLa*, i. *om. La* iperboreis *La* montibus *om. Be*
ADDUCIT capit *BeLa*

356.7-8 ELEPHANTOS INDICOS ORGANICA PERMULSOS DETINERI VOCE

COMPERTUM
PERMULSOS eblanditos *BeLa*
DETINERI capi *La*
COMPERTUM s. est *Be*

.7-8 FISTULIS AVES ALLICI COMPROBATUM, INFANTIBUS CREPITACULA VAGITUS ABRUMPERE
ALLICI provocari *BeLa*
COMPROBATUM s. est *Be*
ABRUMPERE compescere s. comprobatum est *La*

.8 FIDES DELPHINIS AMICITIAM HOMINUM PERSUASERUNT
FIDES cordae *Be*
PERSUASERUNT ipse fides *La*
Sicut de Arione legitur: "inter delfinas Arion"* *BeLa*

/

.9 QUID <QUOD> CANTICIS ALLICI DISRUMPIQUE SERPENTES
QUID s. dicam *BeLa*
ADLICI promoveri *BeLa*
DISRUMPIQUE s. probatum est *La*
provocari, occidi *Leb*

.9-10 GLANDEM FERUNT MESSESQUE TRANSIRE?
GLANDEM quercus *La*
FERUNT In Trachia Orpheo canente dicunt *Be, cf. La* 356.2-3 TRACIUS
TRANSIRE movere *Be*

.10-1 MANES CIERI LUNAMQUE LABORARE NONNE IPSIUS VETUSTATIS PERSUASIONE COMPERTUM?
MANES CIERI s. compertum est *La*
animas mortuorum evocari *BeLa*, evocare *La*
deos infernales *Leb*
LUNAMQUE LABORARE Sicut de Endimione fertur *BeLa*
COMPERTUM s. est *Be*

---

* Verg. Ecl. 8.57, cited in (among others) Isid. Etym. 1.36.7

356.11-2 IN LYDIA NYMPHARUM INSULAS DICI, QUAS ETIAM RECENTIOR ASSERENTIUM VARRO SE VIDISSE TESTATUR
DICI s. quis nesciat *BeLa*
QUAS insulas
RECENTIOR novus *BeLa*
iuvenior Varro *Leb*
iunior *Par*
ASSERENTIUM narrantium *La*

356.12-3 QUAE IN MEDIUM STAGNUM A CONTINENTI PROCEDENTES CANTU TIBIARUM
QUAE insulae *LebPar*
A CONTINENTI PROCEDENTES s. terra *BeLaPar*, s.] a *Par*

929/
356.14-5 IN ACTIACO LITORE MARE CITHARAM SONAT
IN ACTIACO i. continenti vel terra[m]: sic enim sepe vocatur *La*
†CYTHARA videlicet sicut cythara *BeLa*, vedilicet *Be* cithara *Be*

356.15 MEGARIS SAXUM AD ICTUM PULSUS CUIUSCUMQUE FIDICINAT
MEGARIS SAXUM nomen lapidis †DICTUM percussum *BeLa*
MEGARIS SAXUM mons *LebPar*
FIDICINAT Quasi cum fidibus sonat *Be*
Est in mari Mediterraneo saxum, quod si quis percusserit, velut cum fidibus resultat *BeLa*

356.15-6 POSSEM INNUMERA MORTALIBUS A ME COLLATA PERCURRERE
PERCURRERE numerare *La*

356.17 SED INGRATAE HUMANITATIS IGNAVIAM VIDERER IURE DAMNASSE
INGRATAE s. viderer. Non mihi gratias egit de mea arte, nisi mentiam (*sic, lege* mentiar?) *La*
IGNAVIAM stultitiam *Be*
IURE legaliter *BeLa*

356.18-9 SED IAM AD ARTIS PRAECEPTA DESILIAM, UT NUPTURAE VIRGINI PROMISSUM MUNUS IMPENDAM
DESILIAM procedam *BeLa*

UT NUPTURAE VIRGINI i. Philologiae *BeLa*, i.] nupturae *La*
VIRGINI a virgine *La*
†PROXIMUM promissum *La*

.20 ET QUONIAM OFFICIUM MEUM EST BENE MODULANDI SOLLERTIA
difinitio musicae *LebPar*, diffinitio *Leb*
ARS. INCIPIT DIFFINITIO *Par*

.20-1 QUAE RHYTHMICIS ET MELICIS ASTRUCTIONIBUS CONTINETUR
RITHMICIS numeris *La*, numerus *Be*
MELICIS i. in humana et organica voce *BeLa*, umana *La*
  in vocibus *Leb*

.21 PRIUS DE MELICIS DISSERTABO
i. de organicis vocibus disseram *BeLa*, i. *om. La*
in cordis *Leb*
Quae sonant extra humanam vocem *LebPar*

.22-3 DICO QUICQUID RITE SONUERIT AUT TONUM ESSE, AUT HEMITONIUM, AUT QUARTAM <PARTEM> TONI, QUAE DIESIS APPELLATUR
QUID SIT TONUS *BeLa*
cxxvi. diffinitio toni *Leb*
DICO perhibeo *Be*
  adfirmo *La*
AUT TONUM ESSE AUT EMITONIUM duplum, sesqualterum *LebPar*
AUT QUARTAM TONI partem *BeLaPar*, s. *add. Be*
  sesquiquartam *Par*
  i. quarta pars toni *Be*
  sed non semper quarta pars toni *La*
QUAE DIESIS APPELLATUR sesquiquinta *Leb*

.23-1 VERUM TONUS EST SPATIUM CỤM LEGITIMA QUANTITATE
VERUM TONUS EST ubi dupla *Leb*
SPATIUM a corda ad cordam *LebPar*, ad] in *Par*
CUM LEGITIMA QUANTITATE rationabili magnitudine *Be*
  cum rationabili mensura *La*
  Hoc est epogdous: una et octo *La*

Cum epiogdoo id est octava parte *Par*
cxxvi *Par*

357.1 QUI EX DUOBUS SONIS DIVERSIS INTER SE INVICEM CONTINETUR
EX DUOBUS SONIS DIVERSIS i. acuto et gravi *BeLaPar*, i. *om. Par*

357.2 HEMITONIUM DICITUR, QUOD TONI MEDIUM TENET
EMITONIUM emiolia vel sesqualtera *Leb*

357.2-3 DIESEON VERO DISTANTIAE TRES SUNT
†DIESIS quarta pars toni *Leb*
Proprie tamen quarta pars dicitur *Par*
DISTANTIAE i. formae *BeLa*, i. *om. La*

357.3-4 NAM PRIMA BREVIOR, QUAE TETARTEMORIA NOMINATUR EX EO, QUOD QUARTAM PARTEM TONI RECIPIAT
NAM PRIMA distantia *Leb*
TETARTEMORIA prima forma diesis *La*
quarta pars *LebPar*, pars] parte *Par*
NOMINATUR EX EO s. ita nominatur *La*

357.4-5 ENARMONIOS QUOQUE DICITUR PROPTEREA, QUOD ENARMONION MODULANDI GENUS PER HANC MAXIME DIMETIMUR
ENARMONIOS s. secunda forma *BeLa*
Tria sunt genera musicae: cromaticum, diatonum, enarmonion i. adiunctum *BeLa*, adiunctum] adunatum *Be*
Inadunatio, quia non semper aequaliter sonat et quod non una regula ibi est *LebPar*, quia non semper aequaliter sonat et *om. Par*
Vel valde adunatum, quia parta (*sic, lege* quarta) pars ibi tantummodo in differentia est et numquam habet emitonium *Leb*
Vel valde adunatum, tantummodo quia ibi quarta pars est in diferentia. Crommaton in iii parte, diatonus in dimidia. Vocantur a proprietate et qualitate *Par*
QUOQUE DICITUR Ipsa diesis, que vocatur tetartemoria *La*
PER HANC diesin *LebPar*
†DIMITTIMUS Hanc quartam divisionem Boetius non ponit *BeLa*

.5-6 SECUNDA AB ILLA MAIOR EST, NAM TRITEMORIA NOMINATUR
SECUNDA AB ILLA i. enarmonia i. tetartemoria *Be*
  s. distantia diesis s. tetartemoria *La*

.7-8 ITEMQUE CHROMATICE APPELLATUR, QUOD CHROMATICUM MODULANDI GENUS PER IPSAM FUNDITUR
CROMATICAE colorabile *BeLaLeb*
  Cromaticon gen<u>s dicitur colorabile. Croma enim Grece Latine color dicitur.* Sicut enim color dicitur, quicquid est medium inter album et nigrum, sic cromaticum modulandi gen<u>s in medio inter enarmonion et diatonon positum est. *La*
  Cromaticus in tertia parte, tonos in dimidia. Vocantur a proprietate et qualitate, quod non habet nec grave nec gracile i. mixtum de duobus i. de duabus extremitatibus medium, ut aliquis color inter album et nigrum *LebPar*, cromaticus in - et qualitate] colorabile *Par* aliquis] aliqua *Par*
APPELLATUR s. tritemoria *La*
PER IPSAM s. tritemoria *BeLa*
†FINDITUR partitur *BeLa*

.8-9 TERTIA VERO HABET TONI QUARTAM PARTEM AC DIMIDIAM QUARTAE
TERTIA VERO s. diesis distantia *BeLa*
  diatonus *Leb*
  differentia *Par*
†TERTIAM PARTEM AC DIMIDIAM †TERTIAE ut viiii ad vi, xviii ad xii *Leb*, viiii] viii *Leb*[1]
†TERTIAE i. semitonii *Leb*

.9-10 ET VOCATUR HEMIOLIA ENARMONIAE DIVISIONIS, QUONIAM HEMIOLII MODUM COMPLET
EMIOLIA i. semis et tonum *La*
  dimidium et totum *Leb*
†ET ARMONIAE s. ad diatonum *BeLa*, ad] a *La*
  musicae *LebPar*
  Armonia dicitur, quod diversos sonos adunat *LebPar*

---

* Cf. Isid. Etym. 10.45

DIVISIONIS s. est *Be*
EMIOLII MODUM i. sesqualteram proportionem *BeLa*
semipleni *LebPar*

931[*]/
357.11 VERUM SONI SUNT PER SINGULOS QUOSQUE AC PER OMNES TROPOS NUMERO XVIII
†VERO SONI praeposterus ordo *BeLa*, ordo praeposterus *La*
Generaliter xviii sunt soni et unusquisque ... et tropus xv. <Hab>et xviii sonos: non <ta>men alii in isto, <a>lii in illo, sed <i>sti ipsi qui in <i>sto sunt, in illo etiam inveniuntur. <I>nter sonum et tonum dif<f>ert: sonus est <proprie> in voce, <t>onus in spatio <v>oce (*sic, lege* vocis), scilicet inter diversas cordas *La*
TROPOS formas *BeLa*
qui sunt xv *La*
tropus forma *BeLa*
xv *LebPar*
Tropus est forma vocis *LebPar*, est *om. Par*

357.12-3 QUORUM PRIMUS DICITUR APUD GRAECOS ΠΡΟΣΛΑΜΒΑΝΟΜΕΝΟΣ
PRIMUS s. sonus *BeLa*
ΠΡΟCΛΑΜΒΑΝΟΜΕΝΟC i. adquisitus *BeLa*
Secundum Boetium prosmelodos dicitur proslambanomenos[**] *BeLa*, Boetium] boeticum *Be*
Iste est in omnibus *LebPar*

357.14 QUIA EADEM VOCE NOS UTI SUMMUS IUPPITER VETUIT, IDEM DICITUR ADQUISITUS
EADEM VOCE i. Romana *BeLa*
†STATUIT praecepit *BeLa*
IDEM sonus *LebPar*
ADQUISITUS Quia conveniunt illi omnes soni *BeLa*

---

[*] In *Par* we find next to this section a lost gloss, which in fact belongs to the part on rhythm and metre (972-995): Cur versus dividitur in pentimemerim et eptimemerim cum habet sex pedes

[**] Cf. Boeth. Inst.Mus. 1.20

Quia adquisitur ad omnes tropos *LebPar*

15 SECUNDUS ΥΠΑΤΗ ΥΠΑΤΩΝ, HOC EST PRINCIPALIS PRINCIPALIUM
SECUNDUS sonus *La*

15-7 TERTIUS ΠΑΡΥΠΑΤΗ ΥΠΑΤΩΝ, ID EST SUBPRINCIPALIS PRINCIPALIUM, QUARTUS ΥΠΑΤΩΝ ΔΙΑΤΟΝΟΣ, ID EST PRINCIPALIUM EXTENTA
ΠΑΡ- pro sub *BeLa*
ΥΠΑΤΟΝ ΔΙΑΤΟΝΟC Diatonus dicitur, quia abundat tonis *Leb*
PRINCIPALIUM EXTENTA Ipsa est penultima, quia acutiorem sonum habet i. penultimam *Leb*
Quia acutiorem sonum habet i. penultima *Par*

19 SEPTIMUS ΜΕΣΩΝ ΔΙΑΤΟΝΟΣ, ID EST MEDIARUM EXTENTA, OCTAVUS ΜΕΣΗ
MEDIARUM EXTENTA i. penultima *Leb*
OCTAVUS MECH Usque mece sunt grave<s> soni, a mece autem acuti sunt soni *La*

1 NONUS ΤΡΙΤΗ ΣΥΝΗΜΜΕΝΩΝ, ID EST TERTIA CONIUNCTARUM
NONUS TRITE CYNENMENON Secunda est a media, tertia ab ultima *BeLa*
TERTIA CONIUNCTARUM a fine *Leb*

2 HOC EST CONIUNCTARUM EXTENTA i. penultima *Leb*

4 HOC EST PROPE MEDIA iuxta MHCHN *La*

4-5 ID EST TERTIA DIVISARUM a fine *LebPar*

5-6 QUOD EST DIVISARUM EXTENTA i. penultima *Leb*

7 ID EST TERTIA EXCELLENTIUM a fine *LebPar*

8 HOC EST EXCELLENTIUM EXTENTA Quia tono distat ab ultima *BeLa*
a fine *Leb*

358.10 CUM RATIONE i. rationabiliter *BeLa*, i. *om. La*

932/
358.12 GRAVITAS DICITUR, QUAE SONI QUADAM REMISSIONE MOLLESCIT
†MODI i. modulationis *BeLaPar*
vel soni *Leb*
SONI vel modi *Par*
†EMISSIONE* productione MOLLESCIT gravescit *BeLa*
†EMISSIONE laxatione cordarum *LebPar*, cordarum *om. Par*

358.12-3 ACUMEN VERO, QUOD IN ACIEM TENUATAM GRACILIS ET ERECTAE MODULATIONIS EXTENDITUR
ACUMEN VERO s. dicitur *BeLa*
IN ACIEM in acumen *LebPar*, in *om. Par*
ET ERECTAE i. altae *BeLa*

933/
358.15 SYMPHONIAE TRES <SUNT>, QUARUM PRIMA EST DIATESSARON
SYMPHONIAE i. consonantiae s. sunt *BeLa*
DIATESSARON [figure 1] *BeLa*
ex quattuor *Leb*

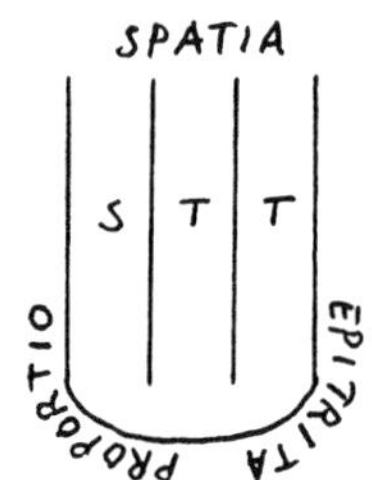

Figure 1

358.16-7 ET RECIPIT SONOS QUATTUOR, SPATIA TRIA, PRODUCTIONES DUAS ET DIMIDIAM
SONOS i. cordas *BeLaLebPar*, i. *om. Par*
SPATIA TRIA cordarum *LebPar*
PRODUCTIONES DUAS i. tonos *BeLaPar*, i. *om. Par*
duos tonos *Leb*

---

* REMISSIONE *Leb*

18 EST AUTEM HEMITONIORUM QUINQUE i. diatessaron *BeLaLeb*, i. *om.* *Leb*

18-9 QUAE AD PRODUCTIONES PLENAS ET INTEGRAS MEDIATENUS VALENT
MEDIATENUS Usque ad mediam quinarii *LebPar*

19-0 DIESIS VERO INTERPRETATIO EST, SICUT SUPRA DIXI, QUARTA PARS TONI
DIESIS ΔIECON s. est *Leb*
QUARTA PARS TONI Non negat Boetius sonum dividi in duas partes, sed in aequas partes dividi non posse dicit* *BeLa*
proprie *Leb*

21 EPITRITUS supertertius *Leb*

21-1 ALIA SYMPHONIA QUINARIA EST, ET DICITUR DIAPENTE
DIAPENTE Diapente nihil est aliud nisi tonus additus ad diatessaron *BeLa*, diatessarons *Be*
[figure 2] *BeLa*

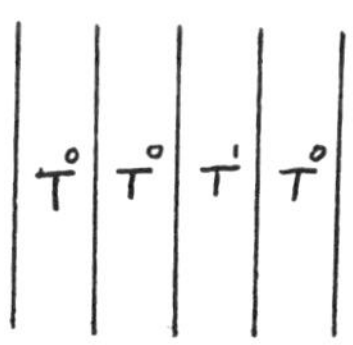

Figure 2

1-2 ATQUE CONSTAT SONIS QUINQUE, QUI INTER SE QUATTUOR SPATIIS DIVIDUNTUR
SONIS cordis *BeLa*
QUI soni *Be*

2 PRODUCTIONES HABET TRES MEDIAMQUE PRAECIDIT
PRODUCTIONES tonos *BeLeb*

* Cf. Boeth. Inst.Mus. 1.16

TRES MEDIAM i. tres et mediam separabit *BeLa*, separabit] separabiliter *La*
†QUE PRAECIDET Vel que discernit diatessaron et diapente *BeLa*

359.3-4 ATQUE HEMIOLII POSSIDET RATIONEM, QUAE FORMA ET EUNDEM NUMERUM CIRCA COLLATA DETINET ET EIUS MEDIUM
EMIOLII sesqualterae *BeLaLebPar*, i. *add. Be* sequaltera *Leb*
QUAE FORMA i. emiolii *La*
diapente *LebPar*
DETINET i. extremis comparatis *Be*

359.5-6 TERTIA DIAPASON, QUAE EX OMNIBUS DICITUR
TERTIA s. symphonia *BeLa*
DIAPASON simphonia diapason omne *La*
EX OMNIBUS DICITUR Ideo ex omnibus, quia veteres non <h>abebant nisi viii cordas *La*

359.6 OCTO SONOS RECIPIT, SPATIA SEPTEM, PRODUCTIONES SEX
SONOS cordas *LebPar*
SPATIA VII tonos *Be*
Hoc est tonos quamvis non simul positos *La*
PRODUCTIONES tonos *LebPar*

359.7-8 ATQUE CONSTAT EX RATIONE DIPLASIA, HOC EST DUPLI
ut i. ad ii. *Be*
Ut sit duo ad unum *La*

935/
359.9 TROPI VERO SUNT QUINDECIM, SED PRINCIPALES QUINQUE, QUIBUS BINI COHAERENT
TROPI i. modi *BeLa*
PRINCIPALES sunt v *Be*
BINI COHERENT i. superior et inferior *BeLa*, i. *om. La*
Primo <facis> materiam in animo simul cum gravitate aut etiam cum acumine. Ergo si libet tibi, ut ex gravioribus tropis alterum formes, praeparandum est opus aut fistula aut fidibus et caetera, aut etiam voce, similiter <facere potes> fide acutis *BeLa*, *difficile lectu in La*
BINI secundum numerum Grecae linguae *LebPar*, Grecae linguae] grecorum linguarum *Par*

ascensiones et descensiones *LebPar*

10 ID EST LYDIUS, CUI ADHAERENT HYPOLYDIUS ET HYPERLYDIUS
ID EST LIDIUS Qui gravius sonat *La*
in Lydia inventus *LebPar*
YPO pro sub *LaLeb*, pro *om. Leb*
i. sub, quia gravius sonat *Be*
in gravitate *LebPar*
YPER super *BeLeb*
in acumine *LebPar*

10-1 SECUNDUS IASTIUS, CUI SOCIATUR HYPOIASTIUS ET HYPERIASTIUS
IACTIUC Iones eo utuntur *Leb*
ANAX rex, asti (civitas). astianax, astianactis *Par*
YΠOIACTIOC nominativus singularis *BeLa*

12-13 QUARTUS PHRYGIUS CUM DUOBUS HYPOPHRYGIO ET HYPERPHRYGIO
quintus Dorius, cum *υποδοριο* et *υπερδοριο* *BeLa*
FRIGIUS a Frigia *Par*

14 VERUM INTER HOS TROPOS EST QUAEDAM AMICA CONCORDIA
VERUM sed TROPOS modos *BeLa*

17 ITEM CONVENIENS APTAQUE RESPONSIO INTER HYPOPHRYGIUM ET HYPOLYDIUM, QUI TAMQUAM DUPLICES COPULANTUR
YΠOΛIΔIOYM i. ypofrigius *BeLa*, i. *om. La* iprofigius *La*
<ypo->eolium *Par*
QUI †TANTUM DUPLICES COPULANTUR Subaudis: in gravioribus tonis vel tonus ad emitonium *Leb*
Germani, ut tonus ad emitonium in duplici proportione *LebPar*

17-1 MEDIAE VERO GRAVIORUM TROPORUM HIS, QUI ACUTIORES SUNT, ΠPOΣΛAMBANOMENAI FIUNT
MEDIAE cordae *LebPar*
HIS tropis *La*
ΠPOCΛAMBANOMENAY i. in gravissimis *BeLa*
Non solum proslambanomenai fiunt in gravissimis, sed etiam

in accutis tropis graves sunt, sed tamen non sic graves in acutis sicut in gravissimis *BeLa*
adquisiti *Leb*

360.1 VERUM TROPI SINGULI QUIQUE TETRACHORDA FACIUNT QUINA
VERUM sed *La*
QUINA i. quinquies in tropo duplicata *BeLa*

360.1-2 TETRACHORDUM AUTEM EST AFFECTIO QUAEDAM SONORUM QUATTUOR PER ORDINEM COMPOSITORUM
†TETRACORDA i. diatessaron *BeLa*
Ideo per diatessaron dividitur, quia elimentum est totius musicae *Par*
AFFECTIO copulatio *Be*
adiuncte *La*
convenientia *Par*
SONORUM cordarum *BeLa*
PER ORDINEM in *La*

360.3 QUORUM EXTREMI SIBI DEBEANT CONVENIRE
EXTREMI s. soni *Be*
DEBEANT CONVENIRE i. primi et ultimi *Be*
Verbi gratia: si quis dixerit mihi quomodo resonare debet ultima corda ad primam, tetracordon dicetur sub epitriti ratione *BeLa**

360.3-4 NUNC AD INCHOAMENTORUM RATUM ORDINEM REDEAMUS
INCHOAMENTORUM armoniae *LebPar*
REDEAMUS i. ad principium musicae artis *BeLa*, i. *om. La*

360.5 [UT ALTIUS TENERENTUR]
ALTIUS firmius s. res *BeLa*
profundius *Par*

---

* In *Be* we find a sign, tying gloss to lemma. In *La*, however, we find the same sign at the top of the gloss, but the scribe did not copy the sign next to the relevant lemma.

6 PRIUSQUAM ME LASUS, EX URBE HERMIONEA VIR, MORTALIBUS DIVULGARET
†PRIMO s. tempore *Be*
†LASSUS laboriosus *Be*
  proprium *Leb*
  auctor *Par*
EX †URSE* opido *Be*
  civitate *La*
  civitas *LebPar*
ERMIONEA (*sic, but the gloss seems to refer to* ENHARMONIA) s. tria genera *BeLa*, tria *om. Be*
†ARMENIAE provinciae. vel ermioniae *Par*
VIR musicus fuit *BeLa*
DIVULGARET pro divulgaverat *BeLa*, di- *om. Be*
  s. cum *LebPar*

7 TRIA TANTUM MEI GENERA PUTABANTUR: ΥΛΙΚΟΝ, ΑΠΕΡΓΑΣΤΙΚΟΝ, ΕΞΑΓΓΕΛΤΙΚΟΝ
GENERA i. musicae *BeLa*
†ΥΔΙΚΟΝ †ΕΝΕΡΓΑCΤΙΚΟΝ †ΕΖΑΝΤΕΑΤΙΚΟΝ in numeris, in forma, in sonis *BeLa*, in forma] i. forma *Be*
  energastidum i. operativum *BeLa*
  Yda forma. ΥΔΙΚΟΝ formabile. Prius enim formatur vox, deinde opus cantilenae, inde interpretatio i. iudicium *LebPar*, yda] ida *Par*
†ΑΠΑΡΓΑCΤΙΚΟΝ activum vel operativum *LebPar*, ΕΝΕRΓΑCΤΙΚΟΝ vel ΑΠΑΡΓΑCΤΙΚΟΝ *Par*

8 QUOD ETIAM ΕΡΜΗΝΕΥΤΙΚΟΝ DICITUR
†ERMENEITICON in verbis ab opere primo *BeLa*
  similiter in verbis *BeLa*
  interpretativum *Leb*

8 ET ΥΛΙΚΟΝ EST
†ΥΔΙCON i. formabile *BeLa*, i.] i. in *La*
[1]Primum genus est idicon, quo genere forme inveniuntur. [2]Forma-

---

* SURSE *La*[2]*Leb*

rum autem tres sunt species, quarum prima in sonis, secunda in numeris, tertia in verbis. [3]Illa autem prima i. quae in sonis invenitur, ad armoniam pertinet. [4]Non etenim in ea verba aliquid significantia, sed soni tantum sibimet ipsis aliqua ratione coniuncti queruntur, ut sunt sequentiae apud cantores. [5]Secunda autem que ad numeros pertinet nichil aliud requirit nisi tantum convenientem numerum sonorum, nichil significantium [aut nichil] aut aliquid significantium absque ullo certo termino, ut est scansio versuum. [6]Tertia vero, quae ad verba pertinet et vocatur metrica, non solum sonos certis temporibus terminatos inquirit, sed etiam cogit, ut aliquid significent et certum numerum terminumque non excedant. [7]Quorum exempla sunt versus et metrum, in quibus duobus certus terminus numerusque pedum ponitur, <in> versu[i] quidem in senario, in metro vero in octonario. [8]Sed notandum, quod hae tres formae in humanis vocibus inveniri possunt. [9]In armonia vero i. in his vocibus, quae carent sensu, non est nisi una i. prima. [10]In ea enim sonus tantum invenitur et non sine ratione numerorum. [11]In hac vero i. humana voce aliquando sonus sine sensu, non tamen sine numero invenitur, si aliqua proportione copuletur; aliquando sonus cum sensu et numero, quae verborum nomine meruit appelari *BeLa*, 2 in numeris] in nu in numeris *Be* in verbis] in menbris *Be*; 3 *post* invenitur] et *add. La*; 4 etenim] enim *Be* queruntur] quierunt *Be*[1] querunt *Be*[2]; 5 aut nichil] aut nihil *La*; 7 certus terminus] certis terminis *Be*; 10 et] tamen *Be*

360.8-9 QUOD EX PERSERVERANTIBUS ET SIMILIBUS CONSONABAT
PERSEVERANTIBUS immutabilibus cordis *LebPar*
SIMILIBUS inmutabilibus i. quod semper uno sonat sonitu *BeLa*, inmutabilis *La*
CONSONABAT tunc temporis *LebPar*

360.9 ID EST SONO, NUMERIS ATQUE VERBIS
in SONO *Be*
Sonus ad instrumenta, numerus ad versus, verba ad singula verba. Ut scias in omnibus verbis qui pes sit *La*
melodia, rithmis, metris *Leb*

360.9-10 SED QUAE EX HIS AD MELOS PERTINENT, HARMONICA
HIS tribus *LebPar*
AD MELOS ad sonum *BeLa*

ARMONICA Hoc est quod dixit sonos *BeLa*, dicit *La*
Ubi vox tantum consideratur *Leb*
vocis *Par*

.10 QUAE AD NUMEROS, RHYTHMICA Hoc est quod dixit numeros *BeLa*

.10-1 QUAE AD VERBA, METRICA Quod ad verba, metrica s. dicuntur *BeLa*

.11-2 ΑΠΕΡΓΑΣΤΙΚΟΝ EST QUIDAM MATERIAE TRACTATUS EFFICIENS EXERCITIUM EIUS
†ΥΠΕΡΓΑCTICON vel energasticon *BeLa*, emergasticon *Be*
MATERIAE vocis *Leb*
scilicet *Par*
†TRACTUS extensio *Be*
productio, confirmatio *La*
EXERCITIUM EIUS s. materiae *BeLa*

.12-3 CUIUS TRES ITIDEM PARTES, ID EST ΜΕΛΟΠΟΙΙΑ, ΛΗΠΣΙΣ, ΠΛΟΚΗ
PARTES s. materiei vel modi *BeLa*, matheriei *La*
ΜΕΛΟΠΟΕΙΑ melodiae factura *LebPar*, factura melodiae *Par*
†ΛΕΧΙC dictio vel narratio *LebPar*, dictio vel *om. Leb*
ΠΛΟΚΕ geminatio *LebPar*

.13-4 ΕΞΑΓΓΕΛΤΙΚΟΝ AUTEM AD EXPOSITIONEM PERTINERE VIDETUR ET HABET PARTES TRES
†ΕΚΑΝΤΕΛΤΙΚΟΝ interpretativum *Par*
EXPOSITIONEM Quia rationem reddit *LebPar*

.14 ΟΡΓΑΝΙΚΟΝ, ΩΔΙΚΟΝ, ΥΠΟΚΡΙΤΙΚΟΝ, QUAE INFERIUS RERUM ORDO DISPONET
ΟΡΓΑΝΙΚΟΝ, ΩΔΙΚΟΝ, ΥΠΟΚRITIKON in musis, in verbis, in numeris *BeLa*, in numeris *om. La*
ΩΔΙΚΟΝ cantabile ΥΠΟΚRITIKON simulativum *LebPar*
†INTERIUS inferius *BeLa*
docebit *BeLa*

.15 NUNC DE PRIMA VOCE VELUT DE SONITUS TOTIUS PARENTE

DICEMUS
VOCE i. sono *BeLa*

937/
360.16 OMNIS VOX IN DUO GENERA DIVIDITUR, CONTINUUM ATQUE DIVISUM
OMNIS VOX Mater musicae vox est *BeLa*
CONTINUUM assiduum. Ubi nulla est distinctio sicut in lectione *BeLa*
Sine colo et commate et periodo vel scansione *LebPar*, commate] comate *Par* periodo vel *om. Par*

360.17 DIVISUM, QUOD IN MODULATIONE SERVAMUS
DIVISUM econtra fit *LebPar*, fit *om. Leb*
MODULATIONE cantilene *BeLa*

361.1 EST ET MEDIUM
s. inter hos duos *BeLa*, inter] in *Be*
Quidam dixerunt, quod continuum ad omnes pertineret, divisus (*sic, lege* divisum) ad laudem deorum, medium ad heroes* *La*

361.1-2 QUOD IN UTROQUE PERMIXTUM AC NEQUE ALTERIUS CONTINUUM MODUM SERVAT NEC ALTERIUS FREQUENTI DIVISIONE PRAECIDITUR
IN UTROQUE genere *LebPar*
PERMIXTUM i. compositum est *BeLa*
NEQUE ALTERIUS i. continui *BeLa*
generis *LebPar*
NEC ALTERIUS i. divisi *BeLa*

361.2-3 HOC PRONUNTIANDI MODO CARMINA CUNCTA RECITANTUR
"Tytire, tu patulae recubans sub tegmine fagi"** *BeLa*

361.3-4 HORUM ILLA, QUAM IN DIVISAS PARTES CERTASQUE DEDUCIMUS, DIASTEMATICA NOMINATUR
ILLA vox *Leb*

---

* Cf. Boeth. Inst.Mus. 1.12
** Verg. Ecl. 1.1

DEDUCIMUS i. dividimus *BeLa*
DIASTEMATICA spatiosa *BeLaLebPar*, i. *add. Be* spaciosa *Par*
Propter tritemoriam et tetartemoriam et tonos et emitonia *BeLa*

4-5 ET EI PARTI, QUAE HARMONICA VOCATUR, APTANDA EST
†ENARMONICA Quae ad melos pertinet *Leb*
melodia *Par*
APTANDA EST iungenda est *BeLa*

6 QUAE QUIDEM HARMONICA HABET PARTES DISPUTATIONIS SEPTEM
Unum genus persequitur i. divisum *BeLa*

7-8 PRIMAM DE SONIS, SECUNDAM DE SPATIIS, TERTIAM DE SYSTEMATIS, QUARTAM DE GENERE, QUINTAM DE TONIS, SEXTAM DE COMMUTATIONIBUS, SEPTIMAM DE MODULATIONE
SECUNDAM DE SPATIIS i. sistematis *Be*
diastematis *LebPar*
DE SISTEMATIS de continuis vocibus *Leb*
Sistema est vox uniformiter stans *Par*
†QUARTO DE GENERE Adverbium posuit pro nominibus *La*

9 AC PRIUS DE SONIS, UBI EST ARTIS ELEMENTUM
Quoniam in corda ponenda est littera, vel qui sonus in ea sit gravis aut acutus *La*
ARTIS ELEMENTUM i. principium s. est dicendum *Be*
s. dicendum est principium *La*
materies *LebPar*

10-1 SONUS QUIPPE TANTI APUD NOS LOCI EST, QUANTI IN GEOMETRICIS SIGNUM, IN ARITHMETICIS SINGULUM
SONUS i. una corda musice *BeLa*
APUD NOS apud musicam *BeLa*
musicos *LebPar*
†LOCIS honoris *LebPar*
IN GEOMETRICIS SIGNUM i. punctus *BeLaLebPar*, i. *om. LebPar*
punctum *Par*

IN ARITHMETICIS SINGULUM i. unus *BeLa*
monos *LebPar*, monas *Leb*

361.11 PTHONGOS SONOS DICIMUS
PTONGOS Inter sonum et tonum differt: sonus est proprie in voce, tonus in spatio vocis inter cordas et cordas *BeLa*

361.11-2 PTHONGUS DICITUR VOCIS MODULATAE PARTICULA UNA INTENTIONE PRODUCTA
UNA INTENTIONE i. uno tenore *BeLa*
Ut sit aut acuta aut gravis *LebPar*

361.12-3 EST AUTEM INTENTIO, QUAM DICIMUS, TASIS, IN QUA VOX CONSISTIT AC PERSEVERAT
INTENTIO i. productio *BeLa*
IN QUA VOX i. finis vocis *La*

361.13-4 SONUS PTHONGOS VEL SPECIATIM VEL GENERALITER APPELLATUR
generalis videlicet et specialis *BeLa*
de uno sono, de omnibus *LebPar*

361.14-5 SED GENERALE VOCABULUM EST, SPECIALIA QUAE IΔIAITATA GRAIA VOCE PERHIBENTUR
Sed generale vocabulum habent specialia, quae ideotetra propriae positionis *Leb*
sed generalia et specialia idio$\theta\eta$ta *Par*
†IDEO TETRA* i. tongus generalis et specialis *La*
quaterna vel tetra i. positionis propriae *BeLa*, vel tetra *om. La*
positionis] pro positione *La*
Antiqui enim singulis sillabis notulas apponebant, quae produceretur, quae corriperetur, quae circumflecteretur, quae acueretur *BeLa, difficile lectu in La*
Intende (*sic, lege* intentio) facit, ut fiat acumen; remissio agit, ut sit gravitas. A quo ideo tetra dicuntur i. quatuor *La*
propriae positionis *LebPar*

---

* TEGRA *La*[1]

15-6 UT SI QUEMADMODUM NOBIS SCRIBENDUM SIT COGITEMUS
SCRIBENDUM SIT verum parvas ac magnas litteras *LebPar*, verum ... ac] utrum ... an *Par*

1 ACUMINIS EXERAMUS VEL LENIMINIS REMITTAMUS
EXERAMUS exaltemus vel pronuntiemus *BeLaLebPar*, vel pronuntiemus *om. LebPar*

.2 FACIUNT INTENTIO VEL REMISSIO, PATIUNTUR ACUMEN ET GRAVITAS
De generali sono dicit, hoc est de toto tropo *BeLa*
INTENTIO extensio cordarum *LebPar*

.3-4 PRODUCTIO AUTEM EST (HOC EST EPITASIS) VOCIS COMMOTIO A LOCO GRAVIORE IN ACUTUM LOCUM
PRODUCTIO i. motus vocis *Be*
intentio *LebPar*

.4-5 ANESIS VERO CONTRA; NAM AB ACUMINIS CULMINE IN GRAVE QUIDDAM SERIUMQUE DESCENDIT
ANESIS i. descensio vel positio deorsum *BeLa*
remissio *LebPar*
SERIUMQUE DESCENDIT i. molle remittit *BeLa*
SERIUMQUE tardum *LebPar*

.5-6 FIT AUTEM SONI GRAVITAS, CUM EX INTIMO QUIDAM SPIRITUS TRAHITUR
EX INTIMO s. pectore *BeLa*
SPIRITUS exemplum de humana voce *LebPar*

.7-8 SED SPECIALITER PER SINGULOS TROPOS VIGINTI OCTO TANTUM POTERUNT CONVENIRE
PER SINGULOS TROPOS In omni tropo tria genera considerantur musicae i. enarmonicum, diatonicum, cromaticum *BeLa*
Omnia ista in cantilena deprehendi possunt *BeLa*
XXVIII Supra xviii posuit, nunc xxviii. Si volueris diatonicum tantum commemorare, tunc x et viii habebis. Si autem cromaticum vel enarmonium intromiseris, x supra habebis et

erunt xxviii. Sub uno tamen nomine v cromatici et v enarmonici adduntur i. pro una corda. Sed subtiliter et obscure iste dixit. Boetius vero enucleatius patefecit* *BeLa*, et erunt xxviii *om. La*

941/
362.9 QUORUM PRIOR EST ADQUISITUS i. proslambanomenos *BeLa*, i. *om. La*

362.9-10 QUONIAM EORUM, QUAE TETRACHORDA NOMINANTUR, NULLI OMNINO CONSENTIT
EORUM s. sonorum *BeLa*
QUE TETRACORDA absolutae *BeLa*
Cythara: pectora vel a pectore cuiusdam animalis** *LebPar*
NULLI OMNINO proprie, sed omnibus generaliter *LebPar*, omnibus *om. Leb*

362.10-2 SED EXTRINSECUS VELUT ADQUISITUS ACCEDIT PROPTER CONSORTIUM MEDIAE, CUI CONCINERE CONSUEVIT
CONSORTIUM iuncturam *Be*
MEDIAE i. mese *Be*
Id est mese, quae omnibus accedit *LebPar*
CONCINERE Duplo enim superatur a media et efficit proportionem diapason *BeLa*

362.12-3 QUI ADQUISITUS UNO TONO A PRINCIPALI PRINCIPALIUM SEPARATUR, QUAE PRINCIPALIS PRINCIPALIUM
A PRINCIPALI PRINCIPALIUM i. ab *υπατε ιπατον BeLa*
PRINCIPALIS PRINCIPALIUM *πηριπατη ιπατον La*

362.14 QUASI CUIUSDAM RECTORIS NOMEN ACCEPIT
ypatos, consul vel rector *Leb*
Pato calco. Quia sicut dux calcat omnes sub se, ita ista omnes cordas premit *Leb*
Unde ypate dicitur: ypatos enim consol vel rector interpretatur *Par*
Macrovius Theodosius *Par*

---

* Cf. Boeth. Inst.Mus. 1.22
** Cf. Isid. Etym. 3.22.2

.14-5 SUBPRINCIPALIS DENIQUE DICITUR, QUAE PRINCIPALI SUBIECTA CONIUNGITUR

SUBPRINCIPALIS περυπατη ιπατον *Be*

QUAE PRINCIPALI SUBIECTA CONIUNGITUR Quia medium locum tenet inter enarmonicum et diatonicum *BeLa*

Prima dicitur mese, quia finis est graviorum et initium acutorum, vel quia quantum superat graviorem (i. adquisitus), tantum superatur ab acutiore (i. nete yperboleo, quae est ultima excellentium) i. proportione diapason *BeLa*, vel *om. La*
quantum] quartum *La* excellentium] excellente *Be*

.15-6 PRINCIPALIUM VERO ENHARMONIUS ET CHROMATICE (QUAM NOS VIX FORSAN RECTE COLORABILEM MEMORAMUS

ENARMONIUS †CROMATICUS Tres penultimae istae movent tria genera modulandi *LebPar*

COLORABILEM MEMORAMUS s. sequitur *BeLa*

.18 COLORABILE GRAIA SIGNIFICATIONE PERHIBETUR) intellectu dicitur *BeLa*

.19-0 ERGO ENHARMONIUS, CHROMATICE ITEMQUE DIATONUS, QUAM EXTENTAM DICIMUS, INDICIA SUNT GENERUM MODULANDI

EXTENTAM Quia tonis abundat *LebPar*

INDICIA i. manifestationes *BeLa*

.20-1 MULTIPLEX QUIPPE TETRACHORDORUM RATIO SIC FORMATUR

MULTIPLEX i. per additionem istarum penultimarum *LebPar*

SIC †FIRMATUR i. ex distinctionibus harum cordarum i. cromaticum, enarmonicum, diatonicum *BeLa*

Si confirmatur in principio tetracorda i. una corda, geminatur in omnibus tetracordis i. illa aut tertia *La*

.22-3 PRINCIPALIS AUTEM MEDIARUM, QUIA PRIMA EST MEDII TETRACHORDI, IDEO ID NOMEN ACCEPIT

PRINCIPALIS AUTEM MEDIARUM i. ypate meson *BePar*, i.] haec dicitur *Par*

lychanos vel ipate meson *Leb*

QUIA PRIMA s. est *Be*

IDEO ID NOMEN ACCEPIT enarrationem *La*

362.22-1 CUI SUBNEXA SUBPRINCIPALIS MEDIARUM ESSE DICITUR
SUBNEXA i. subposita *BeLa*, supposita *Be*
SUBPRINCIPALIS MEDIARUM paripate meson *Be*
pariπατε μεσον *La*

363.1 DENIQUE RELIQUAE TRES SIMILES USU OBOEDIENT
i. enarmonium, cromaticum, diatonicum *Be*
penultimae, enarmonicae, cromaticae, diatonicae, vel simili usu *LebPar*
Quia faciunt ipsae tres penultimae distantiam in tribus generibus modulandi *LebPar*, ipsae *om. Par*

943/
363.4-6 QUAE IDEO MEDIA DICITUR, QUIA TAM GRAVIS MODI FINIS EST, QUAM IN TROPIS OMNIBUS FUTURI ACUMINIS CAPUT
MEDIA Sicut mese est finis gravitatis, sic etiam est initium acuminis. Nam graves cordae praecedunt usque ad mesen, acutae subsequuntur ab illa *LebPar*, mese] in mese *Par* graves] greves *Leb*[1] usque ad mesen *om. Par* illa] illo *Par*
GRAVIS MODI FINIS EST s. in omnibus tropis *La*
soni *Par*
QUAM s. est *Be*
IN TROPIS OMNIBUS xv *Par*

363.6-7 ATQUE IPSIUS QUODAMMODO VINCULO TAM GRAVIS MODULATIO QUAM ACUTA CONECTITUR
IPSIUS s. mese *Be*
s. μησῃ *La*
CONECTITUR coniungitur *BeLa*

363.7 UT IN LYDIO MODO, UBI RECTUM I NOTA EST
IN LIDIO MODO s. apparet *BeLa*
tropo *Par*
RECTUM †IOTA EST littera Greca *La*

363.7-8 VERUM POST MEDIAM EXTENDENTI HEMITONIUM CONIUNCTARUM ERIT TERTIA

MEDIAM mese *LebPar*
EXTENDENTEM elevantem *BeLa*
EMITONIUM CONIUNCTARUM ERIT TERCIA i. trite senemmene i. secunda a prima, tercia ab ultima *BeLa*, senemmene] semmene *Be*
Est enim emitonio in mediam et tertiam coniunctarum a fine *La*
i. trite sinemmenon *Leb*
TERCIA a fine *LebPar*

.8-9 IN EODEM MODO i. tropo *BeLa*

.9 POST HANC i. trite coniunctarum *LebPar*, *post* trite] i. *add. Par*

.11 ΠΑΡΑΝΕΤΕ i. penultima *LebPar*

.12 POST QUAM ULTIMA SOCIATUR
ULTIMA s. tertia coniunctarum *La*
i. nete sineme<non> *Be*
i. nete *LebPar*

.12-3 QUAE IDEO DICTA EST CONIUNCTARUM ULTIMA
Quia mese coniungit *LebPar*

.14 QUIA MEDIAE IPSI, QUAE PERFECTAM SYMPHONIAM PRIMA COMPLEVIT, ADIUNGITUR
MEDIAE i. mese *LaPar*, i. *om. Par*
Ac subprincipali principalium facit plenum sistema *BeLa*
PERFECTAM SYMPHONIAM i. diapason *BeLa*
Octo cordas habet ante se i. diapason *LebPar*, habet ante se] habens *Par* diapason] diapasson *Par*

.15-6 †AC PRODUCENTE MEDIA SEQUENTEM† TONUM
MEDIA mese *Par*
mese, trite, paranete, nete, paramesos *Par*

.16 ILLE QUI SEQUITUR TONUS ΠΑΡΑΜΕΣΟΣ
ILLE i. tetracordum *BeLa*
SONUS ΠΑΡΑΜΕCΟC propemediam *La*
Quasi ad mese, quia iungitur spatio toni ad mesen *BeLa*, mese] messe *La*

iuxta mese *LebPar*

-MECOC Mese ideo media dicitur, quia ibi finitur omnis gravitas et incipit omnium cordarum acutio *Leb*

363.16-7 QUOD HUNC SATIS PROXIMUM SONUS ADIUNCTAE MODULATIONIS OFFENDAT

QUOD HUNC paramesen *LebPar*

ADIUNCTAE i. mediae *BeLa*

i. mese *Be*

MODULATIONIS sonus *Par*

OFFENDAT inveniat *BeLaLebPar*

944/
363.17-8 DEHINC DIVISARUM TERTIA, QUAE PLENUM SYSTEMA DIAPASON FINIT

DEHINC post parameson *LebPar*

DIVISARUM TERCIA s. ab ultima *BeLa*

trite divisarum *LebPar*

PLENUM SISTEMA i. magnitudinem *BeLa*, i. *om. Be*

spatium intervallum *LebPar*, spatium *om. Par*

DIAPASON FINIT A subprincipali principalium facit plenum sistema *La*

genitivus *LebPar*

Quia a trite divisarum usque ad lychanos principalem mediarum octo cordae sunt *LebPar*, lychanos] lichnos *Par*

363.19-1 UNDE DIVISARUM TERTIA PERHIBETUR; POST HANC SONI CETERI CONSEQUUNTUR

†CONIUNCTARUM TERTIA i. ab ultima *Be*

i. a fine tertia et a prima secunda *BeLa*, prima] primo *La*

†PROHIBETUR per<hibetur> *La*

POST HANC triten divisarum *LebPar*

SONI CAETERI enarmonicum, cromaticum, diatonicum *BeLa*

s. trium penultimarum *LebPar*, s. *om. Leb*

364.1-2 GENUS VERO DIVISARUM IDEO HOC NOMEN ACCEPIT

†SONUS tetigit hic diazeumenon *BeLa*, diezeumenon *Be*

i. tetracordum *LebPar*

.2 QUIA UNO ET DIMIDIO TONO A MEDIAE FINE DIVELLITUR
DIVELLITUR i. separatur *BeLa*, i. *om. La*
Quod post mesen sequitur tetracordum coniunctarum, ita ut primum locum mesen teneat, emitonio superatur ipsa media a sequente i. tertia coniunctarum. Quando vero sequitur tetracordum divisarum ita etiam, ut eadem media principium habeat, tunc a proxima corda hoc est a paramese tono superatur. Quando vero tetracordum divisarum inchoat, ita ut mese principium non habeat, sed ipsa paramese, tunc paramese a media uno tono et emitonio separatur atque ideo divisarum tetracordum appellatur. Et ideo dictum est uno et dimidio tono a fine medie divellitur *BeLa*
In uno enim spatio tria emitonia sunt *LebPar*
MEDIAE quae est mese *LebPar*

.3 POST HOC TETRACHORDUM EXCELLENTIUM COLLOCATUR i. post tetracordum divisarum *BeLa*, postetracordum *Be*

.4-5 IN SINGULIS MODULATIONIBUS FASTIGATUR
FASTIGATUR acuminatur *BeLa*
acuitur *LebPar*

.6-7 HORUM IGITUR SONORUM, ID EST ΦΘΟΓΓΩΝ, SUNT ALII, QUOS CONSISTERE ET PERSERVERARE NECESSE EST
HORUM IGITUR SONORUM de omnibus predictis xxviii *La*
xxviii *LebPar*
†PTOGGON genitivus pluralis *Leb*
SUNT ALII sicut proslambanomenos *BeLa*
principalis et subprincipalis et nete *LebPar*, et nete *om. Par*
CONSISTERE ET PERSEVERARE ut mese et nete excellentium *Be*
Sine ulla mutatione eum sustinent *La*

.7 ALII VERO SUNT VAGI
i. mobiles sicut paramese *BeLa*, mobile *La*
Varii ut penultimae, quia una semper modulatio esset, nisi ipsae variarentur *LebPar*, modulatio esset] esset modulatio *Par*

.7-8 DENIQUE ALII ΒΑΡΥΠΥΚΝΟΙ NOMINANTUR, ALII ΜΕΣΟΠΥΚΝΟΙ,

< ALII ΟΞΥΠΥΚΝΟΙ, ALII ΑΠΥΚΝΟΙ > PERHIBENTUR
ΒΑΡΥΠΙΓΝΟΥ i. primi spissi *BeLa*
    Graviter fixi; semper enim stant *LebPar*, stant] sunt *Leb*
†ΜΕCΟΠΙΓΝΟS i. medii spissi *BeLa*
    Medii fixi; per vices gravantur et acuuntur *LebPar*, fixi *om. Leb*
PERHIBENTUR i. ultimi spissi vel alii oky*πικνοι* *BeLa*, difficile lectu in *La*

364.9 SPISSUM VERO DICITUR TRIUM SONORUM COMPOSITIVA QUAEDAM QUALITAS
SPISSUM Unum est quod baripygnoy *Leb*
TRIUM SONORUM COMPOSITIVA i. minoris *BeLa*
    Cum tres soni simul iunguntur *La*
    Cordarum, quia non parum differunt in tribus penultimis *LebPar*

364.10-1 ΒΑΡΥΠΥΚΝΟΙ AUTEM SUNT, QUI VELUT REGIONES PRIMAS SPISSI RETENTANT, ΜΕΣΟΠΥΚΝΟΙ VERO, QUI MEDIA POSSIDENT
ΒΑΡΥΠΙΓΝΟΥ i. spissi in gravioribus cordis i. in tetracordo (*sic, lege* tetracordum) principalium *BeLa*
†ΜΕCΟΠΙΓΝΟS Tetracorda continet mediarum, coniunctarum, disiunctarum *BeLa*
QUI MEDIA Nec tam graviter nec acute sonant *LebPar*, sonant] sonat *Par*

364.11-3 ΟΞΥΠΥΚΝΟΙ, QUI ULTIMA TENENT, ΑΠΥΚΝΟΙ QUI IN POSITIONE SONORUM TRIUM, QUI SUNT SPISSI, NULLO GENERE AUT LEGE IUNGUNTUR
ΟΚΥΠΙΓΝΟΥ Ultimi spissi s. in tetracordo (*sic, lege* tetracordum) excellentium *BeLa*
    acute fixi *LebPar*
ΑΠΙΓΝΟΙ inspissi *BeLa*
    non fixi *LebPar*, fixi *om. Leb*
SPISSI s. soni tres *LebPar*

946/
364.13-4 STANTES AUTEM PERSEVERANTESQUE DICUNTUR VEL ΑΠΥΚΝΟΙ VEL ΒΑΡΥΠΥΚΝΟΙ
VEL ΑΠΙΓΝΟΥ i. non spissi *BeLa*
    A Boetio unisonae et dissonae vocantur. Fortassis

Aristoxenum secutus est *BeLa*
VEL BARYΠΙΙΓΝΟΙ i. spissi *BeLa*

.15 IDEO AUTEM QUIDAM STATARII NOMINANTUR
STATARII i. stabiles *BeLa*

.16-7 QUIA INTERDUM LARGIORA, INTERDUM MINORA SPATIA RECEPERUNT
LARGIORA in gravitate *Leb*
quando gravius *Par*
MINORA in acumine *LebPar*

.17-8 SED HI ALII ΠΑΡΥΠΑΤΟΕΙΔΕΙΣ, ALII ΛΙΧΑΝΟΕΙΔΕΙΣ VOCANTUR
ΠΑΡΙΠΑΤΟΥΔΙC i. subprincipales *BeLa*
iuxta principales vel subprincipales *LebPar*, vel subprincipales *om. Par*
ALII †ΛΥΚΑΟΝΙΔΙC i. dactilici *BeLa*
vigilantes *LebPar*

.19 QUOD STATIM PRINCIPALIBUS PRIORIBUS SUBNECTANTUR
PRIORIBUS s. sonis *La*
principalibus *LebPar*
SUBNECTANTUR post sta<ta>rios *La*

.19-2 ΛΙΧΑΝΟΕΙΔΕΙΣ DICUNTUR A DIGITO, QUI MINISTER ARTIS ET CANTUS SINGULOS MOVET, EST AUTEM PRIMUS A POLLICE
†ΛΥΚΑΟΝΙΔΙC digitales *Leb*
PRIMUS A POLLICE Primo pollice movetur corda *BeLa*

7/

5.2 HI VERO ALII SIBI INVICEM CONGRUUNT, ALII DISCREPANT ET RESULTANT
†HII VERO s. de univoce *La*
soni *LebPar*
DISCREPANT dissonant *BeLa*

5.3 SED ILLI ΣΥΜΦΩΝΟΙ, QUIA SIBI INVICEM CONIUNGUNTUR
†ΟΜΟΦΟΝΙ s. dicuntur *BeLa*
similis vocis *La*

365.3-4 ΔΙΑΦΩΝΟΙ AUTEM, ID EST DISSENTIENTES
ΔΙΑΦΟΝΙ dissoni *Leb*
aφoni id est dissoni *Par*

365.4 ΟΜΟΦΩΝΟΙ, QUI VOCIS QUIDEM ALIAM SIGNIFICATIONEM GERUNT
ΟΜΟΦΟΝΙ i. in simili voce vel aequivoci *BeLa*
aequisoni *LebPar*
QUI VOCIS Id est aliud nomen i. vel mediam, vel gravem, vel acutam *BeLa*, Id est *om. La*
SIGNIFICATIONEM i. non in uno genere intelleguntur *La*
virtutem *LebPar*
GERUNT s. in tono, in emitonio, in diesi *LebPar*

365.5 EUNDEM TAMEN IMPETUM SERVANT
EUNDEM i. unum sonum *La*
IMPETUM vocis *La*
pulsum *LebPar*

365.6 SUNT AUTEM ET ALIAE SONORUM DIVERSITATES
SUNT AUTEM propter singulorum sonorum *BeLa*
ET ALIAE exceptis praedictis *La*

365.6-7 ET PRIMA QUIDEM PER INTENTIONEM, UT AUT ACUMINE AUT GRAVITATE DISSENTIAT
ET PRIMA s. diversitas *La*
INTENTIONEM cordarum *LebPar*
DISSENTIAT s. vox *La*

365.7-8 SECUNDA PER SPATIORUM PERCEPTIONEM, CUM UNI AUT PLURIBUS CONIUNGITUR SPATIIS
SECUNDA s. diversitas *BeLa*
SPATIORUM intervallorum *Leb*
CUM UNI AUT †PLURIMI CONIUNGITUR vox *La*

365.8-9 TERTIA PER CONIUNCTIONEM SYSTEMATIS, QUOD AUT UNUM AUT PLURA RECIPIT
TERCIA s. diversitas *BeLa*
SISTEMATIS i. magnitudinis vocis s. sistema *BeLa*, i. *om. La*

persistens vox *Leb*
Sistema persistens vox vel uniformiter stans *Par*
†QUO AUT UNUM s. spatium *LaLebPar*, s. *om. Par*
AUT PLURA s. sistemata *Be*
Tropus <for>matur, eo quod vo... gens secundum... *La*, difficile lectu

.10 QUOD SECUNDUM MOREM DICITUR, ID EST KATὰ Tὰ ἤTH
SECUNDUM MOREM i. humanam naturam *BeLa*, umanam *La*
KATA EΘOC secundum morem *LebPar*

.10-1 ALIUM QUIPPE MOREM ACUTA SIGNIFICANT, ALIUM GRAVIORA
ALIUM QUIPPE MOREM i. usum *Be*
i. secundum proprietatem uniuscuiusque tropi *La*
ACUTA ex distentis *BeLa*
GRAVIORA ex remissis *BeLa*

3/

.12 NUNC IAM DE DIASTEMATIS EDISSERENDUM
i. de magnitudine vocis *BeLa*
QUID DIASTEMA *BeLa*
Aliud est sistema, aliud diastema. Sistema magnitudo vocis, diastema spatium vocis hoc est locus inter cordam et cordam *BeLa*

5.12-3 DIASTEMA EST VOCIS SPATIUM, QUO ACUTA ET GRAVIOR INCLUDITUR
†DIASTEMATA spatium inter cordam et cordam *Be*
Spatium intervallum, quod fit in viii cordis *LebPar*, spatium intervallum *om. Par*
ACUTA ET GRAVIOR i. corda vel sonus *BeLa*
vox *LebPar*

5.13-4 ALIA <BREVIORA SUNT, ALIA MAIORA>
BREVIORA pro brevissima *BeLa*
Quia nichil est aliud diesis nisi quarta pars toni *BeLa*
spatia *LebPar*
MAIORA i. maxima *BeLa*

5.14-5 BREVIORA ILLA, QUAE SUNT IN DIESI ENHARMONIA, MAIORA

VERO SUNT QUE PER SINGULOS TROPOS BIS EX OMNIBUS FACIUNT

IN †DIESIN i. quarta parte toni *LebPar*, i. *om. Leb*

BIS EX OMNIBUS i. bis diapason *BeLaLebPar*, i.] s. *La*, *om. Leb* diapasson *Par*

365.15-16 QUO NIHIL MAIUS IN TROPIS POSSUMUS INVENIRE

NICHIL MAIUS i. bis diapason *BeLa*

IN TROPIS Exemplum minus in <proslambanomenon et> ipate ipaton, exemplum maius in proslambanomenon et nete excellentium *BeLa*

949/
365.16-7 ATQUE IN SPATIIS ALIA SUNT COMPOSITA, ALIA DISIUNCTA ATQUE ASYNTHETA

COMPOSITA Compositum est, quando multa spatia similiter iunguntur, nulla interposita corda. Adiunctum est, quod, quamvis videatur esse compositum, tamen videtur esse incompositum; interponuntur enim cordae. Unaquaque corda, quae cum alia iungitur, loica dicitur; quae non iungitur aloga *BeLa*

in spatiis *LebPar*

ASINTETA incomposita *BeLaLebPar*

365.17-8 ET COMPOSITA SUNT, QUAE PER ORDINEM CURRUNT, INCOMPOSITA AUTEM, QUAE EX DIVERSIS SIBI INVICEM COPULANTUR

ET COMPOSITA SUNT i. quando una corda consentit alia *BeLa*

PER ORDINEM Sicut ordo cordarum, sic etiam spatiorum *LebPar*

INCOMPOSITA AUTEM s. sunt *BeLa*

i. quando una corda non consequenti[s] <sed> tertiae aut quartae coniungitur in sono *BeLa*, i. *om. La* consequentis] consentientis *La*

EX DIVERSIS ultra duas aut tres cordas *LebPar*

365.18-1 ITEM ALIA LOGICA, ALIA ALOGA MEMORANTUR

†LOYCA rationabilia *BeLaLebPar*

ALOGA inrationabilia *BeLebPar*, irrationabilia *Leb*

<in>rationa<bilia> *La*

.1-2 ATQUE RATIONABILIA ILLA SUNT, QUORUM CONSENSUS POSSUMUS PRAESTARE PROPORTIONEM
RATIONABILIA i. loica *La*
QUORUM s. sunt *BeLa*
genitivus *Leb*
CONSENSUS unitatem *BeLa*, unitate *La*
PRESTARE dicere *BeLa*
ostendere *LebPar*
PROPORTIONEM i. possumus dicere, quam proportionem ad invicem habeant *BeLa*
Utrum dupla an sesqualtera vel sesquitertia sit *LebPar*, vel] an *Par*

.2-3 IRRATIONABILIA, QUIBUS NON SUBEST RATIO
INRATIONABILIA aloga *La*
Nesciens, quod facit *Leb*
Quod nesciens facit *Par*

.3-4 ITEM ALIA CONVENIENTIA, ALIA DISCREPANTIA, TUNCQUE ALIA ENHARMONIA, ALIA CHROMATICA, ALIA DIATONICA
†CONHIBENTIA s. consentanea sibi *BeLa*
coeuntia sibi vel concordantia *LebPar*
DISCREPANTIA inconsona *La*
ENARMONIA quarta pars *LebPar*
CROMATICA spatia *BeLa*
emitonium *LebPar*
DIATONICA toni *LebPar*

.4-5 ITEM ALIA ARTIA, ALIA PERISSA
†ARCHIA i. aequalia *BeLaLebPar*, i. *om. LaLebPar*
APKOC princeps vel equalis *Par*
PERISSA i. inaequalia *BeLa*, i. *om. La*
excurrentia *LebPar*

.6 UT TONUS IN DUO HEMITONIA Non negat Bo<etius> *Be*

.6-7 PERISSA AUTEM, QUAE IN TRIA HEMITONIA DISCERNUNTUR
PERISSA Excurrentia vel discurrentia, quia currunt in tria emitonia, quod fit in cromatico genere *BeLa*, quia] que *La*
IN TRIA EMITONIA tonum et emitonium *LebPar*, et *om. Leb*

950/
366.7-8 DEINDE ALIA DIASTEMATA SPISSA, ALIA RARIORA
DIASTEMATA SPISSA i. spatia acuta *BeLa*, i. *om. La*
Hoc fit in <en>armonio genere *BeLa*, genere *om. La*
RARIORA s. sunt *Be*
graviora *La*

366.8 SPISSA SUNT, QUAE PER DIESIS COLLIGUNTUR, RARIORA, QUAE TONIS
SPISSA SUNT Quia non potest esse minus spatium quam in illis *LebPar*
QUAE PER DIESIS COLLIGUNTUR Cum enim quarta pars differentiae ponitur spissum, non enim multum separatur *La*
QUAE TONIS s. colliguntur *BeLa*

366.9 ET IN HIS ALIA SIBI CONGRUUNT, ALIA DISCREPANT
HIS spatiis *LaPar*

366.10 CONVENIENTIA VERO PER SINGULOS TROPOS SEX SUNT
CONVENIENTIA s. symphoniae *La*
PER SINGULOS TROPOS SEX genera *LebPar*

366.11 ΔΙΑ ΠΕΝΤΕ, QUOD DE QUINQUE s. sonis *La*

366.11-2 ΔΙΑ ΠΑΣΩΝ, QUOD EX OMNIBUS CONTINET Quando duplo spatio vox superat vocem *La*
Hae tres symphonie simplices dicuntur, tres relique compositae vocantur *La*

366.12-4 ILLUD ETIAM, QUOD EX OMNIBUS ET EX QUATTUOR CONSTAT, VEL EX OMNIBUS ET EX QUINQUE, VEL BIS EX OMNIBUS, QUOD ΔΙΣ ΔΙΑ ΠΑΣΩΝ DICITUR
ILLUT s. est *La*
QUOD EX OMNIBUS ET EX QUATTUOR i. ex diapason et diatessaron *BeLa*, ex *om. Be*
EX OMNIBUS ET EX QUINQUE i. diapason et diapente *BeLa*

951/
366.18 IN HEMIOLIA RATIONE in sesqualtera *La*

.1 HEMITONIA GEMINAT i. xii *Be*

.1 IN DIPLASIA RATIONE dupla *Par*

.3-4 SONORUM EST UNDECIM SPATIORUM DECEM, TONOS HABET OCTO SEMIS, †HEMITONIUM EADEM†, DIESISQUE PRAEDICTA RATIONE MULTIPLICAT
SONORUM EST XI Quia finis diapason initium est diatessaron, ideo xi dixit *BeLa*
Cur non xii? Quia octava bis computatur *Par*
†EMITONIA xvii *Par*
PRAEDICTA RATIONE MULTIPLICAT i. multiplex sesqualtera *BeLa*
i. quadruplicat xxxiiii *LebPar*, i. *om. Par*

.4-5 ET EST IN DIPLASIEPIDIMOERA REGULA
†IN DYPLASI i. super duplam duo habet; viii ad iii *LebPar*
†ET DIAMETRI REGULA In du<o>bus semper fit diametrum *Leb*
Ubicumque binarius fuerat, diametrum est *Par*

.5-6 QUOD VERO EX OMNIBUS <ET EX QUINQUE> CONSTAT
EX OMNIBUS et ex v *Be*
i. diapason *La*

.7 HEMITONIA GEMINANTUR, DIESIS TETRAPLANTUR
EMITONIA GEMINANTUR xviiii *La*
DIESIS TETRAPLANTUR i. quadruplicantur *BeLa*, i. *om. La*
i. xxxviii *La*

.8-10 QUOD ΔΙΣ ΔΙΑ ΠΑΣΩΝ DICITUR, HABET SONOS XV, SPATIA BIS SEPTENA, TONOS DECEM AC DUO, HEMITONIA DUPLICATA, DIESIS XLVIII
†ΔΙΕΔΙΑΠΑCON dicitur *Leb*
TONOS DECEM AC DUO i. emitonia iiii *Leb*
EMITONIA DUPLICATA Quod sit in cromatico genere *BeLa*
i. emitonia iiii *LebPar*

.11 EST AUTEM TONUS IN EPOGDOI RATIONE

IN EPOCDOI RATIONE Ut sunt viii ad viiii *Leb*
ut viiii ad viii *Par*

367.11 UBIQUE DIESIN ENARMONION DEBEMUS ACCIPERE
†IN ARMONIO Diesis enim in enarmonio tantum genere ponitur *BeLa*

954/
367.13 NUNC QUID SIT SISTEMA PERHIBENDUM
Quid sit sistema *BeLa*
Sistema non potest esse minus quam in diapason, diastema potest inter duas aut tres aut quattuor cordas esse *LebPar*

367.13-4 MAGNITUDO VOCIS EX MULTIS MODIS CONSTANS, QUAE LICET MULTA DIVISIONUM GENERA RECIPIAT
EX MULTIS MODIS i. ex multis symphoniis: diatessaron, diapente, diapason *BeLa*
i. tonis *LebPar*, i. *om. Par*
GENERA RECIPIAT sicut diastema *LebPar*

367.15 QUIA EADEM ET IN DIASTEMATIS MEMORAVI, PRAETEREO
ET IN DIASTEMATIS MEMORAVI Id est diapason tantum commemorat *BeLa*
†PRAETEREAM dimittam *BeLa*

367.16-8 QUOD AB ADQUISITO, QUEM ΠΡΟΣΛΑΜΒΑΝΟΜΕΝΟΝ DICIMUS, AD MEDIAM, QUAM ΜΕΣΗΝ DIXIMUS, OMNE CONFICITUR
ΠΡΟCΛΑΜΒΑΝΟΜΕΝΟΝ i. primo *BeLa*
[Qui est pro<pe> mediam] *La, cf.* 367.18-9 ΠΑΡΑΜΕCON
MESEN i. octavam *BeLa*
OMNE i. diapason *BeLa*

367.18-9 SECUNDUM, QUOD A PRINCIPALI PRINCIPALIUM IN ΠΑΡΑΜΕΣΟΝ USQUE TENDITUR
SECUNDUM s. est
A PRINCIPALI PRINCIPALIUM ypate ipaton. secundo *BeLa*, ipate *La*
IN ΠΑΡΑΜΕCON i. duodecimum, hoc est prope mediam *BeLa*

367.19-1 TERTIUM, QUOD A SUBPRINCIPALI PRINCIPALIUM IN DIVISARUM

TERTIAM IUNGITUR
TERTIUM s. est *Be*
A SUBPRINCIPALI PRINCIPALIUM i. paripate ipaton. tertio *BeLa*, i.] in *La*
IN DIVISARUM †TERTIUM i. in trite diezimenon i. tertium decimum *BeLa*

.1-2 QUARTUM, QUOD AB EXTENTA PRINCIPALIUM USQUE IN DIVISARUM DIATONUM PROFERTUR
QUARTUM s. est
EXTENTA PRINCIPALIUM i. ipaton diatonos. quarto *BeLa*
IN DIVISARUM DIATONUM i. in quartam decimum *BeLa*, in *om. Be*

.2-3 QUINTUM, QUOD A PRINCIPALI MEDIARUM IN NETEN DIVISARUM USQUE PROGREDITUR
A PRINCIPALI MEDIARUM ipate meson. hoc est quinto *BeLa*
IN NETEM DIVISARUM i. usque in ultimam divisarum *BeLa*
unicam (*sic, lege forte* ultimam?) *La*

.3-4 SEXTUM, QUOD A SUBPRINCIPALI MEDIARUM IN TERTIAM EXCELLENTIUM TRAHITUR
A SUBPRINCIPALI MEDIARUM i. paripate meson. sexto *BeLa*, sexto] quinto *Be*
IN TERTIAM EXCELLENTIUM i. in trite iperboleon. sexto decimo *BeLa*
triten hiperboleon *LebPar*, yperboleon *Par*
TRAHITUR ducitur *La*

.4-5 SEPTIMUM, QUOD A MEDIARUM EXTENTA IN EXCELLENTIUM DIATONON EXCURRIT
A MEDIARUM EXTENTA i. meson diatonon *BeLa*
IN EXCELLENTIUM DIATONON i. septimum decimum *BeLa*

.5-6 OCTAVUM, QUOD A MEDIA EXCELLENTIUM IN ULTIMAM DUCITUR
A MEDIA a mese *LebPar*^{App}, a *om. Par*^{App}
IN †ULTIMUM i. in nete iperboleon *BeLa*

/
.7 QUIBUS EXCURSIS GENERA TETRACHORDORUM MODULANDIQUE

DISCURRAM
QUIBUS EXCURSIS i. sistematibus *BeLa*
GENERA †CORDUM Quartum genus persequitur, quod superius post diastema posuit i. diatonum *BeLa*

368.8 †TETRACHORDUS EST CHORDARUM† CUM CERTA QUALITATE DIVISIO
QUID SIT TETRARDUS* *BeLa*
magnitudine. Qualitas est, utrum principalis (*sic, lege* principalium) an divisarum et ceterorum sit *LebPar*^{App}

956/
368.9-10 ET HARMONIA QUIDEM DICITUR, QUOD PLURIBUS SPATIIS ET ANGUSTIORIBUS SEPARATUR
SPATIIS <inte>gris tonis *Par*^{App}
ANGUSTIORIBUS spatiis *La*
proportionibus *Par*^{App}
SEPARATUR i. distinguitur *BeLa*, distingitur *La*
Propter diesis; sic enim dicitur tonus, tonus, diesis *Par*^{App}

368.10-1 DIATONON VERO, QUOD TONIS COPIOSUM
Quia tonus in primo, tonus in secundo *BeLa*
COPIOSUM s. est *BeLa*
spatiosum *BeLa*
tonus tonus emitonium <...> *Par*^{App}

957/
368.13 AB IMMUTABILIBUS Gravibus, qui<a> numquam transeunt in acumen *LebPar*^{App}

368.14 MODULATIONI OBOEDIET IN ACUMEN PER DIESIN <ET DIESIN> ET DITONON IMCOMPOSITUM
OBOEDIET IN ACUMEN asce<n>det *BeLa*, abscedet *La*
PER DIESIN ET †DIATONOS s. surgit i. per duos tonos *BeLa*, *post* duos tonos] illud gen<u>s armoniae *add. La*
per duos tonos *LebPar*^{App}

---

* This is illustrated by a figure in the upper margin of the next page in *La*: see Plate 3 in Chapter 3 (p. 186).

per diesin et duos tonos *Leb*
INCOMPOSITUM Quia simul sunt duo toni. Si fuerint duo toni in uno spatio, incompositum est. Quicquid vero ex diversis colligitur, compositum est, quicquid per se est in uno spatio, incompositum est *LebPar*^{App}, vero *om. Leb* *post* incompositum] est *om. Leb*

.15[*] IN GRAVEM VERO SONUM E CONTRARIO MODULATIO TOTA SUBMITTITUR
†TOTO s. a *Be*

.16-7 PER HEMITONIUM ET HEMITONIUM ET TRIA HEMITONIA, QUAE INCOMPOSITA PROVENIET
PER EMITONIUM s. ascendendo *BeLa*
ET TRIA EMITONIA Quia in uno spatio tria emitonia habent et non componitur ex multis spatiis *La*
INCOMPOSITA Quia in uno spatio sunt *LebPar*^{App}

.17-1 IN GRAVEM VERO EX DIVERSO RECURRET
IN GRAVEM s. melodiam *La*
EX DIVERSO contrario *LebPar*^{App}

.1-2 AT VERO DIATONUM IPSIS CONTENTUM SONIS IN ACUMEN QUIDEM ITA ACCIPIET MODULATIONEM
†CONVENTUM coniunctum *La*

.4 SED NUNC MAXIME DIATONO UTIMUR i. usui habemus eo, quod pulcrior caeteris sit *BeLa*, pulchrior *La*

.4-5 SED HORUM ALIA MODULAMUR PER ΑΓΩΓΗΝ, ALIA PER ΠΛΟΚΗΝ
SED HORUM generum *LebPar*^{App}
PER AGOGEN ascensionem *LebPar*^{App}
PER PLOCEN copulationem vel geminationem *LebPar*^{App}

---

[*] In the upper margin we find a gloss in a different hand in *La*: Hic tres vel iiii^{or} lineas dimitte. Supposedly it refers to the lines 368.16-369.4 CROMA AUTEM HOC MODO ... MAXIME DIATONO UTIMUR. The meaning of the gloss, however, is not clear.

369.5-6 CUM PER ORDINEM SONUS SEQUITUR De corda in cordam, in acumen de gravi *LebPar*^{App}, de gravi in acumen *Par*^{App}

369.6 CUM DIVERSA SOCIAMUS i. iungimus quartam cordam primae *BeLa*
DIVERSA spatia. Cum una corda geminatur et alia dimittitur *Leb*

369.7 HOC EST CIRCUMSTANS est etiam rotunda *Par*^{App}

369.7-8 EYΘEIA EST, QUAE A GRAVI IN ACUMEN ERIGITUR
EUTIA Quando quis a gravioribus ad acutiora conscendit *BeLa*, acutiora] atiora *Be*
Unde eutigrammos i. recta linea* *LebPar*^{App}, eutigrammos] EYTI ΓPAMMOC *Par*^{App} i. *om.* *Par*^{App}

369.8 PERIFERES i. et ad superiores et ad inferiores plausis (*sic, lege* plausibus?) convertitur *BeLa*
rotunda *Leb*

369.10 AUT COMMODATUR AUT SERVIT i. aut intenditur aut re<la>xatur, i. acuitur aut gravatur *La*
sursum ad acumen, ad gravem *LebPar*^{App}

959/
369.11 QUADRIFIDOS a fidibus *Par*^{App}

369.12 SEX SUNT NOTAE manifeste, usitate *La*
cognitae *LebPar*^{App}

369.12 ENHARMONIA UNA iterum numeravit *La*

369.12-3 CHROMATIS TRES, QUARUM PRIMA QUAE MOLLIS AC SOLUTA, SECUNDA QUAE HEMIOLIA EST, TERTIA QUAE EST TONIAEA
QUARUM trium *LebPar*^{App}
MOLLIS propter tria emitonia *LebPar*^{App}
SECUNDA QUAE EMIOLIA EST i. semis totum *BeLa*
tonum et dimidium, vel tonum et semitonium *Leb*
In qua considera<n>tur tria semitonia ad duo *LebPar*^{App}

---

* Cf. Boeth. Ars Geom. 393.16

i. acuta *Par*^{App}

†TONIA Quia unus tonus dividitur in ii emitonia et ii spatiis (*sic, lege* spatia) *Leb*

duo emitonium *Par*^{App}

.13-4 DIATONICI DUAE, MOLLIS ATQUE ROBUSTA

†ΔΙΑΤΟΝΙΚΕC genitivus *Par*^{App}

†UNA MOLLIS Una mollis, quia <sonat> molliter ubi tonus, mollius ubi tria emitonia *LebPar*^{App}, Una mollis, quia *om. Par*^{App} emitonia] semitonia *Par*^{App}

Quae currit per tonum i. gravis *Leb*

gravis *Par*^{App}

†BIS TRIA s. sunt vi modulationes *La*

ROBUSTA acute <modulationes> *Leb*

.14-5 ET MODOS QUIDEM ACCIPIT HARMONIA ΤΕΤΑΡΤΗΜΟΡΙΑ ΔΙΕΣΕΙ, ID EST TONI PARTE QUARTA

MODOS differentias *LebPar*^{App}

In prima <differentia> spatia, in secunda proportio, in tertia quod tonus dividitur in duo semitonia consideratur *LebPar*^{App}

Eo quod unus tonus divid<itur> *Par*^{App}

TONI PARTE QUARTA A quarta parte incoat *La*

a toni *Par*^{App}

.15-6 ILLUD VERO, QUOD MOLLIUS DIXIMUS, PER ΤΡΙΤΗΜΟΡΙΑΝ ΔΙΕΣΙΝ COPULATUR

ILLUD croma *LebPar*^{App}

ΤΡΙΤΕΜORIAN ΔΙΕCIN i. tertiam partem toni *La*

COPULATUR i. iungitur *BeLa*, i. *om. La*

s. armonia *La*

.16-7 IPSUM VERO CHROMA ΗΜΙΟΛΙΟΝ FIT ΗΜΙΟΛΙΑ ΔΙΕΣΕΙ EIUS, QUAE SIT ENHARMONIAE

IPSUM VERO CROMA [1]Est emiolium, cum tres integri toni ad duos proximos tonos comparantur. [2]Est et emiolium, quando tres partes unius toni ad duas partes quaternas comparantur, de quo nunc loquitur dicendo: *ipsum vero croma emiolium fit.* [3]Ac ne putaremus de emiolio dixisse quod plenis et integris tonis conficitur, ideo subdidit *emiolio diesi* i. emiolio, quod quaternis partibus comparatur. [4]Ideoque sequitur *quae sit enarmonia*<*e*>. [5]Proprie etenim ad <en>armoniam pertinet illa divisio diesis, quae t[r]etartemoria dicitur. [6]Nam modulatio <en>armoniae a tetratemoria inchoat et in tritemoria terminatur. [7]Praedixit enim armonicum genus pluribus, sed angustioribus spatiis dividi. [8]Quid autem angustius potest fieri quam ut a tetratemoria inchoet et in tritemoriam desinatur? [9]Ergo cromaticum genus ab emitonio (*sic, lege* emiolio) diesi inchoat hoc est a tribus quaternis partibus unius toni *BeLa*, 4 ideoque] ideo *La*; 8 et *om. La* desinatur] desinat *La*; 9 diesi inchoat] diesinchoat *La*

modulationem *LebPar*^{App}

KPOMA i. ipsum chromaticum genus *BeLa*, i. *om. La* cromaticum *La*

EMIOLION i. diatonon *LebPar*^{App}, i. *om. Leb*

HEMIOΛIA Quia constat sesqualteris *Par*^{App}

†ENARMONΠ Quia inter primam et ultimam sesqualtera est, in<ter> primam et secundam sesquitertia est *LebPar*^{App}, *difficile lectu in Par*^{App}

960/
370.1 SED NUNC DE TONIS QUID TONUS *BeLa*

370.2-3 HOC EST DE NOTA QUALIBET IN NOTAM
i. de corda ad cordam *BeLa*
Singule enim corde habent notas *La*

370.3-4 UT EST IN LYDIO, SI A IOTA DIRECTO IN SIGMA IACENS SIGNA CONCURRANT
IOTA mese dicit *Leb*
signum mese *Par*^{App}
SIGMA IACENS significat paramese *Leb*
signum parameses *Par*^{App}

370.4 VERUM QUIA HEMITONIA DIESISQUE
VERUM sed *BeLa*
QUIA s. docui *La*

370.8 VERUM HORUM EXTREMOS SIBI APTANDOS ESSE NON DUBIUM EST
HORUM EXTREMOS s. sonos *BeLa*, s. *om. La*

i. primam cordam et ultimam; necesse est enim, ut in his concordia inveniatur *BeLa*
Inter penultimam et ultimam; in omnibus tetracordis duae ultimae <cordae> sesquitertiam proportionem habent *Leb*
Inter penultimam et ultimam sesquitertia est *Par*[App]

/

.12 INCIPIT A PRINCIPALIUM PRINCIPALI ET DESINIT IN SUBPRINCIPALEM MEDIARUM
PRINCIPALIUM PRINCIPALI Id est *υπατε υπατον* *BeLa*
IN PRINCIPALEM MEDIARUM Id est *υπατε μεσον* *BeLa*, inpate meson *Be*
in lych<a>nos *Leb*

.13-4 VERUM PRINCIPALIUM ET MEDIARUM CONEXUM, QUOD CONIUNCTARUM VOCATUR
VERUM sed *BeLa*
CONEXUM Quia unam cordam habent, quae terminat tetracordum praecedens et subsequens <inchoat> *BeLa*
Quia una corda coniungit *LebPar*[App]

.14 A MEDIA IN NETEN CONIUNCTARUM Nete vocatur ultima corda in omni tetracordo *La*

.15-6 A SUBMEDIA IN NETEN SEPARATARUM
SUBMEDIA i. paramese *LebPar*[App], i. *om. Par*[App]

.16 EX DIVERSIS i. contrariis ad graves aliorum tetracordorum *La*
gravi et acuto *Leb*

.18 SED DE TETRACHORDIS SATIS DICTUM s. est *Be*

/

.19-0 AC SIMILITER HAEC QUOQUE QUINQUE ESSE NON DUBIUM EST
QUINQUE s. pentacorda *BeLa*
NON DUBIUM EST In tetracordis est semper diatessaron symphonia, diapente in pentacordis *La*

.21 ET IN PRINCIPALEM MEDIARUM TERMINATUR lich<a>no<s> *Leb*

370.22-3 ET CONSTAT A PRINCIPALIUM EXTENTA IN MEDIARUM ILLAM, QUAE IN LYDIO IOTA RECTUM HABET
PRINCIPALIUM EXTENTA [hy]penultima *Leb*
IN MEDIARUM mesen *LebPar*[App]
IOTA RECTUM mese *Leb*

371.1 QUOD A MEDIARUM EXTENTA IN ULTIMAM CONIUNCTARUM DUCITUR
MEDIARUM EXTENTA meson diatonos *Leb*
ULTIMAM CONIUNCTARUM nete sinemmenon *Leb*

371.2-3 QUARTUM VERO, QUOD PER DIVISIONEM EST, A MEDIA IN NETEN DIVISARUM CONECTI SOLET
QUARTUM VERO paramesos *Leb*
PER DIVISIONEM i. per separationem *BeLa*
Dimittit tetracordum coniunctarum *Leb*
A MEDIA mese *Leb*
†TERTIAM DIVISARUM trite diuzemmennon (*sic, lege* diezeugmenon) *Leb*

371.3 QUINTUM IN ACUMEN ERIGITUR trite hyperboleon *Leb*

963/
371.5-6 NAM NON SUM NESCIA ASSERERE QUOSDAM PENTACHORDA AB HEMITONIIS INITIA POSSE SORTIRI
Iste Martianus ab acutioribus inchoavit. Semper ideo se dicit non ignorare quosdam a gravioribus inchoasse *BeLa*, inchoavit] inchavit *Be*

371.7-9 SANE NOTUM ATQUE INTIMATUM VELIM ADQUISITUM IN OMNIBUS QUIDEM ALIIS COLLOCARI, IN HIS AUTEM, QUAE NON PER SONOS FIANT, IN TETRACHORDIS FIERI NON POSSE
QUIDEM ALIIS Quae a tonis incipiunt *Leb*
IN HIS AUTEM i. in pentacordis *La*
QUAE NON PER SONOS s. per emitonia *La*
tonos *Leb*
FIERI NON POSSE i. semitonium *LebPar*[App], semitonia *Par*[App]

371.9 IN HIS ENIM SEMPER s. tetracordis *La*
diatonicis *Leb*

10-11 IN ALIIS AUTEM, QUAE PENTACHORDA NOMINANTUR NEC TAMEN MODULATIONI SUBICIUNTUR Nec ita ponuntur in tropo, ut solummodo per tropos constare videantur et non per emitonia *BeLa*

11 UT PER SONOS CONSTARE VIDEANTUR, APPONITUR
PER SONOS Quia non habent omnes tropos *La*
APPONITUR s. id est adquisitus *BeLaLeb*, id est *om. La* s. id est *om. Leb*

.12 IN HIS s. pentacordis *BeLaLeb*, s. *om. Leb*

/

.13 NUNC DE TRANSITU MODULANTIUM s. est dicendum *BeLa*
QUID SIT TRANSITUS *BeLa*

.15-6 VEL PER SYSTEMA, UT A PRINCIPALI PRINCIPALIUM SI FORTE IN PRINCIPALEM AUT IN ALIUD FORTE SYSTEMATUM
PER SISTEMA i. per magnitudinem *BeLa*
magnitudinem vocis *Leb*
IN SUBPRINCIPALEM s. facimus transitum *Leb*
IN ALIUD FORTE SISTEMATUM s. transierit *BeLa*

.17 VEL PER TONUM, CUM A LYDIO VEL IN PHRYGIUM VEL IN ALIUM TROPUM
Tonum pro tropo ponit, quia tropi ex tonis constant *BeLa*

.19-0 VEL CUM A VIRILI CANTILENA TRANSITUS IN FEMINEOS MODOS FIT
VIRILI forti *BeLa*
FEMINEOS MODOS molles i. gravis *BeLa*, molles] molliores *La*

/

.21 DE GENERE MODULANDI CONSEQUENTER EDISSERAM QUID MELOPOEIA *BeLa*

.21-2 MELOPOEIA EST HABITUS MODULATIONIS EFFECTAE
MELOPOEIA modulatio *Be*
i. melodiae factura *LaLeb*, i. *om. Leb*
factura melodiae *Par*[App]

MODULATIONIS Utrum de Lidio vel de alio tropo fieri debeat *La*

371.22 MELOS AUTEM EST ACTUS ACUTI AUT GRAVIORIS SONI QUID MELOS *BeLa*
ACTUS operatio *BeLa*

371.23 MODULATIO EST SONI MULTIPLICIS EXPRESSIO QUID MODULATIO *BeLa*

371.23-1 MELOPOEIAE SPECIES SUNT TRES: ΥΠΑΤΟΕΙΔΗΣ, ΜΕΣΟΕΙΔΗΣ, ΝΗΤΟΕΙΔΗΣ
YPATOIDES MESOIDES NETOIDES gravissimi, medii, acutissimi *BeLa*
graves, temperatae, acutae *LebPar*^{App}

372.2 ET ΥΠΑΤΟΕΙΔΗΣ EST, QUAE APPELLATUR TRAGICA
†TRAIKA Tracia *Leb*

372.2-3 ΜΕΣΟΕΙΔΗΣ, QUAE ET DITHYRAMBICA NOMINATUR, QUAE TONOS AEQUALES MEDIOSQUE CUSTODIT
†ΔIPANIAMBICA pinguia *Leb*
Hoc est pinguia iambica *Par*^{App}
AEQUALES temperatos *LebPar*^{App}
†MEDIASQUE inter gravem et acutum *LebPar*^{App}

372.3-4 ΝΗΤΟΕΙΔΗΣ, QUAE ET NOMICA CONSUEVIT VOCARI, QUAE PLURES SONOS EX ULTIMIS RECIPIT
†NOMIKOC Legales, vix enim mutantur *LebPar*^{App}, vix enim] quia vix *Par*^{App}
EX ULTIMIS In acutis sonat vel in ultimis *BeLa*, in ultimis] exultimis *La*^{1}
excellentibus *Leb*

372.4-6 SUNT ETIAM ALIAE DISTANTIAE, QUOD EROTICA MELE DICUNTUR <ALIA, ALIA COMICA,> ALIA ENCOMIOLOGICA
†TROPICA MELE nominativus pluralis *BeLa*
melodia modificata *BeLa*, modificatae *Be*
†COMIOLOGICA i. similis rationis *BeLa*
vel isologica *BeLa*, isologia *La*^{1}
comoediis convenientia *LebPar*^{App}

6 SUBROGANTUR subiunguntur *BeLa*

7 HAE AUTEM SPECIES ETIAM TROPI DICUNTUR In quibus est fictio *LebPar*[App]

9-10 SYSTEMATE, QUIA ALIA EST HYPATOIDES, ALIA MESOIDES, ALIA NETOIDES
†SPECIAE QUOQUE s. dissentiunt *BeLa*
YPATOIDES i. prima *BeLa*

10-12 TROPIS, UT DORIO, LYDIO VEL CETERIS <...> OMNIS TAMEN, QUI MELOS INCHOAT, PRAE CUNCTIS SYSTEMA DEBET ADVERTERE
TROPIS s. dissentiunt *BeLa*
OMNIS TAMEN s. tropus vel homo *La*
MELOS accusativus grecus *La*
SISTEMA i. spatia magnitudinum *BeLa*, i. *om. La*

14 NUNC RHYTHMOS QUID RITHMUS *BeLa*

15-16 RHYTHMUS IGITUR EST COMPOSITIO QUAEDAM EX SENSIBILIBUS COLLATA TEMPORIBUS
EX SENSIBILIBUS in voce *La*
Quia sunt tempora, que intellectu intelliguntur *La*
Quia audiuntur. Sentitur numerus tribus modis: visu, auditu et tactu *Leb*
TEMPORIBUS syllabis et pedibus *Leb*

16-17 AD ALIQUEM HABITUM ORDINEMQUE CONEXA i. aut versus aut metri *La*

17 8 NUMERUS EST DIVERSORUM MODORUM ORDINATA CONEXIO, TEMPORI PRO RATIONE MODULATIONIS INSERVIENS
DIVERSORUM i. gravis vel acuti *BeLa*
TEMPORI PRO RATIONE i. secundum temporum rationem *BeLa*

18-19 PER ID QUOD AUT EFFERENDA VOX FUERIT AUT PREMENDA

i. aut in exaltatione vocis aut in infimatione *BeLa*, infimatione] graviorum *La*
in acumen, in gravem *Leb*

372.19-1 ET QUI NOS A LICENTIA MODULATIONIS AD ARTEM DISCIPLINAMQUE CONSTRINGAT
ET QUI s. numeri *La*
numerus *Leb*
A LICENTIA i. vulgari *La*

373.1 RHYTHMUM numerum *Leb*

968/
373.2 QUIPPE RHYTHMIZOMENON MATERIA EST NUMERORUM
†RITIZOMENON i. vox ipsa naturalis *BeLa*, vox ipsa] ipsa vox *La*
i. forma vocis *Be*
numerabile *Leb*

373.3 NUMERUS rythmus *Leb*

373.5 VISU, SICUT SUNT EA, QUAE MOTU CORPORIS COLLIGUNTUR
MOTU CORPORIS i. saltationes *Be*
ut motu manus vel saltationes *La*

373.6-7 TACTU, UT EX DIGITIS VENARUM EXPLORAMUS INDICIO
EXPLORAMUS probamus *BeLa*

373.7 VERUM NOBIS ATTRIBUITUR MAXIME IN AUDITU VISUQUE
Quia quod ad tactum sit venarum, magis ad medicinam pertinet *BeLa*, quod ad] quod ayd *Be*
VERUM sed *La*
NOBIS musicis *Leb*
ATTRIBUITUR numeris *Leb*

969/
373.8-9 SED RHYTHMICE EST ARS OMNIS IN NUMERIS, QUAE NUMEROS QUOSDAM PROPRIAE CONVERSIONIS ACCIPIAT FLEXUSQUE LEGITIMOS SORTIATUR
†RYTOITIZETE numeri inquisitio *Leb*

Nam zi pro inquisitione ponitur *Leb*[*]
PROPRIAE CONVERSIONIS i. reversionis ad principium *BeLa*
Convertitur pes in pedem *Leb*
FLEXUSQUE LEGITIMOS i. recursiones, quia in infinitum pergit, sed revertitur *BeLa*
reditus, sicut a dispondeo in spondeum *Leb*

.11-2 SED QUIA VISUS AUDITUSQUE NUMERO DICTUS ACCEDERE, HI QUOQUE IN TRIA ITIDEM GENERA DIVIDENTUR
DICTUS s. est *Be*
†HII QUOQUE i. visus vel auditus *BeLa*, i. *om. La*
DIVIDENTUR s. quae accidunt illis *Be*
Accidunt illi *La*

.12-3 IN CORPORIS MOTUM, IN SONORUM MODULANDIQUE RATIONEM ATQUE IN VERBA, QUAE APTA MODIS RATIO COLLIGARIT
IN CORPORIS MOTUM Omnis musica in motu est[**] *BeLa*
ad visum *Leb*
IN SONORUM MODULANDI Hoc est, quod dicit auditus *La*
QUAE APTA De musica humanae vocis dicit *BeLa*
MODIS i. tropis *BeLa*
COLLIGARIT iunxerit *BeLa*

.13-4 QUAE CUNCTA SOCIATA PERFECTAM FACIUNT CANTILENAM
QUAE CUNCTA i. tria *BeLa*

.14-5 DIVIDITUR SANE NUMERUS IN ORATIONE PER SYLLABAS, IN MODULATIONE PER ARSIN AC THESIN
PER SYLLABAS Hoc est quod superius dixit verba i. per tempora syllabarum *BeLa*
MODULATIONE Item recapitulatio, hoc est quod superius dixit "in sonorum modulandi"[***] *BeLa*, item] ite *Be*
PER ARSIN et thesin *Be*

.15-6 IN GESTU FIGURIS DETERMINATIS SCHEMATISQUE COMPLETUR

---

[*] zi < ζητέω, to inquire, to look for.
[**] Cf. Aug. De mus. 2.3.3
[***] Mart.Cap. 373.12

IN GESTU FIGURIS i. in corporis motu *BeLa*, i. *om. La* corporis motu] motu corporis *La*
DETERMINATIS discretis *BeLa*
accentibus. supra figuratis *Leb*

970/
373.17 PRIMUM DE TEMPORIBUS ut in pedibus *La*

373.18-9 QUAE IN NUMERUM CADERE POSSUNT, QUAE RHYTHMOIDE, ID EST SIMILIA NUMERIS
IN NUMERUM CADERE POSSUNT Quae non possunt venire in metrum *La*
in rithmum *Leb*
†RYTMOIDES Formam numerorum habent, quae sisugiae (*sic, lege* synzugiae) sunt *Leb*
SIMILIA NUMERIS Non habent aptum numerum *La*

373.20 HOC EST ERRHYTHMON, ARRYTHMON, RHYTHMOIDES
HOC EST ENRITHMON, ARITHMON, RITHMOIDES Numerabile, sive carens numero; partim consentit, partim dissentit *Leb*
Numeri forma vel medius *Leb*

373.21 QUARTUM DE EORUM GENERE Ut scias cuius generis pes *La*

373.21-2 QUINTUM EST, QUOD AGOGEN RHYTHMICAM NOMINAMUS, ID EST QUO GENERE NUMERUS MODIQUE DUCANTUR
AGOGEN ascendentem numerum *Leb*
QUO GENERE s. metri *La*
DUCANTUR Quia qui ascendit, consequens est, ut descendat *Leb*

373.22-3 SEXTUM DE CONVERSIONIBUS Qui pedes pro quibus ponantur *La*

373.23-4 ID EST QUEMADMODUM PROCREATIO NUMERI POSSIT EFFINGI Ut scias, in quem finem numerus perveniat, utrum ad vii pedes *La*
Verti de prosa in metrum *Leb*

971/
374.1 PRIMUM IGITUR TEMPUS EST
PRIMUM s. dicendum est *La*

TEMPUS breve *Leb*

1-2 QUOD IN MOREM ATOMI NEC PARTEM NEC MOMENTA RECISIONIS ADMITTIT
Non ideo tempora dicuntur in metro, ut illic sint duo tempora in una sillaba, sed ad comporationem illius spatii, quod est brevissimum, illud quasi duo in se habeat *La*
RECISIONIS i. partitionis *BeLa*, partiteonis *La*
ADMITTIT recipit *BeLa*

3 SE IPSA NATURA i. sine ulla arte *BeLa*

3-5 SED [NUMERUS] IN VERBIS PER SYLLABAM, IN MODULATIONE PER SONUM AUT SPATIUM QUOD FUERIT SINGULARE
SED NUMERUS s. athomus *La*
AUT SPATIUM unus pes *Leb*
.5 SINGULARE i. sine productione *BeLa*, i.] unum *La*

.5 IN GESTU EX INCIPIENTE CORPORIS MOTU
†ET INCIPIENTEM i. in ipsa inchoatione motus vel mora unius syllabae inchoatio *La, cf. Be* 374.10-2 UT ENIM ILLE

.6 HOC ERIT BREVISSIMUM TEMPUS s. unius corporis gestus *BeLa*

.7-8 COMPOSITUM VERO, QUOD POTEST DIVIDI ET QUOD A PRIMO AUT DUPLUM EST AUT TRIPLUM AUT QUADRUPLUM
COMPOSITUM VERO Quando multa brevia tempora iunguntur insimul *BeLa*
AUT DUPLUM EST AUT TRIPLUM s. a primo *La*

.8-9 EATENUS ENIM TEMPUS OMNE NUMERI PROFERETUR
EATENUS i. usque ad quadruplum *BeLa*, i. *om. La*
Numerus legitimorum, qui ultra quaternarium non procedit *Lcb*
†PROFERITUR crescet *BeLa*

.9 ATQUE EI FINIS EST, QUI PLENAE RATIONIS EST TERMINUS
ATQUE numero *La*
QUI PLENAE RATIONIS quaternarius *La*

In denario numero, qui continetur in quaternario *Leb*
TERMINUS i. denarius *Leb*

374.9-10 ATQUE IN HOC NUMERUS TONI SIMILIS INVENITUR i. bisdiapason *Be*
i. diapason, quia non transit quaternarium *La*

374.10-2 UT ENIM ILLE PER QUATTUOR SPECIES, HOC EST DIESIS, DIVIDITUR, HIC ETIAM QUATERNARIA TEMPORUM MODULATIONE CONCLUDITUR
UT ENIM ILLE i. in ipsa inchoatione motus vel mora unius syllabae inchoatio *Be, cf. La* 374.5 †ET INCIPIENTEM
tonus *Leb*
HIC ETIAM numerus *BeLaLeb*

972/
374.13-5 ALIA SUNT, QUAE ERRHYTHMA TEMPORA NOMINANTUR, ALIA QUAE ARRHYTHMA, TERTIA QUAE RHYTHMOIDE PERHIBENTUR
†ENRITHMA TEMPORA in numero posita *BeLa*
RYTHMOIDES Partim habent, partim carent *Leb*

374.15-7 ET ERRHYTHMA QUIDEM SUNT, QUAE RATIONE CERTA ORDINEM SERVANT, UT IN DUPLICI VEL HEMIOLIO VEL IN ALIIS, QUAE ALIQUA RATIONE IUNGUNTUR
†ENRYTHMA valde numerabilia *Leb*
IN DUPLICI VEL EMIOLIO iambo, peonibus *Leb*
IN ALIIS QUAE †ALIIS nume<ris> vel equali et epitrita *La*
sesquitertia, sesquiquarta *Leb*

374.17-8 ARRHYTHMA SUNT, QUAE SIBI NULLA OMNINO LEGE CONSENTIUNT AC SINE CERTA RATIONE CONIUNCTA SUNT
NULLA OMNINO LEGE et (*sic, lege* ut) sinsugiis *Leb*
AC SINE CERTA RATIONE Ut sunt sinzigiae in pedibus *BeLa*

374.18-9 RHYTHMOIDE VERO IN ALIIS NUMERUM SERVANT IN ALIISQUE DESPICIUNT
DESPICIUNT i. contemnunt *BeLa*, i. *om. La* contemniunt *Be*

19-0 QUORUM TEMPORUM ALIA ΣΤΡΟΓΓΥΛΑ, HOC EST ROTUNDA, PERHIBENTUR, ALIA ΠΕΡΙΠΛΕΑ

HOC EST ROTUNDA Ubi nusquam pes cum verbo finitur, ut: "immortale nichil m.c.t."* *Leb*

†ΠΕΡΙΠΛΕΟ De his versibus dicit, ubi multitudo consonantium excurrit, ut: "ponite spes sibi quisque"** *BeLa*

.21 ATQUE ORDO LEGITIMUS EXPETIT PRAECIPITANTUR

ORDO LEGITIMUS Quando versus cito ad finem ducitur propter facilitatem sillabarum, ut: "Arma virumque"*** *La*

Ut sunt illi versus, qui ex vocalibus ordiuntur, ut: "Atque ea diversa"**** *BeLa*, atque] adque *Be*

PRAECIPITANTUR cito cantantur *Leb*

.22-1 ΠΕΡΙΠΛΕΑ VERO, QUAE AMPLIUS QUAM DECET MORAS COMPOSITAE MODULATIONIS INNECTUNT

ΠΕRΙΡΛΕC et amplius vel copiosa *Leb*

AMPLIUS Ubi semper pes cum verbo finitur ut: "haec tua sunt b.s.q.t.b.o.c."***** *Leb*

.2 QUAE PODICA ETIAM PERHIBENTUR

PODICA i. pedica *Be*

i. pedalia, que pedibus intel<l>eguntur. Quia sunt quedam, que non possunt in sillabis, ut sunt tempora mensilia ebdomalia *La*

pedalia *Leb*

Legitimos etiam pedes habent *Leb*

---

* "Immortale nichil mundi compage tenetur": Iuv. Praef. 1. Cited in (among others) Isid. Etym. 1.35.5; Sed.Scot. Don.art.mai. 3.358; Ars Laur. Exp.Don.mai. p. 186.19.

** Verg. Aen. 11.309. Cited in, among others, Beda Ars met. 3.1-41; Don. Ars gram. De littera, lin. 799; Sed.Scot. Col.misc. 13.25.20; Ars Laur. Exp.Don.mai. 1.164.19.

*** Verg. Aen. 1.1

**** Verg. Aen. 9.1

***** "Haec tua sunt, bona sunt, quia tu bonus omnia condis": Beda Ars met. 12.7. Cited in: Ars Laur. 1.186.47

974/
375.3-4 PES VERO EST NUMERI PRIMA PROGRESSIO PER LEGITIMOS ET NECESSARIOS SONOS IUNCTA
PES VERO Prima mensura totius versus, quia secundum pedem totus versus metitur *La*
PROGRESSIO motus *Leb*
LEGITIMOS ET NECESSARIOS aptos *BeLa*
ad illum numerum *La*
Illi pedes, qui proportionaliter ponuntur, legitimi dicuntur *BeLa*

975/
375.7-8 COMPOSITOS VERO, UT SUNT PAEONES VEL EORUM PARES i. epitriti *BeLa*, i. *om. La*

375.9-11 ALIOS VERO ALOGOS, HOC EST IRRATIONABILES, NOMINAMUS, QUORUM[QUE] RATIO NULLA PRAESTATUR, SED INCONDITA QUAEDAM COMPOSITIO PROFERTUR
ALOGOS alios etiam *BeLa*
IRRATIONABILES ut amfibracus vel sisugiem *Leb*
SED INCONDITA Ut sunt sinzigiae *BeLa*

976/
375.11-3 ALIA DEINDE DIFFERENTIA EST, QUA PER DIVISIONEM QUAERITUR, QUALIS EXSISTIT, HOC EST ΠOIA, CUM VARIE ET MULTIPLICITER EA, QUAE CONEXA FUERINT, DIVIDUNTUR
†QUAE PER DIVISIONEM Aliud est divisio in simplicibus, aliud in sisugiis i. compositis *Leb*
QUALIS †EST vel existit *La*
ΠOIA i. positio vel factura *BeLa*
VARIAE ET MULTIPLICITER s. nominamus *La*
CONEXA FUERUNT Proceleumaticus et dispondeus per se simplices, ad pyrrhichium et spondeum multiplices *Leb*

375.13-4 ATQUE ILLI, QUI SIMPLICES PEDES <FUERANT, IAM POSSUNT PEDES> ESSE MULTIPLICES
SIMPLICES s. nominamus *La*
PEDES ESSE MULTIPLICES s. possunt *BeLa*
s. nominamus *Be*

.14-5 ALIA EST, QUAE PER DIVISIONEM FIERI CONSUEVIT
PER DIVISIONEM Cui generi metri convenit *La*
  in sisugiis *Leb*
FIERI CONSUEVIT s. incompositis pedibus *BeLa*

.15-7 SEPTIMA, QUAE PER OPPOSITIONEM FIT, ID EST CUM DUOBUS PEDIBUS ACCEPTIS UNUS HABET PROLIXIUS TEMPUS, QUOD PRAECEDIT EX ORDINE
†SEPTIMUS s. modus *La*
†APPOSITIONEM i. aucmentum *BeLa*
QUOD PRAECEDIT EX ORDINE Veluti ipse inferius dicet <de> incompositis pedibus: si duo pedes ponuntur compositi pro uno pede, ut habeat unus et primus pes viii, alter vero iiii. In quibus etiam perhibentur rethores finire suos sermones, quas (*sic, lege* quos) etiam rethorica clausulas appellat *BeLa*

.18 VEL CUM PER CONTRARIUM ORDINEM TEMPORA PRAEDICTA VERTUNTUR Videlicet, ut primus habeat iiii et sequens viii *BeLa*^{1}*La*^{2}

/
.19-0 RHYTHMICA VERO GENERA SUNT TRIA, QUAE ALIAS DACTYLICA, IAMBICA, PAEONICA NOMINANTUR, ALIAS AEQUALIA, HEMIOLIA, DUPLICIA
RITHMICA i. numerorum ordo *BeLa*
DACTILICA s. rihtmica *La*
IAMBICA s. rihtmica *La*
ALIA AEQUALIA Quasi nunc planius loquitur *BeLa*
EMIOLIA alia *Be*
  i. in peonibus *La*
  peonica *Leb*
DUPLICIA in iambis *La*
  iambica *Leb*

5.20-1 DENIQUE ETIAM EPITRITUS SOCIATUR
DENIQUE et postea *BeLa*
EPITRITUS SOCIATUR s. proportionibus supradictis *BeLa*

6.1-2 ETENIM UNUS SEMPER, CUM SIBI FUERIT APTATUS, UT AEQUALIS

CONVENIT

UNUS SEMPER Sicut pyrrichius, qui aequali proportione partitur *BeLa*, pyrichius *La*

CUM SIBI i. unus ad unum *La*

376.2-3 DUPLEX VERO QUI FUERIT AD SINGULAREM, GEMINAM RATIONEM TAM SYLLABARUM QUAM TEMPORUM SERVAT

DUPLEX proportio *Leb*

AD SINGULAREM i. duo ad i *Leb*

GEMINAM RATIONEM TAM SILLABARUM QUAM TEMPORUM In quibusdam sillabe et temporum, in quibusdam temporum numerasque (*sic, lege* numerique), ut in tribracho *La*

Quod non facit in dissyllabis: aequales sunt ibi syllabae. Nam una longa habet duo brevia signa, sicut et duae breves <una longa>. Quod non facit Donatus *Leb*

ut molosus, tribracus *Leb*

376.3 TRIA VERO AD DUO peonica *Leb*

376.4-5 SED QUAE AEQUALIA DIXIMUS, EADEM DACTYLICA ESSE DICEMUS

AEQUALIA esse *Be*

EADEM DACTILICA ESSE Quia perfectus est numerus *Leb*

376.5-6 DENIQUE IN DACTYLICO GENERE SIGNA AEQUALI SIBI IURE NECTUNTUR

IN DACTILICO GENERE SIGNA ∪ ∪ — ∪ ∪ *La*

AEQUALI SIBI IURE NECTUNTUR Quot sunt in arsi, tot in thesi nectuntur. ∪ signa *La*

376.6-7 UNUM AD ALTERUM, VEL AD NUMERUM GEMINUM DUO, VELUT †FORTE AEQUALITAS NUMEROSA DECURRET

UNUM AD ALTERUM i. binarium *Be*

Unum est *La*

De binario dicit, qui secundum Boetium alter numerus dicitur sive geminus, ubi enim aequa proportio est *BeLa*, geminus] minus *La*

ut in pyrrichio *Leb*

VEL AD NUMERUM GEMINUM in spondeum *Leb*

DUO Hoc est diversi *La*

AEQUALITAS i. duo ad duo, unum ad unum *Leb*
NUMEROSA Una figura multiplicetur copiosa *La*

.7 SEQUITUR IAMBICUM GENUS Quae in pedibus fient tali modo *BeLa*

.9 SIVE UNUS AD DUO, SIVE AD QUATTUOR GEMINI
SIVE †UNIUS s. rationis *Par*
GEMINI bini *BeLa*
  duo *LebPar*

.10-1 HEMIOLIUM SANE, QUOD PAEONICUM MEMORATUM, TUNC EST, CUM PEDUM SIGNA HEMIOLII RATIONEM IUSQUE SECTANTUR
EMIOLIUM SANE i. totum semis *BeLa*
TUNC EST i. ibi invenitur *BeLa*, ibi] ubi *Be*

.12-3 ACCEDIT [AUTEM] ETIAM IN EPITRITI RATIONE SAEPE NUMERUS, CUM PES IN EO ACCIPITUR, QUI SIT <UT> TRES AD QUATTUOR
NUMERUS rithmus *LebPar*
IN EO s. numero *BeLa*
QUI SIT †AD III IIII Istum Martianus supertertium vocat *BeLa*, supertercium *La*

/
.15-6 AEQUALE EST IGITUR NUMERI GENUS, QUOD A DISEMO USQUE IN SEDECIM PEDES PROCEDIT
AEQUALE EST Qui aequali proportione gaudent *BeLa*
  dactilicus *Leb*
QUOD A DISSEMO a duorum temporum pede *BeLa*
  pyrrichio. x in simplicibus, vi in sisugiis *LebPar*, pyrrichio] pirrichius *Par* in simplicibus] per symplicibus *Par* in sisugiis] per sinsugiis *Par*
USQUE †AD XVI PEDES †PROCEDUNT Hoc est dicit, quod xvi pedes sunt, qui habent aequam divisionem *La*

.16-7 DISEMUS AUTEM APPELLATUR PES, QUI PER ARSIN ET THESIN PRIMUS CONSTARE DICITUR, UT EST *LEO*
DISSEMUS pyrrichius *La*
  Bitempus, duo tempora brevia habet *Leb*
  semus tempus, bitempus *Par*
  Cronos apud Grecos longum tempus, semus breve dicitur *Par*

UT EST LEO vi sunt in symplicibus *Par*

376.17-8 DUPLUM VERO INCIPIT A TRISEMO, XVIII AUTEM SYLLABAS IN FINEM USQUE DEDUCET

A TRISEMO A pede trium temporum, ut est trocheus *La*
tribrachus. vi in simplicibus, xii in sisugiis *Leb*
SYLLABAS i. pedes. Sillabas posuit pro pedibus *BeLa*, posuit pro pedibus] pro pede posuit *La*

376.18-9 HEMIOLIUM SANE A PENTASEMO DUCIT EXORDIUM, IMPLETUR AUTEM IN XXV NUMERO

A PENTASEMO vii sunt in symplicibus *Par*
IN †XV^{MO} NUMERO i. in xv pedibus *BeLa*, i. *om. La*
vii in simplicibus, viii in sisugiis *Leb*

376.19-1 EPITRITUS AB HEPTASEMO PRINCIPIUM FACIT, XIIII SYLLABIS DESINENS, CUIUS DIFFICILIS EST USUS

AB EPTASEMO vii sunt in simplicibus *Leb*
FACIT pro fit *La*
XIIII i. in ea ratione et proportione *BeLa*, i. *om. La*
Ita tamen ∪ geminatio fiat, ut verbi gratia spondeus cum tribrachio *La*
†SIMILIBUS IDEM PONENS idem gens *La*
epitritus s. proportionem suam *LebPar*, suam *om. Par*
CUIUS generis *La*
DIFFICILIS EST USUS i. raro eo utuntur *Be*

377.2-3 UT SINGULORUM LEGES PER UNIVERSA SERVENTUR

UNIVERSA s. genera *BeLa*

979/
377.4 SED NUMERORUM ALII SUNT COMPOSITI, ALII INCOMPOSITI, ALII PERMIXTI

NUMERORUM s. pedes *La*
rithmorum *LebPar*, ritmorum *Par*
ALII SUNT COMPOSITI i. compositi et incompositi *BeLa*

377.4-5 ET COMPOSITI EX DUOBUS GENERIBUS VEL PLURIBUS COHAESERUNT

DUOBUS GENERIBUS ut spondeis cum trocheo *La*
ut dactilico et iambico *Leb*
VEL PLURIBUS Quando ex iambico et dactilico pes efficitur, vel pluribus i. iambico, dactilico et peonico *BeLa*

.5-6 INCOMPOSITI, QUI UNO PEDE GENERE CONSISTUNT, UT SUNT TETRASEMI
QUI UNO PEDE GENERE CONSISTUNT compositum i. pedibus unius generis *La*
UT SUNT TETRASEMI i. quattuor temporum pedes *BeLa*, i. *om. La* quatuor *La* pedes *om. La*
Ut est proceleumaticus: componitur duobus pedibus unius generis *La*

.6-8 MIXTI VERO, QUI ALIQUANDO IN PEDES, ALIQUANDO IN NUMEROS RESOLVUNTUR, UT IN HEXASEMO NUMERO ACCIPERE DEBEMUS
QUI ALIQUANDO IN PEDES Ut si ponantur pyrrichius et dactilus pro uno pede *BeLa*
ut proceleumaticus et caeteri *LebPar*
UT IN †EXESEMI i. in pede sex temporum *Be*
Ut est ditrocheus *La*
in ditrocheo et diiambo *LebPar*, in *om. Par* ditrocheo] distrocheo *Par*
NUMERO i. in sola tempora *BeLa*

.8-9 ALII PER COPULAS, ALII PER PERIODUM COLLIGANTUR
PER COPULAS i. per compositionem et appositionem *BeLa*
"Arma vi-"*: post unum pedem fieri potest *LebPar*, arma vi *om. Par*

.9-11 ETENIM SYZYGIA, ID EST COPULA, DUORUM PEDUM IN UNUM EST ASTRICTA CONEXIO, QUI [IN] DISSIMILES SIBI POSITI ESSE VIDENTUR
ETENIM SINZIGIA Nunc difinit, quid sit copula *La*
DUORUM pedum *Par*
†QUIQUE DISSIMILES i. secundum genus *BeLa*

---

* "Arma vi<rumque cano>": Verg. Aen. 1.1

377.12 QUI DISSIMILES SIBI PARITER SOCIANTUR
†IMPARES SOCIANTUR si *La*
Secundum aliquam rationem sociantur *BeLa*

980/
377.13 PER OPPOSITIONEM relationem *La*

377.15 VEL QUOD EX PLURIBUS AEQUALITER COPULATUR
s. generibus i. in aequali proportione *BeLa*

377.16-7 CUM AUT PRIMOS DISEMOS PONIMUS, INSEQUENTIBUS LONGE POTIORIBUS
†POSTIORIUS tetrasemis, <posti>oribus vel potioribus *Leb*

377.17-2 UNUM ETIAM PEDEM POSSE SUFFICERE AD COMPLENDAM PERIODON, SI SOLUS CETERIS INAEQUALIS INSERITUR
AD COMPLENDAM PERHIODON Quod vocatur monocolon i. clausulam *BeLa*
ambitum *LebPar*
INAEQUALIS i. divisus *BeLa*

981/
378.3-4 SED EORUM, QUAE IN PEDEM RECIDUNT, DACTYLICUM GENUS PRIMUM EST, IN QUO GENERE PEDES INCOMPOSITI VOCABUNTUR
†PEDUM RECIDUNT resolvuntur *LebPar*
DACTILICUM GENUS PRIMUM EST i. principatum tenet dactilicum genus *La*
PEDES INCOMPOSITI Quia uno pede genere convocant *La*

378.4-5 ID EST PROCELEUMATICUS, ANAPAESTUS, SPONDEUS SIMPLEX ET SPONDEUS MAIOR
ANAΠECTOYC dactilus *BeLa*
SPONDEUS SIMPLEX in pyrrichio *LebPar*
quintus pyrrichius *Leb*

378.6-7 AC PROCELEUMATICUS QUIDEM ET POSITIONEM BREVEM ET ELATIONEM BREVEM RETINET; UTETUR AUTEM HIC IDEM TETRASEMO FREQUENTIUS
AC PROCELEUMATICUS pro pirrichio *Be*

Proceleumaticum pirrichium vocat *BeLa*
Pyrrichius est *LebPar*, pirrichius *Par*
ET POSITIONEM thesin *La*
ET ELATIONEM arsin *La*
HIC IDEM s. pes *La*
i. in hoc dactilico genere iiii habet tempora *LebPar*, i. *om. Par*

8.9 BREVIOR NOMINATUR minor duarum syllabarum *LebPar*

8.9 ILLE VERO maior *La*

8.10-2 AT VERO BREVIOR, ID EST DISEMUS, SYNECHES VOCATUR, QUIA IPSA ASSIDUITAS ET FREQUENTIA COMPREHENDENTIS SE INVICEM SYLLABAE NEC MAGNITUDINEM ALIQUAM NEC MODUM DIVISAE POTESTATIS EXTENDIT
BREVIOR pyrrichius *Leb*
SYNECES i. continentia *Be*
Inde sinoca i. passio continens *BeLa*, sinoca] sonica *Be*
continua *LebPar*
COMPREHENDENTIS i. consequentis *BeLa*, i. *om. La*
coniunctis *La*
SYLLABAE i. brevis post brevem *BeLa*, i. *om. La* brevis post] brevem post *Be*
DIVISAE POTESTATIS EXTENDIT ut "leo" *Leb*

8.13 UTI DECET i. oportet *BeLa*, i. *om. La*

8.14 INCIDAT interimat *La*
violet *LebPar*

8.14-6 IN PERMIXTIONE VERO ALIORUM PEDUM, QUI LONGIORES PONUNTUR, DECENTER APTATUR, UT ILLORUM PROLIXAM MORAM INTERVENIENTE SUA CELERITATE COMPENSET
LONGIORES PONUNTUR proceleumaticus et pyrrichius *Leb*
pirrichius et proceleumaticus *Par*
DECENTER †APTANTUR disemi *La*
COMPENSET i. temperet *BeLa*, i. *om. La*

8.16-7 QUARE PROCELEUMATICUS, QUI AD NUMEROS APTATUR, <A>

QUADRESIMO EXORDIUM DEBET ACCIPERE
AD NUMEROS ad pedes *La*
rithmos *Par*
QUADRISEMO EXORDIUM Quia ille minor [ux] aliquid temperat, u<t> est "ariete"[*] *La*

982/
378.18-9 ANAPAESTUS QUI VOCATUR ACCIPIET ELATIONEM PEDIS UNIUS TEMPORIS, POSTITIONEM VERO DUORUM TEMPORUM FACIET
ANAPESTUS i. dactilus maior *Be*
De dactilo maiore loquitur s. ille pes *La*
Anapestum dactilum appellat. Unde unum minorem, alterum maiorem scribit. Una enim ratione utuntur, hoc distat quia *BeLa*
ex duabus brevibus et longa *LebPar*, duabus] dubus *Leb*
ELATIONEM arsin *LebPar*
UNIUS TEMPORIS i. longioris *BeLa*
Colliguntur duae breves in unam longam *Par*
POSITIONEM finem *BeLa*

378.19 MONOCHRONON i. unius temporis *BeLa*

378.21 VEL CUM TRIA TEMPORA SIMUL BREVIA COLLOCANTUR
s. dicitur monocronon. Ut "sol", "rex" *La*

378.21-2 VEL CUM SUNT QUATTUOR NUMERO s. dicitur monocronon. Ut "strirps" (*lege* stirps)[**] *La*

378.22 QUAE OMNIA AD COMPARATIONEM LONGAE SYLLABAE COMPUTANTUR
QUAE OMNIA s. brevia, nam per se non sunt brevia *La*

378.23 MAIOR ANAPESTUS i. dactilus *BeLaPar*, i. *om. Par*

---

[*] cf. Beda Ars met. 14.15; Ars Laur. 1.168.59

[**] "Stirps" is the most common example for a mono-syllabic word with no less than six letters. Cf., among others, Prisc. 2.44.5-7; Mar.Vict. Ars gram. 5.1; Bon. Ars gram. 1.275 (p. 24); Ars.Laur. 1.161.31; Don.Orth. *De syllaba* 1.17; Mur. Don.art.mai. 1.36.28; Sed.Scot. Don.art.mai. 1.20.28.

Quia a maiori syllaba i. longa incipit *LebPar*, Quia a mai *om. Leb*

9.1-3 QUARE UTRIUSQUE TEMPORIS, QUOD IN POSITIONE FUERIT, AEQUALI SIBI POSITO OPORTET ELATIONIS GEMINUM TEMPUS ACCIPERE
AEQUALI i. sub aequali positione *La*
OPORTET ELATIONIS GEMINUM TEMPUS ACCIPERE Oportet, ut tot tempora sint in elatione, quot in depositione *BeLa*

9.4-5 QUARE ANAPAESTUS APOMIZONOS DACTYLICUS A NOBIS ESSE DICITUR
QUARE i. ut ea, quae sunt in positione, aequalia sint, et quae in elatione <gemina sint> *BeLa*
APOMIZONOS a maiori spatio *BeLa*, spatio] sono *La*
sursum sonans *LebPar*
DACTILICUS i. <a> longiore sillaba *Be*
Eo quod a longa sillaba incoet *La*

9.5-6 AT VERO ANAPAESTUS, QUI APELASSONOS NOMINATUR, EX DUABUS BREVIBUS, QUAE IN ELATIONE SINT, ET UNA, QUAE IN POSITIONE SIT, COPULATUR
AT VERO ANAPESTUS Anapestus apellatur *Leb*
QUI APELLASONOS NOMINATUR i. a minori sono vel spatio *BeLa*, i. *om. La* sono vel *om. Be*
Quia sillabas ii breves habet *La*
deorsum sonans vel descendens sonus *LebPar*, deorsum sonans vel *om. Par*
QUAE IN ELATIONE SINT ionicus maior et minor *Leb*

3/
9.8-9 MAIOR VERO DICITUR, QUI QUATERNARIAM NON SOLUM ELATIONEM, SED POSITIONEM ETIAM VIDETUR ADMITTERE
MAIOR VERO i. dispondeus *BeLaLeb*
Mirabiliter facit in hoc loco: ionicum maiorem et minorem anapestum appellat *Par*
ELATIONEM SED POSITIONEM i. incoationem sed positionem *La*
arsin, thesin *LebPar*

9.10-1 PER COPULAM VERO DUPLICES ACCEDUNT NUMERI, QUORUM

ALTER EX MAIORE ERIT IONICUS, ALTER EX MINORE
PER COPULAM per sinzigiam *La*
sisugiam *LebPar*, sinzugiam *Par*
NUMERI pedes vel rithmi *LebPar*, pedes vel *om. Par*
†QUONIAM ALTER s. ionicorum *La*

379.12 VEL PROCELEUMATICO i. pirrichio *BeLa*, i. *om. La* pyrrihio *La*
de parvo proce<leumatico> *La*
VEL pro et *LebPar*

379.13 QUI VERO EX MINORE EST, CONTRARIUM FACIT
i. constat ex pirrichio et spondeo *BeLa*, pirihio *La*

379.13-5 ATQUE HI QUIDEM IN DACTYLICO GENERE PONENTUR RHYTHMI INCOMPOSITI AC COMPOSITI, QUI VII NUMERO OMNES ERUNT IN DACTILICO GENERE Quoniam aequalis numeri sunt in tempori<bu>s secundum istum. Sunt enim duo ex ista parte, duo ex illa *La*
RITHMI INCOMPOSITI pedes *BeLa*
De pedibus dicit, qui secundum rationem incedunt *BeLa*
Incompositi: ubi non est diversitas pedum *LebPar*, incompositi *om. Leb*
QUI VII NUMERO OMNES ERUNT De his dicit, quos superius nominavit *BeLa*

379.15 DACTYLUS digitus *Par*

984/
379.16 ANAPAESTUS Quasi sursum scandens; "ana" enim "sursum" dicitur *BeLa*, ana] na *Be*
anapesto: sursum peto *LebPar*, anapesto] ANAΠECTO *Par*

379.17-8 PYRRHICHIUS VERO, ID EST PROCELEUMATICUS, QUIA HIC ASSIDUUS VEL IN CERTAMINE VEL IN LUDO QUODAM PUERILI
PIRRICHIUS A Pyrro, qui semper faciebat in scuto. Nunc pedem* *Par*

* Cf. Isid. Etym. 1.17.2; Mur. Don.art.mai. 1.29; Ars Laur. Exp.Don.mai. 1.169.85, etc.

Par
PROCELEUMATICUS A celeuma, i. nauticus clamor[*] *LebPar*, a celeuma i. *om. Leb*
VEL IN LUDO QUODAM PUERILI s. ponebatur *BeLa*

.18-9 SPONDEUS, QUIA PLERUMQUE <SACRIFICIORUM USUI LIBATIONUMQUE> INSERVIT
SPONDEUS Quia sacrificalis est[**] *Be*
sacrificalis[**] *LaLebPar*
spondo, sacrifico[**] *LebPar*, spondo] ΣΠΟΝΔΩ *Par*
INSERVIT Diis inservit vel sacrificat *La*
sacrificiis *Leb*

.19-0 IONICUS SANE PROPTER NUMERORUM INAEQUALEM SONUM
IONICUS inaequalis[***] *LebPar*
INAEQUALEM SONUM s. dicimus *Be*

.20-1 QUO PEDE CARMINE MULTI SAEPE REPREHENSI SUNT
Qui alios vituperabant, haec carmine utebantur *BeLa*, viteperabant *La*[1]

.21 AC <DE> DACTYLICIS SATIS s. dictum *BeLa*

5/
.1-2 NUNC IAMBICA MEMORAMUS; IN QUO GENERE NUMERI INCOMPOSITI ERRANT QUATTUOR, COMPOSITI PER COPULAM DUO; AT VERO PER PERIODUM SUNT DUODECIM
IAMBICA s. divisionem *La*
NUMERI INCOMPOSITI †ERANT simplices variantur *BeLa*
ERRANT discurrunt *Leb*
currunt *Par*
AT VERO PER PERIODUM clausulam *BeLa*
Periodus non potest minus <habere> quam quattuor pedes *LebPar*, periodus *om. Par* quattuor] in quattuor *Par*

---

[*] Cf. Isid. Etym. 1.17.3; Sed.Scot. Don.art.mai. 1.33; Ars Laur. Exp.Don.mai. 1.172.75, etc.
[**] Cf. Isid. Etym. 1.17.2; Sed.Scot. Don.art.mai. 1.31; Ars Laur. Exp.Don.mai. 1.169.85, etc.
[***] Cf. Isid. Etym. 1.17.17

380.3 IAMBUS EX DIMIDIA ELATIONE Quae brevis est *BeLaLebPar*, id est *add. Par*

380.5-6 ORTHIUS VERO PES, QUI EX TETRASEMI ELATIONE [ID EST ARSI] ET OCTASEMI POSITIONE CONSTABIT
ORTIUS Ortius, qui per copulam effectus est. Sic est sicut simplex iambus. Troesus autem sicut trocheus. Ortius autem et troesus per copulam sic sunt, ut bachius per compositionem. Troesus et ortius per copulam imitantur, antibachium per compositionem *BeLa*
honestus, s. est *LebPar*
ORCIUS id est duplus duabus *Par*
QUI EX TETRASEMI s. per copulam isti iungitur *BeLa*
quatuor temporum *La*
POSITIONE CONSTABIT* i. ex proceleumatico et dispondeo *BeLa*

380.7 ATQUE HABET PROPINQUITATEM ALIQUAM CUM IAMBICO PEDE Octo enim ad quatuor sic sunt, quomodo duo ad i. *BeLa*, duo ad i.] i. ad duo *Be*
ut in dup<l>o, diiambo *Leb*

380.8-9 RELIQUIS OCTO TEMPORIBUS ADIUNCTIS Ad similitudinem iambi. Ita enim sunt viii ad quatuor, sic<ut> in iambo unum ad duo *La*

380.9-11 DEHINC TROCHAEUS QUI SEMANTICUS DICITUR, ID EST QUI E CONTRARIO OCTO PRIMIS POSITIONIBUS CONSTET, RELIQUIS IN ELATIONE QUATTUOR BREVIBUS ARTETUR
DEHINC TROCHEUS quasi troheus *La*
SEMANTICUS duplus ditrocheus. semanticus retardans vel remorans *LebPar*, duplus ditrocheus] duplex ditroheus *Par*
IN †ELATIONIBUS arsi *Leb*
in arsin *Par*
†BREVITER thesi *Leb*
in thesin *Par*

986/
380.12-4 BACCHIUS <...> QUI EX TROCHAEO DEDUCIT AUSPICIUM, FINE

---

* CONSTABAT *La*[1]

AUTEM IAMBICI TERMINATUR
BACCHIUS pro coriambo *BeLa*
AUSPICIUM i. initium *Be*
i. initio. Sicut apud Graecos telos significat arche<n> et finem *LebPar*
†TERMINANTUR Incipit ut trocheus, finit ut incipit iambus *LebPar*

).15 ATQUE A CONTRARIO HIS QUOS DIXIMUS PEDIBUS APTABITUR
HIS QUOS DIXIMUS PEDIBUS trocheo et iambo *La*
trocheo et antibachio *Leb*
trocheo vel antibachio *Par*

).15 PERIODUM s. est positio in genere iambico *La*
Est vero perchiodum, quod sequitur viam bachii et antibachii *BeLa*

).16-8 IN HOC GENERE CUM SINT DUODECIM NUMERO, QUATTUOR QUIDEM PER SINGULAS PERIODOS ACCIPERE DOCENTUR UNUM IAMBUM AC TRES TROCHAEOS
IN HOC GENERE i. incipiens iambo *La*
iambico *LebPar*
DUODECIM s. periodi *LebPar*
PER †SINGULOS PERIODOS Si unus externus inter tres similes pedes ponatur, periodus est *LebPar*, pedes ponatur] ponatur pedes *Par*
†DOCETUR imperfectum *LebPar*
UNUM IAMBUM AC TRES TROCHEOS Quatuor sunt pedes semper, sed tamen variantur per xii *La*

).18-9 AC DE ISDEM QUATTUOR PRIMUM QUIDEM, QUI IAMBUM HABERE MONSTRATUR, TROCHAEUS <A> IAMBO NOMINATUR
†HAC DE HISDEM perhiodis *BeLa*, perchiodis *Be*
TROCHEUS IAMBO trocheus a iambo *LebPar*, trocheus] trochei *Leb*
†DENOMINATUR ille periodus *LebPar*

).20-1 QUI VERO RHYTHMUS SECUNDUM IAMBUM RECIPIT, A BACCHIO TROCHAEUS VOCABITUR
RITHMUS i. numerus *BeLa*
A BACHIO pro iambo *Leb*

381.1-2 QUI VERO IAMBUM TERTIUM RECIPIT, BACCHIUS A TROCHAEO POTERIT NOMINARI
BRACHIUS A TROCHEO ex brevi et duabus longis *LebPar*

381.2-3 ILLE VERO, QUI QUARTUM ADMITTET IAMBUM, APPELLATUR EPITRITUS IAMBUS
ILLE VERO s. numerus *La*
s. ut est hoc exemplum: "hoc dotis"[*] *La*
EPITRITUS IAMBUS primus *Leb*
Epitritus ex brevi et tribus longis, peon ex longa et tribus brevibus *LebPar*, longa et tribus] longa ex tribus *Par*

381.3-4 QUI EX UNO TROCHAEO SUNT <ET IAMBIS TRIBUS, PRIMUS EST, QUI PRIORI LOCO TROCHAEUM HABET>, ET IAMBUS A TROCHAEO APPELLATUR
QUI EX UNO TROCHEO i. ex primo *La*
IAMBUS A TROCHEO Habemus hucusque iiii perhiodos. Nunc de his dicit, qui incipiunt a trocheo et <ex> tribus iambis constant. Primus est, qui a trocheo inchoat *BeLa*, periodos *La*[1]

381.5 SECUNDUS IAMBUS s. numerus *La*

381.6 QUI VERO TERTIUM RECIPIT i. iterum trocheum *BeLa*
s. trocheum *LebPar*, s. *om. Par*

987/
381.7 QUI VERO QUARTUM RECIPIT trocheum *Leb*

381.7 OCTO VERO <...> QUATTUOR DE HIS
VIII s. praedicti sunt *Be*
sunt octo *La*
s. praediximus *LebPar*
IIII DE HIS s. praedicti sunt *La*

381.9 QUI BINOS TROCHAEOS ATQUE IAMBOS PER PERIODUM SERVANT
i. secundum proportionem pirrichii *BeLa*

---

[*] I have not been able to locate this example.

1.10-1 ATQUE ILLE, QUI PRIMOS TROCHAEOS RECIPIT, DUPLEX BACCHIUS A TROCHAEO ESSE DICETUR
ATQUE ILLE †PRIMUS s. numerus *BeLa*
periodus qui *Leb*
DUPLEX BACHIUS s. et iambos secundos *BeLa*
ESSE †DICITUR s. qui vero secundos iambos <h>abebit, duplex bachius a trocheo esse dicetur *La*

1.11-2 QUI VERO PRIMOS IAMBOS HABEBIT, DUPLEX BACHIUS AB IAMBO NOMINATUR
DUPLEX BACHIUS AB IAMBO s. et primos iambos *BeLa*, s. *om. La*

1.13-5 OMNES VEL QUI INCOMPOSITI <SUNT, VEL QUI COMPOSITI> PER PERIODON VEL QUI PER COPULAM COLLIGANTUR
PER †SINODON i. per iuncturam simplices videlicet *BeLa*, iuncturam] iunturam *Be*

8/
1.17-1 ITEM HOC NOMEN EST AB EO, QUOD VENENUM MALEDICTI AUT LIVORES INFUNDAT
AB EO QUOD VENENUM Nam ios Grece venenum dicitur *LebPar*, nam *om. Leb* Grece *om. Par*
yos filius *Leb*
ios venenum, yios filius *Par*
LIVORES i. iambatrassa *BeLa*, i. *om. Be*
INFUNDAT s. homines *La*

2.1 TROCHAEUS trochos, rota[*] *LebPar*, trocos *Par*

2.2 ORTHIUS vel ostius *BeLa*

2.3-5 SEMANTUS SANE, QUIA, CUM SIT TARDIOR TEMPORE, SIGNIFICATIONEM IPSAM PRODUCTAE ET REMANENTIS CESSATIONIS EFFINGIT
†SEMANTICUS s. nominatus *La*
Videlicet iiii in capite, viii in fine *Be*
TARDIOR TEMPORE remanens vel retardans *LebPar*, remanens vel *om.*

---

[*] Cf. Isid. Etym. 1.17.3, 4.9.10, etc.

Par

Quia ab octo incipit *LebPar*

ET REMANENTIS i. iiii in fine habet *BeLa*

CESSATIONIS EFFINGIT Quia cum ascendit, per viii tempora ascendit. Videtur enim tunc, quod etiam per alia viii descendat, sed fingit, nam per iiii deponitur. Fingit ergo se <h>abiturum esse moram elationis in depositione *La*

382.5-6 BACCHII VERO SUNT DICTI, QUOD BACCHICIS MAXIME SONIS CONGRUUNT, ISQUE BACHIUS LUDUS EST, QUI ILLIS CARMINIBUS APTATUR

QUOD †BACCHIS bacchantibus *LebPar*, bacantibus *Par*[1] bachantibus *Par*[2]

CONGRUUNT placent *La*

†HISQUE s. pedibus *La*

†BACCHIS Illis mulieribus insanientibus, quae sacerdotes Liberi sunt* *BeLa*, sunt] sunt Bachi *La*

Inde venit verbum bacoi *La*

989/
382.7-8 IN EO VERO GENERE, QUOD PAEONICUM NOMINATUR, INCOMPOSITI DUO RHYTHMI ESSE DICENTUR

IN EO VERO GENERE nunc de peonico *La*

INCONPOSITI Incompositus ex similibus pedibus constat, ut ex peonibus *Par*

382.8 QUORUM ALIUS PAEON ΔΙΑΓΥΙΟΣ APPELLATUR

QUORUM †UNUS alius *La*

PEON †ΔΥΑΓΥ duplex *La*

i. duplex *Leb*

Vel ΛΙΔΙΟC, quod peones in Lydia abundant *LebPar*, vel *om. Par*
quod] quia *Par* lydia] lidio *Par*

382.9-10 EX LONGA POSITIONE ET LONGA ELATIONE AC DUPLICI POSITIONE PRODUCTA ET ARSI LONGIORE IUNGUNTUR

AC DUPLICI i. alter peon epipatois i. in thesi *BeLa*

ARSI LONGIORE Non distinguit quantum *BeLa*, distinguit] distingunt *La*

---

* Cf. Isid. Etym. 1.17.11

2.11 PAEONICI GENERIS NUMERI, QUOS INCOMPOSITOS ESSE
Incompositus ex similibus pedibus constat, ut ex peonibus *Leb*

2.12 NEQUE VERO PER CONIUNCTIONEM Quia rennuunt hoc poete *La*

2.13-4 INDE ΔΙΑΓΥΙΟΣ QUIDEM DICTUS EST, ID EST QUASI DUPLICIA MEMBRA DISCERNAT
†ΔΙΑΤΙΟC ΛΙΔΙΟC Lydius dicitur quasi lysios i. solutus in duplicia membra *LebPar*, lydius] ΛΙΔΙΟC *Par* dicitur *om. Par* lysios] ΛΙCΙΟC *Par*
DUPLICIA MEMBRA i. arsin et thesin *La*
duos peones *Leb*

2.14-5 ΕΠΙΒΑΤΟΣ AUTEM, QUIA MEMBRIS VELUTI UTENS QUATTOR <EX DUABUS ARSIBUS> ET DUABUS DIVERSIS THESIBUS COPULATUR
†ΕΠΙΒΑΤΟΙC Tria tempora in principio et duo in fine, vel contra *La*
a gravioribus *LebPar*
QUATTUOR Duas thesis et duas arsis *Leb*
ET DUABUS i. aequali et duplici *ParLeb*, i. *om. Par*
COPULATUR i. reciprocatur per conversionem, ut quod habet thesis, habeat aliquando arsis *BeLa*, per] pro *La*

0/
2.17 VERUM HAEC GENERA CUM PERMIXTA FUERINT
s. numerorum <genera>, quae praedicta sunt *BeLa*

2.18 PRIMA SPECIES ERUNT ISTAE, QUAE DOCHMIACAE NOMINANTUR
†DOCIMINE ab inventore *BeLa*
approbatae *Leb*
approbativae *Par*
DOKO video, puto, approbo; unde dociminae approbativae. Coles non habent. R *Par*[*]

2.19-0 EX QUIBUS PRIUS QUOD FUERIT, HAC LEGE COMPONITUR, UT SIT

[*] R = Remigius? cf. Rem. 530.5: DOCIMINAE id est approbatae, NOMINANTUR. DOCIMINAE a verbo doco, quod triplicem habet significationem; doco enim video, doco puto, doco approbo, in DOCIMINEA species, i.e. approbativae dicuntur.

EX IAMBO ET PAEONE, QUI ΔΙΑΓΥΙΟΣ VOCATUR
QUIBUS s. speciebus *BeLa*
generibus *Leb*
QUI peon *Par*

382.20-1 HUNC ΔΙΑΓΥΙΟΣ POSTERIORES CRETICUM NOMINARUNT
CRETICUM i. amfimacrum *LebPar*, i. *om. Par* amphimacrum *Par*
sic Agustinus* *Par*

991/
383.2-3 QUI AUTEM DOCHMII NUMERI NOMINANTUR, PROPTER ASSIDUUM ET COMPOSITUM SONUM APPELLARI VIDENTUR
QUI AUTEM †DEDUCTI spondeus molosus *Leb*
producti *Par*
PROPTER ASSIDUUM i. productum *BeLa*
APPELLARI VIDENTUR i. ideo dicti dicuntur *BeLa*, i. *om. La*

383.4 FIUNT AUTEM NUMERI, QUI ET PROSODIACI VOCANTUR
FIUNT AUTEM NUMERI De alio genere periodico nunc loquitur *La*
QUI ET PROSODIACI VOCANTUR Qui unum accentum habent *LebPar*
prosa longa *Leb*

383.5-6 UT HIS TRIBUS PEDIBUS IAMBUS PRIMUS APTETUR
HIS TRIBUS PEDIBUS ter quaterna erunt *La*

383.6-7 ID EST COPULIS, BACCHIO ET IONICO APOMIZONOS CONSTARE CONSUERUNT
BACCHIO ET IONICO APOMIZONOS i. copulam duorum *BeLa*
a maioris sono *La*
Maiore, qui incipit a trocheo *LebPar*
Bachio resolvitur una longa in duas breves *Leb*

992/
383.8 IRRATIONABILES Irrationabiles sunt, qui non componuntur, nisi ex similibus *LebPar*, irrationabiles sunt] s. *Par* nisi *om. Par*

---

* Cf. Aug. Mus. 2.1108.44

.9 QUOS ETIAM CHORIOS APPELLARE CONSUEVIMUS
Sic numeri vocantur, qui ex diversis surgunt et nullam aequalitatem habent *BeLa*
tribrachus et trocheus. choro laudo *LebPar*, tribrachus et trocheus] trocheum et tribracus *Par* laudo] ludo *Leb*

.11-2 ET RHYTHMO QUIDEM EST AD DACTYLICUM SIMILIS, PARTIBUS VERO AD NUMERUM IAMBICO IUNGITUR
SIMILIS similitudinem; arsin et thesin *BeLa*
unus iambus pro una longa syllaba *LebPar*
AD NUMERUM †IONICUM Dupla proportio est *LebPar*

.13 TROCHAEIDES
†TROCHOIDES Quia etiam dupla est *LebPar*

.15 PER CONTRARIUM a contrario *BeLa*, contrarium *La*

/

.18 DACTYLUS IN BACCHIUM INCIDENS, [IS] QUI VENIAT EX TROCHAEO
DACTILUS IN †BACHIO INCIDENS i. iungitur bachio *BeLa*, bahio *La*
i. antibachius *LebPar*, i. *om. Par*

.19 DACTYLUS PER CHORIUM
PER CORIUM i. per trocheum *BeLa*
tribrachus *Leb*
Id est tribrachum *Par*

.20-1 QUI EX SIMILITUDINE TROCHAEI VIDEATUR EXPRESSUS
EXPRESSUS inceptus *La*

.1-2 ET CRETICUS QUIDEM CONSTANS EX TROCHAEI POSITIONE ET <..
CRETICUS amphimacrus *LebPar*, amphimracus *Leb*
CONSTANS EX TROCHAEI POSITIONE i. similitudinem habens trochei *BeLa*, i. *om. La* habens *om. La*

/

.3-4 RHYTHMOPOEIA EST CON>DICIO NUMERI COMPONENDI ET

OMNIUM FIGURARUM PLENA PERCEPTIO
†INDITIO Totius compositionis pedum et omnium figurarum pedum plena perceptio inditio i. ratione muri (*sic*, *lege* numeri) dividitur *BeLa*
iambi *Leb*
PERCEPTIO intellectus *Par*

384.4 DIVIDITUR HAEC IN EAS QUAS ET MELOPOEIA PARTES
IN EAS i. rithmicorum ratio *BeLa*
MELOPOEIA PARTES s. in quibus dividuntur *BeLa*, s. *om. Be*

384.5-6 PERCEPTIO, PER QUAM SCIMUS, QUONAM NUMERO UTENDUM SIT
†QUO QUANTUM NUMERO UTENDUM SIT i. quantum pedes debemus in unoquoque ponere *LebPar*, quantum] quod *Par*

384.7 USUS, PER QUEM POSITIONES AUT ELATIONES DECENTER APTAMUS
DECENTER honeste *BeLa*

384.7-8 PERMIXTIO, PER QUAM, QUOD OPPORTUNUM FUERIT, EX ARTE MISCEMUS
EX ARTE MISCEMUS in copulatione pedum *BeLa*

384.8-9 TROPI VERO, UT IN MELOPOEIA, ET IN RHYTHMOPOEIA TRES SUNT: SYSTALTICOS <...>
MELOPOEIA factura melodiae *BeLa*
ET IN RITHMOPOEIA factura numeri *BeLa*
SISTALTICOS i. in rithmopoeia *BeLa*
aequos i. stantes *LebPar*, aequos i. *om. Par*

996/
384.14-5 QUAE CUM HARMONIA INTENTIS TAM DIVIS QUAM HEROUM POPULIS AUGUSTA QUADAM SUAVITATE PERCURRERET
ARMONIA supradicta *Par*
HEROUM POPULIS nobili *BeLa*

384.15-6 AD CANTUS CARMINUMQUE DULCEDINES DECENTER REGRESSA CONTICUIT

†CONTINUIT abstinuit se *La*
    Festinare voluit *Leb*
    Properare voluit *Par*

.16-8 TUNCQUE IOVE ASSURGENTE DIVISQUE PRAEAMBULIS COEMESIN MODULATA IN THALAMUM QUOQUE VIRGINIS MAGNA CUNCTORUM VOLUPTATE PERVENIT
DIVISQUE PRAEAMBULIS praeviis *La*
CIMESIN nomen tropi *BeLa*
    vel nuptiale carmen *La*
    Species est carminis *Leb*
    Grece carminis *Par*

/

.19 HABES ANILEM, MARTIANE, FABULAM
IAMBICUM SENARIUM TRIMETRUM ACATALECTICUM *BeLa*
IAMBICUM SENARIUM *Leb?Par*
†SENILEM antiquam *BeLa*
    vel similem *Be*
MARTIANE ó fili *BeLa*
    Quasi ad se loquitur *La*
    Verba filii Martiani ad patrem Martianum *LebPar*, Martianum] martianem *Par*
FABULAM Finita est *BeLa*, est *om. Be*

.20 MISCILLO LUSIT QUAM LUCERNIS FLAMINE i. ante suas lucernas *BeLa*
MISCILLO turbulento, vario *Be*
    confuso, turbulento, vario *La*
    Miscilla dea vocatur, quae miscet carminibus falsa *La*
    vario i. mixto, vero cum fabulis *LebPar*
    i. mixtis veris et fabulis *Leb*
QUAM i. te *La*
†QUEM s. te *Leb*
LUCERNIS i. nocturno tempore *La*
    s. suis *BeLa*
FLAMINE spiritu turbido *BeLa*
    spiritu *Leb*
    carmine *Par*

384.21 SATURA, PELASGOS DUM DOCERE NITITUR
†SATYRA poetria *BeLa*
Quando vera et falsa iunguntur, satyra dicitur, eo quod suis nos satiet carminibus *BeLa*
PELASGOS Grecos *BeLaPar*
Grece *Leb*
DUM DOCERE NITITUR cupit *Be*
s. illa Satyra *BeLa*, s. *om. La*

385.1 ARTES †CAGRIS VIX AMICAS ATTICIS
ARTES s. liberales *BeLa*
†CREAGRIS sagacibus acutis *La*
i. ingeniosis, sagacibus, an fractuosis. Dicitur etiam creagre fuscina, unde carnes abstra<h>untur a lebetis coctae *BeLa*, a] et *Be*
acutis vel rusticis. Creos vel crea caro, agro venor, unde creager fuscinula vel creos dicitur hircus, agros rusticus *LebPar*, creos vel *om. Par* creager] creagra *Par*
Orpofago Grece Latine fuscinula *Leb*
VIX AMICAS convenientes *BeLa*
ATTICIS Quia vix intelleguntur ab Atticis *BeLa*
Grecis *Par*

385.2 SIC IN NOVENA DECIDIT VOLUMINA
DECIDIT i. pervenit *BeLa*
fabula *Leb*

998/
385.3 HAEC QUIPPE LOQUAX DOCTA DOCTIS AGGERANS
HAEC s. Satyra *BeLa*
LOQUAX garrula *BeLa*
DOCTA s. praecepta *BeLa*
DOCTIS AGGERANS s. hominibus accumulans *BeLa*

385.4 FANDIS TACENDA FARCINAT, IMMISCUIT
FANDIS pronuntiandis *BeLa*
artibus *LebPar*
TACENDA FARCINAT implet, prodit, accumulat *BeLa*
fabulas implet s. et *Leb?Par*

IMMISCUIT s. et *BeLaLeb*, s. *om. La*
Simul iungit loquenda cum non loquendis *La*

.5 MUSAS DEOSQUE, DISCIPLINAS CYCLICAS
MUSAS i. per sua carmina *BeLa*, i. *om. La*
†DEOQUE Mercurio *Be*
Mater inspirante deo furioso i. Miscillo flammine *La*
DISCIPLINAS CICLICAS i. obscuras, rotundas, involutas *Be*
tortas, obscuras *La*
Aliquando enim carmina componuntur, aliquando artes enucleant illae Musae *BeLa*, carpmina *La*
tortuosas a circo *Leb*
†CALICAS Grecum septem liberales artes i. tortuosas *Par*

.6 GARRIRE AGRESTI CRUDA FINXIT PLASMATE
GARRIRE i. aliquid exponere *BeLa*
AGRESTI cum rustico *BeLaPar*, cum *om. Par*
CRUDA cum *Be*
s. cura *BeLa*
i. nova *LebPar*, i. *om. Leb*
FINXIT simul cum s. se *Be*
Satyra *Leb*
PLASMATE conformatione vel figmenta *Be*
indocta, stulta informatione *La*
inveniens nova carmine *Par*

.7 HAEC IPSA NAMQUE RUPTA CONSCIENTIA
†HAC s. Satyra *BeLa*
†NAUCI i. vili, levi *BeLa*
vili *LebPar*
†RUCTA promta Satyra *BeLa*
RUPTA Satyra *Leb*
corrupta, ipsa Satira *Par*
CONSCIENTIA Quasi aliquid pro mens *BeLa*

.8 TURGENSQUE FELLE AC BILI, 'MULTA CHLAMYDE
TURGENSQUE superbiens *BeLa*
FELLE AC †VILEM de veneno *BeLa*

adverbium pro hic *Be?La*
MULTA CLAMIDE bono <h>abitu *La*
vel cum vile clamide vel de *Be*
Textura sive multo habitu. Ac si diceret: voluit vilia demonstrare, sed tamen ipsum vile *BeLa*
per falsa *Par*

385.9 PRODIRE DOCTIS APPROBANDA CULTIBUS
PRODIRE ostendere vel consurgere *Be*
pronuntiare, consurgere *La*
cum DOCTIS ADPROBANDA laudanda *BeLa*
CULTIBUS ornatibus *La*

385.10 POSSEMQUE COMIS UTQUE E MARTE CURIA
COMIS pulchra *Leb*
†MARTIS sicut *BeLa*
i. de vico Martis Ariopagita *BeLa*
i. sum comes Martis et sum e curia eius *BeLa*, eius *om. La*
CURIA s. comes in Athenis *BeLa*, s. comes] i. *Be*

385.11 FELICIS' INQUIT 'SED CAPELLAE FLAMINE
INQUIT illa Satyra *La*
verba Satyrae *Leb*
Satira *Par*
SED CAPELLAE †FLAMMINE de spiritu vel mente *BeLa*
Quia finem facit *Par*

385.12 INDOCTA RABIDUM QUEM VIDERE SAECULA
INDOCTA RABIDUM iratum *Be*
inde rabulatio i. loquacitas cum ira *BeLa*
sum *Leb*
dementem *Par*
QUEM s. Felicem *BeLa*

385.13 IURGIS CANINOS BLATERATUS PENDERE
†LURGIS lurgis: cum canis glutonibus *BeLa*, lurgis *om. Be*
gluttonibus *LebPar*
BLATERATUS i. latratus *Be*
i. locutiones infructuosas *BeLa*

quia in nuptiis *Par*
PENDERE suspendere *BeLa*
loqui *La*
reddere, solvere *LebPar*, reddere *om. Par*

.14 PROCONSULARI VERBA DANTEM CULMINI
PROCONSULARI una pars est *BeLa*
honorabili *La*
†VERO DANTEM CULMINI i. dat in honorem consulum *BeLa*, i. dat *om. Be*
s. tales latratus *La*
et quem videre saecula *BeLa*
blateratus *Par*

.15 †IPSOQUE DUDUM BOMBINATORE FLOSCULO
IPSOQUE DUDUM parvo *Leb*
cum parvo flore *Par?*
†BOMBINAT ORE i. clamat locutionem *BeLa*
†BOMBINATOREM cantatorem *Leb*
FLOSCULO superfluo vel vili *BeLa*
fructu *Leb*

.16 DECERTUM FULQUEM IAM CANESCENTI ROTA,†
DECERTUM s. videmus *La*
Pro decertatum i. rusticam dictationem *BeLa*
†FALCEM gloriantem de sua falce *La*
CANESCENTI senescenti *BeLa*
aetate senili *BeLa*, senili *om. Be*
cano capite vel aetate *Par*

.17 BEATA ALUMNUM URBS ELISSAE QUEM VIDET
BEATA ALUMNUM filium *BeLa*
URBS ELISSE Cartago Didonis *BeLaPar*, Kartago *BePar* didonis *om. Par*
Quia Martianus Cartaginensis fuit *BeLa*
Cartaginis *Leb*
YRONIA *Leb*

.18 IUGARIORUM MURCIDAM VICINIAM
IUGARIORUM s. iugalium *BeLa*, s.] sub *Be*

Iugaria provincia, unde ipse fuit *Leb*
Ex qua provincia fuit *Par*
MURCIDAM pro marcidam i. tabescentem *BeLa*, pro *om. La*
Nam Cartago inter Mauros est *La*
antiquam *Leb*
parvam *Par*
VICINIAM Murcidam viciniam Iugariorum dicit, fabulas in principio fineque librorum positas *BeLa*

385.19 PARVO OBSIDENTEM VIXQUE RESPERSUM LUCRO
PARVO †OBSIDEM cum *Be*
pauperem divitiarum *Be?La*

VIXQUE RESPERSUM LUCRO i. vix senescentem *Be?La*

385.20 NICTANTE CURA SOMNOLENTUM LUCIBUS —
NICTANTE CURA Que vidit *La*
cum vigilante studio *BeLa*, cum *om. La*
SOMNOLENTUM LUCIBUS s. in nocte *BeLa*
s. et quiescentem in die *Be*
in die *La*
i. die quiescentem et nocte laborantem *BeLa*

1000/
386.1 AB HOC CREATUM PEGASEUM GURGITEM
†OB HOC propter *La*
s. illa cura *BeLa*
PEGASEUM GURGITEM i. facultatem carminum *BeLa*
Pege fons, ubi sunt omnes nimphe *Par*

386.2 DECENTE QUANDO POSSEM HAURIRE POCULO?'
DECENTE orante illo *BeLa*
QUANDO vel prout *BeLa*
POSSEM HAURIRE POCULO cum *Be*
s. ego Satyra fingere *BeLa*

386.3 TESTEM ERGO NOSTRUM QUAE VETERNUM PRODIDIT
TESTEM ERGO NOSTRUM i. musarum *BeLa*
Martianus iunior *Leb*

QUAE VETERNUM PRODIDIT Antiquitatem nostram vituperavit *BeLa*
manifestavit *Par*

.4 SECUTE NUGIS, NATE, IGNOSCE LECTICANS
SECUTAE s. me vel verba Satyre *Be*
s. Satyre *LaPar*, s. *om. Par*
NUGIS s. vilibus carminibus *Be*
s. meis carminibus vilibus *La*
NATE ó fili *BeLa*
IGNOSCE LECTICANS saepissime legens *BeLa*
Quod dicit: "quis mea conscium apud persecutam me" (*sic?*)* *La*
VOX MARTIANI PATRIS *BeLa*

MARTIANI MINEI FELICIS CAPELLAE AFRICARTAGINIENSIS. DE ARMONIA EXPLICIT LIBER NONUS
BeLaLebPar
maciani *Be*[1] marciani *Be*[2]*Par* minei] minnei *Leb* minne *Par* africartaginiensi *La* africarthaginensis *Par* explicit liber nonus] liber nonus explicitus est feliciter. AMHN *Leb* liber viiii explicit feliciter *Par*

* I have not been able to locate this citation.

APPENDIX

CONCORDANCE OF PAGE-NUMBERS IN THE MARTIANUS CAPELLA-EDITIONS OF WILLIS AND DICK

Section numbers	Edition J. WILLIS (page.line)	Edition A. DICK (page.line)
11	6.14	10.11
12	7.6	11.9
19	9.18	15.3
27	12.18	19.8
28	13.3	19.19
105	30.3	44.5
106	30.10	44.15
107	30.15	44.20
108	30.25	45.13
117	33.16	49.6
119	34.16	50.14
120	35.7	51.5
121	35.20	51.18
127	40.20	56.23
131	41.19	57.15
138	42.19	60.1
139	43.6	60.14
169	49.14	69.19
170	49.20	70.7
171	50.3	70.12
181	51.7	72.12
182	51.12	72.18
187	52.11	73.20
194	53.12	75.1
195	53.14	75.3
196	53.14	75.4
197	54.5	75.14
198	54.13	75.23
199	54.15	76.3
213	56.12	78.14
888	337.5	469.9
889	338.15	471.4

Section numbers	Edition J. WILLIS	Edition A. DICK
890	338.25	471.18
891	339.3	471.23
892	339.11	472.7
893	339.18	472.16
894	340.5	473.8
895	340.14	474.5
896	341.3	474.13
897	341.10	474.22
898	342.3	475.18
899	342.14	476.13
900	342.21	477.2
901	343.2	477.7
902	343.7	477.14
903	343.23	478.9
904	344.15	479.3
905	344.21	379.11
906	345.11	480.4
907	345.22	480.18
908	346.11	481.10
909	347.7	482.9
910	347.16	483.4
911	348.2	483.14
912	348.13	484.2
913	349.14	485.3
914	349.24	485.13
915	350.16	486.10
916	350.25	486.19
917	351.10	487.6
918	351.26	487.22
919	352.18	488.20
920	353.10	489.15
921	353.18	490.1
922	353.25	490.10
923	354.6	490.19
924	354.9	491.4
925	354.18	491.14
926	355.7	492.7
927	356.2	493.8
928	356.9	493.17
929	356.14	494.4
930	356.20	494.11
931	357.11	495.5
932	358.12	496.13
933	358.15	496.18

Section numbers	Edition J. WILLIS	Edition A. DICK
934	358.21	497.9
935	359.9	497.20
936	360.6	499.7
937	360.16	500.7
938	361.6	500.16
939	361.10	501.1
940	362.2	501.11
941	362.9	502.7
942	362.15	502.15
943	363.4	503.8
944	363.17	504.8
945	364.6	504.16
946	364.13	505.7
947	365.2	505.17
948	365.12	506.12
949	365.16	506.17
950	366.7	507.11
951	366.18	508.1
952	367.3	508.11
953	367.8	508.18
954	367.13	509.6
955	368.7	510.1
956	368.9	510.4
957	368.13	510.9
958	369.4	511.6
959	369.11	511.13
960	370.1	512.5
961	370.12	512.16
962	370.19	513.10
963	371.5	513.21
964	371.13	514.12
965	371.21	515.1
966	372.7	515.14
967	372.15	516.8
968	373.2	516.15
969	373.8	517.4
970	373.17	517.16
971	374.1	518.8
972	374.13	518.23
973	374.19	519.7
974	375.3	519.15
975	375.7	519.18
976	375.11	520.8
977	375.19	520.19

Section numbers	Edition J. WILLIS	Edition A. DICK
978	376.15	521.20
979	377.4	522.10
980	377.13	523.2
981	378.3	523.13
982	378.18	524.15
983	379.8	525.6
984	379.16	525.17
985	380.1	526.5
986	380.12	527.1
987	381.7	528.5
988	381.17	528.16
989	382.7	529.11
990	382.17	530.5
991	383.2	530.11
992	383.8	531.4
993	383.18	531.15
994	384.3	532.4
995	384.10	532.13
996	384.14	533.5
997	384.19	533.11
998	385.3	533.16
999	385.7	534.3
1000	386.1	535.2

BIBLIOGRAPHY

Abbreviations

CCCM	*Corpus Christianorum: Continuatio Mediaevalis* (Turnhout 1966–)
CCSL	*Corpus Christianorum: Series Latina* (Turnhout 1954–)
CSEL	*Corpus Scriptorum Ecclesiasticorum Latinorum* (Vienna/-Leipzig/Prague 1866–)
CTC	*Catalogus translationum et commentariorum: Mediaeval and Renaissance Latin Translations and Commentaries. Annotated Lists and Guides*, eds. P.O. KRISTELLER, F.E. CRANZ *et alii* (Washington 1960–)
MGG	*Die Musik in Geschichte und Gegenwart: Allgemeine Enzyklopädie der Musik*, ed. F. BLUME, 17 vols. (Kassel 1949-86). New edition of the Sachteile, ed. L. FINSCHER, 9 vols. (Kassel 1994-1999), Personenteile (12 vols.): in progress
MGH	*Monumenta Germaniae Historica* (Hannover/Leipzig/Berlin, etc. 1826–)
New Grove	*The New Grove Dictionary of Music and Musicians*, ed. S. SADIE, 20 vols. (London/Washington 1980)
OLD	P.G.W. GLARE, *Oxford Latin Dictionary* (Oxford 1968-1982)
LEWIS & SHORT	C.T. LEWIS and C. SHORT, *A Latin Dictionary* (Oxford 1879; many reprints)
Pal.Mus.	*Paléographie Musicale: fac-similés phototypiques des principaux manuscrits de chant Grégorien*, ed. Dom A. MOCQUEREAU (Solesmes 1889–)
PL	*Patrologia Latina*, ed. J.-P. MIGNE, 222 vols. (Paris 1841-64)
TLL	*Thesaurus Linguae Latinae* (München 1899–)

Primary sources

Alia musica. Traité de musique du ixe siècle, J. CHAILLEY (ed., transl. and intr.), (Publications de l'Institut de Musicologie de l'Université de Paris 6), Paris 1965

ARISTIDES QUINTILIANUS, *De musica libri tres*, ed. R.P. WINNINGTON-INGRAM, Leipzig 1963

Ars Laureshamensis: expositio in Donatum Maiorem, ed. B. LÖFSTEDT, (Grammatici hibernici Carolini aevi, CCCM 40A), Turnhout 1977

AUGUSTINUS HIPPONENSIS, *Confessionum libri tredecim*, ed. L. VERHEIJEN, (CCSL 27), Turnhout 1981

AUGUSTINUS HIPPONENSIS, *De civitate dei*, eds. B. DOMBART & A. KALB,

(CCSL 47-48), Turnhout 1955

AUGUSTINUS HIPPONENSIS, *De diversis quaestionibus ad Simplicianum*, ed. A. MUTZENBECHER, (CCSL 44), Turnhout 1970

AURELIANUS OF RÉÔME, *Musica disciplina*, ed. L. GUSHEE, (Corpus Scriptorum de Musica 21), 1975

BEDA VENERABILIS, *De arte metrica et de schematibus et tropis*, ed. C.B. KENDALL & M.H. KING, (CCSL 123A), Turnhout 1975

BEDA VENERABILIS, *In Lucae evangelium expositio*, ed. D. HURST, (CCSL 120), Turnhout 1960

BEDA VENERABILIS, *De orthographia*, ed. Ch.W. JONES, (CCSL 123A), Turnhout 1975

PS-BOETHIUS, *Geometriae*, ed. M. FOLKERTS, *Boethius' Geometrie: ein mathematisches Lehrbuch des Mittelalters*, Wiesbaden 1970

BOETHIUS, *De institutione arithmetica libri duo, de institutione musica libri quinque*, ed. G. FRIEDLEIN, Leipzig 1867

BONIFATIUS, *Ars grammatica*, ed. G.J. GEBAUER & B. LÖFSTEDT, (CCSL 133B), Turnhout 1980

CALCIDIUS, *Timaeus a Calcidio translatus commentarioque instructus*, ed. J.H. WASZINK, (Plato Latinus 4), London, Leiden 1962

CASSIODORUS, *Institutiones*, ed. R.A.B. MYNORS, Oxford 1937

CICERO, *Somnium Scipionis*, (*De re publica* VI), transl. C.W. KEYES, Loeb Classical Library, London & New York 1928

DONATUS, ed. L. HOLTZ, *Donat et la tradition de l'enseignement grammatical: Étude sur l'Ars Donati et sa diffusion (IVᵉ-IXᵉ siècle) et édition critique*, Paris 1981

DONATUS ORTIGRAPHUS, *Ars grammatica*, ed. J. CHITTENDEN, (Grammatici hibernici Carolini aevi, CCCM 40D), Turnhout 1982

DUNCHAD, *Glossae in Martianum*, ed. C.E. LUTZ, (Philological Monographs published by the American Philological Association 12), Lancaster, Oxford 1944

FULGENTIUS MYTHOGRAPHUS, *Expositio sermonum antiquorum*, ed. U. PIZZANI, *Fabio Planciade Fulgenzio: "Definizione di Parole Antique"*, (Scriptores Latini 9), Roma 1968

FULGENTIUS MYTHOGRAPHUS, *Mythologiarum libri tres*, ed. R. HELM, Stuttgart 1970

Glossa maior in institutionem musicam Boethii, eds. M. BERNHARD & C.M. BOWER, (Bayerische Akademie für Wissenschaften, Veröffentlichungen der musikhistorischen Kommission 9-12), München 1993, 1994, 1996 — the last Volume is in preparation

GREGORIUS OF TOURS, *Historia Francorum*, ed. ARNDT, MGH Scriptores Rerum Merovingicarum I, pp. 1-450

HIERONYMUS, *In Hieremiam prophetam libri vi*, ed. S. REITER, (CCSL 74), Turnhout 1960

HIERONYMUS, *Tractatus sive homiliae in Marci evangelium*, ed. Dom G. MORIN, (CCSL 78), Turnhout 1958

JOHANNES SCOTTUS, *Annotationes in Marcianum*, ed. C.E. LUTZ, Cambridge

Mass. 1939

JOHANNES SCOTUS ERIUGENA, *Expositiones in hierarchiam celestem*, ed. J BARBET, (CCCM 31), Turnhout 1975

JOHANNES SCOTUS ERIUGENA, *Periphyseon*, ed. I.P. SHELDON-WILLIAMS, (Scriptores Latini Hiberniae 11), Dublin 1981

ISIDORUS HISPALENSIS, *Etymologiarum sive originum libri XX*, 2 Vols., ed. W.M. LINDSAY, Oxford 1911 (reprint 1957)

PS-JUVENCUS, *Praefatio operi Iuvenci postea anteposita*, in: GAIUS VETTIUS AQUILINUS JUVENCUS, *Evangeliorum libri quattuor*, ed. J. HUEMER, (CSEL 24), Wien 1891

LACTANTIUS PLACIDUS, *In Statii Thebaida commentum*, ed. R.D. SWEENEY, Stuttgart, Leipzig 1997

LUPUS OF FERRIÈRES, *Epistolae*, ed. L. LEVILLAIN, *Loup de Ferrières, Correspondance*, (Vols. 1-2), Paris 1927-1935

MACROBIUS, *Commentarium in Somnium Scipionis*, ed. J. WILLIS, Stuttgart, Leipzig 1964

MARIUS VICTORINUS, *Ars grammatica*, ed. I. MARIOTTI, (Testi con commento filologico), Firenze 1967

MARTIANUS CAPELLA, *De nuptiis Philologiae et Mercurii*, ed. J. WILLIS, Leipzig 1983

Musica et scolica enchiriadis una cum aliquibus tractatulis adiunctis, H. SCHMID (ed.), (Bayerische Akademie der Wissenschaften, Veröffentlichungen der Musikhistorischen Kommission 3), München 1981

MURETHACH, *In Donati artem maiorem*, ed. L. HOLTZ, (Grammatici hibernici Carolini aevi, CCCM 40), Turnhout 1977

MYTHOGRAPHI VATICANI I & II, ed. P. KULCSÁR, (CCSL 91C), Turnhout 1987

PLATO, *De Republica*, transl. P. SHOREY, Loeb Classical Library, 2 vols., Cambridge Mass., London 1930-1935

PLATO, *Timaeus*, transl. R.G. BURY, Loeb Classical Library, Cambridge Mass., London 1975

PLINIUS, *Naturalis historia*, transl. H. RACKHAM, W.H.S. JONES & D.E. EICHHOLZ, Loeb Classical Library, 10 vols., Cambridge Mass., London 1938-1962

PRISCIANUS, *Institutiones grammaticarum*, ed. M. HERTZ, (Grammatici Latini 2-3), 2 vols., Leipzig 1855, 1859

RATHERIUS VERONENSIS, *Epistolae*, in: F. WEIGLE (ed.), *Die Briefe des Bischofs Rather von Verona*, MGH, *Briefe der deutschen Kaiserzeit* I, München 1977

REMIGIUS AUTISSIODORENSIS, *Commentum in Martianum Capellam I-II*, ed. C.E. LUTZ, *Libri I-II*, Leiden 1962, *Libri III IX*, Leiden 1965

Scriptores ecclesiastici de musica sacra potissimum, ed. M. GERBERT, 3 vols., Sankt Blasien 1784, repr. 1963

Scriptorum de musica medii aevi nova series, ed. C.E.H. de COUSSEMAKER, 4 vols., Paris 1864-1876, repr. 1963

SEDULIUS SCOTTUS, *Collectaneum miscellaneum*, ed. D. SIMPSON, (CCCM 67), Turnhout 1988

SEDULIUS SCOTTUS, *In Donati artem maiorem*, ed. B. LÖFSTEDT, (Grammatici hibernici Carolini aevi, CCCM 40B), Turnhout 1977

SERVIUS, *In Vergilii Carmina commentarii*, ed. G.I. LIEFTINCK, Amsterdam 1960

SMARAGDUS, *Liber in partibus Donati*, ed. B. LÖFSTEDT, L. HOLTZ & A. KIBRE, (CCCM 68), Turnhout 1986

PS-TERTULLIANUS, *Carmen adversus Marcionem*, ed. R. WILLEMS, (CCSL 2), Turnhout 1954

VERGILIUS, *Aeneas*, transl. H.R. FAIRCLOUGH, Loeb Classical Library, 2 vols, London, New York 1916-1918

VERGILIUS, *Eclogae*, transl. H.R. FAIRCLOUGH, Loeb Classical Library, London, New York 1916

Secondary works

H. ABERT, *Die Musikanschauung des Mittelalters und ihre Grundlagen*, Halle a.d. S. 1905

E. AUERBACH, *Literatursprache und Publikum in der lateinischen Spätantike und im Mittelalter*, Bern 1958

E.M. BANNISTER, *Monumenti Vaticani di Paleografia Musicale Latina*, Leipzig 1913

A. BARBERA (ed.), *Music Theory and Its Sources: Antiquity and the Middle Ages*, (Notre Dame Conferences in Medieval Studies 1), Notre Dame 1990

W. BEIERWALTES, *Eriugena. Grundzüge seines Denkens*, Frankfurt 1994

M. BERNHARD, *Studien zur Epistola de armonica institutione des Regino von Prüm*, (Bayerische Akademie der Wissenschaften, Veröffentlichungen der Musikhistorischen Kommission 5), München 1979

—, *Glossen zur Arithmetik des Boethius*, in: *Scire litteras. Forschungen zum mittelalterlichen Geistesleben*, S. KRÄMER, M. BERNHARD (eds.), München 1988, pp. 23-34

—, *Glosses on Boethius' De institutione musica*, in: *Music Theory and Its Sources. Antiquity and the Middle Ages*, A. BARBERA (ed.), Notre Dame 1990, pp. 136-149

—, *Überlieferung und Fortleben der antiken lateinischen Musiktheorie im Mittelalter*, in: *Geschichte der Musiktheorie*, 3: *Rezeption des antiken Fachs im Mittelalter*, ed. F. ZAMINER, Darmstadt 1990, pp. 7-36

—, *Boethius im mittelalterlichen Schulunterricht*, in: *Schule und Schüler im Mittelalter*, M. KINTZINGER, S. LORENZ and M. WALTER (eds.), Köln etc. 1996, pp. 11-27

M. BIELITZ, *Musik und Grammatik*, (Beiträge zur Musikforschung Bd. 4), München 1977

B. BISCHOFF, *Hadoardus and the Manuscripts of Classical Authors from Corbie*, in: *Didascaliae. Studies in Honor of A.M. Albareda*, New York 1961, pp. 41-57

—, *Lorsch im Spiegel seiner Handschriften* (Münchener Beiträge zur Mediävistik und Renaissance-Forschung, Beiheft), München 1974; second, revised edition: *Die Abtei Lorsch im Spiegel ihrer Handschriften*, Lorsch 1989

—, transl. M.M. GORMAN, *Manuscripts and Libraries in the Age of Charlemagne*, (Cambridge Studies in Palaeography and Codicology 1), Cambridge 1994

—, *Katalog der festländischen Handschriften des neunten Jahrhunderts (mit ausnahme der wisigothischen)*, Teil 1: *Aachen—Lambach*, (Bayerische Akademie der Wissenschaften, Veröffentlichungen der Kommission für die Herausgabe der mittelalterliche Bibliothekskataloge Deutschlands und der Schweiz), Wiesbaden 1998

T.A.M. BISHOP, *The corpus Martianus Capella. Welsh and continental tradition, drawn upon by the revived Latin learning of tenth-century England, united in Corpus Christi College 153 (Martianus Capella)*, in: *Transactions of the Cambridge Bibliographical Society* 4, Cambridge 1949, pp. 257-275

—, *Autographa of John the Scot*, in: *Jean Scot Érigène et l'histoire de la philosophie*, Colloques internationaux du C.N.R.S. 561, Paris 1977, pp. 47-58

C.M. BOWER, *Natural and Artificial Music: The Origins and Development of an Aesthetic Concept*, in: *Musica Disciplina* 25 (1971), pp. 17-33

—, *The Role of Boethius' "De institutione musica" in the Speculative Tradition of Western Musical Thought*, in: *Boethius and the Liberal Arts. A Collection of Essays*, M. MASI (ed.), (Utah Studies in Literature and Linguistics 18), Berne etc. 1981, pp. 157-174

—, *Boethius' De institutione musica: A Handlist of Manuscripts*, in: *Scriptorium* 42 (1988), pp. 205-251

— (intr. and transl.), *Anicius Manlius Severinus Boethius: Fundamentals of Music*, (Music Theory Translation Series), New Haven, London 1989

—, *The grammatical model of musical understanding in the Middle Ages*, in: *Hermeneutics and Medieval Culture*, P.J. GALLACHER & H. DAMICO (eds.), Albany 1989, pp. 133-145

—, *Die Wechselwirkung von philosophia, mathematica und musica in der karolingischen Rezeption der "Institutio musica" von Boethius*, in: *Musik und die Geschichte der Philosophie und Naturwissenschaften im Mittelalter*, ed. F. HENTSCHEL, Leiden, Boston, Köln 1998, pp. 163-183

—, *Transmission of Ancient Theory into the Middle Ages*, forthcoming in: *The Cambridge History of Music Theory*

R. BRAGARD, *L'harmonie des sphères selon Boèce*, in: *Speculum* 4 (1929), pp. 206-213

D. BULLOUGH, *Carolingian Renewal: Sources and Heritage*, Manchester, New York 1991

P.L. BUTZER & D. LOHRMANN (eds.), *Science in Western and Eastern Civilizations in Carolingian Times*, Basel 1993

A. CAMERON, *Martianus and his first editor*, in: *Classical Philology* 81 (1986), pp. 320-328

M. CAPPUYNS, *Capella (Martianus)*, in: *Dictionnaire d'histoire et de géographie ecclésiastiques* 11, Paris 1949, col. 835-847

M. CAPPUYNS, *Jean Scot Érigène, sa vie, son oeuvre, sa pensée*, Diss. Leuven, Brussel 1964

A. CASTAN, *La bibliothèque de l'abbaye de St.-Claude du Jura. Esquisse de son histoire*, in: *Bibliothèque de l'École des Chartes* 50 (1889), pp. 301-354

—, *Catalogue général des manuscrits des bibliothèques publiques de France, Départ. T. 32*, Paris 1897, pp. 352-353

J. CHAILLEY (ed., transl. and intr.), *Alia musica. Traité de musique du ixe siècle*, (Publications de l'Institut de Musicologie de l'Université de Paris 6), Paris, 1965

Y. CHARTIER, *L'Epistola de armonia institutione de Réginon de Prüm*, diss. University of Ottawa 1965

—, "Regino of Prüm", in: *New Grove*, vol. 15, p. 682

—, *L'oeuvre musicale d'Hucbald de Saint-Amand*, (Cahier d'études médiévales, cahier spécial 5), Montreal 1995

J.J. CONTRENI, *A note on the attribution of a Martianus Capella commentary to Martinus Laudunensis (Martinus Scotus)*, 'Martianus Capella. Addenda et corrigenda', in: *CTC* III, Washington 1976, pp. 451-452

—, *Three Carolingian Texts attributed to Laon: Reconsiderations*, in: *Studi Medievali Serie Terza* 17 (1976), pp. 797-813

—, *The Irish "colony" at Laon during the time of John Scottus*, in: *Jean Scot Érigène et l'histoire de la philosophie*, Paris 1977, pp. 59-67

—, *The Cathedral School of Laon from 850 to 930: Its Manuscripts and Masters*, (Münchener Beiträge zur Mediävistik und Renaissance-Forschung 29), München 1978

—, *The Irish in the Western Carolingian Empire (According to James F. Kenney and Bern, Burgerbibliothek 363)*, in: *Die Iren und Europa im früheren Mittelalter*, ed. H. LÖWE, Teilband 2, Stuttgart 1982, pp. 758-789

—, *Carolingian Learning, Masters and Manuscripts*, Aldershot 1992

—, *The Carolingian Renaissance: Education and Literary Culture*, in: *The New Cambridge Medieval History*, Vol. II, *c. 700 - c. 900*, ed. R. MCKITTERICK, Cambridge 1995, pp. 709-757

S. CORBIN, *The Neumes of the Martianus Capella Manuscripts*, in: *Essays on Opera and English Music. In honour of Sir Jack Westrup*, eds. F.W. STERNFELD, N. FORTUNE, E. OLLESON, Oxford 1975, pp. 1-7

—, *Die Neumen*, in: *Palaeographie der Musik*, I: *Die einstimmige Musik des Mittelalters*, ed. W. ARLT, Köln 1979, pp. 3.1-3.230

L. CRISTANTE, *Martiani Capellae De nuptiis Philologiae et Mercurii Liber IX*, (Medioevo e Umanesimo 64), Padova 1987

R. CROCKER, *The Early Medieval Sequence*, Berkeley, Los Angeles, London 1977

—, "Sequence", in: *New Grove*, Vol. 17, London 1980, pp. 141-153

—, *Medieval Chant* and *Frankish Music Theory*, in: *New Oxford History of Music* Vol. II, *The Early Middle Ages to 1300*, Oxford 1990 (2nd edition), pp. 256-264 and 278-283 respectively

P. DEPREUX, *Büchersuche und Büchertausch im Zeitalter der Karolingischen Renaissance am Beispiel des Briefwechsels des Lupus von Ferrières*, in: *Archiv für Kulturgeschichte* 76 (1994), pp. 267-284

A.C. DIONISOTTI, *On the Nature and Transmission of Latin Glossaries*, in: *Les manuscrits de lexiques et glossaires de l'antiquité au moyen âge*, ed. J. HAMESSE, Louvain 1996, pp. 202-252

F. DOLBEAU, *Un nouveau catalogue de Lobbes I/II*, in: *Recherches Augustiniennes* 13 (1978), pp. 3-36, and 14 (1979), pp. 191-248

A.K. DOLCH, *Notker Studien III. Stil und Quellenprobleme zu Notkers Boethius und Martianus Capella*, (Ottendorfer Memorial Series of Germanic Monographs 16), New York 1952, pp. 297-386

P. DRONKE, *Fabula. Explorations into the Uses of Myth in Medieval Platonism*, (Mittellateinische Studien und Texte 9), Leiden, Köln 1974

M.-E. DUCHEZ, *Jean Scot Érigène premier lecteur du De institutione musica de Boèce?*, in: *Eriugena. Studien zu seinen Quellen*, W. BEIERWALTES (ed.), Heidelberg 1980, pp. 165-187

—, *Description grammaticale et description arithmétique des phénomènes musicaux: le tournant du IXe siècle*, in: *Sprache und Erkenntnis im Mittelalter*, (Miscellanea Mediaevalia, Veröffentlichungen des Thomas-Instituts der Universität Köln 13.2), Berlin, New York 1981, pp. 561-579

—, *Vocabulaire musical et conceptualisation musico-scientifique*, in: *Documents pour l'histoire du vocabulaire scientifique* 2 (1981), pp. 1-18

—, *L'émergence acoustico-musicale du terme "sonus" dans les commentaires carolingiens de Martianus Capella*, in: *Documents pour l'histoire du vocabulaire scientifique* 7 (1985), pp. 97-149

—, *Le savoir théorico-musical carolingien dans les commentaires de Martianus Capella. La tradition érigénienne*, in: *Giovanni Scoto nel suo tempo. L'organizzazione del sapere en età carolingia*, Spoleto 1989, pp. 553-592

B.S. EASTWOOD, *The chaster path of Venus in the astronomy of Martianus Capella*, in: *Archives Internationales d'Histoire des Sciences* 32 (1982), pp. 145-158

—, *Heraclides and Heliocentrism: Texts, Diagrams and Interpretations*, in: *Journal for the History of Astronomy* 23 (1992), pp. 233-260

—, *Plato and circumsolar planetary motion in the Middle Ages*, in: *Archives d'Histoire Doctrinale et Littéraire du Moyen Age* 60 (1993), pp. 7-26

—, *The astronomies of Pliny, Martianus Capella and Isidore of Seville*, in: P.L. BUTZER & D. LOHRMANN (eds.), *Science in Western and Eastern Civilization in Carolingian times*, Basel, Boston, Berlin 1993, pp. 161-180

—, *Astronomical Images and Planetary Theory in Carolingian Studies of Martianus Capella*, in: *Journal for the History of Astronomy* 31 (2000), pp. 1-28

E. VON ERHARDT-SIEBOLD & R. VON ERHARDT, *The astronomy of Johannes Scotus Eriugena*, Baltimore 1940

—, *Cosmology in the "Annotationes in Marcianum", more light on Eriugena's astronomy*, Baltimore 1940

R. ERICKSON, (transl. and intr.), *Musica Enchiriadis and Scolica Enchiriadis*, New Haven, London 1995

M. ESPOSITO, *Note on a ninth-century commentary on Martianus Capella*, in: *Zeitschrift für Celtische Philologie* 7 (1910), pp. 499-507

J. FLAMANT, *Macrobe et le Néo-Platonisme latin, à la fin du IVe siècle*, (Etudes Préliminaires aux Religions Orientales dans l'Empire Romain 58), Leiden 1977

R. FLOWER, *Irish commentaries on Martianus Capella*, in: *Zeitschrift für Celtische Philologie* 8 (1912), pp. 566-567

J. FONTAINE, *Isidore de Séville et la culture classique dans l'Espagne wisigothique*, Paris 1959

J.C. FRAKES, *Remigius of Auxerre, Eriugena, and the Greco-Latin circumstantiae-*

formula of accessus ad auctores, in: *The sacred nectar of the Greeks: The study of Greek in the West in the Early Middle Ages*, (King's College London Medieval Studies 2), ed. M.W. HERREN, London 1988, pp. 229-256

D. GANZ, *Corbie in the Carolingian Renaissance*, (Beihefte der Francia 20), Sigmaringen 1990

—, *Heiric d'Auxerre glossateur du Liber glossarum*, in: *L'école carolingienne d'Auxerre. De Murethach à Remi 830-908*, D. IOGNA-PRAT, C. JEUDY, G. LOBRICHON (eds.), Paris 1991, pp. 297-305

—, *Theology and the organisation of thought*, and *Book production and the spread of Caroline minuscule*, in: *The New Cambridge Medieval History*, Vol. II, *c. 700 - c. 900*, ed. R. MCKITTERICK, Cambridge 1995, pp. 758-785 and 786-808 respectively.

S. GERSH, *Concord in Discourse: Harmonics and Semiotics in Late Classical and Early Medieval Platonism*, Berlin, New York 1996

M.T. GIBSON and J.L. NELSON (eds.), *Charles the Bald: Court and Kingdom*, rev. edition Aldershot 1990

G. GLAUCHE, *Schullektüre im Mittelalter*, (Münchener Beiträge zur Mediävistik und Renaissance-Forschung 5), München 1970

S. GREBE, *Martianus Capella "De nuptiis Philologiae et Mercurii". Darstellung der Sieben Freien Künste und ihrer Beziehungen zueinander*, (Beiträge zur Altertumskunde 119), Stuttgart, Leipzig 1999.

L.P. GRIJP & P. SCHEEPERS (eds.), *Van Aristoxenos tot Stockhausen* 1: *Van Oudheid tot Renaissance*, Groningen 1990

J. GODWIN, *Harmonies of Heaven and Earth*, London 1986

—, *Music, Mysticism and Magic*, Arkana 1987

J.P. GUMBERT, *De Egmondse boekenlijst*, in: *In het spoor van Egbert. Aartsbisschop Egbert van Trier, de bibliotheek en geschiedschrijving van het klooster Egmond*, (Egmondse Studiën 3), ed. G.N.M. VIS, Hilversum 1997, pp. 151-177

L. GUSHEE (ed.), *Aureliani Reomensis Musica disciplina*, (Corpus Scriptorum de Musica 21), 1975

J. HANDSCHIN, *Die Musikanschauung des Johannes Scotus (Eriugena)*, in: *Deutsche Vierteljahresschrift für Literaturwissenschaft und Geistesgeschichte* 5 (1927), pp. 316-341

A. HAUG, *Ein neues Textdokument zur Entstehungsgeschichte der Sequenz*, in: *Festschrift Ulrich Siegele zum 60. Geburtstag*, ed. R. FABER, A. FÖRSTER, etc., Kassel 1991, pp. 9-19

B. HAURÉAU, *Commentaire de Jean Scot Érigène sur Martianus Capella*, in: *Notices et extraits des manuscrits de la Bibliothèque Impériale* 20.2 (1862), pp. 1-39

F. HENTSCHEL (ed.), *Musik und die Geschichte der Philosophie und Naturwissenschaften im Mittelalter*, Leiden, Boston, Köln 1998

F. HENTSCHEL, "Augustinus", in: *MGG* (Zweite, neubearbeitete Ausgabe, Personenteil I), Kassel etc. 1999, cols. 1169-1175

J.M.M. HERMANS & G.C. HUISMAN, *De descriptione codicum. Handschriftenbeschijving, tevens Syllabus bij de colleges "Inleiding in de Westerse Handschriftenkunde/Codicologie"*, 3[e] oplage, Groningen 1981

M.W. HERREN, *The Hisperica Famina*, 2 Vols., Toronto 1974 and 1987
— (ed.), *The sacred nectar of the Greeks: The study of Greek in the West in the Early Middle Ages*, (King's College London Medieval Studies 2), London 1988
D. HILEY, *Western Plainchant. A Handbook*, Oxford 1993 (repr. 1995)
L. HOLTZ, *L'école d'Auxerre*, in: *L'école carolingienne d'Auxerre. De Murethach à Remi 830-908*, D. IOGNA-PRAT, C. JEUDY, G. LOBRICHON (eds.), Paris 1991, pp. 131-146
Dom HOURLIER, *Le Domaine de la notation messine*, in: *Revue Grégorienne* 30 (1951), pp. 96-113 and 150-158
H. HÜSCHEN, *Regino of Prüm, Historiker, Kirchenrechtler und Musiktheoretiker*, in: *Festschrift K.G. Fellerer zum 60. Geburtstag*, Regensburg 1962, pp. 205-223
D. IOGNA-PRAT, C. JEUDY, G. LOBRICHON (eds.), *L'école carolingienne d'Auxerre. De Murethach à Remi 830-908*, Paris 1991
J. JAMES, *The Music of the Spheres: Music, Science and the Natural Order of the Universe*, London 1993
É. JEAUNEAU, *Note sur l'École de Chartres*, in: *Studi Medievali* (N.S. 3) 5 (1964), pp. 821-865
—, *Les écoles de Laon et d'Auxerre au IXe siècle*, in: *La scuola nell'occidente latino dell'alto medioevo*, T. II. (Settimane di studio del centro italiano di studi sull'alto medioevo 19), Spoleto 1972, pp. 459-522
—, *Quatre thèmes érigéniens*, (Conférence Albert-le-Grand 1974), Montreal, Paris 1978
—, *Le commentaire érigénien sur Martianus Capella (De nuptiis, Lib.I) d'après le manuscrit d'Oxford (Bodl.Libr. Auct.T.2.19, fol. 1-31)*, in: *Quatre thèmes érigéniens*, Montreal, Paris 1978, pp. 91-166
C. JEUDY, *L'oeuvre de Remi d'Auxerre: État de la question*, in: *L'école carolingienne d'Auxerre. De Murethach à Remi 830-908*, D. IOGNA-PRAT, C. JEUDY, G. LOBRICHON (eds.), Paris 1991, pp. 373-397
M.B. DE JONG, *From Scolastici to Scioli: Alcuin and the Formation of an Intellectual Élite*, in: L.A.J.R. HOUWEN & A.A. MCDONALD (eds.), *Alcuin of York. Scholar at the Carolingian Court*, (Germania Latina 3), Groningen 1998, pp. 45-58
J.C. KING, *Notker der Deutsche. Martianus Capella, De nuptiis Philologiae et Mercurii*, (Die Werke Notkers des Deutschen, Neue Ausgabe 4), Tübingen 1979
R. KLIBANSKY, *The Rock of Parmenides: Medieval Views on the Origin of Dialectic*, in: *Medieval and Renaissance Studies* 1 (1941-1943), pp. 178-186
L. KRUCKENBERG, *The Sequence from 1050-1150: A Study of a Genre in Change*, Diss. Univ. of Iowa 1997
—, "Sequentia", in: *MGG* (Zweite, neubearbeitete Ausgabe), Kassel etc. 1998, col. 1254-1286
L. LABOWSKY, *A new version of Scotus Eriugena's Commentary on Martianus Capella*, in: *Medieval and Renaissance Studies* 1 (1941-43), pp. 187-193
F. LEMOINE, *Martianus Capella: a Literary Reevaluation*, (Münchener Beiträge zur Mediävistik und Renaissance-Forschung 10), München 1972
L. LENAZ, *Martiani Capellae De nuptiis Philologiae et Mercurii Liber secundus*,

Padova 1975

C. LEONARDI, *Nota introduttiva per un'indagine sulla fortuna di Marziano Capella nel Medio Evo*, in: *Bullettino dell'Istituto Storico Italiano per il Medio Evo e Archivo Muratoriano* 67 (1955), pp. 265-288

—, *Raterio e Marziano Capella*, in: *Italia Medioevale e Umanistica* 2 (1959), pp. 73-102

—, *I codici di Marziano Capella I-II*, in: *Aevum* 33 (1959), pp. 443-489, and *Aevum* 34 (1960), pp. 1-99 and 411-524

—, *Intorno al Liber de numeris di Isidoro de Siviglia*, in: *Bullettino dell'Istituto storico italiano per il medio evo e Archivio muratoriano* 68 (1956), pp. 203-231

—, *I commenti altomedievali ai classici pagani: da Severino Boezio a Remigio d'Auxerre*, in: *La cultura antica nell'occidente latino dal 7 all'11 secolo*, (Settimane di Studio del Centro Italiano di Studi sull'alto Medioevo 22), Spoleto 1975, pp. 459-508

—, *Glosse eriugeniane a Marziano Capella in un codice Leidense*, in: *Jean Scot Érigène et l'histoire de la philosophie*, (Colloques Internationaux du C.N.R.S. 561), Paris 1977, pp. 171-182

—, *Martianus Capella et Jean Scot. Nouvelle présentation d'un vieux problème*,in: *Jean Scot écrivain. Actes du IVe Colloque international*, Montreal, Paris 1986, pp. 187-207

H. LIEBESCHÜTZ, *Texterklärung und Weltdeutung bei Johannes Eriugena*, in: *Archiv für Kulturgeschichte* 40 (1958), pp. 66-69

—, *Zur Geschichte der Erklärung des Martianus Capella bei Eriugena*, in: *Philologus* 104 (1960), pp. 127-137

—, *John Eriugena and his cosmological interpretation of Martianus Capella*, in: *The Cambridge History of Later Greek and Early Medieval Philosophy*, ed. A.H. ARMSTRONG, Cambridge 1967, pp. 576-578

—, *The place of the Martianus glossae in the development of Eriugena's thought*, in: *The Mind of Eriugena*, Dublin 1973, pp. 49-57

C.E. LUTZ, *The commentary of Remigius of Auxerre on Martianus Capella*, in: *Medieval Studies* 19 (1957), pp. 137-156

—, *Martianus Capella*, in: *CTC*, Vol. II, P.O. KRISTELLER (ed.), Washington DC 1971, pp. 367-381. "Addenda et corrigenda" in Vol. III, 1976, pp. 449-452, and Vol. VI, 1986, pp. 185-186

M. MANITIUS, *Geschichte der lateinischen Literatur des Mittelalters*, Vol. I-III, München 1911-1931

—, *Zur karolingischen Literatur*, in: *Neues Archiv* 36 (1911), pp. 43-75

J. MARENBON, *From the Circle of Alcuin to the School of Auxerre: Logic, Theology and Philosophy in the Early Middle Ages*, (Cambridge Studies in Medieval Life and Thought, 3rd series, 15), Cambridge 1981

T.J. MATHIESEN, *Aristides Quintilianus On Music in Three Books*, New Haven, London 1983

G. MATHON, *Jean Scot Érigène et le problème de l'âme universelle*, in: *L'homme et son destin d'après les penseurs du Moyen Âge*, (Actes du 1er Congrès International de Philosophie Médiévale), Louvain, Paris 1960, pp. 361-375

—, *Les formes et la signification de la pédagogie des arts libéraux au milieu du IXe siècle. L'enseignement palatin de Jean Scot Érigène*, in: *Arts libéraux et philosophie au Moyen Âge*, Montreal, Paris 1969, pp. 47-64

C. McDONOUGH, *Trier, Bibliothek des Priesterseminars Ms. 100 and the text of Martianus Capella*, in: *Mediaeval Studies* 36 (1974), pp. 56-66

R. McKITTERICK, *Books, Scribes and Learning in the Frankish Kingdoms, Sixth to Ninth Centuries*, Aldershot 1994

— (ed.), *Carolingian Culture: Emulation and Innovation*, Cambridge 1994

C. MEYER, *Métaphore instrumentale et présentation du système acoustique à l'époque carolingienne*, in: *Musik und die Geschichte der Philosophie und Naturwissenschaften im Mittelalter*, ed. F. HENTSCHEL, Leiden, Boston, Köln 1998, pp. 141-149

K.A. DE MEYIER, "Voss.Lat.F. 48 - Martianus Capella, varia," in: *Codices Vossiani Latini, Pars I: Codices in folio*, Leiden 1973, pp. 100-103

B. MÜNXELHAUS, *Pythagoras musicus. Zur Rezeption der pythagoreischen Musiktheorie als quadrivialer Wissenschaft im lateinischen Mittelalter*, Bonn-Bad Godesberg 1976

F. MÜTHERICH, *'De Rhetorica'. Eine Illustration zu Martianus Capella*, in: *Festschrift Bernhard Bischoff*, J. AUTENRIETH, F. BRUNHÖLZL (eds.), Stuttgart 1971, pp. 198-206

D. MUZERELLE, *Pour décrire les schémas de réglure. Une méthode de notation symbolique applicable aux manuscrits latins (et autres)*, in: *Quinio* 1 (1999), pp. 123-170

K.W. NIEMÖLLER, *Die Musik im Weltbild des Johannes Scotus Eriugena*, in: *Musik und die Geschichte der Philosophie und Naturwissenschaften im Mittelalter*, ed. F. HENTSCHEL, Leiden, Boston, Köln 1998, pp. 293-304

G. NUCHELMANS, *Philologie et son marriage avec Mercure jusqu'à la fin du XIIe siècle*, in: *Latomus* 16 (1957), pp. 84-107

B. PABST, *Prosimetrum. Tradition und Wandel einer Literaturform zwischen Spätantike und Spätmittelalter*, I, (Ordo, Studien zur Literatur und Gesellschaft des Mittelalters und der frühen Neuzeit 4/1), Köln, Weimar, Wien 1994

H. PARKER, *The seven liberal arts*, in: *English Historical Review* 5 (1890), pp. 417-461

N. PHILLIPS, *Classical and Late Latin Sources for Ninth-Century Treatises on Music*, in: *Music Theory and Its Sources: Antiquity and the Middle Ages*, ed. A. BARBERA, Notre Dame 1990, pp. 100-135

—, *"Musica" and "Scolica enchiriadis": The Literary, Theoretical, and Musical Sources*, Diss. New York University, 1984

J. PRÉAUX, *Un nouveau texte sur la Vénus androgyne*, in: *Annuaire de l'Institut de philologie et d'histoire orientales et slaves* 13 (1953), pp. 479-490

—, *Le commentaire de Martin de Laon sur l'oeuvre de Martianus Capella*, in: *Latomus* 12 (1953), pp. 437-459

—, *Un nouveau manuscrit de Saint-Gall: le Bruxellensis 9565-9566*, in: *Scriptorium* 10 (1956), pp. 221-228

—, *Deux manuscrits gantois de Martianus Capella*, in: *Scriptorium* 13 (1959), pp. 15-21

—, *Le manuscrit d'Avranches 240 et l'oeuvre de Martianus Capella*, in: *Sacris Erudiri. Jaarboek voor Godsdienstwetenschappen* 17 (1966), pp. 135-149

—, *Les formes et la signification de la pédagogie des arts libéraux au milieu de 9e siècle. L'enseignement palatin de Jean Scot Érigène*, in: *Arts libéraux et philosophie au Moyen Âge*, (Actes du 4e Congrès International de Philosophie Médiévale), Montreal, Paris 1969, pp. 47-64

—, *Securus Melior Felix, l'ultime Orator Urbis Romae*, in: *Corona gratiarum E. Dekkers*, Brugge, Den Haag 1974, pp. 101-121

—, *Jean Scot et Martin de Laon en face du De Nuptiis de Martianus Capella*, in: *Jean Scot Érigène et l'histoire de la philosophie*, ed. R. ROQUES, Paris 1977, pp. 161-170

—, *Les manuscrits principaux du "De nuptiis Philologiae et Mercurii" de Martianus Capella*, in: *Lettres latines du moyen âge et de la renaissance* (Collection Latomus 158), Bruxelles 1987, pp. 76-128

E.K. RAND, *Die Glossen des Johannes zu Martianus Capella*, in: *Quellen und Untersuchungen zur lateinischen Philologie des Mittelalters* I.2, München 1906, pp. 81-82

—, *The supposed autographa of John the Scot*, in: *University of California Publications in Classical Philology* 5 (1918-1923), pp. 135-141

—, *How many leaves at a time?*, in: *Palaeographia Latina* 5, 1927, pp. 52-78

—, *How much of the Annotationes in Marcianum is the work of John the Scot?*, in: *Transactions and Proceedings of the American Philological Association* 71 (1940), pp. 504-523

S. RANKIN, *Carolingian Music*, in: *Carolingian Culture. Emulation and Innovation*, ed. R. MCKITTERICK, Cambridge 1994, pp. 274-316

L.D. REYNOLDS, *Texts and Transmission. A survey of the Latin Classics*, Oxford 1983

P. RICHÉ, *Ecoles et enseignement dans le haut moyen âge: fin du Ve siècle—milieu du XIe siècle*, Paris 1989

Y.-F. RIOU, *Chronologie et provenance des manuscrits classiques latins neumés*, in: *Revue d'Histoire des Textes* 21 (1991), pp. 77-113

R. SCHIEVENIN, *Marziano Capella e il 'proconsulare culmen'*, in: *Latomus* 45 (1986), pp. 797-815

G. SCHRIMPF, *Zur Frage der Authentizität unserer Texte von Johannes Scottus' Annotationes in Martianum*, in: *The mind of Eriugena*, J.J. O'MEARA & L. BIELER (eds.), Dublin 1973, pp. 125-139

D. SHANZER, review of W.H. STAHL & R. JOHNSON, *Martianus Capella and the Seven Liberal Arts*, in: *Beiträge zur Geschichte der deutschen Sprache und Literatur* 104 (1982), pp. 110-117

—, *A Philosophical and Literary Commentary on Martianus Capella's* De nuptiis Philologiae et Mercurii *Book I*, (Classical Studies 32), Berkeley, Los Angeles, London 1986

—, *Review article: Felix Capella: minus sensus quam nominis pecudalis*, in: *Classical Philology* 81 (1986), pp. 62-81

J. SMITS VAN WAESBERGHE, *Die besondere Stellung der* ars musica *im Zeitalter der Karolinger*, in: *Dia-pason. De omnibus*, pp. 48-70; published earlier in

French: *La place exceptionelle de l'Ars Musica dans le développement des sciences au siècle des Carolingiens*, in: *Revue grégorienne* 31 (1952), pp. 81-104

J. SMITS VAN WAESBERGHE, *Dia-pason. De omnibus. Ausgewählte Aufsätze von Joseph Smits van Waesberghe. Festgabe zu seinem 75. Geburtstag*, C.J. MAAS & M.U. SCHOUTEN-GLASS (eds.), Buren 1976.

L. SPITZER, *Classical and Christian Ideas of World Harmony: Prolegomena to an Interpretation of the Word "Stimmung"*, I, *Traditio* 2 (1944), pp. 409-464, and II, *Traditio* 3 (1945), pp. 307-364

B. STÄBLEIN, *Zur Frühgeschichte der Sequenz*, in: *Archiv für Musikwissenschaft* 18 (1961), pp. 1-33

—, "Sequenz", in: *MGG*, Band 12, cols. 522-549

W.H. STAHL, *Commentary on the Dream of Scipio by Macrobius*, New York 1952, reprint 1990

—, *To a better understanding of Martianus Capella*, in: *Speculum* 40 (1965), pp. 102-115

W.H. STAHL & R. JOHNSON, *Martianus Capella and the seven liberal arts*, Vol. I-II, New York 1971, reprint 1977

S. TAFEL, *The Lyons Scriptorium* I-II, in: *Palaeographia Latina* 2 (1923), pp. 40-70, and continued in *Palaeographia Latina* 4 (1925), p. 30-70 (St. Andrews University Publications 16 and 20, ed. W.M. LINDSAY)

M.J. TEEUWEN, *Harmonie der hemelsferen. Een klassiek thema geïnterpreteerd door Johannes Scotus Eriugena*, in: *Millennium* 9 (1995), pp. 117-132

—, *Ninth-century glosses on Martianus Capella: organisation and transmission*, forthcoming in: *Organizing the Written Word: Scripts, Manuscripts and Texts*, ed. M. MOSTERT, (Utrecht Studies in Medieval Literacy 2), Turnhout 2002?

L. TRAUBE, *Computus Helperici*, in: *Neues Archiv* 18 (1893), pp. 73-105

A. VERHULST, *L'activité et la calligraphie du 'scriptorium' de l'abbaye Saint-Pierre-au-Mont-Blandin de Gand à l'époque de l'abbé Wichard (m. 1058)*, in: *Scriptorium* 11 (1957), pp. 37-49

E. VETTER, *Concentrische cirkels. Modus, affect, sfeer en tijd in een middeleeuws muziektheoretisch gedicht*, Proefschrift Universiteit Utrecht, Amsterdam 2000

E.L. WAELTNER, *Organicum melos. Zur Musikanschauung des Iohannes Scottus (Eriugena)*, (Bayerische Akademie der Wissenschaften. Veröffentlichungen der musikhistorischen Kommission 1), München 1977

D.L. WAGNER (ed.), *The Seven Liberal Arts in the Middle Ages*, Bloomington, IN, 1983

J. WARICHEZ, *L'Abbaye de Lobbes, Depuis les origines jusqu'en 1200* (Recueil de travaux publiés par les membres des conférences d'histoire et de philologie), Louvain, Paris 1909

M.L. WEST, *Ancient Greek Music*, New York 1994

H.J. WESTRA, *The commentary on Martianus Capella's De nuptiis Philologiae et Mercurii attributed to Bernardus Silvestris*, Toronto 1986

—, *The Berlin Commentary on Martianus Capella's De nuptiis Philologiae et Mercurii, Book I*, Leiden 1994

—, *The Allegorical Interpretation of Myth: Its Origins, Justification and Effect*, in: *Mediaeval Antiquity* 24 (1995), pp. 277-291

A. WHITE, *Boethius in the Medieval Quadrivium*, in: *Boethius: His Life, Thought and Influence*, ed. M. GIBSON, London 1981, pp. 162-205

J. WILLIS, *Martianus Capella and his Early Commentators*, Dissertation, University of London, 1952

—, *Martianus und die mittelalterliche Schulbildung*, in: *Das Altertum* 19 (1973), pp. 164-174

W. WIORA, *Das vermeintliche Zeugnis des Johannes Eriugena für die Anfänge der abendländischen Mehrstimmigkeit*, in: *Acta Musicologica* 43 (1971), pp. 33-43

K.-A. WIRTH, *Eine illustrierte Martianus-Capella-Handschrift aus dem 13. Jahrhundert*, in: *Städel-Jahrbuch*, N.F. 2 (1969), pp. 43-74

INDEX OF MANUSCRIPTS

INDEX OF ANCIENT AND MEDIEVAL PERSONAL NAMES

MITTELLATEINISCHE STUDIEN UND TEXTE

1. Metellus von Tegernsee. Die "Quirinalien". Untersuchungen und Text 1965. Jacobsen (Ed.).
2. Pseudo-Ovidius. *De vetula.* Untersuchungen und Text. Klopsch (Ed.). 1967. ISBN 90 04 01464 0
3. Lange. *Philologische Studien zur Latinität west-hispanischer Privaturkunden des 9.-12. Jahrhunderts.* Langosch (Ed.). 1966. ISBN 90 04 01465 9
4. Joseph Iscanus. Werke und Briefe. Gompf (Ed.). 1970. ISBN 90 04 01466 7
5. *Numen litterarum:* the Old and New in Latin Poetry. Witke (Ed.). 1971.
6. Theobald. *Physiologus.* Edited with introduction, critical apparatus, translation and commentary by P.T. Eden. 1972. ISBN 90 04 03444 7
7. Geoffrey of Vitry. *The commentary of Geoffrey of Vitry on Claudian* De Raptu Proserpinae. Transcribed by A.K. Clarke and P.M. Giles with an introduction and notes by A.K. Clarke. 1973. ISBN 90 04 03674 1
8. Könsgen, E. *Epistolae duorum amantium.* Briefe Abaelards und Heloïses? Edition und Untersuchungen. 1974. ISBN 90 04 03875 2
9. Dronke, P. *Fabula.* Explorations into the uses of myth in medieval Platonism. Reprint 1985. ISBN 90 04 07715 4
10. Jacobsen, P.C. *Flodoard von Reims.* Sein Leben und seine Dichtung *De Triumphis Christi.* 1978. ISBN 90 04 05407 3
11. Langosch, K. *Donisii Comedia Pamphile.* Untersuchungen und Text. 1979. ISBN 90 04 06007 3
12. Mann, J. *Ysengrimus.* Text with translations, commentary and introduction. 1987. ISBN 90 04 08103 8
13. Petrus Presbyter. *Carmina.* Text und Kommentar. Herausgegeben von M. Rener. 1988. ISBN 90 04 08797 4
14. Nigel of Canterbury. *The Passion of St. Lawrence, Epigrams and Marginal Poems.* Edited and translated by J. M. Ziolkowski. ISBN 90 04 08865 2
15. Orbàn, A.P. *Novus Phisiologus.* Nach Hs. Darmstadt 2780. 1989. ISBN 90 04 08894 6
16. Hilarius Aurelianensis. *Versus et Ludi epistolae. Ludus Danielis Belouacensis.* [Die Egerton Handschrift]. Bemerkungen zur Musik des Daniel-Spiels von Beauvais von M. Bielitz. 1989. ISBN 90 04 09070 3
17. Unterkircher, F. (Ed.). *Hugo von Lüttich: Peregrinarius.* 1991. ISBN 90 04 09325 7
18. Hugo von Mâcon. *Die* Gesta Militum. Ein bisher unbekanntes Werk der Erzählliteratur des Hochmittelalters. Herausgegeben von E. Könsgen. 1990. ISBN 90 04 09201 3
19. Harting-Corrêa, A. *Walahfrid Strabo's Libellus de Exordiis et Incrementis.* 1996. ISBN 90 04 09669 8.
20. Westra, H.J. (Ed.). *The Berlin Commentary on Martianus Capella's* De Nuptiis Philologiae et Mercurii. Book I. With the Assistance of C. Vester. 1994. ISBN 90 04 10170 5
21. Mann, N. & B. Munk Olsen (Eds.). *Medieval and Renaissance Scholarship.* Proceedings of the Second European Science Foundation Workshop on the Classical Tradition in the Middle Ages and the Renaissance (London, Warburg Institute, 27-28 November 1992). 1997. ISBN 90 04 10508 5
22. Haye, T. *Das lateinische Lehrgedicht im Mittelalter.* Analyse einer Gattung. 1997. ISBN 90 04 10668 5
23. Westra, H.J. & T. Kupke (Eds.). *The Berlin Commentary on Martianus Capella's* De Nuptiis Philologiae et Mercurii. Book II. With the Assistance of B. Garstad. 1998. ISBN 90 04 10968 4

24. Ricklin, T. *Der Traum der Philosophie im 12. Jahrhundert.* Traum zwischen Constantinus Africanus und Aristoteles. 1999. ISBN 90 04 11116 6
25. Stein, E. *Clericus in Speculo.* Studien zur lateinischen Verssatire des 12. und 13. Jahrhunderts und Erstedition des "Speculum prelatorum". 1999. ISBN 90 04 11329 0
26. Meckelnborg, C. und B. Schneider. *Opusculum fabularum.* Die Fabelsammlung der Berliner Handschrift Theol. lat. fol. 142. 1999. ISBN 90 04 11333 9
27. Haye, T. *Oratio.* Mittelalterliche Redekunst in lateinischer Sprache. 1999. ISBN 90 04 11335 5
28. Murgatroyd, P. *The Amatory Elegies of Johannes Secundus.* 2000. ISBN 90 04 11603 6
29. Marenbon, P.G. (Ed.) *Poetry and Philosophy in the Middle Ages.* A Festschrift for Peter Dronke. 2001. ISBN 90 04 11964 7
30. Teeuwen, M. *Harmony and the Music of the Spheres.* The *Ars Musica* in Ninth-Century Commentaries on Martianus Capella. 2002. ISBN 90 04 12525 6